Solid State Magnetic Sensors

HANDBOOK OF SENSORS AND ACTUATORS

Series Editor: S. Middelhoek, Delft University of Technology,
The Netherlands

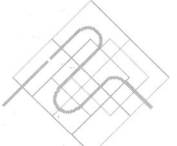

Volume 1   Thick Film Sensors (edited by M. Prudenziati)
Volume 2   Solid State Magnetic Sensors (by C.S. Roumenin)

HANDBOOK OF SENSORS AND ACTUATORS 2

Solid State Magnetic Sensors

Chavdar S. Roumenin
Drujba - I
Sofia, Bulgaria

1994
ELSEVIER
Amsterdam - Lausanne - New York - Oxford - Shannon - Tokyo

ELSEVIER SCIENCE B.V.
Sara Burgerhartstraat 25
P.O. Box 211, 1000 AE Amsterdam,
The Netherlands

ISBN: 0 444 89401 2

© 1994 Elsevier Science B.V. All rights reserved.

No part of this publication may be reproduced, stored in a retrieval system or transmitted in any form or by any means, electronic, mechanical, photocopying, recording or otherwise, without the prior written permission of the publisher, Elsevier Science B.V., Copyright & Permissions Department, P.O. Box 521, 1000 AM Amsterdam, The Netherlands.

Special regulations for readers in the U.S.A. - This publication has been registered with the Copyright Clearance Center Inc. (CCC), Salem, Massachusetts. Information can be obtained from the CCC about conditions under which photocopies of parts of this publication may be made in the U.S.A. All other copyright questions, including photocopying outside of the U.S.A., should be referred to the copyright owner, Elsevier Science B.V., unless otherwise specified.

No responsibility is assumed by the publisher for any injury and/or damage to persons or property as a matter of products liability, negligence or otherwise, or from any use or operation of any methods, products, instructions or ideas contained in the material herein.

This book is printed on acid-free paper.

Printed in The Netherlands

Introduction to the Series

The arrival of integrated circuits with very good performance/price ratios and relatively low-cost microprocessors and memories has had a profound influence on many areas of technical endeavour. Also in the measurement and control field, modern electronic circuits were introduced on a large scale leading to very sophisticated systems and novel solutions. However, in these measurement and control systems, quite often sensors and actuators were applied that were conceived many decades ago. Consequently, it became necessary to improve these devices in such a way that their performance/price ratios would approach that of modern electronic circuits.

This demand for new devices initiated worldwide research and development programs in the field of "sensors and actuators". Many generic sensor technologies were examined, from which the thin- and thick-film, glass fiber, metal oxides, polymers, quartz and silicon technologies are the most prominent.

A growing number of publications on this topic started to appear in a wide variety of scientific journals until, in 1981, the scientific journal Sensors and Actuators was initiated. Since then, it has become the main journal in this field.

When the development of a scientific field expands, the need for handbooks arises, wherein the information that appeared earlier in journals and conference proceedings is systematically and selectively presented. The sensor and actuator field is now in this position. For this reason, Elsevier Science took the initiative to develop a series of handbooks with the name "Handbook of Sensors and Actuators" which will contain the most meaningful background material that is important for the sensor and actuator field. Titles like Fundamentals of Transducers, Thick Film Sensors, Magnetic Sensors, Micromachining, Piezoelectric Crystal Sensors, Robot Sensors and Intelligent Sensors will be part of this series.

The series will contain handbooks compiled by only one author, and handbooks written by many authors, where one or more editors bear the responsibility for bringing together topics and authors. Great care was given to the selection of these authors and editors. They are all well known scientists in the field of sensors and actuators and all have impressive international reputations.

Elsevier Science and I, as Editor of the series, hope that these handbooks will receive a positive response from the sensor and actuator community and we expect that the series will be of great use to the many scientists and engineers working in this exciting field.

Simon Middelhoek

Preface

At present a synthesis is taking place of almost all known technologies with microelectronics. Moreover, the extent to which computers and microsensors are now used is already an indicator of industrial progress. As the development of computers has paved the way for the development of artificial intelligence, the new generation of solid-state transducers complements technology by providing miniature sensing elements. The dramatic improvement in the performance to price ratio of microelectronic components is one of the most important stimuli for those who call themselves members of *"the information society"* and for whom, besides materials and energy, information and data processing play a key role in human progress.

Thanks to the continuous and dynamic development of highly sophisticated technologies like VLSI and ULSI, the "brain" of modern data acquisition and control systems is characterized by a great diversity in terms of architecture and memory, as well as of computing power. For instance, the operating speed of the SX-3 44R supercomputer of the Japanese NEC company has reached 25.6×10^9 operations per second, and most probably this is not the limit by far. But even Man's superb instrument of thinking would not be much use without the eyes, the ears, the nose and the other senses. It was not accidental that a great philosopher said that all that was in our consciousness had initially been in our senses. The analogy with Man is complete in technical systems which consist of a microprocessor and a memory. These systems are helpless, and particularly useless, unless they receive signals from the environment, i.e. unless they have sensors. It would be no exaggeration to assume that the standard and reliability of a system are predetermined by the quality of the primary information transducers.

Unfortunately, at present, the performance to price ratio for sensors considerably lags behind that of other electronic components. This is exactly why their useful application in technology and in everyday life is so limited. However, the growing interest in the field and the purposeful research programmes in many countries of the world are sound premises underlying the author's conviction that, in the high-tech sphere, this decade will be inconceivable without the proper development of microelectronics.

Magnetic field sensors take up a worthy place amidst the great diversity of transducers of nonelectric domains into electronic data. Most probably, the Chinese knew about the existence of a geomagnetic field more than 2000 years ago and tried to measure it using the earliest magnetometer — the compass. Biomagnetism has become a reality thanks to the existence of the latest supersensitive variants of magnetometers — the SQUID (Superconducting Quantum Interference Device). The highly precise data obtained, for instance, by cardiomagnetograms are not the latest curiosity resulting from human innovation, but the groundwork for a life-saving diagnosis. With their help myocardial infarction can be forecast 15–20 days earlier than was possible before, i.e. at a period when the classical electrocardiogram shows nothing out of the ordinary, and what is most important, provided treatment is applied during that period, infarction can be averted altogether. The latest biomagnetic studies will probably lead to a revision of the concept that Man does not possess a sensor for magnetic data. It has been established that to sleep in a bed placed in a North–South direction is good for our well-being and health.

Information concerning the ancient civilizations has survived to our time recorded in stone and gold artifacts; now our life is inconceivable without huge quantities of information recorded by computers, predominantly by magnetic means. The high write and read performance of such systems is impossible without increasing the density of magnetic recording. Replacing the conventional inductive heads with solid-state magnetic sensors, e.g. bipolar magnetotransistors, makes possible a domain resolution of less than one micrometer together with a wide frequency band and optimum sensitivity.

The long and complex evolution of magnetic sensors from the compass to its integrated version and SQUID's, together with their numerous applications, including in general terms geomagnetism and biomagnetism, the identification of magnetic signals, laboratory and industrial measurements, nondestructive material testing, planetary and space research, communication between submarines, and contactless detection of special objects, etc. has always been a challenge to our imagination and production capacities. Various types of magnetoresistors, Hall elements, magnetodiodes, magnetotransistors, carrier-domain magnetometers, SQUID's and other solid-state magnetotransducers of sophisticated design and operation have been developed. The dynamic evolution of this interdisciplinary field has been associated with the continuously growing demands of industry upon magnetometry and with the acknowledged universality of the magnetic field as the physical environment for the bioprocesses in ourselves and for the interrelationships in the environment.

Hundreds of original papers, reviews, monographs and patents have been published, treating various aspects of the properties and operations, device designs, fabrication, technology and packaging, circuitry, metrological problems and applications of solid-state magnetosensors. However, the priority given to specific problems makes a comprehensive and balanced analysis of this field of study more difficult, especially in view of its rapid development over the past few years. It is the purpose of this book to present the overall progress made in the field of magnetotransducers, to sum up the scientific achievements as the groundwork for further development, to follow the evolution of new ideas and facts, and, last but not least, to be a handbook for an ever widening circle of researchers and engineers who are interested in solid-state magnetosensors.

Readers will find, for the first time, collected in one book detailed information regarding the physical mechanisms of the origin of magnetosensitivity, the geometry and design of devices, operating modes, basic parameters and methods for their determination, the incorporation of transducers in circuits and smart solutions, many varied applications and other problems relevant to all the current Hall sensors, magnetodiodes, magnetotransistors, carrier-domain magnetometers, SQUID's and similar transducers of magnetic energy. The attention I have devoted to semiconductor magnetosensitive sensors and especially to their microelectronic versions is, I believe, not only a reflection of my personal point of view. It has been determined by the remarkable rate of development of this class of solid-state devices, and I consider it to have been a dominant trend over a long period of time. I have been highly gratified with the process of research in this field of the "magnetician", as well as with the chance to acquaint myself with the new results, which have also been given due attention in this work. At the same time, upon the discovery of high-temperature superconductivity, the sensors based on that unique phenomenon — SQUID's, have received a powerful boost. They complement magnetic transducers by providing measuring devices which are fantastic in their sensitivity. I am profoundly convinced that notwithstanding the great progress made in solid-state magnetic sensors, they are as yet in their cloudless infancy, whereas there is still so much lying ahead in a world, unlimited in time and space ... Good Heavens! They are a whole Universe into themselves!

The author gratefully acknowledges his good fortune for his personal acquaintance and contacts with a great number of scientists, and the opportunities offered thereby for the discussion of various matters in the domain of solid-state magnetosensors and related topics. I have been especially encouraged by the highly worthwhile consultations with S. Middelhoek (Delft, The Netherlands) and by his human kindness and attention that have stimulated me to undertake the writing of this work. Special acknowledgment is also due to the Grenoble sensor group: G. Kamarinos, A. Chovet, S. Cristoloveanu, F. Balestra and other colleagues for the unique creative atmosphere in which we studied and discussed the physical mechanisms of the magnetosensitivity of integrated devices. I also gratefully acknowledge the help of H.P. Baltes (Zurich, Switzerland) and his collaborators, whom I esteem highly and whose enthusiasm and competence are well known by the community of sensor scientists. My contacts with the Sensor Group in Delft, and especially V. Zieren, S. Kordić, J. Fluitman, J.H. Huijsing, P. Bergveld, G.C.M. Meijer, H.M.M. Kerkvliet, P.P.L. Regtien and P.J.A. Munter, have been a genuine school of science for me. I express grateful acknowledgements to N.B. Brandt (Moscow,

Russia), R. Popović (Lausanne, Switzerland), S. Kataoka and Y. Sugiyama (Electrotechnical Laboratory, Japan), K. Maenaka (Himeji, Japan), I. Igarashi (Toyota Corp., Japan), A. D'Amico (Rome, Italy), Seung-Ki Lee (Seoul, Korea) and J.E. Brignell (Southampton, U.K.). My scientific contacts with them have been a real pleasure to me.

I am highly grateful to the late Professor M. Minkov and to I. Brutchev (Sofia, Bulgaria), who received me in the Laboratory for Geotechnics of the Bulgarian Academy of Science during a difficult period for me in 1982, and without whose protection and democratic spirit I hardly could have achieved anything in sensor electronics.

I would like to mention the fruitful discussions with N.P. Georgiev, of the company INTERQUARTZ (Sofia, Bulgaria), whose comments contributed to the quality of this book.

Finally I would like to thank my wife, my daughter and my parents for their love and care. To them I dedicate this book.

Sofia, Bulgaria
September 1992

Contents

Preface V

Chapter 1. Principles and technologies for the acquisition of magnetic data 1

1.1. General principles 1
1.2. Conversion of magnetic energy by sensors 2
1.3. Nature of the magnetic field and the Lorentz force 6
1.4. Methods and applications of magnetic field measurements 8
1.5. State of the art of magnetosensitive device technology 11
1.6. Solutions employing integrated and smart sensors 13
References 15

Chapter 2. Galvanomagnetic phenomena used in solid-state sensors 17

2.1. Introduction 17
2.2. Conductivity of condensed matter in the crossed electric and magnetic fields . . . 18
2.3. Hall effect 23
    2.3.1. Hall-voltage mode in the case of monopolar conductivity 23
    2.3.2. Hall-current mode in short structures in the case of monopolar conductivity 30
    2.3.3. Hall effect in the presence of high electric fields 34
    2.3.4. Hall effect in medium-length samples (monopolar conductivity) 37
    2.3.5. Hall effect in the case of field and sample inhomogeneities 37
    2.3.6. Hall effect in the case of mixed conductivity 42
2.4. Magnetoresistance effect 46
    2.4.1. Physical magnetoresistance effect (monopolar conductivity) 46
    2.4.2. Geometrical magnetoresistance effect (monopolar conductivity) 47
    2.4.3. Magnetoresistance in the case of mixed conductivity 50
2.5. Magnetodiode effect 52
    2.5.1. Introduction to injection phenomena in semiconductor devices 53
    2.5.2. Magnetoconcentration effect 54
    2.5.3. Magnetodiode effect 58
        2.5.3.1. Magnetodiode effect in long structures with one injecting contact . 58
        2.5.3.2. Magnetodiode effect in structures with two injecting contacts . . . 64
2.6. Related effects 70
2.7. Coherent phenomena associated with superconducting matter . . . 75
    2.7.1. Superconductivity 75
    2.7.2. Coherence effects in superconductors 78
    2.7.3. Resistively shunted junction model 84

	2.7.4.	Principle of operation of SQUIDs	85
		2.7.4.1. RF SQUIDs	86
		2.7.4.2. DC SQUIDs	89
References			94

Chapter 3. Characteristics of solid-state magnetic field sensors 97

3.1. Introduction 97
3.2. Magnetosensor characteristics related to $OUT(B)_C$ 99
    3.2.1. Magnetosensitivity 99
    3.2.2. Nonlinearities 100
    3.2.3. Range of magnetic fields 103
    3.2.4. Directivity and sensor excitation by magnetic fields 103
    3.2.5. Frequency response 104
    3.2.6. Resolution 104
    3.2.7. Error 105
    3.2.8. Accuracy 106
    3.2.9. Hysteresis 106
    3.2.10. Output form 106
    3.2.11. Repeatability 106
3.3. Magnetosensor characteristics related to $OUT(C)_B$ 106
    3.3.1. Noise 106
    3.3.2. Offset 108
    3.3.3. Cross-sensitivity and temperature errors 109
    3.3.4. Drift and creep 110
    3.3.5. Stability, operating life and reliability 110
    3.3.6. Response time 111
3.4. Characteristics of the magnetosensor as a circuit element 111
    3.4.1. Electrical excitation, input and output impedance 111
    3.4.2. Room conditions 111
References 112

Chapter 4. Hall sensors and magnetoresistors 113

4.1. Introduction 113
4.2. Orthogonal Hall sensors 115
    4.2.1. Hall-sensor shapes 115
    4.2.2. Functional relations of the different geometrical shapes of Hall sensors .. 117
    4.2.3. The geometrical correction factor in Hall sensors 120
    4.2.4. Modes of operation of Hall sensors 124
        4.2.4.1. Hall-voltage mode of operation 125
        4.2.4.2. Basic characteristics of Hall sensors 128
        4.2.4.3. Hall-current mode of operation 139
    4.2.5. Requirements as to Hall sensor materials, fabrication technology and packaging 141
    4.2.6. Review of orthogonal Hall sensors 145
        4.2.6.1. Device structure and characteristics of MOSFET Hall sensors ... 145
            Some concluding remarks on MAGFET Hall sensors 152

	4.2.6.2. Bulk bipolar Hall sensors — review of device structures and sensor characteristics	153
	Concluding remarks on bulk bipolar Hall sensors.	159
	4.2.6.3. New trends in the development of orthogonal Hall sensors	160
4.3.	Parallel-field Hall sensors	165
	4.3.1. Conformal mapping and parallel-field Hall devices	166
	4.3.2. Device structures, operating principle and characterization of parallel-field Hall microsensors	169
	4.3.3. Outlook and conclusions	175
4.4.	Magnetoresistance sensors	176
	4.4.1. Magnetoresistor design and materials	176
	4.4.2. Properties and characterization of magnetoresistors	180
	4.4.3. Multiterminal magnetoresistance devices	183
References		186

Chapter 5. Magnetodiode sensors ... 191

5.1.	Device structures, materials and operation of MD sensors	191
5.2.	Effect of device parameters and temperature on the current–voltage characteristics of integrated MDs	200
5.3.	Determination of the MD sensor magnetosensitivity	204
5.4.	Influence of the magnetic field, device parameters, temperature and noise on MD sensor magnetosensitivity	207
5.5.	MD sensor review	213
	5.5.1. Orthogonal MD devices	213
	5.5.2. Parallel-field MD devices	218
5.6.	Concluding remarks	225
References		225

Chapter 6. Bipolar magnetotransistors and related sensors ... 229

6.1.	General approach to BMT design	229
6.2.	General principles of BMT operation	231
	6.2.1. Conditions for the occurrence of magnetosensitivity	231
	6.2.1.1. Emitter region magnetosensitivity	232
	6.2.1.2. Base-region magnetosensitivity	234
	6.2.1.3. Magnetosensitivity associated with the collector–base depletion region	241
	6.2.2. Functional connection of BMT devices with Hall sensors	242
6.3.	Device structures and operation of BMT sensors	244
	6.3.1. Orthogonal BMT sensors	244
	6.3.2. Parallel-field BMT sensors	248
6.4.	Figures of merit of BMT sensors	255
	6.4.1. Magnetosensitivity	255
	6.4.2. Noise	257
	6.4.3. Offset	258
	6.4.4. Linearity	259
	6.4.5. Temperature coefficient of magnetosensitivity	259

	6.4.6. Frequency response	260
6.5.	Unijunction magnetotransistor sensors	260
	6.5.1. UJMT device and operation	261
	6.5.2. Review of UJMT sensors	264
6.6.	Carrier-domain magnetometers	266
	6.6.1. Three-layer carrier-domain magnetic-field sensors	266
	6.6.2. Four-layer carrier-domain magnetic-field sensors	270
	6.6.2.1. Carrier-domain formation in four-layer structures	270
	6.6.2.2. Review of four-layer CD devices	271
6.7.	Concluding remarks	275
References	276

Chapter 7. SQUID sensors — 279

7.1.	The SQUID system	279
	7.1.1. Josephson junctions and their characteristics	279
	7.1.2. Signal–input coupling	283
	7.1.3. Read-out schemes	284
	7.1.4. SQUID sensor periphery	285
7.2.	Noise characteristics and sensitivity of SQUIDs	288
7.3.	DC SQUID sensors	291
	7.3.1. DC SQUID structures	291
	7.3.2. Characteristics of the $1/f$ noise in DC SQUIDs	294
7.4.	RF SQUID sensors	295
	7.4.1. RF SQUID structures	295
	7.4.2. Noise properties of RF SQUIDs	296
7.5.	Review of SQUID sensors	297
	7.5.1. DC SQUID sensors	297
	7.5.2. RF SQUID sensors	301
7.6.	High-T_c SQUID sensors	302
7.7.	SQUID gradiometer sensors	305
	7.7.1. Basic configurations of SQUID gradiometers	305
	7.7.2. Requirements of SQUID gradiometer configurations and their characteristics	307
7.8.	Concluding remarks	309
References		309

Chapter 8. Functional magnetic-field sensors — 313

8.1.	Functional multisensors	314
	8.1.1. Multisensors for magnetic fields and temperature	314
	8.1.2. Multisensors for magnetic fields, temperature and light	319
8.2.	Functional gradiometer sensors	325
	8.2.1. Hall-effect gradiometer sensors	326
	8.2.2. BMT gradiometer sensors	328
8.3.	Magnetic-field vector sensors	329
	8.3.1. Methods for the registration of the components of the magnetic vector	330
	8.3.2. 2-D and 3-D Hall-effect vector magnetometers	331

8.3.3. 2-D and 3-D BMT vector magnetometers 333
        8.3.3.1. BMT vector sensors for simultaneous 2-D and 3-D registration . . 333
        8.3.3.2. BMT vector sensors for successive 3-D registration 338
8.4. Concluding remarks 341
References 341

Chapter 9. Interface and improvement of solid-state magnetosensor characteristics . . 343

9.1. Biasing circuitry 343
9.2. Signal-processing electronics for solid-state magnetosensors 346
9.3. Magnetic flip-flop based sensors 354
9.4. Interfacing solid-state magnetic sensors with digital systems 357
9.5. Improvement of solid-state magnetosensor characteristics 359
    9.5.1. Offset reduction 359
    9.5.2. Cross-sensitivity reduction 370
    9.5.3. Compensation of nonlinearity 377
9.6. Concluding remarks 379
References 379

Chapter 10. Applications of solid-state magnetosensors 383

10.1. Measurement of the magnetic field 384
10.2. Measurement of the current, power and related electrical quantities 388
10.3. Measurement of nonelectrical and nonmagnetic quantities 393
    10.3.1. Principles of operation 393
    10.3.2. Magnetic sensor systems 394
    10.3.3. Selected contactless mechanical sensing devices 397
References 406

Conclusion 409

Appendices 411

I. Magnetic terms and units 411
II. Important quantities and contstants used in galvanomagnetism 412

Subject index 413

Chapter 1

Principles and technologies for the acquisition of magnetic data

1.1. General principles

At the present stage of technological development the data gathered from the environment are processed by means of electronic devices mainly because of the favorable performance to price ratio of electronic components. However, an alternative future method of data acquisition by biological means as achieved by Man and the animals should not be excluded. In general any modern information-processing system, including magnetic data acquisition, is a *triptych* [1–3] made up of the following elements:

(i) an input transducer, which in scientific publications is most often referred to as a *sensor*;

(ii) a signal processor (modifier);

(iii) an output transducer (a memory, display, actuator, transmitter, etc.).

Nonelectric quantities, such as the components of the magnetic-field vector, light, chemical concentration, pressure, temperature, etc. are selectively converted by means of sensors with the maximum possible accuracy and reliability into an adequate electrical signal. In other words, sensors give the key to the physical reality by transforming into electrical signals phenomena associated with the laws of nature. The signal processor, termed modifier, transforms the signal (amplified, filtered and converted from an analog into a digital form or from a digital into an analog form) and improves the sensor characteristics (linearity, offset, temperature degradation, drift, noise, etc.). The last element of the *triptych* displays the final results of information processing in a form which can be perceived by the human senses. In the output transducer, the electrical signals are transformed back into nonelectrical domains, i.e. sound, light, etc. The output signal may be stored by means of memory devices or can be used for the performance of different operations: control-valve opening and closing, control of canal-locks, doors, jets, printing devices, etc. If necessary, by means of a transmitter, the signal can be sent elsewhere. Figure 1.1 shows the functional block-diagram (*triptych*) of a modern electronic measuring system.

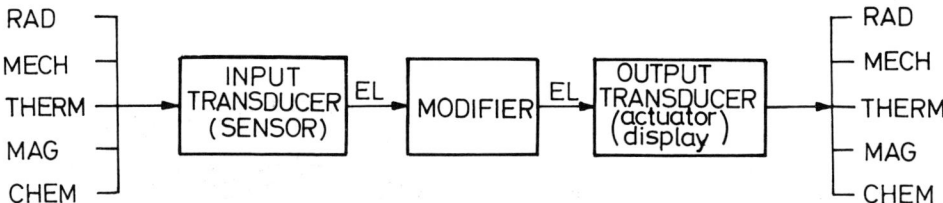

Fig. 1.1. Functional block diagram (triptych) of an electrical measurement system.

Signal transformation in transducers is always based on the adequate conversion of a certain amount of energy. Only by means of mass or energy transport can the information be transferred from the source to the respective receiver [1]. There are six signal domains including the following physical parameters:

– Radiant signals: light intensity, wavelength, polarization, phase, reflectance, transmittance.

– Mechanical signals: force, pressure, torsion, vacuum, levels, position, velocity, acceleration, slope, roughness, acoustic wavelength, amplitude.

– Magnetic signals: magnetic field strength, flux density, magnetization, permeability, momentum, direction of field propagation.

– Electrical signals: electric charge, current, voltage, frequency, inductance, resistance, capacitance, dielectric constant, electric polarization, pulse duration.

– Thermal signals: temperature, heat, specific heat, entropy, thermal flux.

– Chemical signals: concentration, toxicity, composition, pH, oxidation-reduction potential, reaction law.

Input transducers (sensors) necessarily convert one of the nonsignal domains into an electrical domain i.e. the external influence (e.g. magnetic field) acquires its adequate electronic image. Figure 1.2 illustrates the converting functions of sensors. This is possible owing to the dependence of the physical properties of some materials on appropriate external forces.

The input and output signals of the modifier are electrical in their essence, therefore no signal conversion is accomplished in this block. In the output transducer (actuator) the electrical signal is converted into one of the other five signal domains.

1.2. Conversion of magnetic energy by sensors

From a phenomenological point of view, a sensor is a component whose main input is its entire surface, or a portion of it, which is subjected to an external influence by the environment. The sensor includes output terminals which generate an electrical signal functionally connected with the input by a definite physical phenomenon (or phenomena). The main input of solid-state magne-

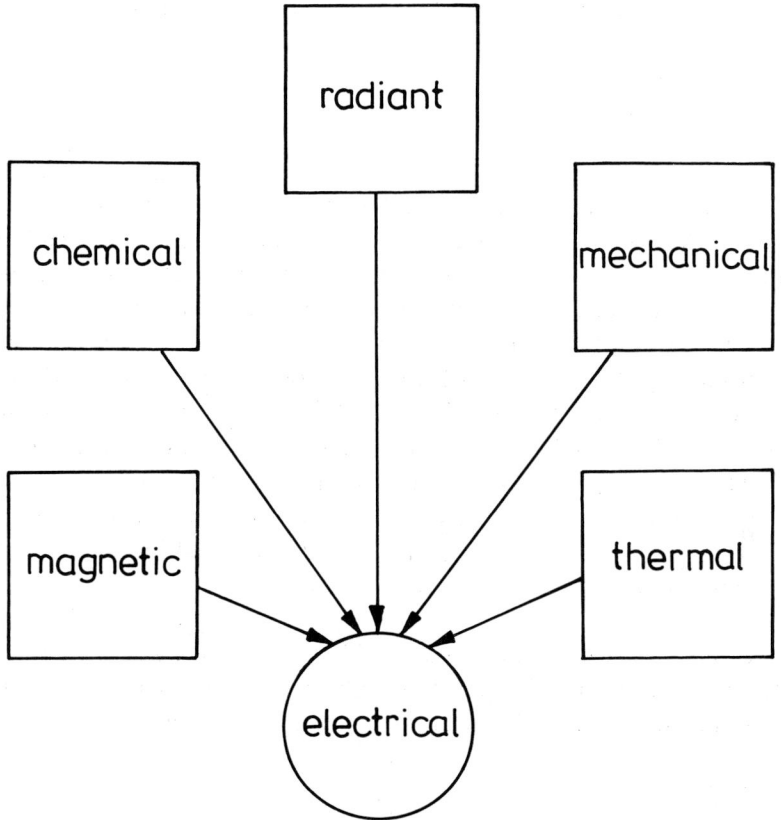

Fig. 1.2. Diagram indicating the five possible signal conversions in input transducers.

tosensitive transducers in particular is the plane of the structure to which the vector of the magnetic fields is applied and the output signal is taken from two or more terminals. There are two principal types of sensors depending on whether their operation requires external energy led through an additional input [1–3].

For instance, solar cells which utilize the photovoltaic effect, thermocouples based on the Seebeck effect, etc. generate the output electrical signal themselves without any additional power supply. The converters of this type are called *self-generating*. The energy of a magnetic field can be converted into a kinetic or internal potential energy provided it has a gradient. On the other hand electrons rotating around an axis which is fixed in space possess the well-known spin properties. This is why the application of an external magnetic field always changes the internal chemical energy of materials. This fact can be observed experimentally in the magnetization of certain materials. The modification of their properties can be registered by optical methods due to the Zeeman effect. If the sample is magnetized by an external magnetic field, it is subjected, as a whole, to the action of a force and therefore acquires a potential energy. The self-generating principles of magnetic energy conversion described above, applied to the construction of magnetic-field sensors, are associated with magnetostriction (change of the length of a magnetic bar in a longitudinal magnetic field) and magnetoelastic (change of Young's modulus of a ferromagnetic material) effects. Detailed information on these phenomena and on sensors based on them is given, for instance, in [4]. The self-generating conversion of magnetic energy can also be accomplished by means of Faraday's induction in magnetic coils. The temporal variation of the magnetic flux density leads to its conversion into the kinetic energy of moving electric charges.

The conversions which belong to the second type are called *modulating* and in them the output energy of the sensor is fed by an external power source through an additional input. The nonelectrical signal exerts an influence on a certain amount of the energy coming from external sources, which determines the output information. There are four possible manners in which such an influence can be exerted according to the law of conservation of energy:

(i) The first possibility is when the external energy source gives rise to a kind of energy different from that generated at the output. In this case there is a self-generating conversion from external energy into output energy. The input energy influences the extent to which this conversion takes place.

(ii) The second possibility occurs when the energy of the external source and the output energy are of the same type. The action at the input determines the self-generating conversion of the external energy into a type which is not measured at the output.

(iii) The third possibility is when a nonelectrical input signal modulates the direction of the external energy in the sensor structure, the result being the generation of an adequate amount of that energy at the output.

(iv) The fourth case is observed when the input sensor signal determines the moment at which the external source energy reaches the output. This mechanism is typical of digital magnetometers, e.g. carrier-domain magnetosensors.

In the first two cases, (i) and (ii), the input energy (the nonelectrical signal) modulates the self-generating conversion, whereby the modulated energy is determined only by the number of material particles. The last two possibilities, (iii) and (iv), occur only if the external energy is spatially confined, e.g. within the volume of the sensor device. Such energies are kinetic energy (of moving electric charges), and internal energy, chemical energy or heat. The solid-state magnetic field sensors discussed in this work are based on the modulation principle and, generally, cases (ii)–(iv) are characteristic of them. The input signal is a magnetic field, an electrical power source is connected to the second (additional) input and the output is also electrical. The external energy is localized within the magnetosensitive structure and the magnetic field, depending on the particular galvanomagnetic effect, deter-

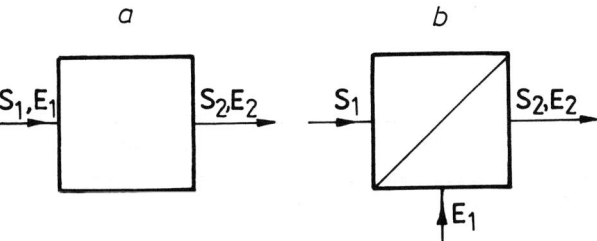

Fig. 1.3. Graphic symbols for (a) self-generating transducer with an input signal S_1 and an energy E_1, and at the output a signal S_2 and an energy E_2; and (b) a modulating transducer with a signal S_1 at one input and an energy E_1 at the other input, and at the output a signal S_2 and an energy E_2 [1].

mines the direction of the electrical power in the device or the moment at which the energy reaches the output (current-domain magnetometers).

Therefore, the general conclusion is that in the modulating-type magnetic-field sensors, the magnetic energy only influences the magnitude and direction of the velocity of the mobile electric charges. By means of the Lorentz force, the magnetic energy interacts with the kinetic energy of the moving particles, which depends on the magnitude and direction of the particle velocity, determined, in their turn, by the external supply of electrical power. The Hall devices based on the Hall effect, magnetodiodes (Suhl magnetoconcentration), magnetotransistors, etc. are examples of such modulating sensors.

Figure 1.3 is a graphical illustration of the two basic types of sensors: self-generating and modulating. Refs [1] and [3] present a convenient sensor classification employing a three-dimensional orthogonal diagram of physical effects, the so-called *cube of sensor effects*, in which the effects are divided into self-generating and modulating. The domains of the input signal, the output signal and the additional source energy have been chosen as classification indices (Fig. 1.4).

The vectors along the X and Y axes indicate the form of the input energy and the output energy, respectively, while the vector along the Z axis indicates the form of the input signal. Every effect can be placed in at least one cube. For instance, the effect that can be placed in the cube depicted here is the temperature dependence of magnetic susceptibility. A similar diagram can be constructed for the position and time derivatives of the measured variables. In compliance with this diagram, when a very small signal is to be measured, for instance in cardiomagnetometry, the proper choice is the utilization of a modulation effect and the pertinent sensor (in this case the superconductivity effect and a SQUID). If the registered signal is not strong enough, and the use of a battery poses serious problems, or if a zero-offset sensor is needed, a self-generating effect and the relevant sensor are to be preferred.

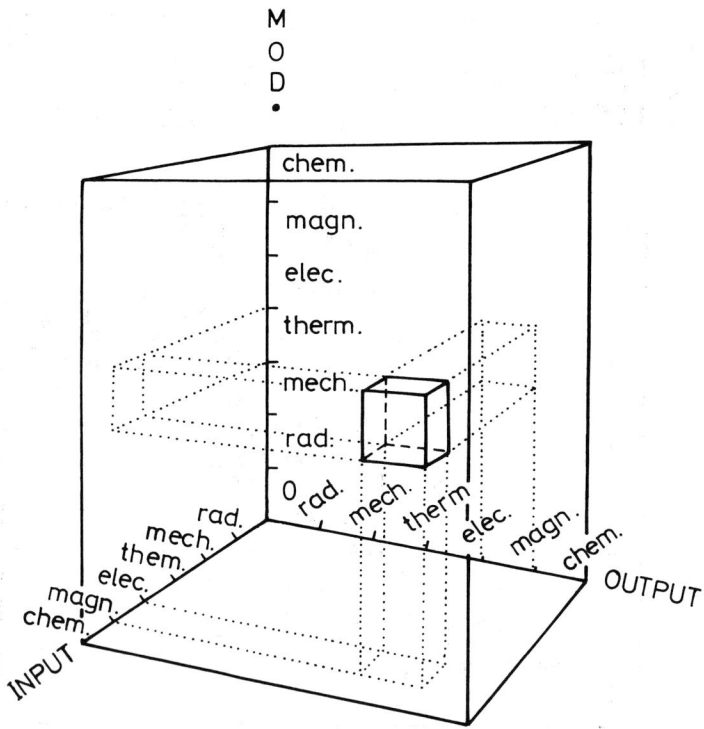

Fig. 1.4. Three-dimensional diagram of physical effects: the *sensor effect cube* [2].

1.3. Nature of the magnetic field and the Lorentz force

The origin of magnetosensitivity in most solid-state sensors is the Lorentz force, which is the connection between the electric and magnetic forces. Such categories as the magnetic and electric fields, velocity of charge carriers, etc. in the analysis of this type of sensor are often assumed a priori, without the necessary discussions. Unlike other nonelectrical scalar parameters like temperature, for instance, the magnetic field B is a vector and is characterized by its magnitude and direction. The choice of the reference system, with respect to which the components of the magnetic field B can be correctly introduced and measured, is of principal importance. Another point of interest is the so-called *assigned velocity*, which exists only in our imagination, as the methods and instruments for its determination have not been specified, nor has the coordinate system within which the experiment is to be performed. Therefore, the connection between the electric and magnetic fields, the properties of the Lorentz force and the significance of magnetosensitive devices as measuring instruments should be explained within a general physical concept.

The force F which is applied to the electric charge q depends on its location as well as its velocity v. Each point in space is characterized by two vector quantities which determine the total force F acting upon an electric charge q.

The first one is the electric force $F_E = qE$ which results from the electric field E. It does not depend on the motion of the charge q. The second force has a magnetic origin and is actually a function of the charge velocity v. Unlike all other forces, this magnetic force F_M possesses a number of surprising properties: at any point in space its direction and magnitude is determined by the direction of the vector v; at any moment it is perpendicular to the vector v; in any region of space the magnetic force F_M is always perpendicular to the fixed direction and its magnitude is proportional to that component of the velocity v which is perpendicular to the fixed direction. All these properties can be described by the introduction of the magnetic-field vector B which determines the fixed spatial direction and at the same time serves as a proportionality factor between the magnetic force F_M and the velocity v, i.e. $F_M = qv \times B$. Therefore the total electromagnetic force F_L, termed also the Lorentz force, exerted upon a charge q will be:

$$F_L = F_E + F_M = qE + qv \times B = q(E + v \times B). \qquad (1.1)$$

It follows from the above definition of the vector B that it changes if the reference system with respect to which the charge velocity is determined is replaced by another one. In Einstein's special theory of relativity magnetism and electricity are not independent phenomena. They determine a uniform electromagnetic field and serve as a method of describing the forces acting upon the moving charge q. The separation of the electric from the magnetic component of that interaction, i.e. the relation of E to B, is to a large extent dependent on the choice of the (reference) system. Therefore no absolute vectors B and E exist in nature, as the parameters which describe the magnetic and the electric fields are relative, depending on the choice of the reference system with respect to which the velocity v of charge carriers is determined. In order that the vectors B and E might have physical meaning, the reference system must be inertial, i.e. its motion must be uniform along a straight line without any acceleration. Only in such inertial systems can the previously unknown components of the field B which is generated by the moving charges be correctly measured.

Assuming that space is isotropic and three-dimensional (Euclidean), it can be described in terms of an orthogonal cartesian coordinate system $(x \perp y \perp z)$ and the time t. The Lorentz force F_L (1.1) acting upon an electric charge q at a point, whose coordinates are (x_0, y_0, z_0, t), is determined by the vectors E and B, and is invariant, if the source of the fields E and B is uniform in its motion along a straight line. The inertial reference system usually used for the investigation of magnetic sensors, for instance the laboratory, which is fixed with respect to the Earth, includes the sensor, the apparatuses which generate the magnetic field, the power sources, any instruments needed for the measurement, and the researcher himself. As regards the magnetic sensor, the charge carriers in it are in quasi-uniform motion with a drift velocity $v = v_{\text{dr}}$, which results from the electric field E. If the initially unknown direction and magnitude of the vector B are to be determined, this information can be obtained from the Lorentz force which generates the respective galvanomagnetic effect in the sensor, exposed to a field E. For this purpose three mutually perpendicular magnetic transducers are employed (or an integrated three-dimensional vector sensor for magnetic fields), which measure the components B_x, B_y and B_z at a given point. Then the magnitude of B is calculated from the formula $|B| = (B_x^2 + B_y^2 + B_z^2)^{1/2}$, and the direction of the magnetic field is determined from the vector components B_x, B_y and B_z in the cartesian coordinate system. In this case the magnetic sensors serve as analyzers of the field B.

1.4. Methods and applications of magnetic field measurements

Regardless of its character, information is usually encoded as a magnitude of a physical quantity, and measurement is necessary in order to determine it. In the measurement process the magnitude of the respective parameter is expressed by a number which is compared with an appropriate scale for that same parameter. Only two techniques for comparison are known. The first is an analog technique using a classical measuring instrument with a pointer. The position of the pointer is compared with a scale. The actual comparison is made by an observer or by taking a photo. The second technique is based on an analog/digital (A/D) converter. The electric voltage, current or signal frequency is compared to a reference voltage, current or frequency by means of electronic digital techniques. The result of the comparison is a numerical value which is presented in a digital code. The great variety of physical parameters requires a great number of alternative techniques of comparison. If one is to measure such quantities, their magnitudes must be converted into magnitudes of a reference voltage, current or frequency, or into a magnitude of the displacement of the pointer of an instrument. All these cases of conversion involve energy exchange. Therefore the magnitude of the respective physical quantity, the magnetic field, for instance, is converted into a certain kind of energy, a definite relationship existing between it and the basic parameter. If the energy registered is not suitable for comparison with a given scale, it is converted into another type allowing the application of at least one of the two possible techniques of measurement.

In magnetometry the magnetic properties are determined by:

(a) the force exerted upon the magnetized material; and

(b) the magnetic flux density or the variation of the magnetic field B in the neighborhood of the sensor or of the sample investigated.

There are various indirect methods based on a well-known and reproducible relationship of the detected phenomenon as a function of the magnetic properties of the sample. The great variety of particular cases makes it impossible to develop a technique and a magnetic field sensor for universal application. All the familiar techniques and sensors have their own advantages and limitations. Magnetic field theory is described by several fundamental parameters: the magnetization M, magnetic field strength H and the density of the magnetic flux, termed the induction B. They are connected by the formula:

$$B = \mu_0(H + M), \qquad (1.2)$$

where μ_0 is the magnetic permeability of vacuum.

This book deals primarily with nonmagnetic media (nonmagnetic semiconductors and materials), for which the vector $M \approx 0$ and the measurement by sensors boils down to the determination of the parameter B (in SI units the magnetic induction B is measured in T, Tesla).

$1\,\text{T} = 1\,\text{Wb m}^{-2} = 1\,\text{V s m}^{-2}$, and $1\,\text{T} = 1 \times 10^4\,\text{G}$, $1\,\gamma = 10^{-5}\,\text{G}$.

(Appendix I).

Like other input transducers, magnetic sensors have two main groups of application [1,5,6]:

– *Direct applications*. The registration is accomplished by a single galvanomagnetic effect. The sensor converts the *magnetic field into an electrical signal*, which is the final goal of the measurement. The determination of the geomagnetic field by a Hall-effect based sensor serves as an example.

– *Indirect applications* by means of tandem transducers. The nonmagnetic signal is measured with the help of an intermediate conversion into a magnetic field which is then followed by a conversion of the *magnetic field into an electrical signal*. Two or more separate physical effects are used for this purpose. The mechanical displacement of a permanent magnet, for instance, can be detected by the variation of the magnetic flux density registered by a magnetotransistor or a magnetodiode.

As seen in Fig. 1.5, which complements Fig. 1.2, five tandem conversions by means of magnetic energy conversion are possible in principle. The tandem conversions are represented in Fig. 1.5 by dotted lines.

The purpose of direct applications (by a single effect) is to gather information about the direction and magnitude of the magnetic vector itself [5,6]. It includes:

(a) Measurement of the natural and artificially generated magnetic fields in the surrounding natural and technical environments, altitude control of satellites, submarines, balloons and aircraft, geophysical and geological investigations, classical and electronic compasses, prediction of volcanic eruptions and earthquakes by making use of geomagnetic-field variations, etc.

(b) Retrieval of data stored on magnetic tapes or disks.

(c) Electromagnetic control of electrical machines, AC and DC motors, transformers, acceleration systems in nuclear physics, MHD generators, etc.

Indirect application includes:

(d) Contactless keys and DC motors, remote-control valves and door locks, keyboards, panels etc.

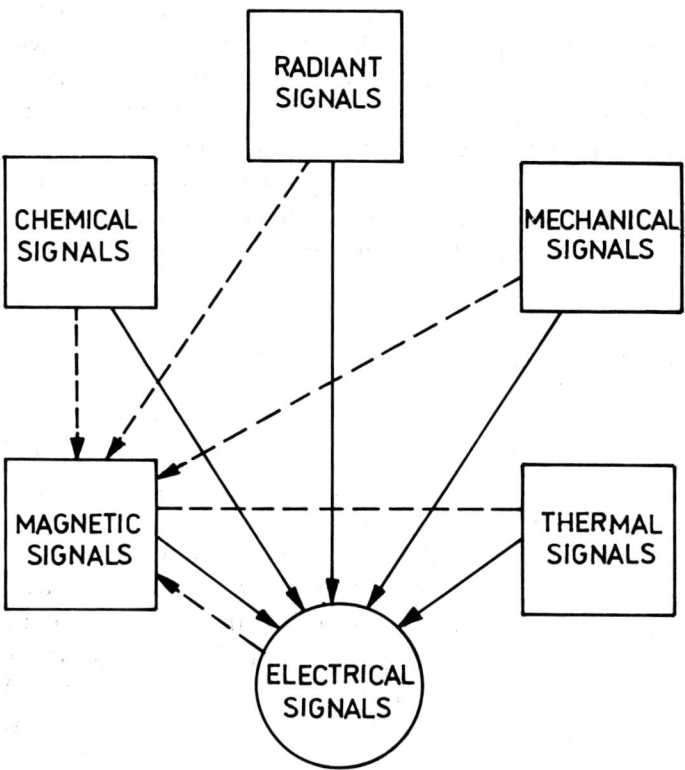

Fig. 1.5. The five possible tandem conversions using an intermediary conversion to magnetic signals (broken lines), see also Fig. 1.2.

(e) Transducers for linear detection: proximity keys, position sensors, magnetic levitation, etc.

(f) Detection of angular displacements: angular decoders, synchroresolvers, tachometers, crankshaft position sensors for ignition control in motor vehicles, etc.

(g) Traffic detection: determination of submarine locations in seas and oceans by the perturbations of the geomagnetic field caused by the submarine, determination of the spatial positions of objects made of ferromagnetic materials, etc.

(h) Current detection: contactless measurements of the magnitudes of direct and alternating currents without interrupting the current-carrying conductors, Wattmeters, power measurements, etc.

(i) Biomedicine (biomagnetism): measurement of magnetic fields generated by human and animal organs and systems, e.g. heart, brain, nerves, muscles, etc.

At this stage groups (b), (d), (f) and (h) seem to hold the best promise for the application of solid-state magnetic sensors. On the other hand each group has its own requirements as to sensors, mainly because of the enormous differences in the magnitudes of the magnetic fields to be detected. For instance, the magnetic induction on the earth's surface in equatorial regions is approximately 0.035 mT (0.35 G) and galvanomagnetic sensors (Hall elements, magnetotransistors, etc.) can be successfully applied. A resolution of 1 nT (1 nT = 1 γ) is required for the detection of geomagnetic-field variations. In the field of biomagnetism the typical values of the magnetic induction measured are several fT. This is why superconductivity quantum interference devices (SQUIDs) should be used. In particle acceleration systems, the magnetic field induction is several T. Therefore this extremely wide range of magnitudes of the induction can be covered by different types of magnetic sensor. The diversity of analog and digital applications also calls for magnetic transducers based on different principles of operation. That is why the search for a single type of sensor of universal applicability is irrelevant in magnetometry.

1.5. State of the art of magnetosensitive device technology

The operation of a great number of conventional magnetic-field sensors is based on the Lorentz force (1.1). Depending on the approach adopted in [5] and [7] two classes of magnetic-field sensors can be distinguished in view of the relationship:

$$\boldsymbol{B} = \mu\mu_0\boldsymbol{H}, \tag{1.3}$$

where μ is the magnetic permeability of the sensor material.

The first class includes ferromagnetic and ferrimagnetic materials, whose $\mu \gg 1$ contributes to the enhancement of sensitivity. The NiFe-film magnetoresistors, magnetostriction of Ni cladding of optic fibres, magneto-optic effects in garnets and different magnetic-field transducers, which use flux concentration implements, serve as examples of this class of sensors. The second class includes diamagnetic and paramagnetic materials whose $\mu \sim 1$ does not lead to significant enhancement of the magnetic field in the active sensor region. Typical representatives of this class are galvanomagnetic sensors based on nonmagnetic semiconductors and materials like Si, Ge, GaAs, InSb, InAs, InP, etc.

Optoelectronic magnetosensitive technologies use light as an intermediate signal carrier. Such devices are the transducers based on the Faraday rotation of the polarization plane of linearly polarized light as a result of the action of the Lorentz force upon bound electrons. Useful magnetic field sensors can be manufactured by making use of optical-fiber coils with a long light path and an adequate rotation per unit magnetic field. On this basis, high-voltage, magneto-optical current transducers can be constructed [8,9]. In addition optically transparent ferrimagnetic garnet materials have higher values of the angle of the Faraday rotation per unit path length [10]. Optical fibers with magnetostriction jacketing, Ni, for instance, or mechanically strained optical fibers coiled around a cylinder of a magnetostriction material are among the most promising optoelectronic magnetic-field sensors.

A strain transferred to a magnetostriction fiber changes the optical path length which in turn modulates the detected phase of an optical-fiber interferometer. A resolution better than 1 nT has been achieved [11].

A resolution unsurpassed among all kinds of magnetic sensors is manifested by devices based on nuclear magnetic resonance (NMR) and especially by SQUID magnetometers, which are the most sensitive of all magnetometers [12]. SQUIDs are capable of magnetic-field registration in the pT and fT ranges. Quantum galvanomagnetic effects which occur in some metals (Nb, Pb, etc.) and compounds when cooled down to helium temperatures are made use of. High-temperature superconductivity in Ba–La–Cu–O metal–ceramic-like materials at temperatures near the boiling point of liquid nitrogen ($T = 77$ K) has recently been discovered [12,13]. Periodic oscillations of the current induced in a superconducting ring interrupted by a weak link (tunnel contact) are a function of the density of the magnetic flux which pierces the ring. This phenomenon is known as the Josephson effect [14]. In RF SQUID sensors, for instance, the ring is inductively coupled with a RF circuit and the current induced modulates the resonant frequency of the circuit. If a suitable feedback circuit is used, the feedback current can be monitored and used as a measure of magnetic flux density. The superconducting ring may be coupled with a superconducting search coil to increase the sensitivity. An on-chip integration of the SQUID and the superconducting coil together with the feedback and signal processing circuitry has been successfully achieved experimentally [15].

Magnetosensitive devices based on thin metal films or wires use ferromagnetic materials. The low magnetostriction $Ni_{81}F_{19}$ alloy is suitable for this purpose. The most pronounced sensor effect is observed in the magnetosensitive switching of anisotropic NiFe or NiCo films [16–18].

There are applications which employ transducers based on Faraday coils. They register temporal variations of magnetic flux density. The use of coils in integral circuits, however, is not an easy technological task. Another serious drawback of this sensor method is the impossibility of direct measurement of static magnetic fields [5].

Solid-state magnetosensitive devices based on well-known effects, for instance the Hall effect, magnetoresistivity, magnetoconcentration, the magnetodiode effect and related phenomena in solid-state materials, connected mainly with the deflection caused by the Lorentz force F_L are very popular and have reached a relatively mature stage. Some of the advantages of solid-state magnetic sensors are the flexibility of their design and application, their small dimensions and the stability of their characteristics. Si, Ge, GaAs and other $A^{III}B^{V}$ compounds are used as materials for their fabrication. Silicon microsensors and, over the short term, GaAs devices will remain much cheaper than their rivals because of the perfection of IC technology. This group of devices includes bulk and surface channel Hall elements, magnetoresistors, magnetotransistors, carrier-domain magnetometers and other related kinds of transducers [1,5–7,16,19–23]. Extended references on this subject can be found in the respective chapters of this book.

Galvanomagnetic effects, on the other hand, are a universal instrument for the investigation of the fundamental properties of materials. In spite of the existence of new methods for the detection of structural defects or impurities, galvanomagnetic phenomena have no rival in the extraction of information about the electron transport parameters, i.e. in the dynamic characterization of materials.

1.6. Solutions employing integrated and smart sensors

The current progress in silicon sensor technology has led to the combination of integrated circuits with the transducer itself, i.e. monolithic silicon chips which include both the sensor and the signal processing circuits. Depending on the particular application, three levels of sensor integration (including magnetic sensors) with integrated circuits can be distinguished:

(i) discrete sensors and IC circuitry;

(ii) hybrid packaging of a sensor (sensors) and IC chip(s);

(iii) a monolithic IC which incorporates a sensor (sensors) and the processing circuitry on a single chip.

An on-chip or off-chip scheme for processing the sensor signal is chosen depending on the particular application and the resulting price. Generally, the one-chip combination of sensors with analog and/or digital ICs provides for maximum improvement in the metrological characteristics of the system, maximum reliability and low price as a result of mass production. Among the commercially available magnetometry devices, the magnetosensitive integrated circuits which include a Hall element and an appropriate analog or trigger-function circuit occupy dominant positions. Thus the weak signal from the Hall sensor is amplified at the source without being transferred to another location. Thereby the overall signal to noise ratio of the measuring system is improved. Magnetic fields are not the only external factor that influences the properties of silicon. Silicon is sensitive to mechanical stress, temperature, radiation, etc. This is why the ideal magnetosensitive transducer, whose output signal is a linear function of the field B and whose value is zero in the absence of magnetic fields ($B = 0$), is only a desirable approximation. There are a lot of imperfections in solid-state transducers, including magnetosensitive devices: e.g. cross-sensitivity, nonlinearity, noise, parameter instabilities, etc. The solution of these problems within the framework of analog electronics is extremely difficult and perhaps impossible [24]. Traditional approaches to the application of linear systems demand a linear output signal from the transducer. For instance, the compensation of the drift of sensor parameters by means of continuous electronics is practically impossible, since the components of analog circuits are subjected to parameter drift themselves. The introduction of low-priced and compact digital signal processing is an entirely new approach in the development of sensor electronics. The transducer defects considered above can be easily avoided if digital circuitry is employed.

A prime example of a problem which has greatly diminished in importance is the linearization of input/output characteristics which exhibit simple nonlinearity. The realization of inverse nonlinearity in continuous electronics is not a trivial task. Conversely, the *programmatic* realization is simply a stored table of ideal outputs corresponding to equally spaced values of actual sensor outputs. Thus the digitized form of the actual outputs can be used as an increment pointer to the value of the output that an ideal linear transducer would have. This technique of the *reference table* is very powerful in a number of applications [25].

The natural extension of digital techniques into sensor electronics has led to the development of a new generation of transducers, i.e. *intelligent sensors*. In spite of the existence of different points of view about the meaning of the term smart sensor, the definition given in [26] seems to represent most adequately the tasks tackled by these devices. The principal criteria refer to the ability of the sensor system to perform logical functions, to take decisions and to take part in two-way communication. Smart sensors dramatically promote such applications as self calibration, computation and multisensing. The possibility of independent and simultaneous measuring of more than one variable, e.g. magnetic field and temperature, by means of one and the same microstructure can eliminate, in a wide temperature range, such shortcomings as the temperature dependence of magnetosensitivity, offset, parameter drift, etc. [27]. The simultaneous measurement of magnetic field, temperature and light flux by one and the same sensor structure is a sophisticated solution. Multisensing helps to avoid cross-sensitivity by the sensor itself in combination with digital circuitry and appropriate software.

Figure 1.6 presents in a generalized form the problem of sensor compensation within a smart solution. The sensor experiences a variety of physical inputs, of which one, say $x(t)$, is the desired one, while the rest, $y(t)$ and $z(t)$, are unwanted. Furthermore, the sensor itself adds further noise and drift signals, $n(t)$ and $d(t)$. The output of the sensor is some complicated function, not necessarily linear and

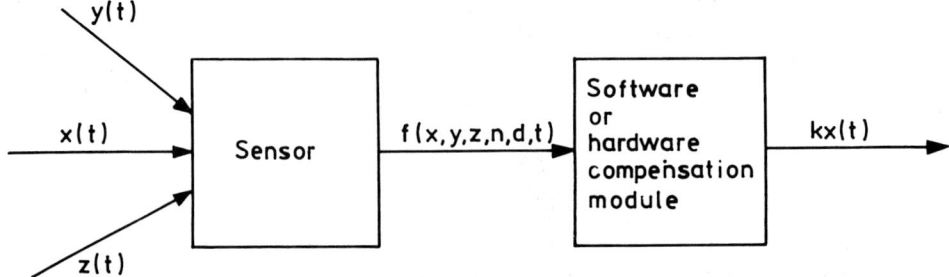

Fig. 1.6. Principle of the problem of sensor compensation.

time-invariant, of all of these variables. The compensation element is a conceptual box representing the hardware and/or software operations which produce an output that is proportional to $x(t)$ and nothing else. In the general case this is a tall order, but fortunately in many specific cases a reasonable solution is achievable. Some sensor defects may pose few problems individually but may prove very difficult in combination, with a nonideal frequency response. Even in this case, however, if the defects can be modelled as two cascaded processes, then their compensating elements can be cascaded in the reverse order [28].

The basic metrological parameters of the smart sensor constructed in such manner can be considerably improved by a compromise distribution of the electronic processing power among the Hall transducer, interface electronics and the central processor [29]. At the present stage the "intelligence" of the individual transducer and the one-chip digital module integrated with it is limited primarily by the price, by the possible incompatibility of fabrication processes for the circuitry and the sensor, (e.g. concentration of doping impurities, epitaxial layer thickness, etc.) and by specific protection of the IC from environmental influences like temperature, humidity, light, radiation, magnetic field, etc., which nevertheless must be "seen" by the sensor [30]. It is the author's opinion that the most probable application of smart magnetosensors will be to measure very weak magnetic fields, for instance the geomagnetic field, when the level of the output signal is near the threshold value determined by the noise.

References

[1] S. Middelhoek and S.A. Audet, Silicon Sensors, Academic Press, London, 1989, Ch. 1.
[2] S. Middelhoek and A.C. Hoogerwerf, Sensors and Actuators, 10 (1986) 1–8.
[3] S. Middelhoek and D.J.W. Noorlag, Sensors and Actuators, 2 (1981/82) 29–41.
[4] Sensors, Vol. 5, VCH, Weinheim, 1989, R. Bolk and K.J. Overshott (eds.): K.J. Overshott, 34–42; G. Hinz and H. Voigt, 98–152.
[5] V. Zieren, Integrated silicon multicollector magnetotransistors, Ph.D. dissertation, Delft Univ. Techn., The Netherlands (1983).
[6] S. Foner, IEEE Trans. Magn., MAG-17 (1981) 3358–3363.
[7] H.P. Baltes and R.S. Popović, Proc. IEEE, 74 (1986) 1107–1132.
[8] S.C. Rashleigh and R. Ulrich, Appl. Phys. Lett., 34 (1979) 768–770.
[9] A. Papp and H. Harms, Appl. Opt., 19 (1980) 3729–3834.
[10] J.P. Castera and G. Hepner, Device for modulating optical radiation by a variable magnetic field, US Patent 4236782 (1980).
[11] A. Dandridge, A.B. Tveten, G.H. Sigel, Jr., E.G. West and T.G. Giallorenzi, Electron. Lett., 11 (1980) 408–409.
[12] H. Koch, in: Sensors, Vol. 5, R. Bolk and K.J. Overshott, (eds.), VCH, Weinheim, 1989, 381–445.
[13] J.G. Bednorz and K.A. Muller, Z. Phys., B64 (1986) 189–193.
[14] B.D. Josephson, Phys. Lett., 1 (1962) 251–253; Adv. Phys., 14 (1965) 419–451.
[15] N. Fujimaki, H. Tamura, T. Imamura and S. Hasuo, IEEE Trans. Electron Devices, ED-35 (1988) 2412–2418.
[16] J.P.J. Groenland, C.J.M. Eijkel, J.H.J. Fluitman and R.M. de Ridder, Sensora and Actuators, A, 30 (1992) 89–100.

[17] A.W. Vinal, IEEE Trans. Magn., MAG-20 (1984) 681–686.
[18] International Magnetics (INTERMAG) Conf. Proceed., B.E. MacNeal, R.E. Fontana, Jr., and J.C. Smits (eds.) IEEE Trans. Magn., Session BE-Magnetic Sensors, MAG-20 (1984) 954–974.
[19] S. Kordić, Sensors and Actuators, 10 (1986) 347–378.
[20] R.S. Popović, Sensors and Actuators, 17 (1989) 39–53.
[21] Ch. S. Roumenin, Sensors and Actuators, A, 24 (1990) 83–105; A, 30 (1992) 77–87.
[22] H.P. Baltes and A. Nathan in: Sensors, Vol.1, T. Granke and W. Ko (eds.), VCH, Weinheim, 1989, 195–215.
[23] S. Cristoloveanu, J. Korean Inst. Elect. Eng., 10 (1986) 86–95.
[24] J.E. Brignell and A.P. Dorey, J. Phys. E: Sci. Instrum., 16 (1983) 952–958.
[25] J.E. Brignell, Phys. Bull., 34 (1983) 342–343.
[26] J.M. Giachino, Sensors and Actuators, 10 (1986) 239–248.
[27] Ch.S. Roumenin and P.T. Kostov, Sensors and Actuators, 8 (1985) 307–318.
[28] J. E. Brignell, J. Phys. E: Sci. Instrum., 20 (1987) 1097–1102.
[29] A.R. Cooper and J.E. Brignell, Sensors and Actuators, 7 (1985) 189–198.
[30] G.C.M. Meijer, J. van Drecht, P. de Jong and H. Neuteboom, Sensors and Actuators, A, 35 (1992) 23–30.

Chapter 2

Galvanomagnetic phenomena used in solid-state sensors

2.1. Introduction

The static and dynamic behavior of devices exposed to external electric and magnetic fields can be described by a system of fundamental equations based on concepts from different fields of physics. The electrophysical theory of solid-state devices and the corresponding fundamental equations are included in [1–6] while the applications of numerical techniques are given in [7] and [8]. Unfortunately, with the exception of [1], in these works, the effects generated by the applied magnetic field have been ignored or have played a negligible role in the considerations. The theory of charge-carrier transport in the presence of magnetic fields is consistently developed in [9–12] with the emphasis laid on galvanomagnetic phenomena in semiconductors. The theoretical aspects of galvanomagnetic phenomena are discussed in publications on applications of Hall sensors [13–16]. The Hall effect has been considered in equilibrium conditions, i.e. assuming a negligible deviation of the carrier concentration from its equilibrium value. An analysis of the Hall effect and an attempt to interpret the experimental results is presented in [17]. In a later publication a systematic theory and the fundamental equations of transport phenomena in semiconductor devices under nonequilibrium conditions in the presence of a magnetic and an electric field are presented in [18]. Magnetoconcentration and the magnetodiode effect are considered in [21,22] for example. A consistent theory of the Hall effect and related phenomena in intricate structures with $p-n$ junctions and in magnetodiodes is presented in [23,24]. The Hall effect has recently been considered in detail and its manifestations in particular sensor devices are discussed in [27].

The theory of superconductivity as the coexistence of infinite conductivity and the Meissner effect (expulsion of the magnetic field from the sample volume) arising in most metals and alloys at temperatures lower than a certain critical value, as well as the fundamental equations describing, it can be found in monographs [28–33], which have become classics. The Josephson effect (quantum interference associated with tunnelling of an electron pair through a weak link which divides two superconductors) is discussed in [31–36]. In these works superconductivity is considered at helium temperatures ($T = 4.2$ K). Josephson-effect-based cryoelectronic sensors for magnetic-field measurements, the so-called SQUIDs (Superconducting Quantum Interference Devices) are described in [31–33,37–40]. High-temperature superconductivity and related problems are discussed in [39–42].

In addition to the above review of works on galvanomagnetic phenomena, there are hundreds of other original articles and monographs. Their contribution to the understanding of transport phenomena in condensed matter in the presence of a magnetic field is indisputable. The purpose of all these works and the citations in them is to serve as a reference point in the theory of solid-state magnetic sensors. This chapter is dedicated to the fundamental physical principles of the galvanomagnetic phenomena used in interpreting the operation and the application of solid-state magnetic-field transducers.

2.2. Conductivity of condensed matter in the crossed electric and magnetic fields

Galvanomagnetic effects arise as a result of the influence of an external magnetic field B upon the directed motion of quasi-free charge carriers in condensed matter. Most often such a directed motion is obtained by applying an external electric field E. Ohm's law for the current density j in the case of one type of charge carrier, spherical Fermi surfaces and a moderate electric field under isothermal conditions is given by the expression:

$$j = \sigma E. \tag{2.1}$$

The connection between the electrical conductivity σ in the absence of a magnetic field and the drift mobility μ_{dr} of carriers is expressed by the formulae:

$$\sigma_n = -q\mu_n n \text{ for electrons, and}$$
$$\sigma_p = q\mu_p p \text{ for holes}, \tag{2.2}$$

where

$$\mu\left(\frac{q}{m^*}\right)\langle\tau\rangle \equiv \mu_{dr}. \tag{2.3}$$

On the other hand the drift velocities of electrons and holes are expressed in terms of the drift mobilities μ_n and μ_p in the familiar relationships:

$$v_{dr}^{(n)} = \mu_n E; \quad v_{dr}^{(p)} = \mu_p E,$$

the corresponding current densities being:

$$J_n = q\mu_n n E; \quad J_p = q\mu_p p E.$$

The notations introduced so far are as follows: q is the magnitude of the charge of the electron, the positive sign "+" is used for holes, and "−" is used for electrons. The concentration of the carriers is denoted by n or p, m^* is their effective mass, and the time to cover their mean free path is $\langle\tau\rangle$, known also as the relaxation time. The symbol $\langle\ldots\rangle$ stands for averaging over the energy of an ensemble of quasi-free carriers. In the general case the conductivity is a tensor with nonzero diagonal components:

$$\sigma = \begin{pmatrix} \sigma_{xx} & & 0 \\ & \sigma_{yy} & \\ 0 & & \sigma_{zz} \end{pmatrix}, \tag{2.4}$$

where $\sigma_{ii} = qn\mu_i$; $\mu_i = (q/m_i^*)\langle\tau\rangle$. In an isotropic medium the respective conductivities are equal ($\sigma_{xx} = \sigma_{yy} = \sigma_{zz}$). The permittivity of the material is time independent, the polarization due to mechanical stress is negligible and there are no piezo-electric and ferroelectric phenomena. The permittivity of the material is assumed to be uniform and is treated as a scalar quantity.

In vacuo (no scattering of carriers) in the absence of an electric field ($E = 0$), the application of a homogeneous external magnetic field B leads to the appearance of a magnetic force $F_M = q(v \times B)$ (§1.3). The force F_M is always perpendicular to the vector v of carrier velocity. The physical result is that the force F_M does not affect the magnitude of the carrier velocity and changes only its direction, thus making the trajectory circular, the cyclotron frequency being:

$$\omega_c = \left| q\frac{B}{m^*} \right|. \tag{2.5}$$

The corresponding cyclotron radius is:

$$r_c = \frac{v}{\omega_c}. \tag{2.6}$$

If mutually perpendicular electric, E, and magnetic fields, B, are assumed, i.e. $E \perp B$, whose directions are chosen, for convenience, to be parallel to the x-axis and to the z-axis of a cartesian coordinate system, with respect to that system the two fields will be $E = (E_x, 0, 0)$ and $B = (0, 0, B_z)$ respectively. In this electromagnetic field the well-known Lorentz force will be applied to each carrier (§1.3):

$$F_L = qE + qv \times B. \tag{2.7}$$

In a vacuum, as a result of the force F_L, the carrier drift is oriented perpendicularly to both the electric and magnetic fields, and not in the direction of the field E by contrast with the case of $B = 0$. The trajectory in the xy plane is a cycloid (Fig.

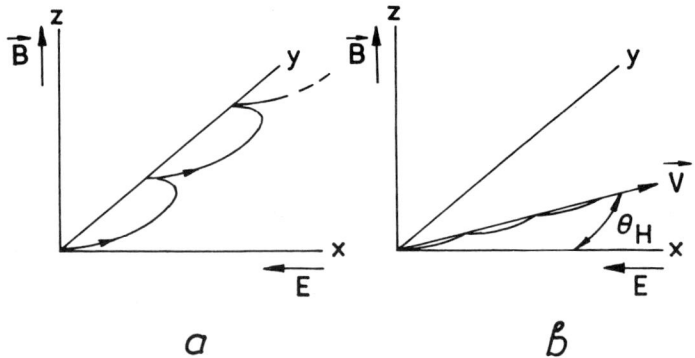

Fig. 2.1. Motion of an electron in crossed electric and magnetic fields; (a) in vacuo, (b) in a solid-state sample.

2.1a). In this case the drift velocity is determined from the expression:

$$v_y = \frac{E_x}{B_z}. \tag{2.8}$$

In real conditions, in semiconductors for instance, the mean free path of carriers is limited owing to electron (hole) scattering by the crystal lattice. A more accurate consideration yields a dependence of the transport mechanism on scattering by the crystal lattice, by ionized or neutral impurities, and electron–electron scattering. The average time interval $\langle \tau \rangle$ between two collisions also depends on temperature and energy. In this case, when there is a weak magnetic field B the collisions do not allow the cycloid motion to be completed. As a result the starting point of every cycloid is shifted in the direction of E_x. Thus a longitudinal drift component v_x arises (Fig. 2.1b). This process is repeated many times, and hence the resultant trajectory is a series of small arches. From the average time $\langle \tau \rangle$ is obtained an average velocity oriented at an angle Θ_H with respect to the vector E_x. This macroscopic parameter is generally known as the Hall angle.

The equation which describes the motion of the charges in the presence of crossed electric and magnetic fields in the hydrodynamic approximation will be:

$$\frac{\partial (m^* v)}{\partial t} = qE + q(v \times B) - \frac{m^* v}{\langle \tau \rangle}$$

therefore:

$$\frac{v_y}{v_x} = \left(\frac{q}{m^*}\right) \langle \tau \rangle B_z = \mu B_z = \omega_c \langle \tau \rangle. \tag{2.9}$$

From equation (2.9) can be determined the Hall angle:

$$\tan \Theta_H = \mu B_z \equiv \omega_c \langle \tau \rangle. \tag{2.10}$$

The value of the Hall angle $\Theta_H = \pi/4$, i.e. $\mu B = 1$, is accepted to be the boundary between low and high magnetic fields:

$$\mu B \ll 1 \text{ (for low magnetic fields)} \tag{2.11}$$

$$\mu B \gg 1 \text{ (for high magnetic fields)} \tag{2.12}$$

In high fields without quantization the Hall angle $\Theta_H \to \pi/2$ and the carrier trajectory coincides with that for a vacuum (no collisions happen in a vacuum); for instance, if $E_y = 0$, for electrons $v_x = 0$ and $v_y = -E_x/B_z$. This ideal case is characterized by a maximum value of ω_c and a minimum radius r_c (Fig. 2.1a).

In low magnetic fields, (2.11) is valid and the influence of the Lorentz force upon carrier trajectories can be neglected. The galvanomagnetic phenomena in a homogeneous medium are determined mainly by the scattering mechanisms which predominate at a given temperature. In the case of high magnetic fields (2.12), the carrier trajectory is controlled completely by the Lorentz force, and galvanomagnetic phenomena are determined mainly by the shape of the Fermi surface [43].

In the presence of an external magnetic field the nondiagonal components of the conductivity tensor differ from zero since the directions of the electric field and the resulting electric current do not coincide. The relationship between the current density and the electric field is represented by the equations:

$$j_i = \sum_j \sigma_{ij}(B) E_j. \tag{2.13}$$

If the directions of the fields $\mathbf{E}$ and $\mathbf{B}$ are assumed to be along the x and z axes respectively, and if the medium is homogeneous and isotropic, equation (2.13) can be rewritten in the following detailed form:

$$\left. \begin{array}{l} j_x = \sigma_{xx}(B) E_x + \sigma_{xy}(B) E_y; \\ j_y = \sigma_{yx}(B) E_x + \sigma_{yy}(B) E_y; \\ j_z = \sigma_{zz}(B) E_z. \end{array} \right\} \tag{2.14}$$

In an isotropic medium with cubic symmetry the following relations are valid:

$$\sigma_{xx}(B) = \sigma_{yy}(B) = \langle \mu(B) \rangle q n; \tag{2.15}$$

$$\sigma_{xy}(B) = -\sigma_{yx}(B) = \langle \mu^2(B) \rangle q n B_z; \tag{2.16}$$

$$\sigma_{zz}(B) = \langle \mu \rangle q n;$$

where $\langle \mu(B) \rangle \equiv \langle \mu/(1 + \mu^2 B^2) \rangle$; $\langle \mu^2(B) \rangle \equiv \langle \mu^2/(1 + \mu^2 B^2) \rangle$.

In the considered case the low-field condition is $\mu^2 B^2 \ll 1$ and the high-field condition is $\mu^2 B^2 \gg 1$.

An important property of the diagonal components of the conductivity tensor follows from the theorem of Onsager : they are even functions of the magnetic field **B** [44]:

$$\left.\begin{array}{l}\sigma_{xx}(B) = \sigma_{xx}(-B);\\ \sigma_{yy}(B) = \sigma_{yy}(-B);\\ \sigma_{zz}(B) = \sigma_{zz}(-B).\end{array}\right\} \quad (2.17)$$

The nondiagonal tensor components are odd functions of the field:

$$\left.\begin{array}{l}\sigma_{xy}(B) = -\sigma_{xy}(-B);\\ \sigma_{yx}(B) = -\sigma_{yx}(-B).\end{array}\right\} \quad (2.18)$$

The current density j is often referred to as an independent parameter. In such cases it is convenient to replace the tensor of conductivity in equation (2.13) by the resistivity tensor:

$$E_i = \sum_k \rho_{ik}(B) j_k. \quad (2.19)$$

By analogy with (2.13), equations (2.19) can be presented in an extended form:

$$\left.\begin{array}{l}E_x = \rho_{xx}(B) j_x + \rho_{xy}(B) j_y;\\ E_y = \rho_{yx}(B) j_x + \rho_{yy}(B) j_y;\\ E_z = \rho_{zz}(B) j_z.\end{array}\right\} \quad (2.20)$$

Similar to (2.17) and (2.18), the following relations hold:

$$\rho_{xx}(B) = \rho_{xx}(-B);$$
$$\rho_{yy}(B) = \rho_{yy}(-B);$$
$$\rho_{zz}(B) = \rho_{zz}(-B);$$
$$\rho_{xy}(B) = -\rho_{xy}(-B);$$
$$\rho_{yx}(B) = -\rho_{yx}(-B).$$

The relationship between the components of the conductivity tensor and the tensor of resistivity can be obtained from the equation system (2.14) and (2.20) by taking into account (2.15) and (2.16):

$$\sigma_{xx} = \frac{\rho_{xx}}{\rho_{xx}^2 + \rho_{xy}^2};$$

$$\sigma_{xy} = -\frac{\rho_{xy}}{\rho_{xx}^2 + \rho_{xy}^2};$$

$$\sigma_{zz} = \frac{1}{\rho_{zz}}.$$

Analogous formulae are valid for the reverse dependence $\rho_{ij} = f(\sigma_{ij})$ [12].

2.3. Hall effect

The Hall effect is a generation of transverse electromotive force in a sample carrying an electric current and which is at the same time exposed to a perpendicular magnetic field. Depending on the structural geometry this force can cause an electric field (voltage) with a transverse orientation with respect to the sample or a current deflection. The transverse voltage is the most widespread criterion for the occurrence of the Hall effect. This phenomenon was discovered in 1879 at the John Hopkins University (USA) by Edwin H. Hall [45,46], a graduate physics student at that time. The effect can be analyzed in two modes of operation: Hall voltage in long samples and Hall current in short structures.

2.3.1. Hall-voltage mode in the case of monopolar conductivity

Let us consider the carrier trajectories in a sample with the shape of a long and thin parallelepiped of dimensions l, w and t (Fig. 2.2). The sample is oriented in such a way that its thickness t is parallel to the direction of the applied magnetic field B. The sample is made of a semiconductor material with only one type of charge carrier (monopolar conductivity). By a long structure is meant the case when the length l is much greater than the width w, i.e. $l \gg w$. Ideal ohmic contacts are assumed so that their influence on the transport process may be neglected. When an external magnetic field B_z is applied, the charge carriers deflect, as a result of the Lorentz force (2.7), in the direction of the y-axis towards one of the sample boundaries (the upper or the lower boundary), depending on the direction of the vector B_z, on the current direction j_x, as well as on the type of conductivity (n or p; Fig. 2.2). If the two boundaries are not shunted, i.e. if the current component $j_y = 0$, a charge of electrons accumulates upon one of them, the upper boundary for instance, when the material is n-type and the opposite boundary remains positively charged. Moreover the carrier concentration on the upper boundary increases, and therefore the current density is increased. On the lower boundary the carrier concentration and the current density decrease. This is why charge neutrality is no longer preserved. Hence an electric field E_y parallel to the y-axis is generated between the two boundaries. This transverse field, known as the Hall field, always has such a direction so as to balance the Lorentz deflection:

$$q(v_{\text{dr}} \times B) + qE_H = 0. \tag{2.21}$$

On the other hand, the Hall field E_H acts upon the ensemble of free carriers in such a way as to neutralize the excess charge upon the respective boundaries. This dynamic process leads to the straightening of the trajectories curved by the deflection. As a result the current j again becomes parallel to the x-axis. Equation (2.21) needs special comment. It is absolutely valid only if all carriers have one

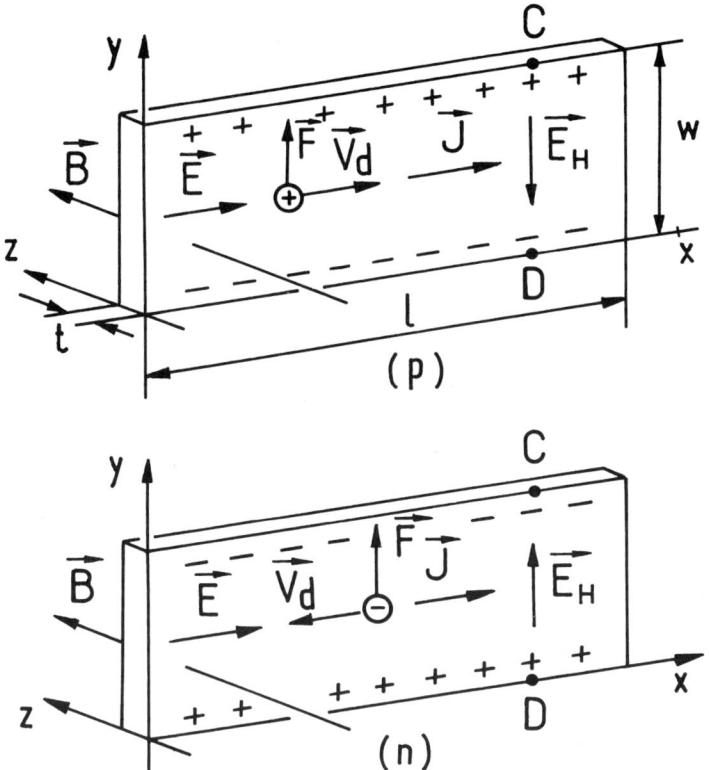

Fig. 2.2. Hall effect in long samples of *p*-type and *n*-type materials. The vectors presented are: E — applied external electric field; B — magnetic field; v_d — drift velocity of carriers; $F \equiv F_M$ — magnetic part of the Lorentz force; J — current density; and E_H — Hall electric field.

and the same drift velocity or equal mobilities. In fact (2.21) is not valid for all carriers at the same time since their drift velocities differ both in magnitude and direction. On the other hand, stationary conditions do not mean the balancing of the magnetic deflection of each individual carrier by the electric force (this never happens). Stationary conditions occur when the charge accumulation on the respective sample boundary ceases. Equilibrium is established when the current generated by the Hall field E_H and the current which results from the Lorentz deflection compensate each other. Therefore a balance of currents instead of a balance of forces is needed for stationary conditions to be fulfilled. The spatial confinement of the structure is a necessary condition for the generation of the Hall field.

When there is no boundary in the direction of the *y*-axis and the carrier mean free path l^* is not limited by scattering ($l^* \to \infty$), all carriers will drift in the direction of $E \times B$ and no Hall effect will occur.

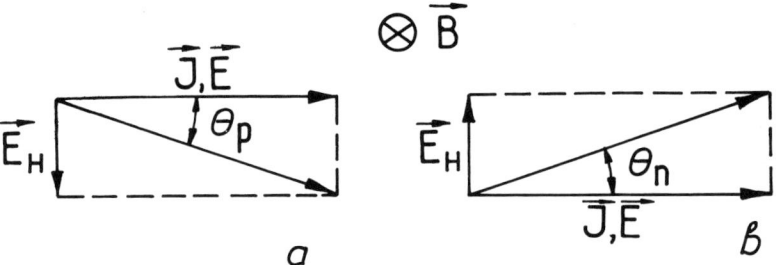

Fig. 2.3. The Hall angle in the case of *p*-type conductivity (a) and *n*-type conductivity (b).

In the presence of a magnetic field the total electric field in the sample $E = E_x + E_H$ is no longer colinear with the external electric field E_x applied by the current source. As seen from Fig. 2.3, the field E is rotated to an angle Θ_H (the Hall angle) with respect to the current $j_x \parallel x$. The value of the Hall angle is obtained from the expression:

$$\tan \Theta_H = \frac{E_y}{E_x} = \frac{\sigma_{xy}(B)}{\sigma_{yy}(B)} = -\frac{\rho_{xy}(B)}{\rho_{xx}(B)} = \qquad (2.22)$$
$$= \mu_H(B) B_z,$$

where

$$\mu_H(B) = \frac{\langle \mu^2(B) \rangle}{\langle \mu(B) \rangle} \qquad (2.23)$$

is the so-called Hall mobility. The parameter $\mu_H(B)$ is connected with the drift mobility $\langle \mu \rangle$ of the carriers by the factor $r_\mu(B)$:

$$r_\mu(B) = \frac{\mu_H(B)}{\langle \mu \rangle}. \qquad (2.24)$$

From the expression (2.22) it follows that:

$$E_y = \mu_H(B) E_x B_z. \qquad (2.25)$$

Therefore the Hall mobility $\mu_H(B)$ can be regarded as a factor of proportionality between the Hall field $E_y \equiv E_H$ and the applied external fields $E \equiv E_x$ and $B \equiv B_z$. The Hall field E_y is proportional to the supplied current j_x as well to the external magnetic field B_z. (This is exactly the relationship empirically found by E. Hall.)

From (2.14) and from the relation

$$\sigma(B) = \frac{1}{\rho(B)} = \frac{j_x}{E_x} = \frac{\sigma_{xx}^2(B) + \sigma_{xy}^2(B)}{\sigma_{xx}(B)} = \frac{1}{\rho_{xx}(B)} \qquad (2.26)$$

an expression for the Hall field can be derived as follows:

Table 2.1

Values of the galvanomagnetic coefficients depending on the scattering mechanism

Scattering	Value of s in (2.32)	Hall factor r_H in (2.30)	Value of r_{M_0} in (2.72)	Value of r_M in (2.73)
Acoustic phonons, deformation potential	$-1/2$	$3\pi/8 = 1.18$	0.616	$3/4(\pi)^{1/2} = 1.33$
Acoustic phonons, piezoelectric scattering	$1/2$	$45\pi/128 = 1.105$	0.33	$(27\pi)^{1/2}/8 = 1.15$
Nonpolar optical phonons	$-1/2$	$3\pi/8 = 1.18$	0.616	$3/4(\pi)^{1/2} = 1.33$
Ionized impurities	$3/2$	$315\pi/512 = 1.93$	1.47	$(15\pi/8)^{1/2} = 2.43$
Neutral impurities	0	1	0	1

$$E_H \equiv E_y = \frac{\sigma_{xy}(B)}{\sigma_{xx}^2(B) + \sigma_{xy}^2(B)} j_x =$$
$$= \rho_{yx} j_x = R_H(B) j_x B_z, \tag{2.27}$$

where

$$R_H(B) = -\frac{1}{qn} \frac{\langle \mu^2(B) \rangle}{\langle \mu(B) \rangle^2} \frac{1}{(1 + \mu_H^2 B_z^2)} = -\frac{r_H(B)}{qn} \tag{2.28}$$

is a factor of proportionality, termed the Hall coefficient. The expression (2.28) refers to electrons and in the case of holes the sign "−" is replaced by "+". R_H is measured in m^3 C^{-1} or in cm^3 C^{-1}. The Hall coefficient is an important parameter of a given material since its value determines the magnitude of the Hall effect and its sign determines the type of conductivity. The parameter

$$r_H(B) = \frac{\langle \mu^2(B) \rangle}{\langle \mu(B) \rangle} \frac{1}{(1 + \mu_H^2 B_z^2)} \tag{2.29}$$

is called the Hall factor.

It is physically dependent on the scattering mechanism and its value is usually within the interval $1 \leq r_H \leq 2$. For low magnetic fields:

$$r_\mu(B) = r_H(B) = r_H = \frac{\langle \tau^2 \rangle}{\langle \tau \rangle^2}. \tag{2.30}$$

The values of the Hall factor r_H (2.30) for different scattering mechanisms are given in Table 2.1 [12].

In classical high magnetic fields (fields which do not cause quantization):

$$r_\mu(B) = \left(\frac{\langle \mu^{-1} \rangle}{\langle \mu \rangle} \right)^{-1}; \quad r_H(B) = 1, \tag{2.31}$$

i.e. r_H does not depend on the scattering mechanism.

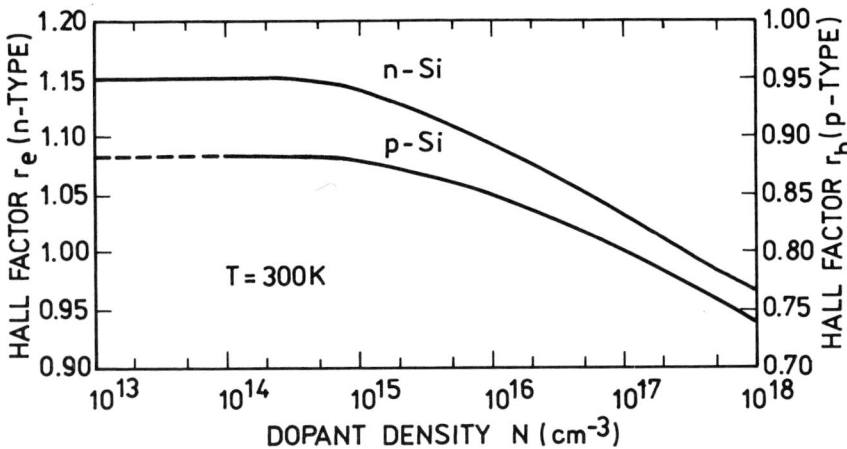

Fig. 2.4. Theoretical Hall factors for *n*-type and *p*-type silicon [47,48].

The parameter τ is characterized by an energy dependence of the type:

$$\tau = \tau_0 E^s, \tag{2.32}$$

where E is the energy of the particles and the power s is a function of the scattering mechanisms.

The values of s for different scattering mechanisms are given in Table 2.1. Figure 2.4 presents the theoretically determined Hall factors for *n*-Si and *p*-Si according to [47,48]. The computed values of r_H for different single crystal semiconductor materials do not exceed 1.3 [49]. For degenerate materials $r_H = 1$ and $\mu_H \approx \mu_{dr}$ can be assumed with a reasonable degree of accuracy. In a heavily doped semiconductor (*n*-type or *p*-type), to a first approximation the Hall coefficient will be:

$$R_{H_n} = -\frac{1}{qn}; \quad R_{H_p} = +\frac{1}{qp}. \tag{2.33}$$

From (2.33) it can be understood why the intensity of the Hall effect in semiconductors is, in general, much more pronounced than in metals. The obvious answer follows from the inversely proportional dependence of R_H on the carrier concentration. In metals the concentration of carriers is higher by several orders of magnitude than that in semiconductors. For instance, in Au, Cu or Bi the Hall coefficient is $R_H \sim 5$ to 10 cm^3 C^{-1}, whereas in silicon R_H is about 10^4 times as big.

In the practical applications of the Hall effect in long samples it is convenient to operate with entirely macroscopic integral parameters, which can be measured directly on the structures. These are the generated potential differences (voltages) and the magnitude of the current forced through the sample, rather than the electric field and the current density. Therefore the following relationship can be used:

$$I_x = j\omega t;$$
$$V_x = lE_x, \quad V_H \equiv V_y = wE_y;$$
$$V_H = \frac{w}{l}\mu_H V_x B_z = \left(\frac{1}{t}\right) R_H I_x B_z.$$
(2.34)

There are important details which must be taken into account in the analysis and application of the Hall effect:

(a) By contrast with the conventional direction of the current accepted in electrical engineering, the Lorentz force is determined by the actual direction of carrier velocity, i.e. from "−" to "+" for electrons and from "+" to "−" for holes.

(b) The external magnetic field is always directed from the north pole (N) to the south pole (S) and is easily found for instance by use of a compass for each particular magnet or magnetic system.

(c) Provided the information from points (a) and (b) is available, the direction of the Lorentz force can be found by the left-hand rule: if the thumb, the first finger and the second finger are mutually perpendicular (similar to the axes of a cartesian coordinate system) in such a way that the first finger is oriented in the direction of the magnetic field B_z and the second finger represents the actual direction of the carrier velocity v_x, then the direction of the thumb will coincide with that of the Lorentz force F_L for holes. If the carriers are electrons, the direction of the Lorentz force will be just the opposite to that of the thumb. By means of this useful rule, from (2.34) the concentration of carriers can easily be determined as well as their mobility and the type of conductivity.

There is an important fact that must be taken into consideration in the practical applications of formulae (2.34) for the investigation of the dependence of the Hall effect on external factors — on the temperature for instance. For a fixed direction and magnitude of the magnetic field B, the Hall voltage can be measured in two modes: the forcing of a constant current (I_x = const) through the sample or the application of a constant voltage V_x = const to the sample. In the first case (I_x = const), from the expressions $V_y = R_H I_x B_z/t$ and $V_y = (w/l)\mu_H V_x B_z$ it follows that the Hall voltage $V_H \equiv V_y$ is a function of the carrier concentration, and in the second case (V_x = const) the voltage V_H is a function of the carrier mobility μ_H. The validity of the Hall measurements is determined by the x-coordinates of the Hall terminals. The conclusions drawn so far are correct if C and D (Fig. 2.2) are point contacts and if they are located in the middle part of the long sample ($l \gg w$). As the distance between the low-resistance current contacts is reduced, the condition $j_x = 0$ is no longer fulfilled, owing to the shunting of V_H by the carrier contacts. Thus inhomogeneities of the electric field arise in the neighborhood of the ohmic contacts, the field E_y being smaller than in the middle part of the sample (Fig. 2.5). The analysis in [50,51] yields a length of about 1–1.5 times the sample width w for the regions of field inhomogeneity. From Fig. 2.5 it can also be seen

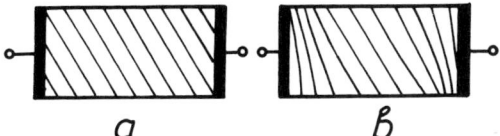

Fig. 2.5. Distribution of equipotential lines in a semiconductor plate when a magnetic field is applied: (a) contacts and semiconductor material with equal resistivities; (b) the resistivity of contacts is much lower than the resistivity of the semiconductor material.

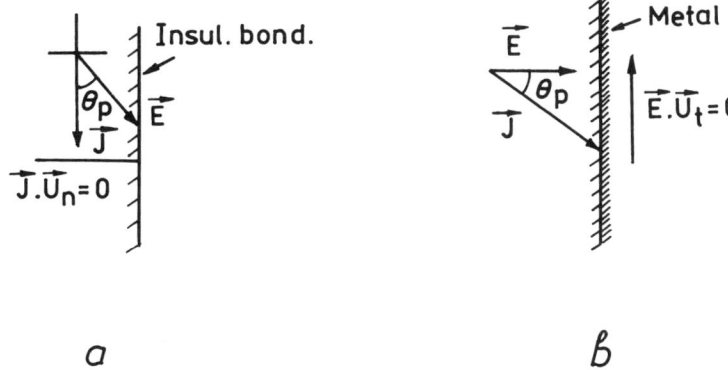

Fig. 2.6. The electric field and current density vectors on the boundary of a p-type semiconductor: (a) floating surface and (b) metallized surface. u_n is a unit normal vector and u_t a unit tangential vector.

that if the current contacts have the same resistivity as the Hall sample they do not change the equipotential picture in the presence of a magnetic field. This fact manifests the substantial role of boundary conditions in the analysis of the Hall effect. In the case of high-conductivity metallized surfaces the electric field (E_x in the considered case) is always perpendicular to the surface, and in the presence of a magnetic field the current j enters the sample at a Hall angle Θ_H (Fig. 2.6a). With respect to the free (floating) surfaces, the electric field E_x is oriented at a Hall angle Θ_H and the current is parallel to those surfaces (Fig. 2.6b).

The amount of charge accumulated by the Lorentz force upon the respective side of the Hall sample plays a significant role. The surface density Q_s of these charges is given by the expression $Q_s = C_s V_H$, where C_s is the capacitance per unit area. The evaluation of C_s is based on the assumption that the permittivity of the material of the Hall device is much greater than the environmental permittivity and there are no other conducting materials near the sample [27]. In this case $C_s \simeq \varepsilon_0 \varepsilon_s / \omega$. For instance, silicon has $\varepsilon_s \simeq 12$ and typically $C_s \sim 1 \times 10^{-6}$ F m^{-2}, thus yielding for Q_s a value of about 0.46×10^{-6} C m^{-2}. This value corresponds to a very low electron density $N_s = Q_s/q \simeq 2.9 \times 10^{12}$ m^{-2}. From this result it can be concluded that in spite of the negligible small additional charge concentration the measured voltage V_H may reach tens or hundreds of millivolts.

The time necessary for the Hall effect to set in is important too. The characteristic time for the electrical charging of the sides of the Hall sample is the relaxation time of the conductivity $\tau_D \approx \varepsilon\varepsilon_0/\sigma$. In typical semiconductor sensor materials, the values of the parameters τ_D are in the range 10^{-14} to 10^{-12} s [16].

Usually in investigations of the Hall effect in ferromagnetic materials the Hall resistivity is considered:

$$\rho_H = \frac{E_y}{j_x} = \frac{\sigma_{xy}}{\left(\sigma_{xx}^2 + \sigma_{xy}^2\right)} = \rho_{yx}. \tag{2.35}$$

Since the magnetic induction for this class of materials is the sum of the external and internal magnetic fields, i.e. $\boldsymbol{B} = \mu_0(\boldsymbol{H} + \boldsymbol{M})$, the corresponding equation is valid:

$$\rho_H = R_H \mu_0 H_z + R_{HA} \mu_0 M = R_H B_z + R_{HS} \mu_0 M, \tag{2.36}$$

where $R_{HA} = R_H + R_{HS}$ and R_{HS} are, respectively, the anomalous and the spontaneous Hall coefficients. The coefficient R_{HS} is related to the spin–orbital interaction and to the potential of scattering by impurities, phonons, magnons, etc. [52].

There is also the so-called longitudinal Hall effect, where the field $\boldsymbol{E}_H \parallel \boldsymbol{B}_z$ [44,53]. This effect is not observed in materials whose Fermi surfaces possess a spherical symmetry, nor is it observed in high magnetic fields. A square and a planar Hall effect are also known. The voltage V_H generated by these effects is an even function of the magnetic field and results from the magnetoresistance effect [53]. The dimensions of the physical quantities and the values of some fundamental constants commonly used in Hall-effect investigations are given in Appendix II.

2.3.2. Hall-current mode in short structures in the case of monopolar conductivity

The Hall-current mode of operation sets in when the boundary surfaces are electrically shunted and the field $\boldsymbol{E}_H \equiv E_y = 0$. This condition is fulfilled in the so-called *infinite sample*. Its practical implementations are:

(a) Short Hall structures in which the distance l between the contacts to the current source is much smaller than the distance w between the insulating boundary surfaces, i.e. $l \ll w$ (Fig. 2.7).

(b) The well-known two-terminal Corbino disk. One of the terminals is located at its center and the second one at its periphery. When $\boldsymbol{B} = 0$ the current has a radial symmetry (Fig. 2.8) [3,14,15,27].

Since the Corbino disk has no insulating surfaces where electric charges could accumulate, no Hall field can be generated. Therefore the magnetic force $\boldsymbol{F}_M$ is not balanced and the carriers deflect from their initial minimum length trajectories. When a potential difference between the center and the periphery is applied

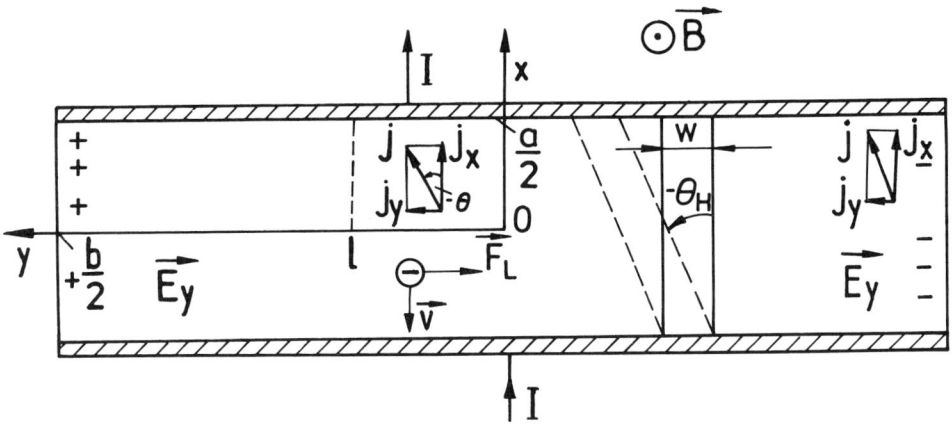

Fig. 2.7. The origin of a Hall current $j_y \equiv j_H$ in a short semiconductor sample

(Fig. 2.8) the equipotential lines will be concentric and both the electric field and the current will have a radial direction. A magnetic field $\boldsymbol{B}$ whose direction is perpendicular to the plane of the Corbino disk will lead to the generation of a tangential component of the current density.

When crossed electric and magnetic fields are applied to a short sample (Fig. 2.7), the carriers are deflected by the Lorentz force in the direction of the y-axis. Owing to the absence of a transverse Hall field $\boldsymbol{E}_y$, a transverse current component $j_y \equiv j_H$ termed the Hall current arises, $E_y(x = \pm a/2) = 0$. In this case, as can be seen from Fig. 2.7, the Hall angle Θ_H is determined from the ratio j_x/j_y, and from equations (2.14)–(2.16) and (2.20) is obtained the following expression:

$$\tan \Theta_H = \frac{j_y}{j_x} = \frac{\sigma_{yx}}{\sigma_{xx}} = \frac{\rho_{xy}}{\rho_{yy}} = -\mu_H B_z. \tag{2.37}$$

With the same type of carrier and with the same directions of the current and the field $\boldsymbol{B}_z$, the vector of the resultant current in a short sample and the vector of the total electric field (2.22) for $\boldsymbol{B}_z \neq 0$ are deflected in opposite directions with respect to the x-axis. A relationship similar to (2.25) can be obtained from equation (2.37):

$$J_y = \mu_H j_x B_z. \tag{2.38}$$

In the Hall-current mode the expressions for the conductivity and the resistivity in a magnetic field differ from (2.26):

$$\sigma(B) = \frac{1}{\rho(B)} = \sigma_{xx}(B) = \frac{\rho_{xx}(B)}{\rho_{xx}^2(B) + \rho_{xy}^2(B)}.$$

In all cases the conductivity of short samples is reduced by the magnetic field.

Detailed information on the relationships between the Hall current and the vector of the electric field can be obtained from the solution of the following system

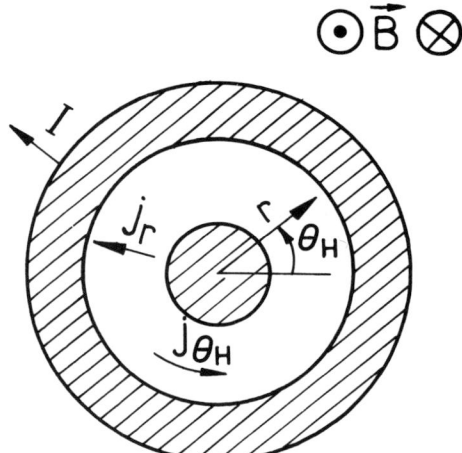

Fig. 2.8. A Corbino disk sample. The current density J is tilted relative to the electric field by the Hall angle Θ_H.

of equations:

$$\operatorname{div} \boldsymbol{j} = 0; \tag{2.39}$$

$$\operatorname{rot} \boldsymbol{E} = 0; \tag{2.40}$$

$$\boldsymbol{j} = \sigma \boldsymbol{E} - \mu_H \boldsymbol{j} \times \boldsymbol{B}. \tag{2.41}$$

The boundary conditions are:

$$E_t = 0; \tag{2.42}$$

$$j_n = 0. \tag{2.43}$$

The first condition (2.42) expresses the absence of a tangential component of the field $\boldsymbol{E}$ upon the metallized regions of the sample surface. The second condition (2.43) represents the absence of a normal component of the current through the free (floating) regions of the sample surface. Figure 2.9 presents in a graphic form the lines of flow of the conduction current j_n as obtained from (2.39)–(2.43) (solid lines) and the lines of flow of the Hall current j_H (dashed lines). In order to define the considered case, monopolar n-type conductivity and $\mu_n B = 1$ are assumed [54]. Since upon the contacts $E_t = 0$, (2.42), from (2.41) it follows for the Hall current that $j_H = 0$, i.e. this current does not directly enter or leave the sample via the contacts.

When the field $\boldsymbol{B} = 0$, the current through the free boundary surfaces is $j_n = 0$ (2.43). If an external field $\boldsymbol{B} \neq 0$ is applied, the Hall current j_H leads to charge accumulation upon the floating boundary surfaces and a conductivity

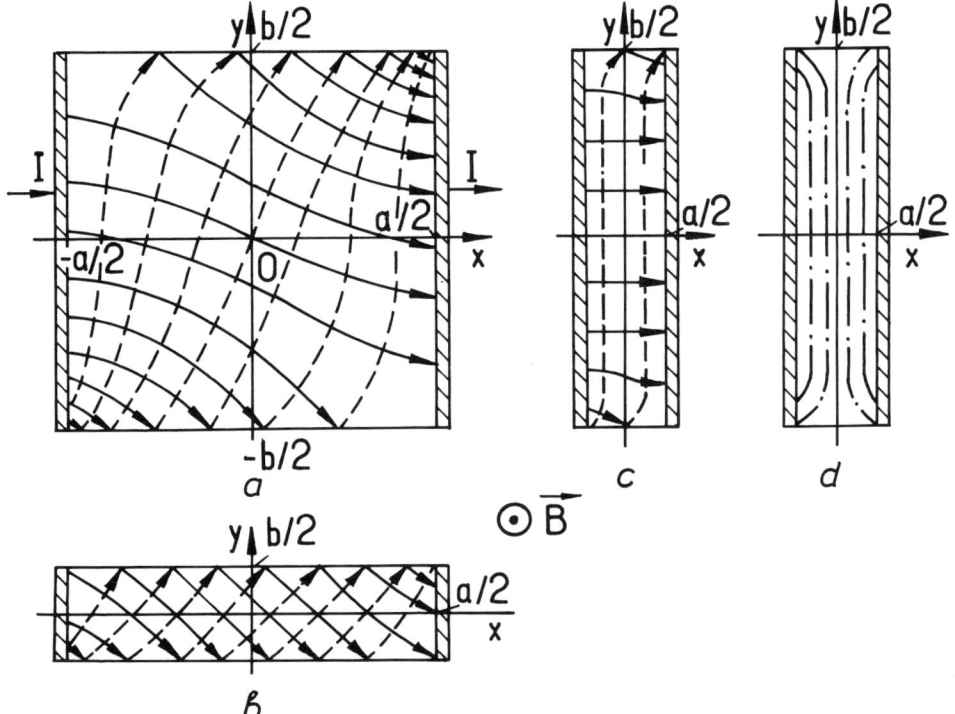

Fig. 2.9. The behavior of the lines of flow of the conductivity current j_n, the Hall current J_H and the current J_B in samples with different geometries [54].

current $j_n \neq 0$ arises. From the condition (2.43) and from equation (2.41) it follows that $j_n = -j_H$, i.e. a mutual conversion of the Hall current and the conduction current takes place on these surfaces. Having been converted into a conduction current, the Hall current is closed by the ohmic contacts and by the supply circuit of the sample. This is why the current density increases in the two corner regions near the points $(-l/2, -w/2)$ and $(l/2, w/2)$ and decreases near the points $(-l/2, w/2)$ and $(l/2, -w/2)$ (Fig. 2.9a). The current density is constant in the central part of the short sample in Fig. 2.9c and it varies only in the regions of width w adjacent to the free (floating) surfaces $y = \pm l/2$. According to (2.41) the lines of the conduction current coincide with the lines of the electric field and the lines of the Hall current are perpendicular to E and coincide with the equipotential lines in the structure. In regions with high values of j_n the magnitude of E is also high. Such regions are those round the points $(-l/2, -w/2)$ and $(l/2, w/2)$.

In the low-field case ($\mu_H B \ll 1$), the total current density j can be presented as $j = j_0 + j_B$, where j_0 is the current for $B = 0$ and j_B is the sum of the magnetic-field induced increment of the conduction current and the Hall current j_H.

The lines of the current j_B are shown in Fig. 2.9d. In the central part of the sample $j_B = j_H$ and near the corners of the structure the current j_B is closed by the metal contacts. The picture of the lines of the current j_B is symmetrical with respect to the y-axis. The total Hall current through the cross-section of the central part of a short sample with a plane $y = $ const is obtained from the expression (2.44) by taking into consideration (2.38), i.e. ($j_y = j_H$):

$$I_H = -ltj_H = -\mu_n B I_n \left(\frac{l}{w}\right). \tag{2.44}$$

When $l \ll w$ the Hall current I_H is generated only by the deflection due to the Lorentz force. For $l \gg w$ (a long sample), the Hall voltage is reduced and the generated Hall current has a maximum value [54], which is determined from the formula:

$$I_H = -0.742 \mu_H B I_n. \tag{2.45}$$

In most practical cases it is easier to determine the carrier mobility and the type of conductivity of the material from the Hall current rather than from the Hall voltage V_H in long samples. For this purpose the ohmic contacts for the current are split by narrow slits and the Hall current is measured between the contacts at $\boldsymbol{B} \neq 0$. More details on this promising technique can be found in refs. [12,16,51,54].

Information concerning the kinetic parameters of the material, e.g. the carrier mobility, obtained from the Hall effect in long or short samples is always equivalent. This result can be verified experimentally, but its theoretical explanation is based on the Van der Pauw dual theorem. According to [55], if the solid-state material has a cubic crystal symmetry, when the positions of the metallized surfaces are exchanged with those of the isolated (floating) surfaces and the direction of the external magnetic field is reversed, i.e. $(+\boldsymbol{B}) \to (-\boldsymbol{B})$, the same data are obtained from these two Hall configurations concerning the kinetic parameters. In particular, there is a duality between the parallelepiped samples from Fig. 2.9b, c. But in solid-state materials whose symmetry is not cubic, these two Hall arrangements give different values of the mobility [56].

The galvanomagnetic behavior of the Corbino disk is analogous to that of the short sample described above since the latter can be regarded as a Corbino disc section with an infinitely large radius [3,27,57].

2.3.3. Hall effect in the presence of high electric fields

The theory of galvanomagnetic phenomena is well developed only for small values of the electric field. As seen in [2], in a low electric field E the drift velocity of carriers v_{dr} in semiconductors is proportional to the magnitude of the electric field E, the factor of proportionality μ_{dr} being field independent, i.e. $v_{dr} = \mu_{dr} E$, where μ_{dr} is the drift mobility. For high fields, however, the relationship $v_{dr} = f(E)$

becomes nonlinear and manifests saturation after a certain value of v_{dr} is exceeded, i.e. $v_{dr} \to v_{dr.sat}$. In thermodynamic equilibrium the number of acoustic phonons generated by carrier scattering equals the number of absorbed acoustic phonons.

The additional energy acquired by the quasi-free carriers owing to the application of an external electric field E is transferred to the crystal lattice. In this case the mean energy of the subsystem of carriers exceeds the equilibrium energy and the subsystem energy distribution is characterized by an effective electron temperature T_e, which for $E \neq 0$ is higher than the crystal lattice temperature T_l. The dependence of the drift velocity v_{dr} on the parameters T_l, T_e and E is given by the expression $v_{dr} = \mu_0 E (T_l/T_e)^{1/2}$, where μ_0 is the low-field drift mobility. In low fields the difference between T_l and T_e is negligible, i.e. $T_e \approx T_l$. In electric fields that fulfill the condition $\mu_0 E \simeq 8v_s/3$, where v_s is the velocity of sound in the respective semiconductor material, the electron temperature is two times as high as that of the crystal lattice, the mobility μ_{dr} decreasing by 30% [2]. The magnitude of the electric field at which the velocity saturation begins is given by the formula $E_{sat} \simeq v_{dr.sat}/\mu_0$, where in most materials $v_{dr.sat}$ is in the range 10^7 cm s^{-1}. For instance, in n-type silicon the saturation of v_{dr} begins at $E_{sat} \simeq 10^4$ V cm^{-1}. There are depletion layers in silicon magnetotransistors and carrier-domain magnetometers that assist the establishment of the high-electric-field mode of operation of such devices [58,59]. It can be intuitively assumed that an increase in the field E will lead to an increase in the magnetic deflection of the current since the drift velocity v_{dr} is increased and $F_M \sim v_{dr} \times B$; therefore an increased angle θ_H is expected as well. If the average time interval $\langle \tau \rangle$ between two successive collisions is considered to be independent of the electric field, i.e. if $\langle \tau \rangle = $ const; from (2.3), (2.5) and (2.10) it follows that the Hall angle θ_H is invariant with respect to the field E, i.e. $\theta_H = $ const. This model is based on the assumption that as a result of a collision a given charge carrier loses the total amount of energy acquired due to its acceleration by the electric field (§2.3.1). After each collision the carrier velocity is zero and the motion starts again along a new cycloid in the direction of the field E. This is why the velocity vector is inclined at a constant angle $\theta < \pi/2$ with respect to E.

Contrary to this scenario, the measured sensitivity of magnetotransistors indicates that both the Hall angle and the transducing efficiency of the device are reduced when the space charge is modulated by the collector bias, and a high electric field E arises in it [58,59]. These results can be explained by the dependence of the relaxation time $\langle \tau \rangle$ on the carrier energy. In addition the Hall angle depends on the frequency of collisions $f = 1/\langle \tau \rangle$, (2.5) and (2.10). Hence for two different transit times τ_1 and τ_2, where $\tau_1 > \tau_2$, the part of the cycloid which corresponds to τ_1 is longer than the part which corresponds to τ_2, and $\theta_{H_1} > \theta_{H_2}$. Figure 2.10 shows the effect of reducing the relaxation time $\langle \tau \rangle$, by increasing the carrier energy as a result of an increase in the external electric field, i.e. $E_2(\langle \tau_2 \rangle) > E_1(\langle \tau_1 \rangle)$, and the subsequent decrease in the Hall angle.

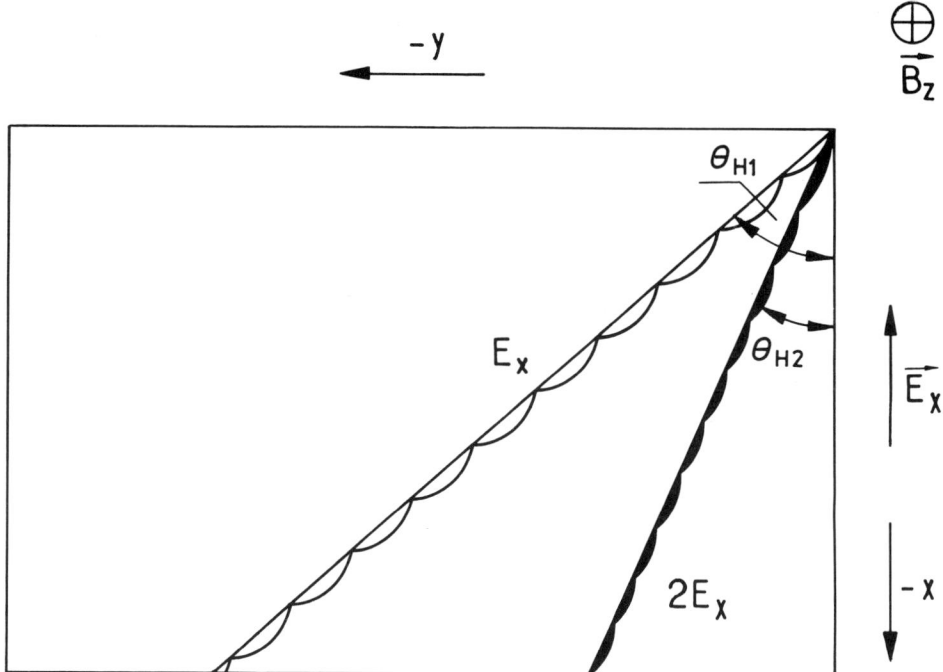

Fig. 2.10. Decreasing the Hall angle Θ_H by reducing the time $\langle \tau \rangle$; $\Theta_H(\tau_1) > \Theta_H(\tau_2)$ at $\tau_1 > \tau_2$.

The drift mobility is also dependent on the average time between two collisions as well as on the effective mass, (2.4), (2.5), i.e. $\mu_{dr} = v_{dr.sat}/|E|$ [2,27]. According to [2,60] the decrease in the carrier mobility in silicon starts at fields $|E| \gtrsim 10^3$ V cm^{-1}, which is an indication that the relaxation time $\langle \tau \rangle$ is reduced by the electric field E. The region of the $\mu_{dr}(E \gtrsim 10^3$ V cm$^{-1})$ curve is called the warm region and for $|E| \sim |E_{sat}|$ the mobility reaches its minimum value.

The analysis of the Hall effect at high electric fields is further complicated by the fact that the field itself is not homogeneous in the depleted layers typically found in complex solid-state magnetic sensors. This leads to a functional dependence of the Hall angle on the position within the depletion layer. Furthermore for $E = $ const, the relaxation time $\langle \tau \rangle$ is not constant and can acquire random values. A pronounced dependence of the Hall mobility (2.23) on the field magnitude is observed in high fields. Since the drift mobility μ_{dr} decreases with increasing E, the Hall factor (2.29) approaches unity, hence μ_H is also reduced by E. The Hall field E_H is also saturated with increasing E. This is explained by the saturation of the drift velocity $v_{dr}(E)$, and the magnitude of the saturated Hall field is $E_{H\,sat} \simeq v_{dr.sat}B$. Since in high electric fields the Hall factor r_H tends to 1 and the drift velocity is in the range 10^7 cm s^{-1}, the physical limit for the highest possible magnitude of the Hall field with a fixed field B is $|E_{H\,sat}|/|B| \lesssim 10^3$ V cm^{-1} T^{-1}

[27]. This value of $E_{H\,sat}$ is almost the same for many semiconductor materials and must be taken into consideration when optimizing Hall sensors, for instance.

2.3.4. Hall effect in medium-length samples (monopolar conductivity)

The importance of the Hall effect in medium-length samples, i.e. when neither of the conditions $l \gg w$ or $l \ll w$ is fulfilled, follows from the fact that most real Hall sensors belong to this category. It can quite naturally be expected that the behavior of such devices with intermediate geometry will be somewhere between that of very long and that of very short samples. In the neighborhood of the insulated boundary, the current flow is parallel to it, owing to the Hall field; thus a Hall voltage V_H is generated. Near the high conductivity ohmic contacts, the Lorentz deflection of the current dominates and the voltage V_H is reduced (§2.3.2). According to (2.34), the voltage V_H decreases with a decrease in the ratio l/w. The influence of the geometry upon the Hall voltage in samples of arbitrary shape can be taken into account by introducing a geometrical correction factor G [14,15]:

$$V_H = G V_{H\infty}. \tag{2.46}$$

$V_{H\infty}$ denotes the Hall voltage which corresponds to an infinitely long sample with point contacts. The correctness of (2.46) is determined by the identical characteristics of the two structures: equal Hall coefficients and thicknesses, equal supplied currents and the same homogeneous magnetic field.

In fact the factor G is introduced in order to represent the decrease in the Hall voltage V_H in structures with finite dimensions as a result of the imperfect spatial confinement of the current. The geometrical correction factor G is a numerical value within the interval [0–1], i.e. $0 \leq G \leq 1$. In very long samples $G \to 1$ and in very short ones $G \to 0$. The value of the geometrical factor G depends on the Hall angle θ_H and on the width s of the Hall terminals which determines the s/w ratio, as shown in Fig. 2.11 [61,62]. This is why in the general case the factor G is a function of s/w, l/w and θ_H, i.e.

$$G = f\left(\frac{l}{w}, \frac{s}{w}, \theta_H\right). \tag{2.47}$$

The following methods can be used to obtain the Hall voltage in structures with different shapes: conformal mapping [63,64], boundary element methods [65,66], finite-difference [67,68], or finite-element approximations [69]. The influence of the G-factor upon the Hall effect is discussed in detail in §4.2.3.

2.3.5. Hall effect in the case of field and sample inhomogeneities

The considerations of galvanomagnetic properties are based on the assumption of monocrystalline materials with ideal homogeneous structures and a homoge-

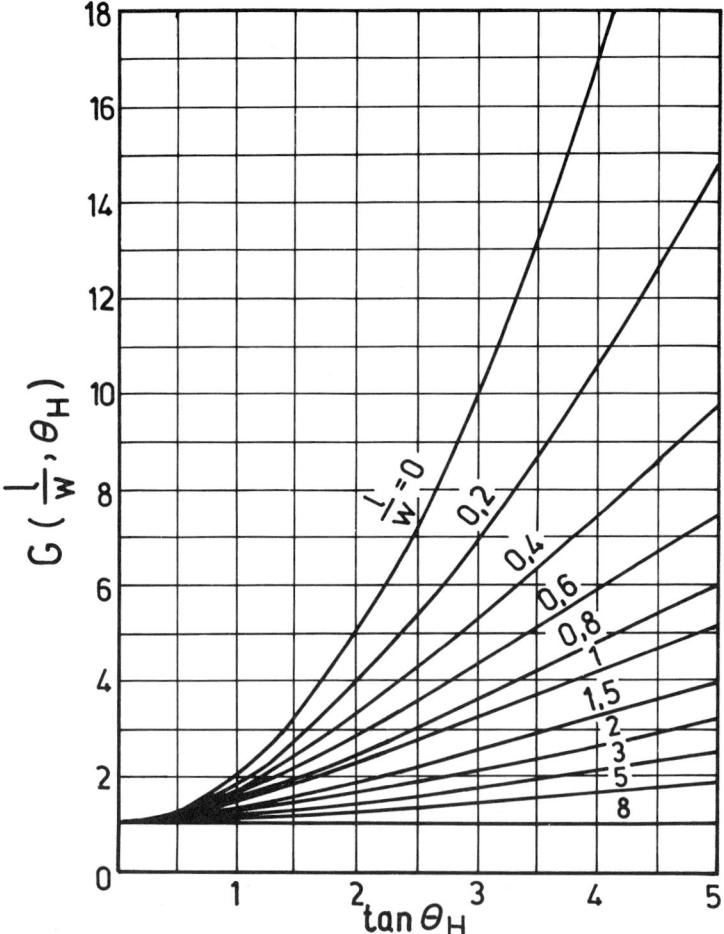

Fig. 2.11. The dependence of the correction factor $G(l/w, \Theta_H)$ on the parameter $\tan \Theta_H$ [61,62].

neous external magnetic field. But even in structures with ideal homogeneity (such structures do not really exist), perturbations arise owing to the effect of the finite dimensions upon the free movement of carriers. Real crystals always include inhomogeneities, even those synthesized in space by means of crystal-growth technology. The inhomogeneities are regular and random spatial fluctuations in the concentration of doping impurities and its gradient, crystal defects, high conductivity or insulating regions, etc. An analysis of the Hall-effect related phenomena in inhomogeneous structures can be found in [70,71].

The importance of the problem is determined also by the fact that modern integrated microsensors for magnetic-field registration, magnetotransistors for instance, include purposely created inhomogeneities like abrupt p–n junctions, where a large concentration gradient of doping impurities arises. The relative variation of this concentration in such junctions is about 10^5 times within a distance of $1\mu m$. This high concentration gradient entails the occurrence of a built-in electric field which can alter considerably the magnetosensitivity of the device.

One type of inhomogeneity that results from crystal growth is the exponential gradient of carrier concentration:

$$n(x) = n_0 \exp(k_x, x), \tag{2.48}$$

where $k_x \equiv n^{-1} \partial n / \partial x = \text{const.}$

If monopolar conductivity, a carrier mobility which is independent of the concentration of doping impurities, a homogeneous magnetic field $(0, 0, B_z)$ and a long and thin sample are assumed, the density of the current along the x-axis is given by the expression:

$$j_x(y) = \frac{I_x}{wt} \cdot \frac{\gamma/2}{\sinh(\gamma/2)} \exp\left[-\gamma \left(\frac{y}{w}\right)\right], \tag{2.49}$$

where $\gamma \equiv k_x w \mu_H B_z$ and w and t are the width and thickness of the sample respectively (see Fig. 2.12).

It can be seen from (2.49) that the external magnetic field moves the current towards one of the Hall sides, i.e. a transverse exponential gradient of the current density j_x arises; thus the corresponding Hall field $E_H \equiv E_y$ gradient is generated as well. It should also be pointed out that the current distortion predicted by (2.49) has to do with the longitudinal component of the current density j_x and should not be confused with the Hall current. With an increase in $\boldsymbol{B}_z$ the current density upon the corresponding Hall side is further increased. The reversing of the direction of $\boldsymbol{B}_z$ moves the current towards the opposite side. Another characteristic of this galvanomagnetic effect is the generation of a circulating current j_x^* that envelops the sample. Upon one of the Hall sides j_x^* is added to the current j_x and upon the other side it is subtracted from it [53,72].

Figure 2.13 shows a structure with a region of abrupt inhomogeneity pierced by the x-axis, which is perpendicular to the current flow $\boldsymbol{j}_x$ [3,53]. In a magnetic field $\boldsymbol{B}_z$ the region of the sample with higher conductivity has a shunting effect upon the Hall field in the lower conductivity region. As a result of the different magnitudes of the Hall field $E_H^{(1)}$ and $E_H^{(2)}$ a compensating circulation current j^* appears in the boundary region between the two parts with different carrier concentrations n_1 and n_2. The current j^* flowing in a transverse direction (along the y-axis) is, actually, a Hall current, $j^* = j_H$, which reduces the Hall field in the boundary region. Furthermore, the current j^* flows also in the direction of the x-axis thus yielding an algebraic sum with the current j_x. This is why the ohmic voltage drop between contacts 3 and 4 is increased and between contacts 5 and 6 the voltage drop is reduced (see Fig. 2.13). If the abrupt inhomogeneities are arranged along the x-axis in an $n_1 - n_2 - n_1 - n_2 - n_1 - \ldots$ sequence, in such a structure the current flows in a zig-zag manner [3]. According to [73] the effective Hall coefficient measured

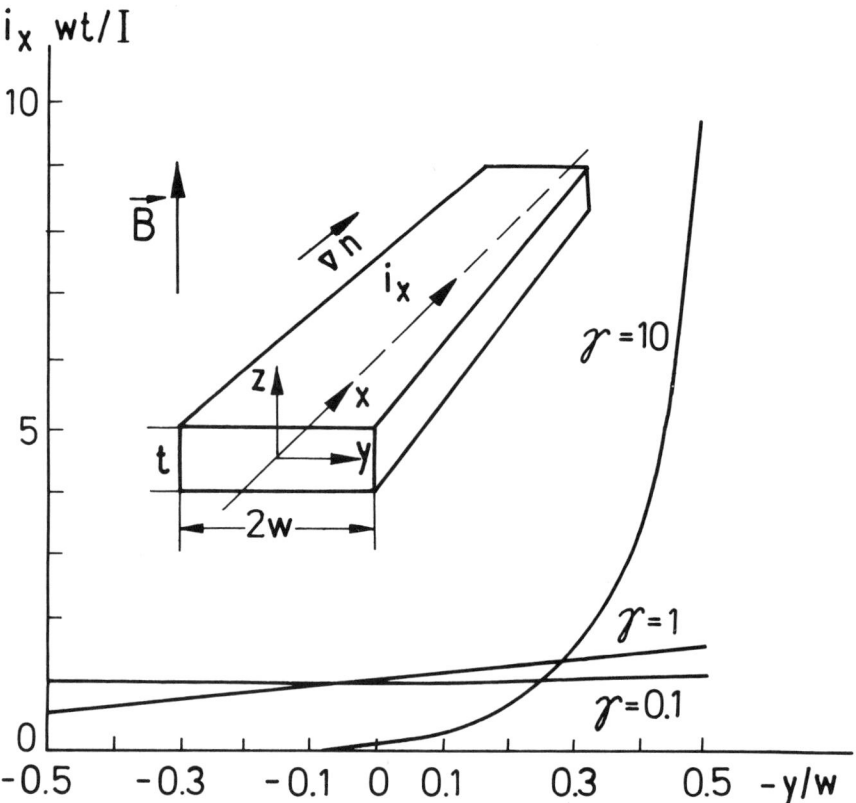

Fig. 2.12. The longitudinal current density distribution in a nonhomogeneous semiconductor sample. The parameter γ is related to the inhomogeneity and the Hall angle.

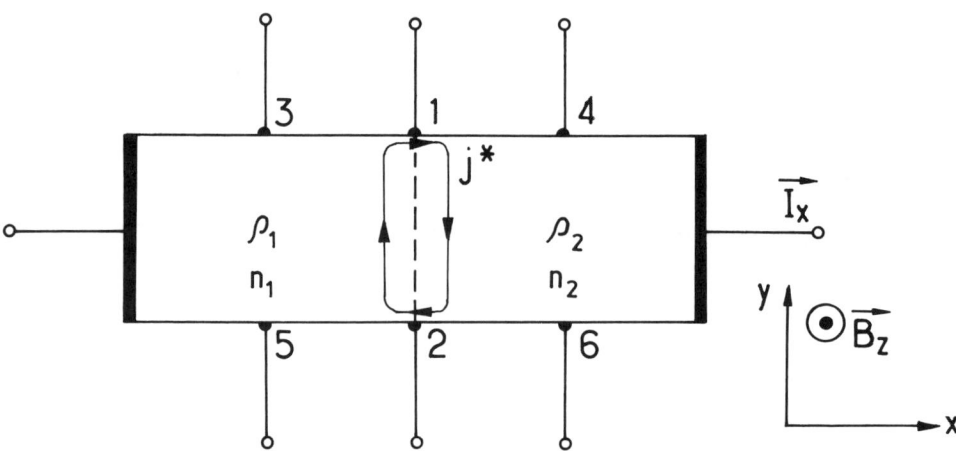

Fig. 2.13. The arising of a circulation current j^* in a sample with an abrupt inhomogeneity in the direction of the x-axis.

between contacts 1 and 2 is given by the expression:

$$R_{H_{\text{eff}}} = \frac{(R_{H_1}\rho_2 - R_{H_2}\rho_1)}{(\rho_1 + \rho_2)}, \quad (2.50)$$

where ρ_1 and ρ_2 are the resistivities of the regions with carrier concentrations n_1 and n_2, respectively (Fig. 2.13).

The effective mobility is obtained from the formula:

$$\mu_{H_{\text{eff}}} = \frac{R_{H_{\text{eff}}}}{\langle\rho\rangle} = \mu_H \frac{4(n_1/n_2)}{[1 + n_1/n_2]^2}, \quad (2.51)$$

where $\langle\rho\rangle = (\rho_1 + \rho_2)/2$.

It follows from (2.50) and (2.51) that $R_{H_1} > R_{H_{\text{eff}}} > R_{H_2}$ and $\mu_{H_{\text{eff}}} < \mu_H$, when $n_1 > n_2$.

Besides these two types of regularly inhomogeneous conductivity there are other inhomogeneities along the y-axis and along the z-axis. Random, local, polycrystalline, surface and other inhomogeneities generated by the process of sample preparation are common. The peculiarities of the galvanomagnetic phenomena described above can also be observed in such samples. In general the layers with higher carrier concentration have a shunting effect upon the Hall field and always give rise to circulating currents which influence the distribution of the main current j_x over the cross-section of the sample. A nonuniform heating of inhomogeneous semiconductor structures is also possible [51].

Like the material inhomogeneities, the inhomogeneities of the external magnetic field generate spatial variations in the conductivity of the Hall field and distortion of the lines of current flow. The influence of an inhomogeneous magnetic field of the type $k_{xB} = B_z^{-1} \partial B/\partial x$, where k_{xB} is analogous to the parameter k_x from (2.48), is discussed in [53]. It has been proved that under such conditions the current distribution in the sample is described by the expression (2.49). For instance, in the case of large Hall angles, when $\mu_H B_z \approx 30$, the relative gradient of the field magnitude whose value is $k_{xB} = 2.5\%$ cm^{-1} leads to such a redistribution of the current density in the structure that the current densities upon the two opposite sides of the sample differ by a factor which exceeds 2.

Therefore at high values of the parameter $\mu_H B_z$ insignificant inhomogeneities of the magnetic field can entail a considerable distribution of the lines of the current flow. In [53,62] the Hall effect is analyzed in a long homogeneous sample when the magnetic field is applied only to a part of the structure. It has been found that the regions which are not exposed to the field **B** have a shunting effect upon the Hall field in that part of the sample to which the field is applied.

The consideration of the Hall effect under inhomogeneous conditions leads to the conclusion that the discussed peculiarities of galvanomagnetic phenomena should influence the operation of semiconductor sensors for magnetic-field registration. Many of them include p–n junctions and operate in the conductivity-gradient mode. Unfortunately in most cases the properties of complex magnetic sensors are interpreted without taking into account the impact of inhomogeneities on their operation.

2.3.6. Hall effect in the case of mixed conductivity

So far our considerations of galvanomagnetic phenomena assume a simple model of the band structure: for instance, only one energy minimum or maximum in the band diagram and an isotropic effective mass of carriers. But the accuracy of this model is not satisfactory for such important semiconductors like Si and Ge. There are several equivalent minima of energy in the conduction band of these materials. The electron effective mass related to these energy valleys depends on the direction of the applied electric field. This is why in semiconductors with multi-valley band structures the effective mass is anisotropic, i.e. it is a tensor quantity. The scattering mechanism is also anisotropic. In this case electrical conductivity arises in several bands together and several types of charge carriers participate in it. Usually these are electrons from the conduction band and holes from the valence band, as well as injected or generated nonequilibrium carriers. The charge transport may include carriers of only one type: for instance the electrons from the three energy valleys of GaAs. In such a case each band contributes to all components of the tensor of electrical conductivity [12,44], and the tensor (2.14) acquires the form:

$$\left.\begin{aligned} j_x &= \left[\sum_\alpha \sigma_{xx\alpha}(B)\right] E_x + \left[\sum_\alpha \sigma_{xy\alpha}(B)\right] E_y; \\ j_y &= \left[\sum_\alpha \sigma_{yx\alpha}(B)\right] E_x + \left[\sum_\alpha \sigma_{yy\alpha}(B)\right] E_y; \\ j_z &= \left[\sum_\alpha \sigma_{zz\alpha}(B)\right] E_z. \end{aligned}\right\} \qquad (2.52)$$

The index α is used to denote the type of charge carrier that participates in the process of electrical conductivity. In the summation with respect to α the sign of σ_{xx} is always positive, whereas σ_{xy} is negative for electrons and positive for holes.

As in §2.2 the analysis of the tensor (2.52) yields general expressions for the effective electrical conductivity $\sigma_{\text{eff}}(B)$, for the effective Hall mobility $\mu_{\text{eff}}(B)$, and for the effective Hall coefficient $R_{H_{\text{eff}}}(B)$ which can be applied in the case of an arbitrary magnetic field:

$$\sigma_{\text{eff}}(B) = \frac{\left[\sum_\alpha \sigma_{xx\alpha}(B)\right]^2 + \left[\sum_\alpha \sigma_{xy\alpha}(B)\right]^2}{\sum_\alpha \sigma_{xx\alpha}(B)} =$$

$$= [1 + \mu_{H_{\text{eff}}}^2(B)B^2] \sum_\alpha q_\alpha n_\alpha \langle \mu_\alpha(B) \rangle; \tag{2.53}$$

$$\mu_{H_{\text{eff}}}(B) = \frac{\sum_\alpha \sigma_{xy\alpha}(B)}{\sum_\alpha \sigma_{xx\alpha}(B)} B_z^{-1} = \frac{\sum_\alpha q_\alpha n_\alpha \langle \mu_\alpha^2(B) \rangle}{\sum_\alpha q_\alpha n_\alpha \langle \mu_\alpha(B) \rangle}; \tag{2.54}$$

$$R_{H_{\text{eff}}}(B) = \frac{\sum_\alpha \sigma_{xy\alpha}(B)}{\left[\sum_\alpha \sigma_{xx\alpha}(B)\right]^2 + \left[\sum_\alpha \sigma_{xy\alpha}(B)\right]^2} B_z^{-1} =$$

$$= \frac{\sum_\alpha q_\alpha n_\alpha \langle \mu_\alpha^2(B) \rangle}{\left[\sum_\alpha q_\alpha n_\alpha \langle \mu_\alpha(B) \rangle\right]^2} [1 + \mu_{H_{\text{eff}}}^2(B)B^2]^{-1}. \tag{2.55}$$

The dependence of the effective coefficients on the field B has a complex behavior with extremum points and more than one inversion of sign in spite of the fact that σ_α, μ_{H_α} and R_{H_α}, expressed as functions of B, may have simplified forms for the separate energy bands and for all values of the effective mass m^* they may be invariant. Two types of carrier ($\alpha = 1, 2$) are often used in the analysis of the Hall effect and the following expressions can be conveniently applied:

$$\sigma_{\text{eff}}(B) = [\sigma_1^2(1 + \mu_{H_2}^2 B^2) + \sigma_2^2(1 + \mu_{H_1}^2 B^2) + 2\sigma_1\sigma_2(1 + \mu_{H_1}\mu_{H_2}B^2)] \times$$

$$\times [\sigma_1(1 + \mu_{H_2}^2 B^2) + \sigma_2(1 + \mu_{H_1}^2 B^2)]^{-1}; \tag{2.56}$$

$$\mu_{H_{\text{eff}}}(B) = \frac{\mu_{H_1}\sigma_1(1 + \mu_{H_2}^2 B^2) + \mu_{H_2}\sigma_2(1 + \mu_{H_1}^2 B^2)}{\sigma_1(1 + \mu_{H_2}^2 B^2) + \sigma_2(1 + \mu_{H_1}^2 B^2)}; \tag{2.57}$$

$$R_{H_{\text{eff}}}(B) = [R_{H_1}\sigma_1^2(1 + \mu_{H_2}^2 B^2) + R_{H_2}\sigma_2^2(1 + \mu_{H_1}^2 B^2)][\sigma_1^2(1 + \mu_{H_2}^2 B^2) +$$

$$+ \sigma_2^2(1 + \mu_{H_1}^2 B^2) + 2\sigma_1\sigma_2(1 + \mu_{H_1}\mu_{H_2}B^2)]^{-1}. \tag{2.58}$$

The relationships (2.56)–(2.58) are slightly modified in comparison with those usually used in [53,74], but nevertheless they clearly illustrate the dependence on the magnetic field **B**.

The expressions (2.53)–(2.55) can be written in a simpler form for a low magnetic field, (2.11):

$$\sigma_{\text{eff}} = \sum_\alpha q_\alpha n_\alpha \langle \mu_\alpha \rangle = \sum_\alpha \sigma_\alpha; \tag{2.59}$$

$$\mu_{H_{\text{eff}}} = \frac{\sum_\alpha q_\alpha n_\alpha \langle \mu_\alpha^2 \rangle}{\sum_\alpha q_\alpha n_\alpha \langle \mu_\alpha \rangle} = \frac{\sum_\alpha \sigma_\alpha \mu_{H_\alpha}}{\sum_\alpha \sigma_\alpha}; \tag{2.60}$$

$$R_{H_{\text{eff}}} = \frac{\sum_\alpha q_\alpha n_\alpha \langle \mu_\alpha^2 \rangle}{\left[\sum_\alpha q_\alpha n_\alpha \langle \mu_\alpha \rangle\right]^2} = \frac{\sum_\alpha \sigma_\alpha \mu_{H_\alpha}}{\left[\sum_\alpha \sigma_\alpha\right]^2}; \tag{2.61}$$

where $\langle \mu_\alpha \rangle = \mu_{\text{dr}.\alpha}$ and $\langle \mu_\alpha^2 \rangle = \mu_{\text{dr}.\alpha}^2 r_{H_\alpha} = \mu_{\text{dr}.\alpha} \mu_{H_\alpha}$.

The expressions which correspond to high magnetic fields, (2.12), are:

$$\sigma_{\text{eff}} = \mu_{H_{\text{eff}}}^2 \sum_\alpha q_\alpha n_\alpha \langle \mu_\alpha^{-1} \rangle = \mu_{H_{\text{eff}}}^2 \sum_\alpha \sigma_\alpha \langle \mu_\alpha^{-1} \rangle^2; \tag{2.62}$$

$$\mu_{H_{\text{eff}}} = \frac{\sum_\alpha q_\alpha n_\alpha}{\sum_\alpha q_\alpha n_\alpha \langle \mu_\alpha^{-1} \rangle} = \frac{\sum_\alpha \sigma_\alpha \langle \mu_\alpha \rangle^{-1}}{\sum_\alpha \sigma_\alpha \langle \mu_\alpha \rangle^{-1} \langle \mu_\alpha^{-1} \rangle}; \tag{2.63}$$

$$R_{H_{\text{eff}}} = \left(\sum_\alpha q_\alpha n_\alpha\right)^{-1} = \left(\sum_\alpha R_{H_\alpha}^{-1}\right)^{-1}. \tag{2.64}$$

As seen from (2.62)–(2.64), there is no dependence on B_z.

In the presence of two types of carrier, the expressions (2.59)–(2.61) can be presented in the form:

$$\sigma_{\text{eff}} = q_1 n_1 \mu_1 + q_2 n_2 \mu_2; \tag{2.65}$$

$$\mu_{H_{\text{eff}}} = \frac{r_{H_1} q_1 n_1 \mu_1^2 + r_{H_2} q_2 n_2 \mu_2^2}{q_1 n_1 \mu_1 + q_2 n_2 \mu_2}; \tag{2.66}$$

$$R_{H_{\text{eff}}} = \frac{r_{H_1} q_1 n_1 \mu_1^2 + r_{H_2} q_2 n_2 \mu_2^2}{(q_1 n_1 \mu_1 + q_2 n_2 \mu_2)^2}. \tag{2.67}$$

From (2.67) are obtained the well-known formulae for the Hall coefficient [10]. In the case of bipolar conductivity when $r_{H_1} = r_{H_2} = r_H$ we specify that:

$$q_1 \equiv q_p; \quad q_2 \equiv q_n; \quad |q_p| = -q_n = q;$$

$$\mu_1 \equiv \mu_p > 0 \text{ and } \mu_2 \equiv \mu_n < 0;$$

and the effective Hall coefficient is given by the expression:

$$R_{H_{\text{eff}}} = \frac{r_H}{q} \frac{p \mu_p^2 - n \mu_n^2}{(p \mu_p + n \mu_n)^2} = \frac{r_H}{qp} \frac{(1 - ab^2)}{(1 + ab)^2}. \tag{2.68}$$

In a material with intrinsic conductivity, $n = p = n_i$, the Hall coefficient is:

$$R_{H_{\text{eff}}} = \frac{r}{qn_i} \frac{(1-ab^2)}{(1+ab)^2}, \tag{2.69}$$

where $a = n/p$ and $b = \mu_n/\mu_p$.

It is evident from (2.60), (2.61), and (2.66)–(2.69) that the behavior of $\mu_{H_{\text{eff}}}$ and $R_{H_{\text{eff}}}$ is to a great extent determined by the weight contributions of the second powers of the mobilities [12]. For instance, in narrow bandgap semiconductors with $b \gg 1$, the n-type carriers with a relatively low concentration can predetermine negative signs of $\mu_{H_{\text{eff}}}$ and $R_{H_{\text{eff}}}$. The criterion for the inversion of the sign of the Hall voltage is the condition $\mu_{H_{\text{eff}}} = R_{H_{\text{eff}}} = 0 = 1 - ab^2$; hence $a = b^{-2}$.

An important feature of the Hall effect in the case of mixed conductivity is the fact that electrons and holes are deflected by the Lorentz force F_L in the same direction because of the opposite drift velocities and opposite signs of their electric charges. If the mobilities and concentrations of electrons and holes in the semiconductor structures are equal, their charges completely compensate each other when in a magnetic field, and the Hall effect does not occur. If there is no such symmetry in the concentration or the mobility of the type of carrier, for instance electrons, a negative charge is assumed upon one of the Hall sides and an uncompensated positive charge is assumed on the opposite side. As a result a transverse Hall field arises which retards the movement of electrons towards the negatively charged surface and at the same time accelerates holes towards it. The process of charge accumulation continues until the hole and electron currents balance each other. Thus the Hall voltage is reduced. Moreover, a carrier concentration gradient arises and transverse diffusion currents flow in the sample. Their magnitudes are determined by the rate of surface recombination and by the decay rate of the excess of carriers [17]. An analysis of these and other related effects is given in §2.5.

According to (2.69), in intrinsic conductivity samples the sign of the Hall voltage corresponds to the charge sign of the carriers with the higher mobility. Typically these are the electrons.

In long samples with mixed conductivity determined by two types of carrier (electrons and holes) the current density vectors j_n and j_p have such directions that the resulting electric field E is at the same time deflected in a negative direction at a Hall angle θ_{H_n} with respect to j_n and in a positive direction at an angle θ_{H_p} with respect to j_p. Figure 2.14 illustrates that the resulting current density $j = j_n + j_p$. The resultant Hall angle which is observed experimentally is the angle of deviation of the vector E with respect to the current j. The angle θ_H may be negative or positive, depending on the magnitudes and directions of j_n and j_p, i.e. depending on the ratio of the concentrations n and p, and that of carrier mobilities μ_{H_n} and μ_{H_p}.

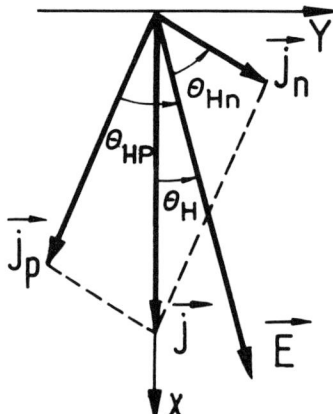

Fig. 2.14. The Hall effect in the case of mixed conductivity.

2.4. Magnetoresistance effect

2.4.1. Physical magnetoresistance effect (monopolar conductivity)

The original cause of the physical magnetoresistance effect is that the velocity (the energy) of the ensemble of carriers, instead of being the same for each carrier, is distributed around the average value $\bar{v}_x$. The effect of the Lorentz force $\boldsymbol{F}_L$ in long samples is compensated by the Hall field $E_H \equiv E_y$ only for carriers whose average velocity is $\bar{v}_x$. That is why the carriers whose velocities exceed $\bar{v}_x$ will deflect from their straight line trajectories in the direction of the Lorentz force $\boldsymbol{F}_L$, and those whose velocities are smaller than the average velocity will deflect in the direction of the Hall field $\boldsymbol{E}_H$. As a result of the overall deflection of the trajectories of carriers with velocity $v_i \lessgtr \bar{v}_x$, their displacement in the direction of the external electric field $\boldsymbol{E}_x$ covered within the time between two collisions decreases, thus reducing the mobility and the electrical conductivity σ. This is the reason why the resistance of long samples increases when they are exposed to a magnetic field. The above effect is called physical magnetoresistance.

In the case of monopolar conductivity, (2.20) or (2.14) can be used to prove that in a homogeneous, transverse magnetic field $\boldsymbol{B}_z$ in an isotropic medium with a current $j_x = $ const (i.e. $E_x = $ const) and a Hall current $j_y = 0$, the change in ρ (hence in σ) is determined from the expression:

$$\left[\frac{\Delta\rho}{\rho}\right]_{j_y=0} = \frac{\rho(B) - \rho(0)}{\rho(0)} = \frac{\sigma_{xx}(B)\sigma_{xx}(0)}{\sigma_{xx}^2(B) + \sigma_{xy}^2(0)} - 1. \tag{2.70}$$

At very low fields, $\mu B \to 0$, the average mobility $\langle\mu(B)\rangle$ and $\langle\mu^2(B)\rangle$ can be developed in a series with respect to $\mu^2 B^2$ and, by taking into consideration only the first and the second terms, the following is obtained:

$$\left[\frac{\Delta\rho}{\rho}\right]_{j_y=0} = \mu_{M_0}^2 B_z^2 = r_{M_0}^2 \mu^2 B_z^2, \tag{2.71}$$

where $\mu_{M_0}^2$ is the physical magnetoresistance mobility, and r_{M_0} is a coefficient which is similar to r_H and depends on the scattering mechanism. In compliance with (2.30) the following formula results:

$$r_{M_0} = \frac{\mu_{M_0}}{\langle\mu\rangle} = \left(\frac{\langle\tau^3\rangle}{\langle\tau\rangle^3} - \frac{\langle\tau^2\rangle}{\langle\tau\rangle^4}\right)^{1/2} - (r_M^2 - r_H^2)^{1/2}. \tag{2.72}$$

The values of the coefficient r_{M_0} for different scattering mechanisms are given in Table 2.1. An important conclusion concerning the square function dependence of the physical magnetoresistance effect on the field $\boldsymbol{B}$ can be drawn from (2.71). The phenomenon is characterized by an even function of the polarity of $\boldsymbol{B}$, and the change in the resistance does not depend on the sign of the carrier charges. In semiconductors with an energy-independent relaxation time, $\langle\tau\rangle =$ const, no physical magnetoresistance effect occurs. In this case all types of mobility (drift mobility, Hall mobility and magnetoresistance mobility) coincide.

For high magnetic fields ($\mu B \gg 1$) the magnetoresistance is saturated and becomes independent of B:

$$\left[\frac{\Delta\rho}{\rho}\right]_{j_y=0} = \langle\tau\rangle\langle\tau^{-1}\rangle - 1 = \text{const}.$$

Nevertheless the magnetoresistance is a function of the scattering mechanisms [12].

The physical magnetoresistance effect is relatively weak in materials with a low carrier mobility. For instance, in silicon the relative change in conductivity for $|\boldsymbol{B}| = 0.1$ T is less than 10^{-4} and cannot be used in sensors for practical applications. In metals the effect is insignificant, since the electron gas is degenerate and only electrons close to the Fermi surface can contribute to conductivity. They have almost equal energies, the same relaxation time $\langle\tau\rangle$ and the dispersion of their velocities is negligible.

2.4.2. Geometrical magnetoresistance effect (monopolar conductivity)

In short samples the Hall current is the dominating mode of operation and all carriers deflect in the direction of the Lorentz force $\boldsymbol{F}_L$. Moreover, in such structures ($l \ll w$) a velocity component v_y arises along the y-axis which is perpendicular to the electric field $\boldsymbol{E}_x$. The Lorentz force associated with v_y generates an extra horizontal component $-v_x^*$ directed in the opposite direction to the carrier velocity. The result is a reduced mobility of all carriers. This is the original cause of the geometrical magnetoresistance effect in short samples. The change in the resistance due to this phenomenon in an electric field $E_H = 0$ is described by the formula:

$$\left[\frac{\Delta\rho}{\rho}\right]_{E_y=0} = \frac{j_x(0)}{j_x(B)} - 1 = \frac{\sigma_{xx}(0)}{\sigma_{xx}(B)} - 1 = \mu_M^2(B)B_z^2, \qquad (2.73)$$

where $\mu_M(B) = \left[\langle\mu^3(B)\rangle/\langle\mu(B)\rangle\right]^{1/2} = r_M\mu$ is the geometrical magnetoresistance mobility, and $\langle\mu^3(B)\rangle \equiv \langle\mu^3/(1+\mu^2B^2)\rangle$ [3,11,12,74].

The coefficient $r_M(B) = \mu_M(B)/\langle\mu\rangle$ is analogous to the Hall factor r_H. Their ratio is $\xi = \mu_M/\mu_H = r_M/r_H$ and differs from unity by less than 26% [12]. The values of r_M are determined by the scattering mechanism (Table 2.1). In low magnetic fields $r_M = (\langle\tau^3\rangle/\langle\tau\rangle^3)^{1/2}$. In the range of high magnetic fields $r_M = (\langle\tau\rangle\langle\tau^{-1}\rangle)^{-1/2}$. Therefore with an increasing field B_z the geometrical magnetoresistance can increase without limit owing to the drift of carriers along the y-axis and, at the same time, due to their rotation with a diminishing radius.

As a result of the geometrical magnetoresistance effect [61], there is a square function relationship between the resistance and the magnetic field B at small values of the Hall angle ($\theta_H \leq 0.45$):

$$\frac{R(B)}{R(0)} = \frac{\rho(B)}{\rho(0)} = \left(1 + C_1(\mu_M B)^2\right), \qquad (2.74)$$

where $R(0)$ is the sample resistance when $B = 0$, $\rho(B)/\rho(0)$ is the relative resistance in an infinitely long rod and C_1 is a geometry-dependent constant.

For high values of the Hall angle ($\theta_H \to \pi/2$), the relationship is linear:

$$\frac{R(B)}{R(0)} = \frac{\rho(B)}{\rho(0)}(C_2\mu_M B + C_3), \qquad (2.75)$$

where C_2 and C_3 are geometry-dependent factors.

In agreement with the results in [61], the geometry-dependent factors C_1, C_2 and C_3 from (2.74) and (2.75), which determine the curve $R(B)$ for $l/w \leq 0.35$, can be expressed as $C_1 = 1 - 0.54l/w$; thus (2.74) acquires the form:

$$\frac{R(B)}{R(0)} = \frac{\rho(B)}{\rho(0)}\left[1 + (\mu_M B)^2\left(1 - 0.54\frac{l}{w}\right)\right]. \qquad (2.74')$$

If $l/w = 0$, the ratio $R(B)/R(0)$ from (2.74') has its maximum value. This is the case with the Corbino disk.

When $C_2 = w/l$, the following expression for large Hall angles ($\theta_H \to \pi/2$) can be obtained from (2.75):

$$\frac{R(B)}{R(0)} = \frac{\rho(B)}{\rho(0)}\left(\mu_M B\frac{w}{l} + C_3\right). \qquad (2.75')$$

The dependence of the geometrical factor C_3 on l/w is shown in Fig. 2.15a [61].

The term geometrical magnetoresistance effect is used to express the fact that the phenomenon is associated with changing the geometry of the current flow, this

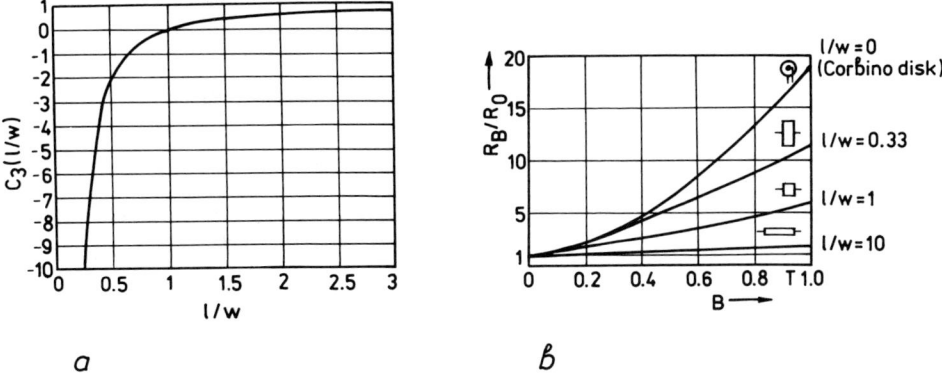

Fig. 2.15. Dependence of the constant C_3 in equation (2.75') on the length/width ratio, l/w (a), and (b) dependence of the relative resistance $R(B)/R(0)$ on the magnetic field B, for various values of l/w; n-InSb has been used as an example [61].

change, in turn, being determined by the geometry of the device. Both the physical and the geometrical magnetoresistance effects are even functions of the field B. Figure 2.15b illustrates the magnetic-field dependence of the relative change of the resistance of n-type InSb samples with equal electron concentrations at $T = 300$ K and different shapes [62]. The effect is most pronounced in the Corbino disk and the smallest effect is observed in a long ($l \gg w$) structure. Among all semiconductors, the narrow-bandgap materials InSb, InAs, InP, as well as GaAs, because of their high carrier mobility, most strongly manifest both the Hall effect and the magnetoresistance effect.

The geometrical magnetoresistance effect is also a convenient instrument for the investigation of carrier mobility in semiconductors with a complex band structure.

The two magnetoresistance effects are transverse (orthogonal), since the field B is perpendicular to the current forced through the structures. There is, however, a longitudinal magnetoresistance effect, the direction of B coinciding with that of the current j. Galvanomagnetic phenomena of this type occur when the constant energy surfaces are anisotropic [53,74], and are not employed in sensor design. If the constant energy surface is spherical and the magnetic field does not cause quantization, there is no longitudinal magnetoresistance effect.

By contrast with the positive magnetoresistance effect (increasing resistance when a field B is applied), there are particular cases, especially at low temperatures, when a negative magnetoresistance occurs. Such anomalous behavior is most often observed in complex compounds As the magnetic field is increased, the function $\rho(B)$ becomes zero, and after that acquires normal positive values. The negative magnetoresistance effect is interpreted in terms of the so-called weak localization, which is characterized by the absence of a spin–orbital scattering [12]. Usually the negative physical magnetoresistance is caused by magnetic impurities.

2.4.3. Magnetoresistance in the case of mixed conductivity

(a) Physical magnetoresistance effect. In long samples with mixed conductivity and Hall current $j_y = 0$, the following expressions can be derived from formula (2.70) and from the tensor (2.52):

$$\left[\frac{\Delta\rho}{\rho}\right]_{j_y=0} = \frac{\sum_\alpha \sigma_{xx\alpha}(B) \sum_\alpha \sigma_{xx\alpha}(0)}{\left[\sum_\alpha \sigma_{xx\alpha}(B)\right]^2 + \left[\sum_\alpha \sigma_{xy\alpha}(B)\right]^2} - 1 = \mu^2_{M_{0,\text{eff}}}(B) B_z^2; \quad (2.76a)$$

$$\mu^2_{M_{0,\text{eff}}}(B) = \left[\frac{\sum_\alpha q_\alpha n_\alpha \langle\mu_\alpha(0)\rangle}{\sum_\alpha q_\alpha n_\alpha \langle\mu_\alpha(B)\rangle}\right]^{1/2} \left[1 + \mu^2_{H_{\text{eff}}}(B) B^2\right]^{1/2} B_z^{-1}; \quad (2.76b)$$

where $\mu_{H_{\text{eff}}}(B)$ is determined from formula (2.54).
The expression for $\sigma_{\text{eff}}(B)$ is identical with (2.53).
For low magnetic fields, the following relationship for the mobility and for $R_{M_{0,\text{eff}}}$ is obtained by taking into account (2.60):

$$\mu_{M_{0,\text{eff}}} = \left[\frac{\sum_\alpha \sigma_\alpha \mu^2_{M_\alpha}}{\sum_\alpha \sigma_\alpha} - \left(\frac{\sum_\alpha \sigma_\alpha \mu_{H_\alpha}}{\sum_\alpha \sigma_\alpha}\right)^2\right]^{1/2} \equiv$$

$$\equiv \mu^2_{M_{\text{eff}}} - \mu^2_{H_{\text{eff}}}, \quad (2.77a)$$

and

$$R_{M_{0,\text{eff}}} = \mu_{M_{0,\text{eff}}} \left(\sum_\alpha \sigma_\alpha\right)^{-1}. \quad (2.77b)$$

If formula (2.56) is used for two types of carrier and if the accuracy is limited to the first-order terms with respect to B^2 by taking into account the magnetic-field independence of σ_1 and σ_2 and $\langle\tau\rangle = $ const, the following expression is obtained for the physical magnetoresistance:

$$-\left[\frac{\Delta\sigma}{\sigma}\right]_{j_y=0} = \left[\frac{\Delta\rho}{\rho}\right]_{j_y=0} =$$

$$= \left[\frac{\sigma_1\sigma_2(\sigma_1 R^2_{H_1} + \sigma_2 R^2_{H_2})}{\sigma_1 + \sigma_2} - \frac{\sigma_1^2\sigma_2^2(R_{H_1} + R_{H_2})^2}{(\sigma_1 + \sigma_2)^2}\right] B_z^2. \quad (2.78)$$

In spite of the fact that, for $\langle\tau\rangle = $ const, in semiconductors with spherical constant energy surfaces and one type of carrier $\Delta\rho/\rho = 0$; for materials with several types of carrier with different values of the effective mass, $\Delta\rho/\rho \neq 0$. There is no longitudinal magnetoresistance effect in such semiconductors.

In a high magnetic field the following expression is valid for the magnetoresistance:

$$\left[\frac{\Delta\rho}{\rho}\right]_{j_y=0} = \frac{\sum_\alpha \sigma_\alpha}{\sum_\alpha \sigma_\alpha \langle \mu_\alpha^{-1}\rangle^2} \left[\frac{\sum_\alpha \sigma_\alpha \langle \mu_\alpha\rangle^{-1}}{\sum_\alpha \sigma_\alpha \langle \mu_\alpha\rangle^{-1}\langle \mu_\alpha^{-1}\rangle}\right]^{-2}, \qquad (2.79)$$

i.e. the physical magnetoresistance effect does not depend on the field B [12].

(b) *Geometrical magnetoresistance effect.* In short structures ($l \ll w$) with a Hall field $E_H = 0$ and mixed conductivity, the magnetoresistance effect is given by the formulae:

$$\left[\frac{\Delta\rho}{\rho}\right]_{E_y=0} = \frac{\sum_\alpha \sigma_{xx\alpha}(0)}{\sum_\alpha \sigma_{xx\alpha}(B)} - 1 = \mu^2_{M_{\text{eff}}}(B) B_z^2; \qquad (2.80)$$

$$\sigma_{\text{eff}}(B) = \sum_\alpha \sigma_{xx\alpha}(B); \qquad (2.81)$$

$$\mu_{M_{\text{eff}}}(B) = \left[\frac{\sum_\alpha q_\alpha n_\alpha \langle \mu_\alpha^3(B)\rangle}{\sum_\alpha q_\alpha n_\alpha \langle \mu_\alpha(B)\rangle}\right]^{1/2}; \qquad (2.82)$$

$$R_{M_{\text{eff}}}(B) = \frac{\mu_{M_{\text{eff}}}(B)}{\sigma_{\text{eff}}(B)} = \left\{\frac{\sum_\alpha q_\alpha n_\alpha \langle \mu_\alpha^3(B)\rangle}{\left[\sum_\alpha q_\alpha n_\alpha \langle \mu_\alpha(B)\rangle\right]^3}\right\}^{1/2}. \qquad (2.83)$$

For low magnetic fields the expressions (2.82) and (2.83) can be simplified:

$$\mu_{M_{\text{eff}}} = \left[\frac{\sum_\alpha \sigma_\alpha \mu_{M_\alpha}^2}{\sum_\alpha \sigma_\alpha}\right]^{1/2}; \qquad (2.84)$$

$$R_{M_{\text{eff}}} = \left[\frac{\sum_\alpha \sigma_\alpha \mu_{M_\alpha}^2}{\left(\sum_\alpha \sigma_\alpha\right)^3}\right]^{1/2}. \qquad (2.85)$$

For high magnetic fields the following expressions are valid:

$$\left[\frac{\Delta\rho}{\rho}\right]_{E_y=0} = \frac{\sum_\alpha \sigma_\alpha}{\sum_\alpha \sigma_\alpha \langle\mu^{-1}\rangle\langle\mu\rangle^{-1}} B_z^2 - 1; \qquad (2.86a)$$

$$\mu_{M_{\text{eff}}} = \left[\frac{\sum_\alpha \sigma_\alpha}{\sum_\alpha \sigma_\alpha \langle\mu^{-1}\rangle\langle\mu\rangle^{-1}}\right]^{1/2}; \qquad (2.86b)$$

$$R_{M_{\text{eff}}} = \left[\frac{\sum_\alpha \sigma_\alpha}{\left(\sum_\alpha \sigma_\alpha \langle\mu^{-1}\rangle\langle\mu\rangle^{-1}\right)^3}\right]^{1/2} B_z^2. \qquad (2.86c)$$

It can be concluded from the above relations that the geometrical magnetoresistance is no longer proportional to B_z^2 and does not saturate [12].

When the electron and hole concentrations are equal, i.e. $n \approx p$, the resultant Hall field E_H differs from the respective fields $E_H^{(n)}$ and $E_H^{(p)}$ owing to the inevitably different mobilities. In this case the Lorentz force does not efficiently balance the two types of carrier and they deflect at the corresponding angles $\theta_{H_n} \sim -\mu_n B$ and $\theta_{H_p} \sim \mu_p B$ from the direction of the applied external electric field $\mathbf{E}$. Therefore, even if the sample is very long ($l \gg w$), there is a geometrical magnetoresistance effect.

The magnetoresistance effect in inhomogeneous structures with an exponential concentration gradient of carriers $n(x) = n_0 exp(k_x x)$ is characterized by the following features [12,51,72]: for the two directions of the external magnetic field ($+\mathbf{B}$ and $-\mathbf{B}$) the magnetoresistance $\Delta\rho/\rho$ has different values and is a function of B_z; the value and the sign of $[\Delta\rho/\rho]$ depend on the location of the sensing contacts upon the sample surface and $[\Delta\rho/\rho]$ is not saturated in high magnetic fields. When there are abrupt inhomogeneities in the sample, or when the direction of the magnetic field is reversed, the circulation current j^* arising produces an inequality in the voltage drops $V_{3,4}$ between terminal 3 and terminal 4 and $V_{5,6}$ between terminals 5 and 6 (Fig. 2.13) and no saturation of $[\Delta\rho/\rho]$ is observed in a high field B_z.

2.5. Magnetodiode effect

The magnetodiode effect is a galvanomagnetic phenomenon which is actually a superposition of magnetoconcentration and injection of carriers. On the other hand, by means of the Hall effect manifested in the form of a current deflection, the magnetoconcentration controls the concentration in semiconductor structures with ambipolar conductivity when $n_0 \approx p_0$, or in the case of extrinsic conductivity when the relationship $n|\mu_n| \approx p\mu_p$ is valid. Magnetoconcentration was discovered by Suhl and Shockley and is also known as the Suhl effect [1,75]. The importance of the magnetodiode effect is illustrated by the fact that a whole family of highly sensitive semiconductor sensors (magnetodiodes), including integrated versions have been developed on its basis. This phenomenon also plays an important role in the operation of bipolar and unijunction magnetotransistors.

2.5.1. Introduction to injection phenomena in semiconductor devices

The magnetic control of the carrier concentration in equilibrium and nonequilibrium plasma in semiconductors lies at the basis of magnetoconcentration and of the magnetodiode effect. A brief description will therefore be given of the basic properties of the plasma in solid-state structures. Knowledge of these properties is necessary in order to analyze these two galvanomagnetic effects.

A plasma is a system of charged particles whose concentration is so high that the Coulomb interactions between them are stronger than the forces which result from external fields. This is why, whereas individual electrically charged particles in an electric or magnetic field move in accordance with the sign of their charges, the reaction of a plasma and its components to external fields is entirely different and under certain conditions manifests "collective" properties. The electrical origin of the interaction forces within the plasma determines the trend towards preservation of electroneutrality as much as possible under these particular conditions. Thus the plasma counteracts the separation of the positive from the negative charges. This property is manifested as a screening effect; hence the electric field does not penetrate into the plasma and the spatial separation of its charges becomes impossible. There are two types of nonequilibrium plasma in semiconductors: (a) uncompensated (electrically charged), which consists of electrons and holes interacting in the presence of doping impurities; and (b) compensated (neutral) plasma, which consists of equal amounts of quasi-free electrons and holes. This type of plasma is observed in intrinsic semiconductors and is responsible for the magnetoconcentration effect.

A nonequilibrium plasma in solid-state materials consists of electrons and holes introduced into the sample via one or two junctions, or obtained as a result of irradiation, exposure to an external field or to heat. The nonequilibrium plasma determines the magnetodiode effect. When nonequilibrium carriers are introduced into the semiconductor material, their charge in the region around the junction is neutralized by carriers with opposite charges that come from the more distant parts of the structure. The plasma state, established in such a manner in the neighborhood of the junction, then spreads all over the sample. The spatial distribution of the plasma is determined by the injection current, by the properties of the contacts and by the surface recombination. The surface recombination rate is of particular importance when considering the magnetodiode effect.

The properties of the injecting contacts have a considerable influence on the processes of formation of the nonequilibrium electron–hole plasma. In conventional diodes with one injecting p^+–n junction and one ohmic contact, the injection of minority carriers (holes) into the n-base generates a solid-state plasma. The majority carriers (electrons) needed to compensate the charge of the nonequilibrium holes come from the ohmic contact to the base, and their number is sufficient for the preservation of charge neutrality [1–3]. The formation of n^+–n or p^+–p junctions can considerably improve the efficiency of the contact responsible for the delivery of majority carriers to the semiconductor sample. These junctions are a potential barrier which does not allow the injected minority carriers to leave the base and lets in the majority carriers needed to maintain the charge neutrality. The amount of majority carriers depends on the value of the voltage applied to the p^+–n–n^+ or p^+–p–n structure. Current excitation, termed the double injection mode, occurs in diodes with two injecting contacts. Double injection introduces holes and electrons into the sample at the same time. This is the most efficient method for obtaining a solid-state plasma of sufficiently high concentration by the application of a bias of several volts. The most important characteristics of the double injection mode associated with the magnetodiode effect are: (a) rapid mixing of the carriers with opposite charges throughout the entire volume of the sample so that an electron–hole quasi-neutral plasma is obtained; (b) the injected electrons and holes recombine within the bulk of the structure. This process is usually controlled by deep impurity levels, while, owing to the concentrating influence of the Lorentz force upon electrons and holes, the recombination process may be enhanced or retarded depending on the quality of the surface. The electric and galvanomagnetic effects related to conductivity modulation are actually bulk effects and this fact determines the usefulness of their application in solid-state magnetic sensors.

2.5.2. Magnetoconcentration effect

An isotropic semiconductor sample with a parallelepiped shape, ohmic contacts for the supplied current and quasi-intrinsic conductivity $n_0 \simeq p_0 \approx n_i$ is shown in Fig. 2.16. The sample is activated by an external electric field $\boldsymbol{E}_x$ directed parallel to the x-axis and a magnetic field $\boldsymbol{B}_z$ directed parallel to the z-axis ($\boldsymbol{E}_x \perp \boldsymbol{B}_z$). As a result of the Lorentz force, the quasi-free electrons and holes deflect towards one and the same surface in the direction of the y-axis (§2.3.6). Since a Hall voltage is prevented from building up (a consequence of electrical neutrality), the Lorentz force is compensated by a carrier concentration gradient perpendicular to the magnetic and electric field vectors. An important condition for the occurrence of the magnetoconcentration effect is that the surface recombination rate upon one of the two surfaces Σ_1 and Σ_2 should be finite. In the ideal case, the recombination rate upon Σ_1, for instance, is $s_1 \to 0$ and upon Σ_2 is $s_2 \to \infty$. The theoretical basis of magnetoconcentration includes the following assumptions [24–26,76]: a

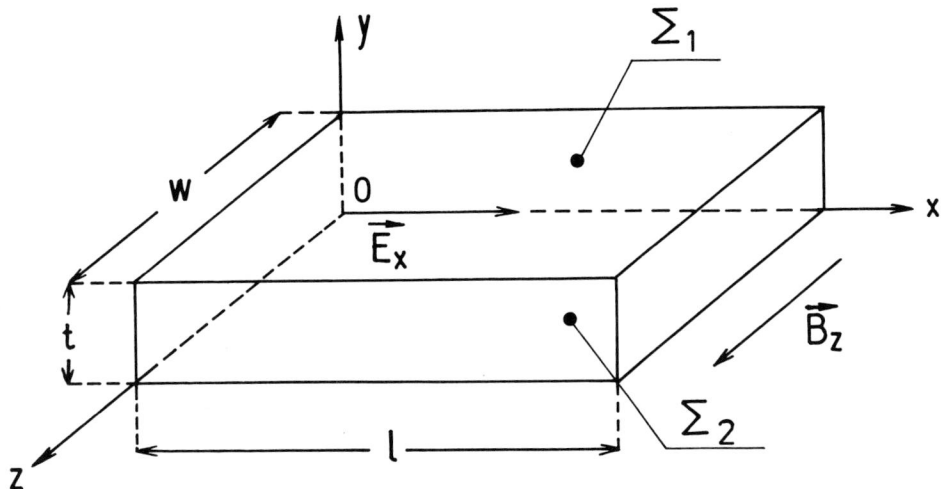

Fig. 2.16. Configuration of the magnetoconcentration effect.

small external magnetic field $\mu B \ll 1$; all effects associated with hot carriers and high field are neglected in order to eliminate the magnetoresistance effects; the condition of quasi-neutrality is fulfilled both in the bulk and on the two surfaces Σ_1 and Σ_2, and hence $n_0 \sim p_0$ and $\delta_n \approx \delta_p$; infinite dimensions of the sample on Fig. 2.16 along the x-axis and the z-axis are assumed in order to eliminate recombination upon the sides which are perpendicular to these axes and to establish a linear law of recombination upon the two surfaces Σ_1 and Σ_2. Under such conditions, in the equilibrium mode of operation the densities of the electron and hole currents j_n and j_p are presented by the expression:

$$j_{n,p} = (j_{n,p})_0 + (j_{n,p})_0 \times \mu_{H_{n,p}} B; \tag{2.87}$$

$$(j_{n,p})_0 = \sigma_{n,p} E - q_{n,p} D_{n,p} \nabla n, p; \tag{2.88}$$

where $D_{n,p}$ is the respective diffusion coefficient, the other notations being given in the preceding paragraphs of this chapter.

The continuity equation is used to determine the transverse carrier distribution $\delta n(y)$ in the presence of the magnetic field. The number of charge carriers is preserved; that is why:

$$\text{div}\, j_{n,p} = -q_{n,p} \frac{\partial n}{\tau_V}; \tag{2.89}$$

$$(j_{n,p})_{\pm t/2} = \pm q_{n,p} s_{1,2} \delta n(\pm \frac{t}{2}). \tag{2.90}$$

In (2.89) and (2.90) τ_V is the minority-carrier lifetime in the bulk, and $s_{1,2}$ are the respective surface recombination rates of Σ_1 and Σ_2 when $y = \pm t/2$. The

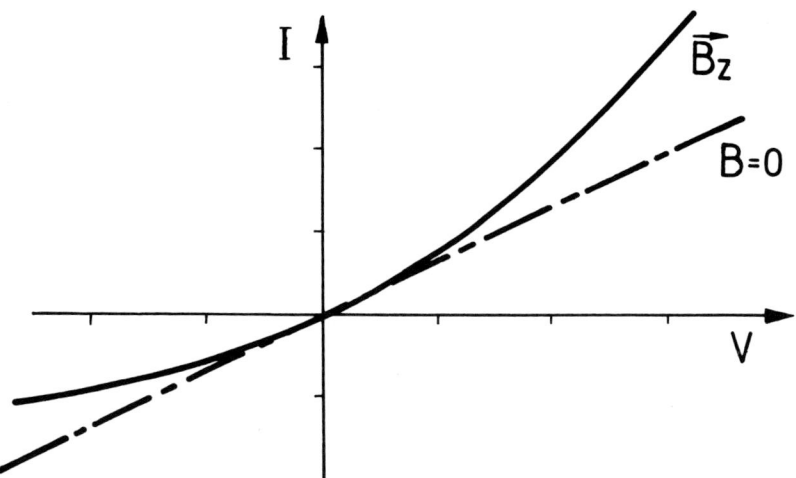

Fig. 2.17. Nonlinearity of the $I(V)$ curve when $\boldsymbol{B}_z \neq 0$, associated with the establishing of a transverse carrier concentration gradient due to the magnetoconcentration effect.

transverse distribution $\delta n(y)$ is obtained from the differential equation:

$$\frac{\partial^2 n}{\partial y^2} + \frac{1}{l_c^*}\frac{\partial n}{\partial y} - \frac{1}{L_{\text{amb}}^2}\delta n = 0, \tag{2.91}$$

where L_{amb} is the ambipolar diffusion length, and the physical meaning of l_c^* is a characteristic length which is determined by the electric and magnetic fields applied:

$$l_c^* = \frac{2k_B T}{q}\left[(|\mu_{H_n}| + \mu_{H_p})\boldsymbol{E}_x \times \boldsymbol{B}_z\right]^{-1}. \tag{2.92}$$

According to [76] the distribution $\delta n(y)$ is given by an expression of the type:

$$\delta n(y) \approx n_0 \exp\left(-\frac{\frac{t}{2} - y}{l_c^*}\right) + n_0 \exp\left(-\frac{\frac{t}{2} + y}{l_c^*}\right).$$

When $E_x = 0$ or $B_z = 0$, and $1/l_c^* = 0$, the nonuniform distribution of carriers along the y-axis disappears i.e. $\delta n(y) = 0$. But when $\boldsymbol{E}_x \times \boldsymbol{B}_z \neq 0$, transverse currents arise [25,26,76]. Under such conditions the current–voltage curve of the sample in Fig. 2.16, which is linear when $B = 0$, becomes nonlinear and resembles a "diode" curve (Fig. 2.17). The excesses δn and δp of carriers over their equilibrium concentrations depend on the surface recombination rates upon Σ_1 and Σ_2. As a result of the application of an external field $\boldsymbol{B}_z$ the carriers deflect towards Σ_2, and the carrier excess is small owing to the high rate of surface recombination. Just the opposite situation is observed near Σ_1. The insufficiency of carriers dominates, since that surface cannot generate a sufficient amount of carriers (Fig. 2.18a). This is why the mean concentration of the electron–hole plasma is reduced, thus

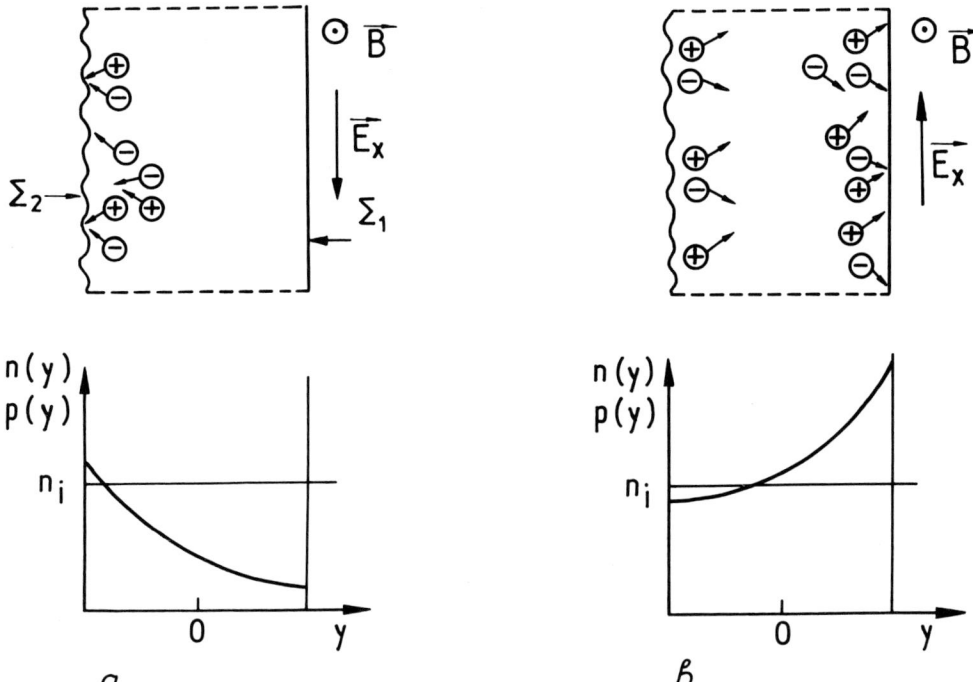

Fig. 2.18. Carrier density distributions as a result of the magnetoconcentration effect in an intrinsic semiconductor plate: (a) the plasma is deflected by the field B_z toward the surface with a high recombination rate; (b) the plasma is deflected by B_z toward the surface with a low recombination rate.

reducing the mean conductivity of the sample. When the polarity of the magnetic field or of the electric field is reversed, the Lorentz force causes deflection of carriers towards the surface Σ_1. In such a way the mean carrier density and the mean conductivity increase. In this case the concentration of electrons and holes generated by Σ_2 is much higher than the concentration of carriers which recombine upon the surface Σ_1 (Fig. 2.18b). Under the conditions described above there is a large asymmetry of the $I(B_z)$ curves. If the two active surfaces Σ_1 and Σ_2 are subjected to identical processing, the reversing of the fields E_x or B_z leads to an equivalent carrier distribution and the $I(V)$, $I(B_z)$ or $V(B_z)$ curves of the device are symmetrical. The general behavior of the $I(B)$ relationship in the different possible configurations $s_1 \ll s_2$, $s_1 = s_2$ and $s_1 \gg s_2$ is shown in Fig. 2.19.

Summarizing the magnetoconcentration effect yields the optimum conditions for its occurrence:

– The sample thickness t must be of the order of magnitude of the ambipolar diffusion length L_{amb}. In thick structures only the regions near the two surfaces Σ_1 and Σ_2 suffer perturbations due to this effect, and the middle part remains in equilibrium. Conversely, when the samples are very thin, only a limited deviation of

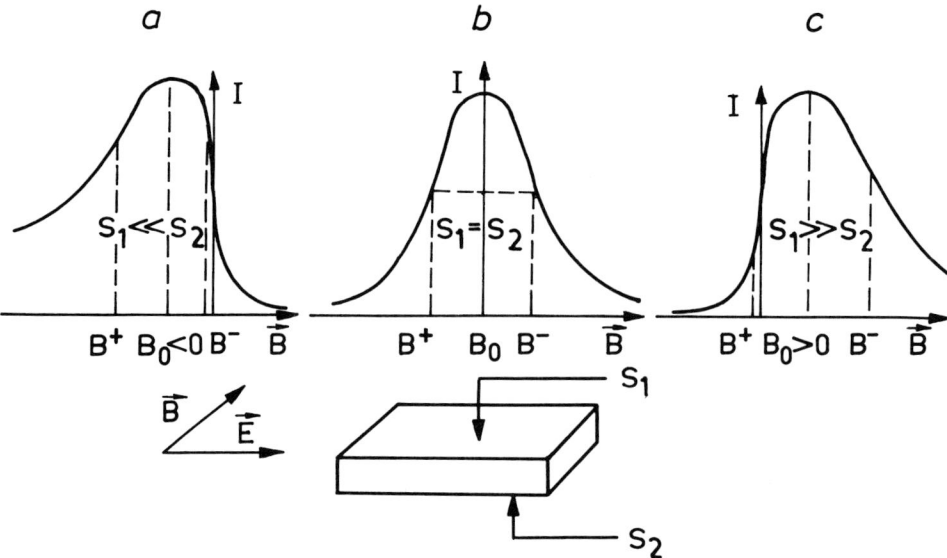

Fig. 2.19. The $I(B_z)$ relationship due to magnetoconcentration with different values of the s_1/s_2 ratio: (a) $s_1/s_2 \ll 1$; (b) $s_1/s_2 = 1$; (c) $s_1/s_2 \gg 1$.

the carrier concentration from its equilibrium value is induced by the field $\boldsymbol{B}_z$ [76].

– The conductivity of the semiconductor material must be bipolar and its value must be near that of an intrinsic sample with $n_0 \approx p_0$. It has been shown in ref. [22] that magnetoconcentration reaches a maximum in p-type semiconductors with a low concentration of doping impurities.

– A low surface recombination rate must be characteristic for at least one of the surfaces, Σ_1 or Σ_2, and a maximum effect is achieved when the conditions for recombination upon Σ_1 and Σ_2 are highly asymmetric.

2.5.3. Magnetodiode effect

In its physical essence the magnetodiode effect is a superposition of magnetoconcentration and injection. In the structures where this phenomenon is observed, there is a highly nonequilibrium electron–hole plasma obtained by injection by one or two p–n junctions in semiconductor samples with a low concentration of doping impurities. The following cases will be analyzed in detail: (a) when the injection is by a single junction, i.e. the case of a long diode; (b) when the magnetodiode structure includes two injecting p–n junctions, so that double injection is accomplished.

2.5.3.1. Magnetodiode effect in long structures with one injecting contact

The theory of this manifestation of the magnetodiode effect was first described in [77,78]. Figure 2.20 shows a long diode structure ($l > L_D$) with an abrupt

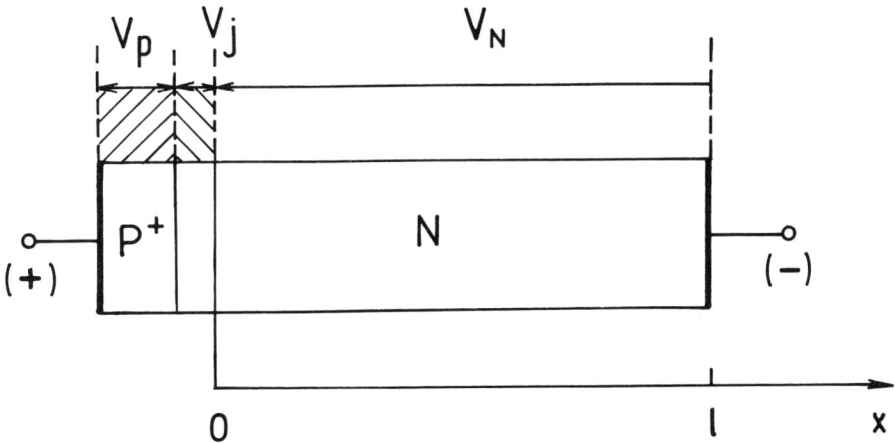

Fig. 2.20. Configuration of the magnetodiode effect in long structure with one injecting p–n junction.

forward biased p^+–n junction ($p \gg N \sim n$). In the n-type base region near the p^+–N junction the current consists exclusively of holes. The analysis has been achieved in two stages: first the equation of the current–voltage curve of a long diode has been derived. Secondly, the influence of an external magnetic field **B** upon the macroscopic parameters determined by the $I(V)$ characteristic has been considered. In the n-base, when there is no magnetic field, the carrier behavior can be described by the following equations:

$$I_p = \underbrace{q\mu_p p E_x}_{\text{drift of holes}} - \underbrace{qD_p \frac{\partial p}{\partial x}}_{\text{diffusion of holes}} ; \qquad (2.93)$$

$$I_n = \underbrace{q\mu_n n E_x}_{\text{drift of electrons}} + \underbrace{qD_n \frac{\partial n}{\partial x}}_{\text{diffusion of electrons}} . \qquad (2.94)$$

In the stationary mode the following relationship is valid:

$$\frac{\partial p}{\partial t} = 0 = \underbrace{-\frac{p - p_0}{\tau} - \frac{1}{q}\frac{\partial I_p}{\partial x}}_{\text{stationary mode}} . \qquad (2.95)$$

The solution of the system (2.93)–(2.95) at low injection levels ($p \ll n$) and at high injection levels ($p \gg n$) is of the conventional type:

$$p = p_0 + A \exp\left(-\frac{x}{L}\right) + B \exp\left(\frac{x}{L}\right), \qquad (2.96)$$

where the parameter L is the longitudinal minority-carrier (hole) diffusion length $L = L_{D,p}$ for $p \ll n$; $L = L_{\text{amb}}$ is the ambipolar diffusion length at high injection levels $p \gg n$. The expression $L_{\text{amb}} = L_{D,p} 2\mu_n/(\mu_n + \mu_p)$ is valid for L_{amb}.

The following boundary conditions are adopted in order to obtain an analytical expression for the current–voltage characteristic of the long diode:
- the total voltage is $V_0 = V_P + V_J + V_N$ (Fig. 2.20), where $V_P \approx 0$ is assumed;
- for $x = 0$ the distribution is of the Boltzmann type since the voltage V_J is applied to the junction;
- for $x = l$ an infinite rate of recombination $s \to \infty$ upon the surface of the contact is assumed.

In accordance with the boundary conditions so defined, the profile of carrier concentration in the n-type base is described by the expressions:

$$p(x) = p_0 + p_0 \left[\exp\left(\frac{qV_J}{k_BT}\right) - 1\right] \frac{\text{sh}\,[(l-x)/L]}{\text{sh}\,(l/L)} \tag{2.97}$$

and

$$\frac{p - p_0}{p_0} = \left[\exp\left(\frac{qV_J}{k_BT}\right) - 1\right] \frac{\text{sh}\,[(l-x)/L]}{\text{sh}\,(l/L)}. \tag{2.98}$$

The current injected into the base by the p^+–N junction can be described by making use of (2.97). Since $I_n \approx 0$, by setting the left-hand side of (2.94) to zero is obtained:

$$q\mu_n n E_x + qD_n \frac{\partial n}{\partial x} = 0,$$

where

$$E_x = \frac{D_n \frac{\partial n}{\partial x}}{\mu_n} = \frac{k_BT}{qn}\frac{\partial n}{\partial x}.$$

Taking into account that $n = N + p$, $\delta n \approx -\delta p$, $E_x = -[k_BT/q(N+p)]\partial p/\partial x$ and the charge-neutrality condition, the substitution of p and $\partial p/\partial x$ in (2.93) yields the following expression for the hole current at $x = 0$:

$$I_p = -k_BT\mu_p \left(1 + \frac{p(0)}{N + p(0)}\right)\left(\frac{\partial p}{\partial x}\right)_{x=0}.$$

According to (2.97), the gradient $\partial p/\partial x$ is obtained from the expression:

$$\left(\frac{\partial p}{\partial x}\right)_{x=0} = -p_0\left[\exp\left(\frac{qV_J}{k_BT}\right) - 1\right]\frac{\text{cth}\,(l/L)}{L}.$$

As a result of the above transformation, the current across a p^+–N junction as a function of the applied voltage V_J is given by the formula:

$$I = \beta I_s \left[\exp\left(\frac{qV_J}{k_BT}\right) - 1\right], \tag{2.99}$$

where

$$I_s = \frac{k_BT\mu_p p}{L}\,\text{cth}\,\frac{l}{L}, \tag{2.100}$$

and its physical meaning is the saturated value of the reverse current of the diode.

The coefficient β is presented by the expression:

$$\beta = 1 + \frac{p(0)}{p(0) + N}. \tag{2.101}$$

The voltage drop in the base region of the diode is evaluated in order to obtain the current–voltage curve $I(V_l)$. This is accomplished by the integration of the electric field in the base region, where $n(x) = N + p(x)$ and $N = \text{const}$:

$$V_0 = \int_0^l E_x dx,$$

where

$$E_x = \frac{I_n + I_p}{q(n\mu_n + p\mu_p)} - \frac{k_B T}{q} \frac{\mu_n - \mu_p}{n\mu_n + p\mu_p} \frac{\partial p}{\partial x}.$$

In the general case the integration yields a complicated expression for V_0, but only the two extreme cases (low level of injection and high level of injection) are of practical importance.

At low injection levels the current–voltage characteristic of the long diode is of the type:

$$I = I_s \left[\exp\left(\frac{q(V_0 - IR_T)}{k_B T} \right) - 1 \right], \tag{2.102}$$

where

$$R_T = \rho_0 l \left[1 + \left(\frac{\mu_n - \mu_p}{\mu_p} \right) \frac{L}{l} \text{th} \frac{l}{L} \right];$$

$$\rho_0 = \frac{1}{q(n_0 \mu_n + p_0 \mu_p)}.$$

The expression (2.102) is the standard formula for a diode except for the effective voltage across the p^+–N junction, which is the difference between the total voltage applied and the product IR_T which is the ohmic voltage drop in the base region. The diffusion length L in this case is a parameter of holes. When $l \ll L$, the following expressions are valid:

$$I_s \sim q \frac{D_p p_0}{l} \quad \text{and} \quad R_T \sim \rho_0 l \frac{\mu_n}{\mu_p},$$

I_s is the typical saturated reverse current of a short diode.

At high injection levels the expression (2.102) for the long-diode current–voltage curve is of the type:

$$I = I_C \left[\exp\left(\frac{qV_0}{Ck_B T} \right) - 1 \right], \tag{2.103}$$

where

$$I_C = 2(I_s)^{\frac{1}{C}} (I_{s_1})^{\frac{2\,\text{ch}(l/L)}{C}\frac{\mu_p}{(\mu_n+\mu_p)}} (I_{s_2})^{\frac{(\mu_n-\mu_p)}{(\mu_n+\mu_p)}\frac{1}{C}} \tag{2.104}$$

and

$$C = 2\frac{\mu_n + \mu_p\,\text{ch}(l/L)}{\mu_n + \mu_p} \tag{2.105}$$

is called the ideality factor. The currents I_s, I_{s_1} and I_{s_2} are functions of the electron and hole mobilities and depend on the parameters l, L and ρ_0 [79].

In this case a diode-type of relationship is derived again but the voltage V_J across the p^+–N junction, instead of the total voltage V_0, is the ratio V_0/C. The voltage V_J is obtained from (2.99):

$$V_J = \frac{V_0}{C}\frac{\log(I/2I_s)}{\log(I/I_C)}.$$

At high injection levels V_J acquires the following form:

$$V_J \sim \frac{V_0}{C}, \quad \text{since} \quad \frac{\log(I/2I_s)}{\log(I/I_C)} \to 1.$$

In the two extreme cases of a short diode and a long diode the following well-known relationships are obtained:
– for a short diode: $l/L \ll 1$; $C \sim 2$; $I_C \sim 2qD_p n_i/l$; $I = I_C \exp(qV_0/2k_BT)$;
– for a long diode $l/L \gg 1 \Longrightarrow \exp(l/L) \gg 1$; $C \to \infty$; $I_C \to 2I_{s_1}$; $V_0 \to 0$ and $I \sim 2I_{s_1} \exp(qV_0/Ck_BT)$.

Therefore in the long diode the conductivity of the base region dominates, and the influence of the injection across the junction is rather small.

When an external magnetic field B is applied to the long diode in Fig. 2.20, the effective minority-carrier lifetime τ_{eff} in the base region is controlled by B. Depending on the polarity of B, the Lorentz force causes deflection of carriers towards one of the two active sides of the semiconductor structure. The recombination rates on the two surfaces are not equal. For instance, the deflection of the electron–hole plasma towards the upper surface, whose recombination rate is assumed to be high, is shown in Fig. 2.21. The physical result of such a deflection is a decrease in the excess of carriers δn and δp and a reduction in the effective minority carrier lifetime and, therefore, in the nonequilibrium concentrations of electrons and holes. Reversing the polarity of B produces the opposite effect.

The variation of δn and δp and/or τ_{eff} directly modulates the diffusion length L of minority carriers. By decreasing δp the total concentration of holes is reduced. This is accomplished by the transition of L from L_{amb} to L_p. This dependence of L on δp is weak. On the other hand, a decrease in τ_{eff} has a more direct influence upon L, since $L = \sqrt{D\tau_{\text{eff}}}$.

The magnetic field B has a considerable effect on the current–voltage characteristic of the diode in the high injection mode. It follows from (2.103) that:

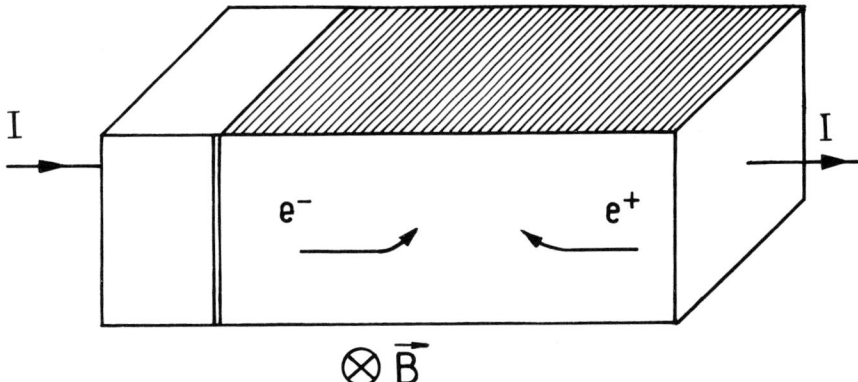

Fig. 2.21. Deflection of both electrons (e^-) and holes (e^+) towards the surface with a high recombination rate when a magnetic field is applied to a long diode.

$$I \sim I_C \exp\left(\frac{qV_0}{Ck_BT}\right), \quad \text{when}$$
$$V_0 = V_J + V_N, \quad \text{and} \quad V_J = \frac{V_0}{C}. \tag{2.106}$$

If the MD structure is operating at the high injection level and the total voltage applied is V_0, the diffusion length L is the ambipolar diffusion length $L = L_{\text{amb}}$. In a magnetic field $\boldsymbol{B}$, the ambipolar diffusion length L_{amb} is reduced.

This entails, however, an increase in the factor C (2.105). If $l/L \geq 3$, the condition $\operatorname{ch} l/L \approx \operatorname{sh} l/L$ is fulfilled. Then $dC/C = (-l/L)(dL/L)$.

It can be seen from (2.106) that an increase in the ideality factor C leads to a reduction in the voltage V_J across the junction when $V_0 = \text{const}$.

Therefore:
$$\frac{dV_J}{V_J} = -\frac{dC}{C} = \frac{l}{L}\frac{dL}{L}.$$

It follows from (2.99) that a decrease in V_J brings a decrease in the current I in such manner that:
$$\frac{dI}{I} \sim \frac{qV_J}{k_BT}\frac{dV_J}{V_J} = \frac{l}{L}\frac{qV_J}{k_BT}\frac{dL}{l}. \tag{2.107}$$

If $l/L \geq 3$ is assumed, the corresponding value of $(qV_J)/(k_BT)$ will be ~ 30 to 40, and as a result $dI/I \sim 100 dL/L$. This is why a small variation of L, caused by the magnetic field $\boldsymbol{B}$, leads to a large variation of the current I, which on the other hand, again influences L; hence a positive feedback is established by modulating the injection level. In fact, decreasing or increasing the current I leads to a lower or higher level of injection, and therefore the diffusion length L tends to L_p, i.e. $L_p < L_{\text{amp}}$, or to L_{amp}, i.e. $L_{\text{amb}} > L_p$. The derived expressions concerning the magnetodiode effect in a long diode with one injecting contact lead to the

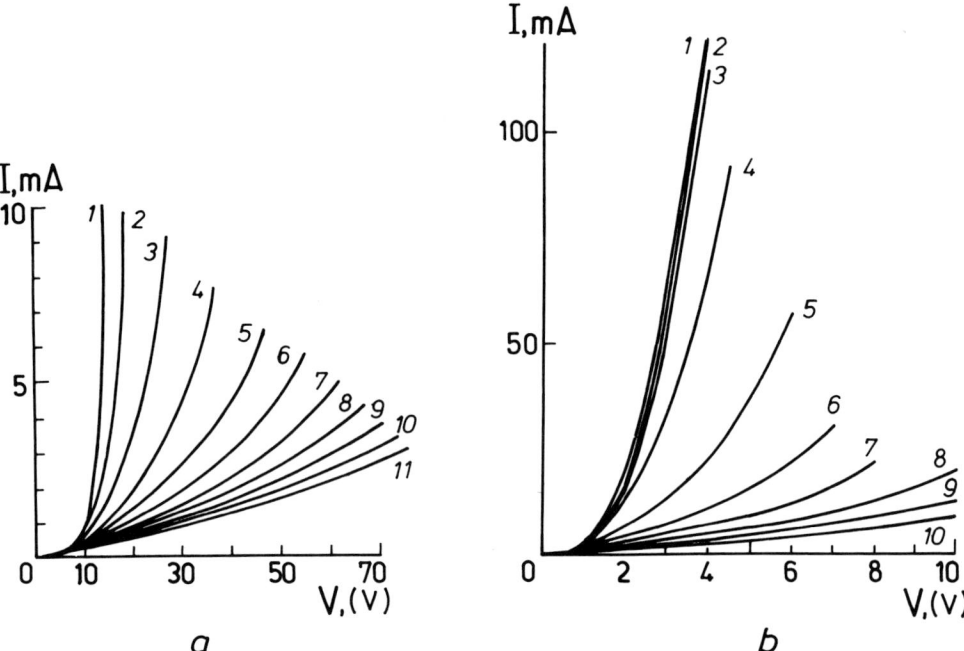

Fig. 2.22. Typical case of magnetic control of the $I(V)$ characteristics of long magnetodiodes with one injecting junction at T = 300 K. (a) Silicon device for various B. *1*: 0 T; *2*: 0.1 T; *3*: 0.2 T; *4*: 0.3 T; *5*: 0.4 T; ...; *11*: 1 T. (b) Germanium device for various B. *1*: 0 T; *2*: 0.05 T; *3*: 0.1 T; *4*: 0.2 T; *5*: 0.3 T; *6*: 0.4 T; *7*: 0.5 T; *8*: 0.6 T; *9*: 0.7 T; *10*: 0.8 T.

conclusion that a very small variation in the system entails self-enhancement by a factor greater than unity (a positive feedback) and to large variation of the current I in a magnetic field B [79].

Ge, Si, GaAs and other materials are used for the implementation of long magnetodiode sensors [78,79], and by optimizing the design and the operating conditions ($l/L \geq 3$, high level of injection) a high transducer efficiency is achieved. Figure 2.22 illustrates the magnetic control of the current–voltage characteristics of typical silicon and germanium long magnetodiodes with one injecting contact [78].

2.5.3.2. Magnetodiode effect in structures with two injecting contacts

The principal ideas and equations of the theory of the magnetodiode effect in semiconductor structures with two injecting p–n junctions are discussed in [24–26,76,80,81]. Figure 2.23 shows a semiconductor sample with the shape of a parallelepiped. The majority-carrier concentration in the lightly doped base region is n_0. The left-hand end of the structure is a p^+–n junction which injects holes into the n region, while at the other end of the structure an n^+–n contact is responsible for the delivery of nonequilibrium electrons. The stationary formulation of the magnetodiode effect is analogous to that in the case of a long magnetodiode (cf. §2.5.3.1)

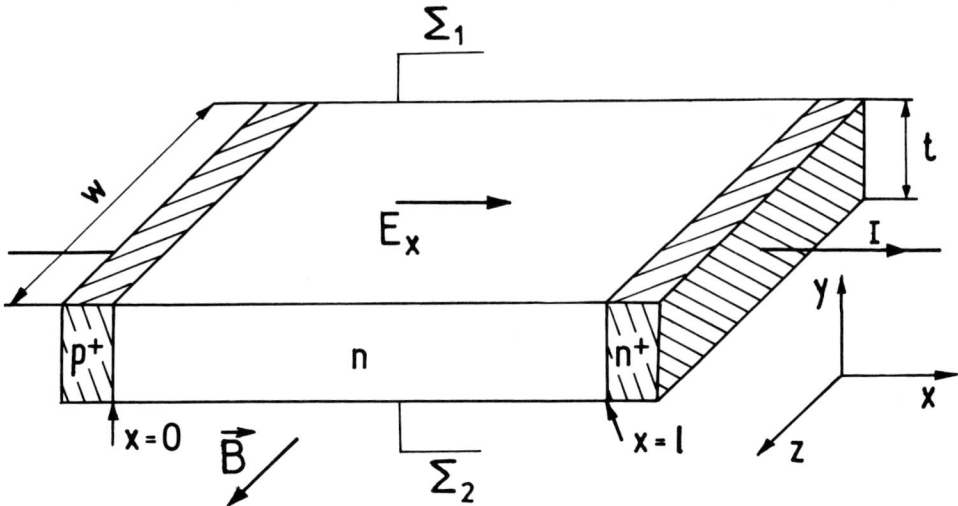

Fig. 2.23. Configuration of the magnetodiode effect in p^+–n–n^+ structure.

and includes the equations for double injection, and for magnetoconcentration. The densities of electron and hole currents $j_{n,p}$ under isothermal conditions are described by the equations:

$$j_{n,p} = (j_0)_{n,p} + (j_0)_{n,p} \times (\mu_H)_{n,p} B, \qquad (2.108)$$

where

$$(j_0)_{n,p} = \sigma_{n,p} E - q_{n,p} D_{n,p} \nabla n, p. \qquad (2.109)$$

The Poisson equation in the considered case is:

$$\operatorname{div} E = \frac{q}{\varepsilon}(\delta p - \delta n), \qquad (2.110)$$

where ε is the dielectric constant of the material.

Also valid is the assumption of a low recombination center density which prevents a negative differential resistance arising. Moreover, the excess of carriers δn and δp is considerably larger than n_0, or the charge density in the space-charge region:

$$\delta n \simeq \delta p \gg n_0 \gg p_0. \qquad (2.111)$$

Charge neutrality of the plasma is assumed both in the bulk and on both surfaces Σ_1 and Σ_2 of the structure. The continuity equation is:

$$\operatorname{div} j_{n,p} = \pm \frac{\delta n}{\tau_V} q. \qquad (2.112)$$

The following relationships are valid on both surfaces Σ_1 and Σ_2 ($y = \pm t/2$):

$$(j_{n,p})_{y=\pm t/2} = \pm q_{n,p} s_{1,2} \delta n \left(\pm \frac{t}{2}\right). \tag{2.113}$$

Linear recombination laws which determine the lifetime in the bulk and the surface recombination rates s_1 and s_2 are assumed for simplicity. The z-dimension of the sample in Fig. 2.23 is long enough and the direction of the magnetic field $\boldsymbol{B}$ is the same as that of the z-axis, i.e. $\boldsymbol{B} = (0, 0, B_z)$. In the general case, the carrier distribution $\delta n(x, y)$ is two dimensional. By combining equations (2.108)–(2.112), and making use of the assumptions already made, an expression which describes the carrier distribution is obtained:

$$D^* \frac{\partial^2 n}{\partial y^2} + \frac{D}{l^*} \frac{\partial n}{\partial y} - \frac{\delta n}{\tau_V} =$$
$$= -\frac{D^*}{2\beta}(n_0 - p_0)\frac{dE_x}{dx} + \frac{D^* \varepsilon}{2\beta} \frac{d}{q \, dx}\left(E_x \frac{dE_x}{dx}\right) - D^* \frac{\partial^2 n}{\partial x^2} = -Q, \tag{2.114}$$

where D^* is the ambipolar diffusion constant:

$$D^* = \beta \frac{|\mu_n|\mu_p(n+p)}{n|\mu_n| + p\mu_p}; \quad \beta = \frac{k_B T}{q},$$

and, similar to (2.92), the parameter l^* is the characteristic length which is used in order to take into account the effect of the fields $\boldsymbol{E}$ and $\boldsymbol{B}$. This length is described by the expression:

$$\frac{1}{l^*} = \frac{1}{2\beta}\left(|\mu_{H,n}| + \mu_{H,p}\right)\frac{BV}{l}, \tag{2.115}$$

where $E_x = V/l$ (Fig. 2.23).

The first two terms on the left-hand side of equation (2.114) represent the magnetoconcentration effect; the right-hand side of the equation includes an expression which corresponds to the longitudinal injection of carriers. The term $(-\delta n/\tau_V)$ in (2.114) stands for the loss of carriers due to recombination. The meaning of the symbol Q, introduced in (2.114), is a source of electrons and holes whose density is almost constant along the y-axis. In a magnetic field $\boldsymbol{B}$ the electrons and holes delivered by the source are subjected to magnetoconcentration. Q is a function of the concentration of doping impurities in the base region, the space-charge region and the longitudinal diffusion. The solution of (2.114) for the transverse distribution of carriers under boundary conditions (2.113) is of the type [24–26]:

$$\delta n(x, y) = f_1(y) + \tau_V Q(x) f_2(y),$$

where f_1 and f_2 are expressions used to take into consideration the parameters τ_V, s_1, s_2 and B [80].

The mean current density in the sample as a function of the applied voltage V is obtained by the integration:

$$V = \int_0^l E_x(x)dx. \tag{2.116}$$

In (2.116) the electric field E_x is expressed as a function of I, and the integration of $E_x(x)$ directly yields the current–voltage relationships. The voltage drop across the p^+–n and n^+–n junctions is considered negligible. Only a numerical solution of (2.114) yields an accurate current–voltage characteristic of the sample. But an analytical solution of (2.114) with a certain degree of approximation is also possible. Such a solution offers the opportunity for a clear physical interpretation of the effect. The analysis of the particular type of I vs. V relationship in a magnetic field is made by successively choosing the double-injection operating mode [24–26,76,80].

(i) The magnetodiode ohmic regime. This mode of operation is observed at low values of forward bias, applied to the p^+–n–n^+ structure. The resulting concentration of the carriers introduced into the sample is lower than their equilibrium concentration. The electrical behavior of the structure is that of a resistor with an ohmic $I(V)$ characteristic:

$$I = qn_0|\mu_n|\frac{V}{l}. \tag{2.117}$$

In this mode of operation the current is practically independent of the magnetic field.

(ii) The magnetodiode semiconductor regime. Such behavior is characterized by a moderate level of double injection, the longitudinal diffusion and the space charge being negligible in the expressions for Q and I. A square relationship [25,26], is derived from (2.116) by the integration of the field $E_x(x)$ under the boundary condition $E_x(x=0) = 0$:

$$I = \frac{9}{8}q|\mu_n|\mu_p(n_0 - p_0)\tau_{sc}^* \frac{V^2}{l^3}. \tag{2.118}$$

The effective lifetime τ_{sc}^* which represents the magnetic control of the current in the semiconductor regime is a function of the recombination rate and the fields E_x and B_z [24,82]. A detailed analysis of the parameter τ_{sc}^* leads to the following physical comments:

(a) If the magnetic field leads to deflection of carriers towards a surface with a low recombination rate (piling up of carriers upon that surface plays an important role), the carrier recombination upon the opposite side has only a slight influence on the performance of the device. The current I as well as τ_{sc}^* gain their maximum values when $\tau_{sc}^* = \tau_V$. At high magnetic fields, $I \sim 1/B^2$.

(b) If, in a magnetic field B, carriers deflect from a surface with a low recombination rate towards a surface with a high recombination rate, a considerable insufficiency of carriers results, i.e. $\langle \delta n \rangle$ is reduced and an ohmic $I(V)$ relationship is observed:

$$\tau_{sc}^* \simeq \frac{l^*}{s_2} \sim \frac{l}{s_2 V B}.$$

With increasing B the current becomes proportional to $1/B^3$.

(c) If the recombination rates s_1 and s_2 on the two surfaces Σ_1 and Σ_2 are infinitely high ($s_{1,2} \to \infty$), the excess of carriers "immediately" recombines ($\tau_{sc}^* \to 0$). In this case the effect of the double injection is replaced by ohmic conductivity.

(d) At low magnitudes of the magnetic field B, the current in the structure is proportional to B. If there is a great difference between the recombination rates on the two surfaces ($s_1 \to 0, s_2 \to \infty$), the slope of the $I(B)$ characteristic acquires its maximum value and is proportional to V^3. In this case the optimum sample thickness is 1 to 3 times L_amb. If the recombination rates on the surfaces Σ_1 and Σ_2 are equal ($s_1 \approx s_2$), the current I is an even function of B. This is why the processing of the two surfaces Σ_1 and Σ_2 with the purpose of achieving the desired values of the recombination rates s_1 and s_2 is of considerable importance for the occurrence of the magnetodiode effect.

(iii) The magnetodiode insulator regime. The insulating mode of conductivity is observed when the injection level is sufficiently high and the space charge dominates over the initial doping of the base region and over the longitudinal diffusion:

$$I \sim \langle \delta n \rangle_y E_x. \tag{2.119}$$

A cubic $I(V)$ relationship is obtained by the integration of (2.116) under the boundary conditions $E_x(0) = E_x(l) = 0$ [80,81]:

$$I = \frac{125}{18} \varepsilon |\mu_n| \mu_p \tau_{is}^* \frac{V^3}{l^5}, \tag{2.120}$$

where τ_{is}^* is the effective lifetime in the insulating mode and $\tau_{is}^* \sim \tau_V \langle f_2 \rangle$.

The comments in (ii) are valid in the case of the isolating mode of conductivity as well. Particularly, when there is a great difference between the recombination rates on both surfaces, the $I(V)$ relationship in (2.120) becomes a square function, since $\tau_{is}^* \sim 1/V$.

(iv) The magnetodiode diffusion regime. Enhancement of the injection by increasing the applied voltage V leads to the domination of the longitudinal diffusion throughout the entire length of the structure. (In the other regimes diffusion takes place only in the neighborhood of the p^+–n and n^+–n junctions). Two types of $I(V)$ curves can be observed in this regime, depending on the injection level [80]. At moderate injection levels, $\langle \delta n \rangle \sim I$, the current increases at a constant voltage, cf. (2.119). With such a steep current–voltage curve, magnetoconcentration modulates

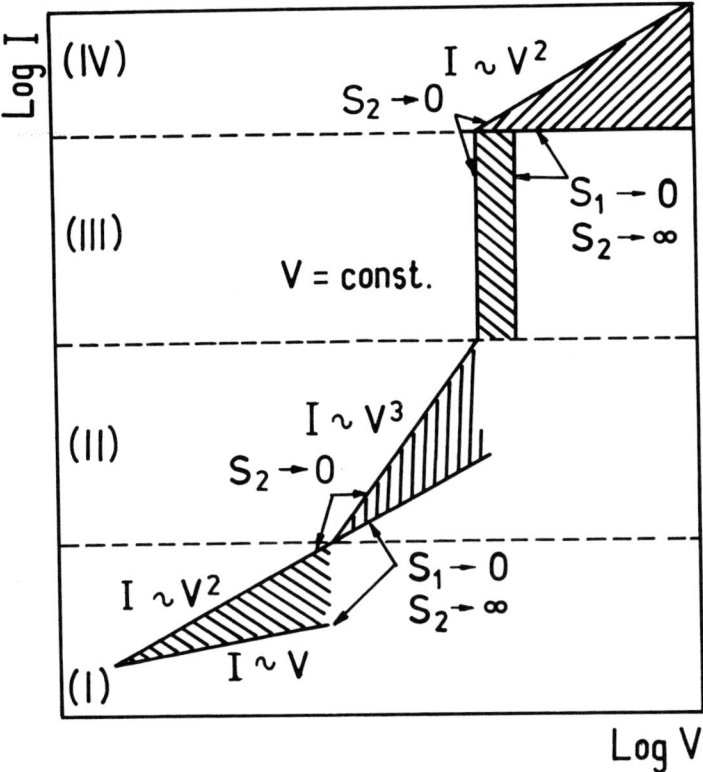

Fig. 2.24. Influence of a magnetic field on the $I(V)$ characteristics of p^+-n-n^+ magnetodiodes, according to the double-injection regime: (I) semiconductor regime; (II) insulator regime; (III) diffusion regime at a moderate level; (IV) diffusion at a very high level. Note that the extreme evolutions correspond to $s_1 \to 0$ and $s_2 \to \infty$.

only the voltage. Further increasing of the current leads to a very high injection level and $\langle \delta n \rangle \sim I^{1/2}$. The square function law is observed again, thus illustrating the increased influence of magnetoconcentration: $I \sim (\langle f_2(y) \rangle V)^2$ [25,26,80].

Figure 2.24 shows a summarized sequence of the particular conductivity regimes of a p^+-n-n^+ magnetodiode in the case of double injection (with the exception of the ohmic conductivity regime). The principal modification of the laws of conductivity due to the influence of the surfaces Σ_1 and Σ_2 and of the magnetic field have been presented too. According to the analysis already given and to Fig. 2.24, magnetoconcentration can completely change the classical modes of double injection. In a particular p^+-n-n^+ structure, however, one or more magnetodiode modes of operation may be impossible because of limitations associated with the sample length, concentration of doping impurities, dissipated power, the quality of the injected contacts, etc. Experiments with silicon and germanium magnetodiodes with two injecting contacts have manifested excellent sensitivity and the behavior of the device has been explained in detail by the presented theory [26,80]. Thus the establishment of the principles of magnetodiode operation has led to the creation of the modern family of SOS magnetodiodes with wide functional capabilities [80–83]. The magnetodiode phenomenon is also a useful instrument for the investigation of recombination parameters in semiconductors and especially in narrow bandgap materials [76]. The following conclusion can be drawn from the origin of magnetosensitivity in long diodes with one injecting contact and p^+-n-n^+ structures with double injection: although magnetoconcentration occurs in both types of devices, the behavior of magnetodiodes with two injecting contacts is more strongly influenced by the magnetic field. The main reason for this is the fact that the active volume of the sensor is almost completely occupied by a region of strongly nonequilibrium quasi-intrinsic conductivity at $\boldsymbol{B} = 0$. When $s_1 \gg s_2$ or $s_1 \ll s_2$ the magnetoconcentration effect very effectively "suppresses" or "releases" the injection of carriers. A large increase or decrease in the current I is observed depending on the direction of the magnetic field $\boldsymbol{B}$. In the long diode, however, the strongly nonequilibrium bipolar conductivity is not established in the entire active region and the magnetic control of the current I is more closely related to the indirect feedback due to potential redistribution between the base region and the injecting junction rather than directly to magnetoconcentration. Therefore in most cases the current $I(\boldsymbol{B})$ is a decreasing function of the field magnitude for both field directions. Substituting an n^+-n junction for the ohmic contact in the long diode is equivalent to reducing the length of the base region by a factor of 2 [78]. This is why the miniaturization of magnetodiode sensors requires double injection, and this is accomplished in integrated SOS magnetodiodes

2.6. Related effects

Transverse effects occurring in a magnetic field, the Hall effect for instance, are influenced by their coexistence with a transverse magnetoconcentration gradient of the carrier distribution. This problem is directly related to the performance of injection-based magnetosensitive devices, like magnetodiodes, bipolar magnetotransistors, etc. The galvanomagnetic picture in them is further complicated by the nonuniform distribution of carriers along the length of the structure as well as the generation of circulating currents (cf. §2.3.5). An advance in the understanding of the phenomena in this class of solid-state transducers has been achieved by the extrapolation of the Hall effect beyond the limits of its trivial manifestations and by the analysis of its role in such effects like emitter injection modulation in bipolar magnetotransistors, the filament magnetosensitivity effect, electrical control of the sign of magnetosensitivity, etc. This is why the perturbation of the Hall effect by magnetoconcentration in complex structures with nonequilibrium bipolar conductivity is an interesting problem.

The peculiarities of the Hall effect under nonequilibrium conditions have been analyzed on the basis of a diode structure with a long base region ($L > L_{\text{amb}}$), the field B being parallel to the z-axis and with the current flow along the x-axis (Fig. 2.25). The boundary condition that in all cases the transverse current does not leave the two surfaces Σ_1 and Σ_2 means that $\int_x [j_{n,y}(x,y) + j_{p,y}(x,y)] dx = 0$. With such an assumption the following expression is obtained for the resultant electric field along the y-axis [76]:

$$E_y(y) = (n|\mu_{H,n}| + p\mu_{H,p})^{-1} \left[(p\mu_{H,p}^2 - n\mu_{H,n}^2) B E_x + \frac{k_B T}{q} (\mu_{H,n} + \mu_{H,p}) \frac{\partial n}{\partial y} \right]$$
$$= E_H(y) + E_D(y). \tag{2.121}$$

The physical meaning of the two terms in expression (2.121) for $E_y(y)$ is the following: the first term represents the Hall field in the case of ambipolar conductivity, while the second is the Dember field E_D, which is associated with the existence of a transverse gradient of carrier concentration. This gradient is responsible for the charge neutrality of the plasma. According to (2.121) the transverse voltage V_y will be:

$$V_y = V(+\frac{t}{2}) - V(-\frac{t}{2}) = -\int_{-t/2}^{+t/2} E_y(y) dy =$$
$$= V_H + V_D. \tag{2.122}$$

At low injecting levels $\delta n(y) \approx \delta p(y) \ll n_0; p_0$, are valid for the Hall voltage and for the Dember voltage the following expressions apply:

$$V_H = \frac{n_0 \mu_{H,n}^2 - p_0 \mu_{H,p}^2}{n_0 |\mu_{H,n}| + p_0 \mu_{H,p}} B E_x t; \tag{2.123}$$

$$V_D = \frac{k_B T}{q} \frac{\mu_{H,n} - \mu_{H,p}}{n_0 |\mu_{H,n}| + p_0 \mu_{H,p}} \left[n\left(+\frac{t}{2}\right) - n\left(-\frac{t}{2}\right) \right]. \tag{2.124}$$

It follows from (2.121) that a higher Dember field will be generated in semiconductor samples with intrinsic conductivity than in extrinsic samples. The analysis of this effect leads to the conclusion that with infinite rates of surface recombination ($s_{1,2} \to \infty$, $\delta n(y) \approx 0$) the Dember voltage is $V_D = 0$ and $V_y = V_H$, i.e. in this case the "pure" Hall effect can be determined experimentally, since $V_y/V_H \simeq 1$. The perturbation of the Hall effect caused by the Dember field can be evaluated from the formula:

$$\frac{V_y}{V_H} = 1 - \frac{1}{2} \frac{\text{th}(t/2L_{\text{amb}})}{t/2L_{\text{amb}}}. \tag{2.125}$$

However, if the sample is thin enough ($t < L_{\text{amb}}$), under boundary conditions $s_1 \to 0$ and $s_2 \to \infty$ at high injection levels the Dember field reduces the Hall voltage by a factor of 2. The lower the recombination rate is upon the two surfaces

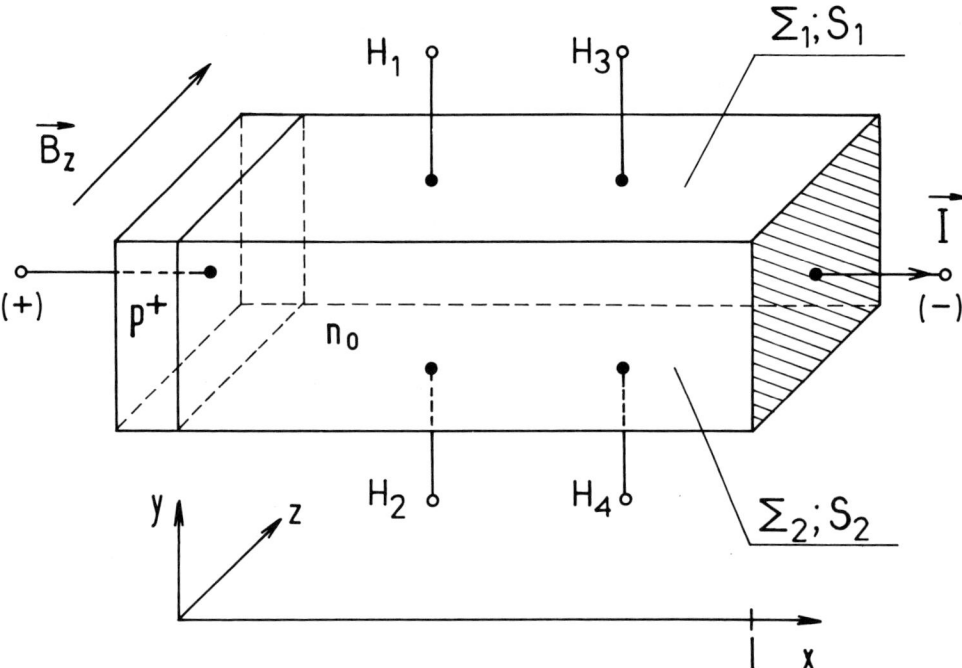

Fig. 2.25. A long diode structure used for the analysis of the influence of nonequilibrium conditions upon the Hall effect.

Σ_1 and Σ_2, the more pronounced is the reduction of the Hall voltage due to the transverse magnetoconcentration gradient of carrier density (Dember field).

Another peculiarity of the galvanomagnetic properties of diode structures is associated with the longitudinal inhomogeneity of the injected carrier density along the x-axis. Until recently the Lorentz deflection of the nonequilibrium electron–hole plasma was assumed to dominate and the role of the Hall voltage associated with majority carriers used to be neglected [78]. It was found from experiments with bipolar samples that the Hall voltage really exists and is characterized by a longitudinal profile along the x-axis. At sufficiently long distances from the p^+–n junction, the Hall voltage reaches considerable values under this condition [84]. The concept of the diode Hall effect is illustrated by Fig. 2.25. The selected rectangular geometry of the structure is not determining the phenomenon, since the results are valid also for samples obtained by conformal mapping, which have planar contacts (cf. §4.3). An infinitely high recombination rate was assumed for the two Hall surfaces so that the gradient of carrier density along the y-axis due to magnetoconcentration could be neglected. Another way to fulfill this condition is to prepare a sample whose dimension along the z-axis is $t \gg L_p$, where L_p is the hole diffusion length. Moreover, a low concentration of doping impurities in the base region $n_0 \approx N$ with a length $l > L_p$ is considered. A low or moderate injection level is assumed. A large portion of the nonequilibrium carriers recombine during their drift and diffusion in the positive direction of the x-axis. The heavy doping of the injecting p^+ region makes the Hall effect in it negligible (cf. §2.3.5). When the diode shown in Fig. 2.25 is forward biased, equal amounts of holes and electrons are introduced into the base in order to preserve the quasi-neutrality of the base region. As a result of recombination, the distribution of injected holes $p(x)$ and the corresponding current $I_p(x)$ in the general case are given by the well-known relationship:

$$I_p(x) \sim p(x) \sim \left[\exp\left(\frac{qV}{k_B T}\right) - 1\right] \exp\left(-\frac{x}{L_p}\right). \tag{2.126}$$

According to (2.126) the hole current decreases exponentially with increasing distance $x > 0$ but the total current across the diode remains the same, i.e. $I = I_p(x) + I_n(x) = \text{const}$. Therefore, sufficiently far from the p^+–n junction, the majority carrier current dominates and $I_n(x) \sim qn_0 v^*_{dr}(x)$, the drift velocity v^*_{dr} being a function of x. It is exactly the majority-carrier current $I_n(x)$ that generates the Hall voltage in the base region of the diode when a magnetic field B_z is applied. Owing to the high concentration of nonequilibrium electrons and holes near the p^+–n junction, the Hall coefficient R_H and the Hall voltage will be considerably reduced, according to §2.3.6 and (2.69). As a result of shunting effects, near the low resistivity n^+ region the Hall voltage will be $V_H \sim 0$. This is why a nonuniform distribution of the Hall voltage sets in along the x-axis of the structure and its maximum is located far enough from the p^+–n and n^+–n junctions. The Hall field $E_H(x)$ generated by majority carriers influences the minority carriers located in a cross-section which is perpendicular to the x-axis. In a magnetic field, because of the Lorentz force, the holes deflect at an angle $\Theta_{H,p} \sim \mu_p B$. In addition, they drift in the direction of the resulting electric field $E(x)$ which is tilted at a Hall angle $\Theta_{H,n} \sim -\mu_n B$ relative to the x-axis. Thus the total angle of deflection of the minority carriers with respect to the x-axis is $\varphi = \Theta_{H,p} - \Theta_{H,n} = (\mu_{H,p} - \mu_{H,n})B$. Since $\mu_{H,n} < 0$ and $\mu_{H,p} > 0$, the two angles add up. Because of the assumption that $s_{1,2} \to \infty$ the deflected carriers are expected to recombine upon Σ_1 or Σ_2. This phenomenon is known as the Suhl effect [1].

As seen from §2.3.5, the nonuniform distribution of the Hall field generates a circulating current j^* that leads to a transverse inhomogeneity of the current density in the direction of the y-axis as the magnetic field is increased. The increase in the injection level will obviously reduce the Hall voltage $V_H(x)$. Figure 2.26 presents experimental results which prove the presence of the Hall voltage generated by majority carriers in a planar silicon diodes and transistors [84,85].

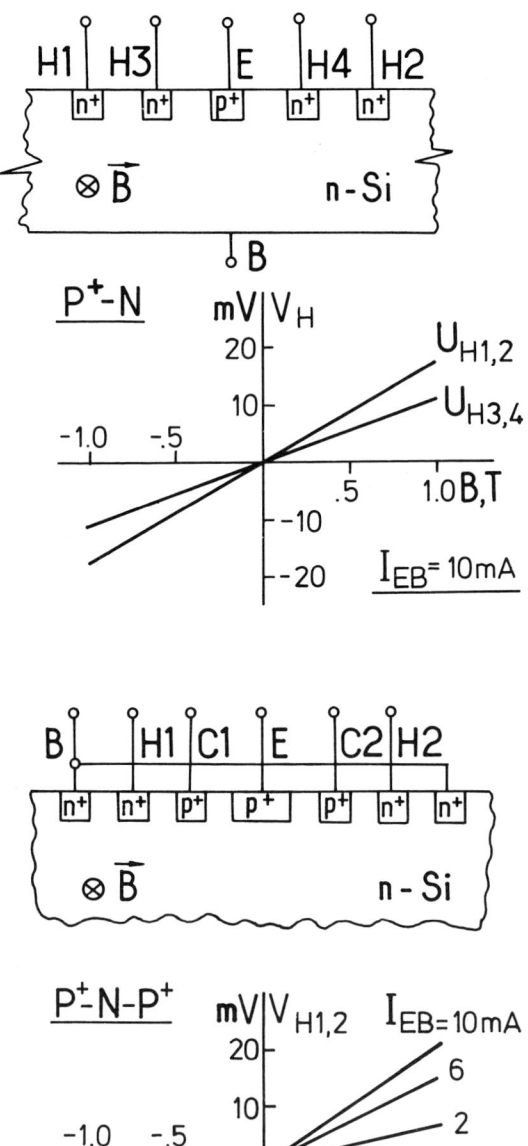

Fig. 2.26. Cross-section of parallel-field diode and transistor structures with Hall probes for the registration of $V_H(B)$.

2.7. Coherent phenomena associated with superconductivity in condensed matter

2.7.1. Superconductivity

At a temperature lower than a certain critical value T_c, a transition into a superconducting state is observed in almost half of the metal elements and in many metal and metal–ceramic alloys, cf. Table 2.2. There are four specific features of the superconducting state [28–33]:

(a) *Infinite conductivity*, i.e. zero-resistance state;

(b) *The Meissner effect*. The essence of this effect is the expulsion of the magnetic field from the superconducting volume. Nevertheless, the field penetrates to a very small depth λ (the so-called London penetration depth) which in most superconductors is in the 10^{-5}-cm range. When the field exceeds a certain critical value B_c, i.e. $B \geq B_c$, superconductivity is no longer observed and the field penetrates into the volume of the material which is already normal.

(c) *Flux quantization*, which forces the total flux inside a superconducting loop to be a multiple of the flux quantum Φ_0;

(d) *The Josephson effect*, which concerns the tunnelling of Cooper pairs between two superconducting materials.

These four important characteristics are systematized in Fig. 2.27. The theoretical basis of this unique phenomenon can be found, for instance, in [29,30]. The superconductive properties of condensed matter are associated with the fact that, instead of there being discrete (individual) electrons as in the case of normal materials, the charge carriers may be the so-called Cooper pairs. These pairs consist of electrons with opposite spin and quasi-momentum orientations. The reason for this coupling is that under certain conditions the electrons effectively attract each other owing to the exchange of acoustic phonons. In superconductors these forces of attraction exceed the classical Coulomb interaction. At high temperatures the thermal motion prevents electron coupling. This is why electron pairs appear only at temperatures $T \leq T_c$, and the stronger the attraction, the higher the value of T_c, and at $T \ll T_c$ all the electrons that determine the conductivity of a given material are coupled.

Table 2.2

Selected superconducting materials

Material	Critical temperature T_c (K)	Material	Critical temperature T_c (K)
Al	1.18	Nb	9.2
In	3.4	NbN	17
Sn	3.7	BaLaCuO	30
Hg	4.15	YBaCuO	90
V	5.3	BiSrCaCuO	110
Pb	7.2	TlCaBaCuO	120

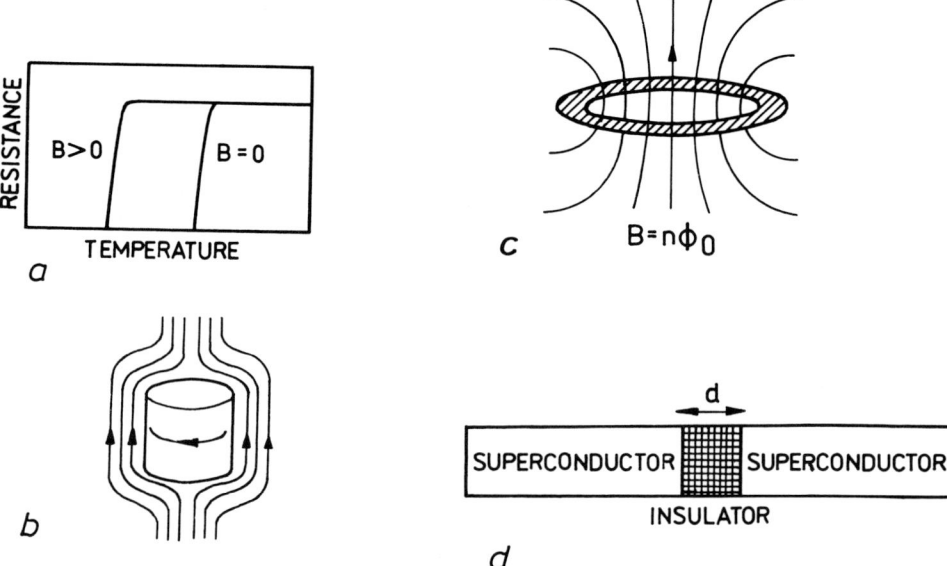

Fig. 2.27. The four basic phenomena of the superconducting state: (a) zero-resistance state; (b) Meissner effect; (c) flux quantization; and (d) Josephson effect.

The spin of electrons is $\pm 1/2$ and they obey Fermi–Dirac statistics, whereas the total spin of the Cooper pair is zero and Bose–Einstein statistics must be applied. The latter type of statistics is known to allow the "condensation" of all particles in the lowest (ground) energy level when $T \to 0$. It is this condensation that determines the difference between superconductors and normal materials.

The behavior of charge carriers in each metal is described by the equations of quantum mechanics. In the case of the weak particle interaction and negligible spin effects, Schrödinger's nonstationary equation can be used:

$$i\hbar \frac{d\Psi(\boldsymbol{r},t)}{dt} = \hat{H}\Psi(\boldsymbol{r},t), \tag{2.127}$$

where Ψ is the wave function of the particle and $\hat{H}$ is Hamilton's operator.

Under stationary conditions, $\hat{H}\Psi(\boldsymbol{r},t)$ can be replaced by the particle energy E. Then, when the complex function Ψ is presented in the form:

$$\Psi = |\Psi| \exp(i\varphi), \tag{2.128}$$

with a well-defined amplitude $|\Psi|$ and phase angle φ, the following formula is obtained:

$$\hbar \frac{d\varphi}{dt} = -E. \tag{2.129}$$

It can be concluded from (2.129) that the phase φ of the particle wave function is time dependent and its velocity is proportional to the energy E. Fermi particles have different values of energy E because of the Pauli principle, and hence their phase velocities differ as well. But in the case of Bose–Einstein statistics, the phases of particles are uniformly distributed within the interval $[0, 2\pi]$. An increase in the phase φ of 2π leads to a state which is physically equivalent to a state with a phase φ that fulfills the condition $0 \leq \varphi \leq 2\pi$. Bose particles that are condensed in the lowest energy level have the same energy E and therefore their phases can differ only by a constant. Another important property of the Cooper pairs is that they have relatively large dimensions. The theory of superconductivity yields the following formula for the mean distance between the two electrons (the so-called coherence length) in a Cooper pair:

$$\xi_0 = \frac{\hbar v_F}{\pi \Delta(0)}, \qquad (2.130)$$

where v_F is the velocity of electrons upon the Fermi surface (v_F is in the 10^8 cm s^{-1} range) and $\Delta(T)$ is the bound energy of the electrons in Cooper pairs. It is often termed the width of the energy gap.

In compliance with the Bardeen, Cooper and Schrieffer microscopic model of the superconducting state, [29,30], $\Delta(0)$ is determined by the critical temperature T_c in the expression:

$$2\Delta(0) \simeq 3.5 k_B T_c. \qquad (2.131)$$

For typical superconductors like Pb and Nb, from (2.131) is obtained $\Delta(0) \approx 3 \times 10^{-22}$ J (about 2 meV), and ξ_0 is in the 10^{-5} cm range. Therefore the coherence length of Cooper pairs is much bigger than the distance between neighboring pairs, which is comparable with interatomic distances. Hence there is a considerable spatial overlapping between the pairs, i.e. a strong interaction between them. This is the reason why the Cooper pairs are stable; they are always synphase and constitute a unified superconducting condensate. All pairs located at a given point have identical phases φ and wave functions Ψ. Owing to this fact, the function $\Psi(r, t)$ does not depend on the pair index and describes the condensate of pairs as an unified quantum system. The consistent theory [29,30], gives a satisfactory explanation of the existence of the quantum condensate and of the characteristics (a)–(d) of the state of superconductivity. Until recently, the practical applications of superconductivity were limited because of the need to use expensive liquid helium ($T = 4.2$ K) or explosive liquid hydrogen ($T = 20$ K) in order to achieve the critical temperatures T_c of the known superconductors. The discovery of high-T_c superconductivity in the ceramic system Ba–La–Cu–O (cf. Table 2.2) by Bednorz and Müller (IBM) gave a powerful impetus to investigations in this field. As a result materials with $T_c \sim 120$ K have been synthesized [40–42]. The first experiments with such metal-oxide ceramics with a complex structure have shown that, unlike the low-temperature superconductors, the charge carriers in them are holes. At temperatures $T < T_c$ the current is due to pairs with a charge $+2q$ and a small coherence length $\xi_0 \leq 10$ to 15 Å. The most probable reason for the mutual attraction of holes and for the formation of Cooper pairs is the exchange interaction of their spins with the magnetic fluctuations of the spin of Cu^{2+} ions. Both the theory of high-T_c superconductivity and the experimental methods are currently in a stage of rapid development.

The operation of SQUID sensors is based on the so-called quantum macroscopic (coherent) effects discovered by B. Josephson [34–36].

2.7.2. Coherence effects in superconductors

The reference point in the analysis of this group of characteristics of superconductivity is the effect of quantization of magnetic flux. Its essence is that the magnetic flux $\Phi = \oint \boldsymbol{B} \cdot d\boldsymbol{A}$ inside a superconducting ring acquires a discrete value:

$$\Phi = n\Phi_0, \quad n = 0, \pm 1, \pm 2, \pm 3 \ldots, \tag{2.132}$$

and Φ_0 is the universal value $\Phi_0 = h/2q = 2.07 \times 10^{-15}$ Wb $= 2.07 \times 10^{-7}$ G cm^{-2} (cf. Fig. 2.28). The constant Φ_0 is termed the quantum of the magnetic flux and q is the elementary electric charge [31–33]. It will be shown that this effect results from the existence of a coherent condensate of Cooper pairs in the superconductor. Actually, if two points 1 and 2 located inside two separate superconductors are chosen (Fig. 2.29a), and if they are characterized by their energy E_1, E_2 and phases φ_1, φ_2, the corresponding equations of the type (2.129) can be written and subtracted to yield:

$$\hbar \frac{d}{dt}(\varphi_2 - \varphi_1) = E_2 - E_1. \tag{2.133}$$

It follows from §2.7.1 that the energy values E_1 and E_2 of the two ensembles of Cooper pairs may differ only if a voltage V is applied between points 1 and 2, i.e. $E_2 - E_1 = 2qV$. In this case the fundamental relationship

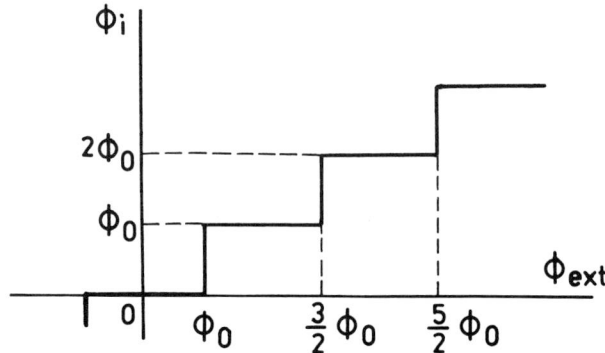

Fig. 2.28. The quantization effect inside a superconducting ring; Φ_i vs. the externally applied flux Φ_{ext}.

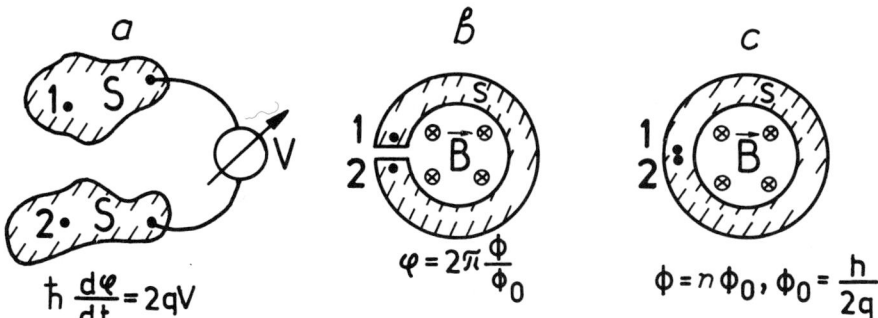

Fig. 2.29. Coherence effects in superconductors. The phase difference $\varphi = \varphi_2 - \varphi_1$ of the wave functions at two arbitrarily selected points *1* and *2* inside the superconductors is: (a) always related to the Josephson equation (2.134); (b) synonymously determined by the magnetic flux $\varphi = 2\pi(\Phi/\Phi_0)$; (c) associated with the magnetic flux quantization in accordance with (2.132).

$$\hbar\frac{d\varphi}{dt} = 2qV \text{ or } V = \left(\frac{\hbar}{2q}\right)\frac{d\varphi}{dt} \qquad (2.134)$$

is valid for the phase difference $\varphi_2 - \varphi_1$, $\hbar = h/2\pi$.

If points 1 and 2 are located on both ends of a superconducting ring with a narrow slit (Fig. 2.29b), the only method of generating a voltage V is to vary the magnetic flux Φ that pierces the ring:

$$V = \frac{d\Phi}{dt}. \qquad (2.135)$$

Comparing equations (2.134) and (2.135) and integrating with respect to time yield the following equation:

$$\varphi = 2\pi\frac{\Phi}{\Phi_0}. \qquad (2.136)$$

It can be seen from (2.136) that there is a simple relationship between the phase difference of the wave functions of the superconducting condensate at the two points (1 and 2) and the magnetic flux inside the ring. If points 1 and 2 coincide (Fig. 2.29c), the validity of (2.136) is preserved and in this case the phases φ_1 and φ_2 differ by a multiple of 2π:

$$\varphi = 2\pi n; \quad n = 1, 2, 3, \ldots \qquad (2.137)$$

The replacement of φ in (2.136) with its equivalent from (2.137) leads to the familiar formula (2.132) for the effect of magnetic flux quantization.

Josephson's coherence effects are of fundamental importance for cryosensor magnetoelectronics [34]. If there is a weak link between two superconductors by means of a so-called tunnel junction (about 10 Å thick oxide layer) (cf. Fig. 2.30a), a quantum mechanical tunnelling may occur. The tunnel current has two

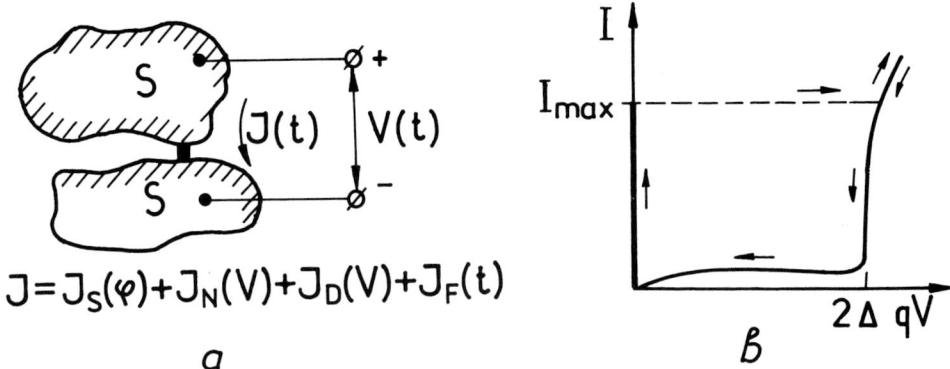

Fig. 2.30. (a) A weak link between two superconducting electrodes and components of the current across it. (b) Current–voltage curve of a Josephson tunnel junction.

components: single electrons that tunnel across the potential barrier and generate a normal current J_N and the current J_S due to the tunnelling of Cooper pairs. The superconductivity tunnel current (supercurrent) was predicted theoretically in [34]. As the current J_S is flowing across the tunnel junction, two entirely different types of behavior, known as the stationary (d.c.) and nonstationary (a.c.) Josephson effect, are observed. In the stationary case the supercurrent J_S may reach a certain critical value J_0 without generating a voltage drop across the junction when $J_S \leq J_0$. In this case the normal current J_N does not arise, i.e. $J_N = 0$. Therefore the thin insulator behaves as a superconductor. However, the critical value J_0 is much smaller than the maximum current that would flow in the superconducting ring in the absence of a tunnel junction. Moreover, the current J_S is destroyed in a magnetic field lower than the field B_c necessary for the transition of the homogeneous material from superconductivity to the normal state (cf. §2.7.1). With a voltage $V = 0$, the current J_S is a periodic function, with a period 2π, of the phase difference φ of the superconducting condensate wave functions:

$$J_S = J_0 \sin \varphi, \tag{2.138}$$

where J_0 is the critical current, strongly dependent on the properties of the insulating layer, on the dimensions and on the material of the contacts.

Convincing arguments which prove the validity of (2.138) can be adduced without resorting to the strict theoretical derivation of this result. In fact the current of Cooper pairs J_S across the weak link must be dependent on the wave functions Ψ_1 and Ψ_2 of the condensate in both electrodes. If there is a variation of J_S, $|\Psi_{1,2}|$ must remain constant, since the small current does not change the physical state of the superconducting electrodes. Therefore the current must depend on the arguments of Ψ_1 and Ψ_2, i.e. on the phases $\varphi_{1,2}$. The phases are determined in such a way that their difference is an arbitrary constant, and the current J_S can be dependent only on their difference $\varphi = \varphi_2 - \varphi_1$. An increase in φ_1 or φ_2 by a multiple of 2π does not change the physical state of the system or of the current J_S. This is why the supercurrent must be a 2π periodic function of φ (c.f (2.138)). According to (2.138) it is this "balancing" current that establishes a constant phase difference between the superconducting electrodes:

$$\varphi = \arcsin \frac{J_S}{J_0} + 2\pi n. \qquad (2.139)$$

Since $d\varphi/dt = 0$, it follows from formula (2.134) that $V = 0$. The direction of the current J_S depends on the values of the phases φ_1 and φ_2. At temperatures $T \geq T_c$, the currents J_S and J_0 are zero. The nonstationary Josephson effect is observed when an external voltage V is applied to the weak link. In this case the Cooper pairs undergo a transition from the higher Fermi level of the first superconductor to the lower Fermi level of the other one, the energy difference between the two levels being qV. Owing to this fact, when tunnelling, Cooper pairs change their energy by $2qV$. This energy is emitted as an electromagnetic wave. According to (2.134), under such conditions the phase φ includes a linear component:

$$\varphi = \omega_V t + \tilde{\varphi}; \quad \omega_V = \frac{2q}{\hbar} V. \qquad (2.140)$$

The meaning of (2.140) is that the supercurrent J_S (2.138) is oscillating in time with a frequency ω_V. Owing to the small value of Planck's constant $\hbar$, this frequency is rather high: $\omega_V = 2qV/\hbar \sim 483.6$ MHz per 1 μV applied voltage. This generation of electromagnetic waves has been observed experimentally and was first described in [86–88]. It occurs in almost all devices based on the Josephson effect. When $V \neq 0$, there is tunnelling of both single electrons and Cooper pairs. The current–voltage curve of a Josephson tunnel junction between two identical superconductors at $T_0 < T_c$ is shown in Fig. 2.30b. The presented I–V relationship was determined on the basis of the above analysis. As the current across the junction is increased, the current I_S reaches $I_0 \equiv I_{max}$, $(V = 0)$ and the tunnel current consists of Cooper pairs only. At $I > I_0$ an abrupt transition of the working point to I_N is observed accompanied by single-particle tunnelling and a voltage across the junction, together with generated oscillations with frequency ω_V. Figure 2.30b illustrates the hysteresis of the $I(V)$ characteristic. Within the voltage interval $0 \leq V \leq 2\Delta/q$ the tunnel junction is working as a high-frequency generator.

The reaction of the high-frequency Josephson junction to a small electrical signal is an interesting problem. If this signal excites a sufficiently small additional voltage $\tilde{V}$ across the junction, the corresponding phase increment will also be small:

$$\tilde{\varphi} = \frac{2q}{\hbar} \int \tilde{V} dt, \quad \text{when } |\tilde{\varphi}| \ll 1. \qquad (2.141)$$

The developing of $\sin \varphi$ in a series with respect to the increment $\tilde{\varphi}$ yields:

$$\sin(\varphi + \tilde{\varphi}) = \sin\varphi + \cos\varphi \times \tilde{\varphi} + \ldots \qquad (2.142)$$

Taking into consideration only the linear dependence on $\tilde{\varphi}$ in (2.142), the corresponding increase in the supercurrent is obtained:

$$\tilde{J}_S = \frac{1}{L}\int \tilde{V}dt, \qquad (2.143)$$

where

$$\frac{1}{L} = \frac{1}{L_0}\cos\varphi; \quad L_0 = \frac{\hbar}{2qJ_0}. \qquad (2.144)$$

The meaning of (2.143) is that the small supercurrent leads to the appearance of an inductive type of reactance. According to (2.144) this inductance has two basic properties:

(a) In the Josephson generation mode the inductance is modulated by a frequency ω_V. From the standpoint of parametric systems this is equivalent to self-excitation of oscillations in such an element when $V \neq 0$.

(b) The inductance L may achieve negative values as well.

The practical applications of the Josephson effect in superconductors are based on the possibility of achieving an unsurpassed sensitivity (several fT) in the measurement of a magnetic flux or of electric quantities that can be directly converted into a magnetic field. The magnetosensitive SQUID is, generally, a superconducting ring closed by one or two Josephson tunnel junctions [38,40,86–89]. Most often the tunnel junction is a cross-like structure of two superconducting films placed crosswise and separated by a thin insulating layer whose thickness is in the 1- to 10-nm range. When a current I whose magnitude does not exceed a certain critical value I_0 is forced across such a device, Cooper pairs are not scattered, i.e. the system superconductor–insulator–superconductor (S–I–S) manifests superconductivity. When $I > I_0$, the $I(V)$ curve of the tunnel junction begins with a transitional resistive region where $R_t < R_n$ (R_n is the normal resistance of the junction at a given temperature). With increasing current I, the curve enters a region where the resistance is entirely determined by the normal component $R_t = R_N$, Fig. 2.31.

SQUID magnetometry employs the magnetic field dependence of the critical current I_0 (or the voltage V) across the junction in the resistive region of the $I(V)$ curve. The relationship $I_0(B)$ is presented in Fig. 2.32 and is described by the expression:

$$I_0 = I_{0\max}\frac{|\sin(\pi\Phi/\Phi_0)|}{\pi\Phi/\Phi_0}. \qquad (2.145)$$

Formula (2.145) is often called a diffraction equation by analogy with the expression for the intensity of light diffracted by a narrow slit. The period of oscillation of the relationship $I_0(B)$ is:

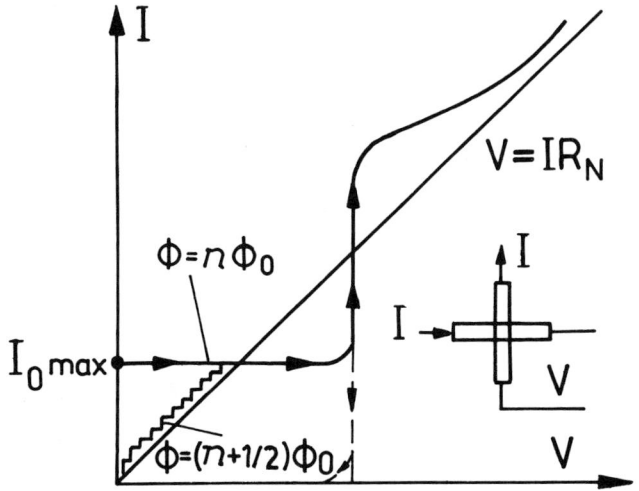

Fig. 2.31. Current–voltage curves of a Josephson tunnel junction.

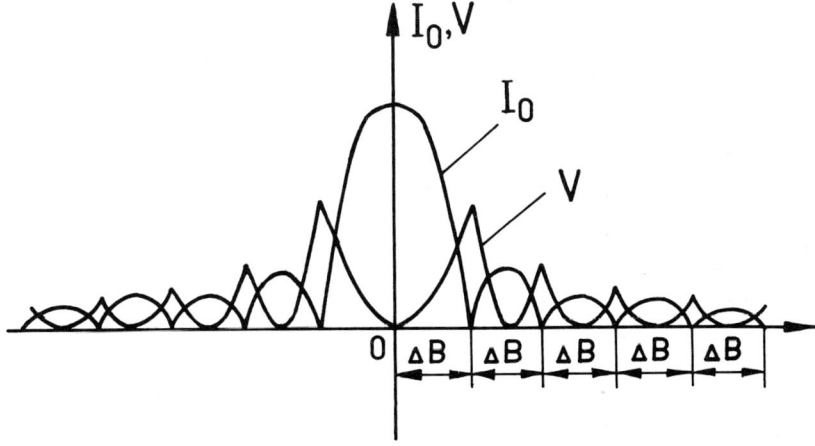

Fig. 2.32. Dependence of the critical current and the voltage across a Josephson tunnel junction on the magnetic field for $I > I_0$. The junction is shunted by a normal resistor in order to eleminate the hysteresis of the current–voltage characteristic.

$$\Delta B \sim \Phi_0. \tag{2.146}$$

The amplitude of oscillations is reduced with an increase in the field B and becomes negligible at a magnitude of $B \sim 1.3 \times 10^{-4}$ T. The normal resistance itself depends exponentially on the thickness t of the insulating layer and is inversely proportional to the area of the junction cross-section [89].

2.7.3. Resistively shunted junction model

In addition to the Cooper-pairs current $J_S = J_0 \sin \varphi$, a normal leakage current J_N is observed in real junctions due to the tunnelling of single electrons and a displacement current due to the capacitance of the tunnel junction. This presentation of the total current across the junction is called the Resistively Shunted Model (RSJ) [90,91]. This approach is illustrated in graphical form by Fig. 2.33, the RSJ model being the optimum and an adequate description of the real junction. An ideal Josephson element JJ shunted by a resistor R and a capacitor C is assumed. According to Fig. 2.33 the bias current in the device can be written as:

$$J = \frac{V}{R} + I_0 \sin \varphi + C \frac{dV}{dt}. \tag{2.147}$$

The capacitance is often neglected in the analysis of the behavior of the Josephson junction, i.e. $C = 0$. In practical cases SQUID junctions are shunted by resistors in order to reduce the adverse effect of the capacitance. In this case $V/R \gg C dV/dt$. By taking into consideration (2.134) the following relationship can be deduced:

$$\frac{I}{I_0} = \frac{\hbar}{2qRI_0} \frac{d\varphi}{dt} + \sin \varphi. \tag{2.148}$$

With an average value of the voltage $V = \langle V \rangle = \hbar/2q \langle d\varphi/dt \rangle$, the integration of (2.148) yields:

$$\left. \begin{array}{l} V = 0, \quad \text{when } I < I_0 \\ V = RI_0 \left[\left(\frac{I}{I_0}\right)^2 - 1 \right]^{1/2}, \text{ when } I > I_0. \end{array} \right\} \tag{2.149}$$

The expression (2.149) describes the current–voltage characteristic of the Josephson junction at $C = 0$. The curve is illustrated in Fig. 2.34 for different values of the current. Figure 2.34 shows that if the magnitude of J is less than the critical value J_0 there is no voltage across the junction, since the entire current $|J| < J_0$ is due to Cooper pairs. This is why this region of the curve is called superconduct-

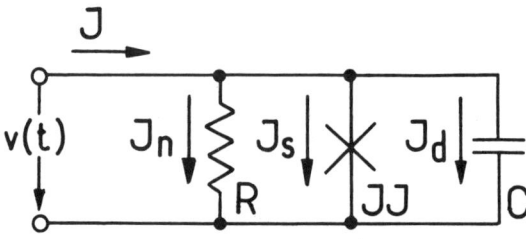

Fig. 2.33. RSJ (resistively shunted junction) model.

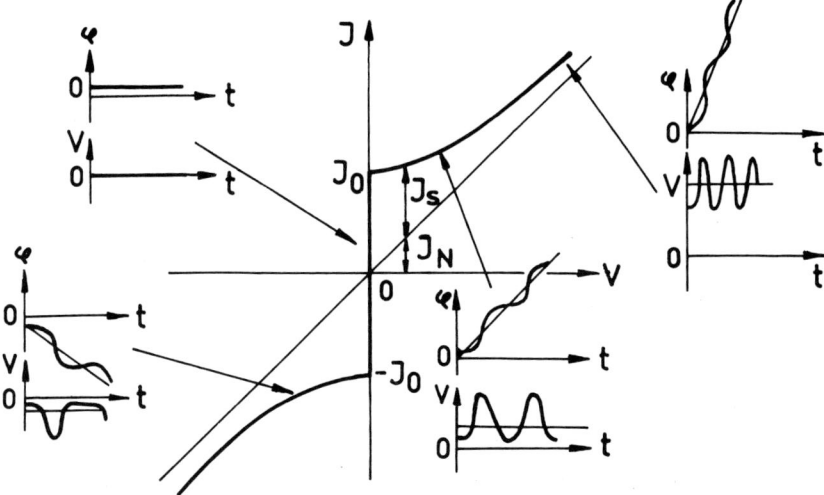

Fig. 2.34. Current–voltage characteristic of a Josephson junction after the RSJ model and time dependence of the phase φ at different magnitudes of the current across the junction.

ing. If the current magnitude is $|J| > J_0$, single (normal) electrons contribute to the transport process, thus generating a voltage across the junction and changing $d\varphi/dt$. The contribution of the superconducting condensate decreases as the total current is increased. Therefore the current–voltage curve of the Josephson device tends asymptotically to the straight line $V = JR_N$ of a conventional resistor. This is why when $|J| > J_0$, the $I(V)$ curve is usually termed resistive.

The $I(V)$ characteristics observed in real Josephson devices differ from the curve in Fig. 2.34. Firstly, the fluctuations and the noise in real junctions smooth the abrupt region of the curve near the critical value J_0. Secondly, the deviation in the behavior of the Josephson junction from ohmic conductivity is the reason for using the straight line instead of the parallel line determined by the normal resistance R_N for the approximation of the $I(V)$ curve:

$$J = \frac{V}{R_N} + J_e, \qquad (2.150)$$

where the "displacing" current J_e is always smaller than J_0.

The fact that, at a fixed value of the total current $J = $ const and $J \geq J_0$, the variation of J_0 leads to a very large variation of the voltage across the junction is rather important for the operation of SQUIDs.

2.7.4. Principle of operation of SQUIDs

The interference of light waves is known from optics and is the basis of the operation of optical interferometers. The interference of electron-wave processes can be considered by analogy with optical interference. The main parameter in optical interference is the wavelength of light, whereas for electrons this is the length of the de Broglie waves. It is known from quantum mechanics that a wave of characteristic length λ^* corresponds to each particle. In low-temperature superconductors, the charge carriers are Cooper pairs and in high high-T_c materials they are hole pairs. If the waves of such pairs interact in a given spatial region, interference should be expected. The observation of the interference picture is possible in superconductors since all charge carriers move coherently (cf. §2.7.1) and possess the same wave function energy and phase; hence the same de Broglie wavelength λ^* corresponds to all of them. SQUIDs are devices that can be used for the observation of the macroscopic quantum interference of Cooper-pair waves. These devices functionally unite two physical phenomena: the effect of magnetic-flux quantization and Josephson tunnelling. The output voltage oscillates periodically depending on the magnetic flux applied, the period being equal to the quantum Φ_0.

Any SQUID includes a superconducting ring of arbitrary shape which is interrupted by one or two Josephson junctions. The SQUIDs with one junction are known as RF SQUIDs and the double-junction devices are called DC SQUIDs. This classification is associated with the type of the read-out electronics used for the RF and DC devices.

2.7.4.1. RF SQUIDs

The RF SQUID is a magnetic-field transducer which consists of a superconducting ring and one Josephson junction *JJ*. The junction interrupts the flow of the current I_S which maintains a constant phase difference φ (cf. §2.7.2). The activation of the sensor is accomplished by means of an additional RF magnetic flux, hence the name of the device.

If an external magnetic field is applied to a superconducting ring which includes a weak link (Fig. 2.35), according to §2.7.2 there will be a quantization of the magnetic flux:

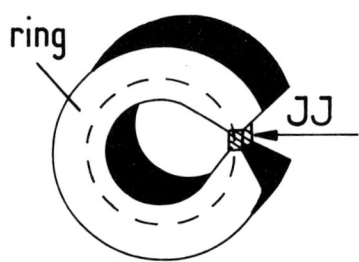

Fig. 2.35. Superconducting ring containing one weak link.

$$2\pi n = 2\pi \frac{\Phi}{\Phi_0} - \delta; \quad \delta = \varphi_2 - \varphi_1; \tag{2.151}$$

where the phase difference δ is associated with the magnetic flux inside the ring.

The relationship between the magnetic flux inside the ring and the externally applied flux Φ_{ext} is the following:

$$\Phi = \Phi_{\text{ext}} + L_0 J_S, \tag{2.152}$$

where L_0 is the self-inductance of the ring and J_S is determined from (2.138). The result is:

$$\Phi = \Phi_{\text{ext}} + L_0 J_0 \sin\left(2\pi \frac{\Phi}{\Phi_0}\right). \tag{2.153}$$

The characteristic corresponding to (2.153) is presented in Fig. 2.36 for three different values of the parameter $\pi\alpha$, where $\alpha = 2L_0 I_0/\Phi_0$. In the general case, the relationship $\Phi(\Phi_{\text{ext}})$ (2.153), is nonlinear and has been chosen in such a way as to describe the different modes of (2.153). The value $\pi\alpha = 1$ is a boundary between the nonhysteresis ($\pi\alpha = 0.5$) and the hysteresis ($\pi\alpha = 5$) modes. The hysteresis mode is characterized by the presence of a negative slope. If the externally applied magnetic flux Φ_{ext} is increasing, the internal flux Φ follows the curve only to the point where the slope becomes infinite. After that point the slope is energy-unfavorable and a flux jump is observed (Fig. 2.36). A similar situation occurs when the external flux Φ_{ext} is decreasing. The flux jumps associated with the decreasing external flux are observed at another value Φ'_{ext} which differs from that of the increasing Φ_{ext}. As a result the SQUID configuration in Fig. 2.35 manifests a

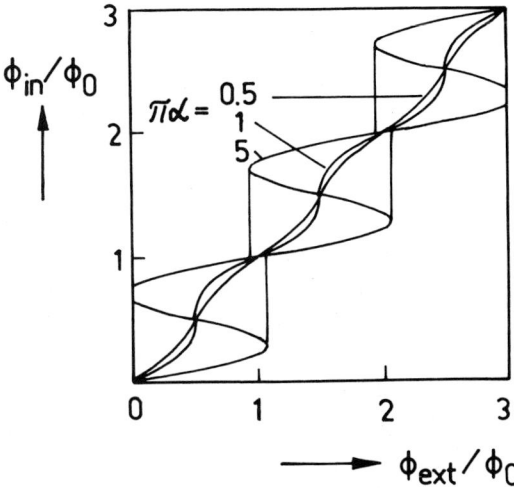

Fig. 2.36. Relation between the magnetic flux inside a superconducting ring from Fig. 2.35 and an externally applied flux Φ_{ext}.

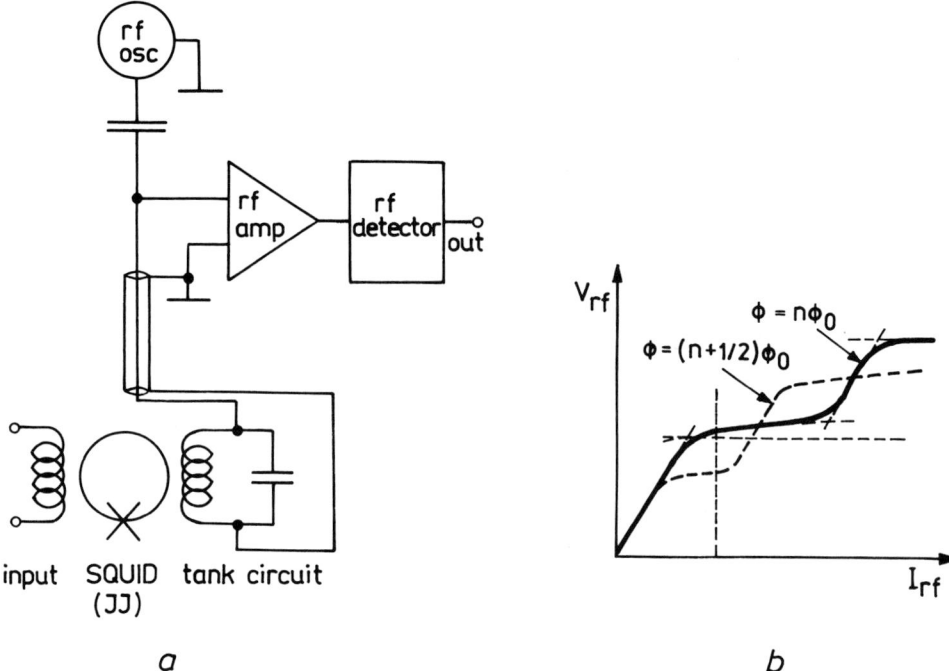

Fig. 2.37. (a) Electronic circuits for obtaining the RF SQUID characteristics. (b) Demodulated voltage V_{rf} as a function of the demodulated current I_{rf} at $\Phi = n\Phi_0$ and $\Phi = (n+1/2)\Phi_0$.

hysteresis behavior. The case $\alpha = 0$ cannot be observed when the conducting ring is normal. The relationships between the input and output signals of a RF SQUID can be obtained by means of the electronic circuits shown schematically in Fig. 2.37a [40].

The SQUID ring together with the Josephson junction is inductively coupled to a LC tank circuit that is tuned to the frequency of an RF signal provided by an RF oscillator. The peak value of the voltage drop V_{rf} across the tank circuit is measured by means of a RF amplifier and a detector. V_{rf} serves as an output signal. The value of V_{rf} depends on the low-frequency input signal, i.e. on the flux Φ_{ext} applied perpendicularly to the SQUID ring by means of an inductively coupled input coil. The general principle of operation of the circuitry in Fig. 2.37a is the following: according to (2.136) the variation of the magnetic flux Φ_{ext} leads to a variation of the phase φ of the current across the Josephson junction. Since this current is a strongly nonlinear function of Φ, (2.138), the variation of the phase φ changes impedance of the Josephson junction and therefore the state of the RF interferometer as well. This is exactly the variation registered by the output elements of the circuit. The entire circuit in Fig. 2.37a is analogous to a parametric amplifier. A very small low-frequency input signal is mixed with the relatively strong pump signal across the nonlinear Josephson junction. As a result of the parametric up-conversion the detected signal appears at the output of the device as an amplified side-band signal adjacent to the RF bias signal. The operation of RF SQUIDs can be understood by tracing the amplitude of I_{rf} and V_{rf} (Fig. 2.37b) obtained as a result of RF signal demodulation. When $\Phi = n\Phi_0$, a sequence of steps (the uninterrupted line) is observed. When the magnetic flux Φ_{ext} is increased or decreased, each step is split into two and the maximum voltage difference between them is achieved at $(n + 1/2)\Phi_0$. Further increasing or decreasing the flux Φ_{ext} to $(n + 1)\Phi_0$ or $(n - 1)\Phi_0$ leads again to an uninterrupted current–voltage curve $V_{rf}(I_{rf})$ (Fig. 2.37b). This is why at a given value of the current I_{rf} (the vertical line in Fig. 2.37b) the demodulated voltage is a periodic function. The choice of the working point which corresponds to $\Phi_{ext} = n\Phi_0$ is accomplished by means of an orthogonal DC magnetic field via the input coil [40]. It has been found experimentally that in RF SQUIDs the transition from a given quantum level to the neighboring level can take place only if the condition $I_0 L_0 \leq \Phi_0$ is fulfilled. When $I_0 L_0 > 10\Phi_0$, quantum transitions of the flux can take place via several states to a random final state with many quanta of the flux Φ inside the superconducting ring. This disorder limits to a certain extent the parameters of SQUIDs as magnetic sensors. Owing to this fact and because of the higher sensitivity of DC SQUIDs, only modest progress has been made in the field of RF SQUIDS in recent years.

2.7.4.2. DC SQUIDs

A DC SQUID consists of two Josephson junctions JJ_1 and JJ_2 connected in series to a closed superconducting loop which operates at a constant dc supply current I^*; hence the name of this type of sensors. A DC SQUID is presented schematically in Fig. 2.38. The superconducting current of electron pairs in the two symmetric branches of the device can be presented as a propagation of de Broglie waves of length λ^*. The increase in phase of the de Broglie waves for one complete traverse of the left-hand branch is $\delta\varphi_1 = 2\pi \int_{(1)} dl/\lambda^*$, and for the right-hand branch the corresponding phase increase is $\delta\varphi_2 = 2\pi \int_{(2)} dl/\lambda^*$, where dl is an infinitely small linear element from the SQUID loop. The SQUID area of quantization is A_0. To a first approximation A_0 is determined by the effective area of the opening in the superconducting ring. The de Broglie relationship $1/\lambda^* = p/h$ is well known from quantum mechanics. p is the total quantum mechanical momentum of the Cooper pair; $h \simeq 6.63 \times 10^{-34}$ J s is Planck's constant; $\boldsymbol{p} = m_e \boldsymbol{v}_s + (q/c)\boldsymbol{A}$; c is the velocity of light in vacuo; $\boldsymbol{A}$ is the vector potential; and m_e is the mass of a free electron.

The density of the current of superconducting electrons can be written in the form $\boldsymbol{J}_s = n_s q \boldsymbol{v}_s$, where n_s is the electron-pair concentration. The above relationships can be used in order to write $\delta\varphi_1$ and $\delta\varphi_2$ in the form:

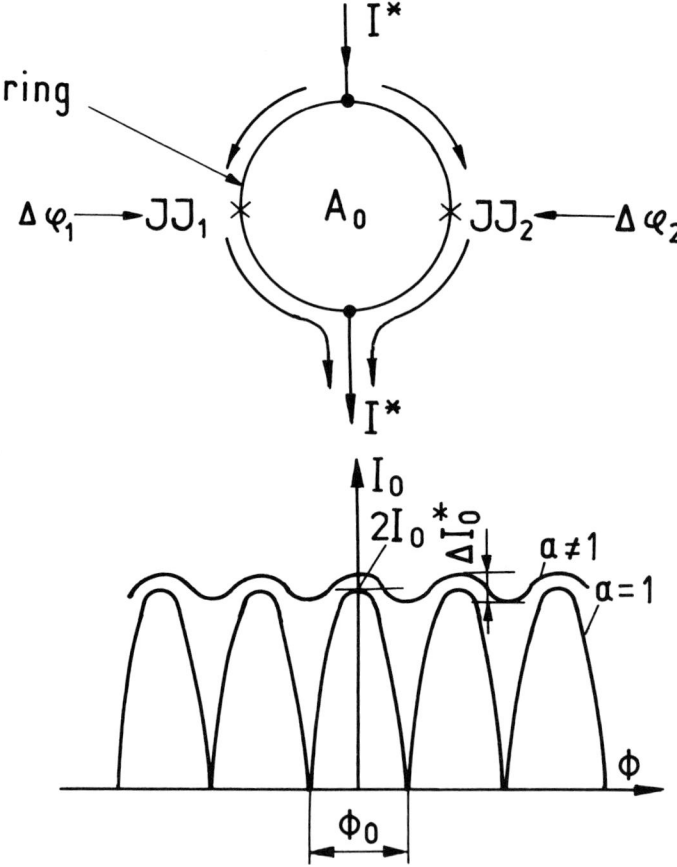

Fig. 2.38. A DC SQUID and the dependence of the critical current on the magnetic flux Φ at $I_0^* L_0 < \Phi_0$.

$$\delta\varphi_1 = (4\pi m_e/n_s hq) \int_{(1)} J_s dl + \left(\frac{4\pi q}{h}\right) \int_{(1)} A dl;$$

$$\delta\varphi_2 = (4\pi m_e/n_s hq) \int_{(2)} J_s dl + \left(\frac{4\pi q}{h}\right) \int_{(2)} A dl.$$

The contour of integration is chosen in such a way as to be located in the metal ring of the SQUID at a depth that exceeds the depth of penetration, λ, of the magnetic field into the superconductor. The two junctions abruptly change the phases by $\Delta\varphi_1$ and $\Delta\varphi_2$ respectively. Under such conditions, the integration across the enclosed loop of the ring yields a phase difference which is a multiple of 2π, i.e. $\Delta\varphi_1 + \Delta\varphi_2 + (4\pi q/h) \oint A dl = 2\pi n$. In the integration, the fact that the integrals $\int J_s dl$ along the contour equal zero everywhere except for the junctions JJ_1 and JJ_2 has been taken into account. The relationship between the vectors $\boldsymbol{B}$ and $\boldsymbol{A}$ and

Stokes' theorem are used to obtain:

$$\Delta\varphi_1 + \Delta\varphi_2 + \left(\frac{2\pi}{\Phi_0}\right)\int_{A_0} BdA = 2\pi n.$$

Since $\int_{A_0} BdA = \Phi$ is the flux quantization of the magnetic field B that pierces the loop of the SQUID and $\Phi_0 = h/2q$, the following formula is valid: $\Delta\varphi_1 + \Delta\varphi_2 = 2\pi(n - \Phi/\Phi_0)$. When $n = 0$ this reduces to $\Delta\varphi_1 + \Delta\varphi_2 = -2\pi\Phi/\Phi_0$. In the case when the current I^* is zero, the jump of the two phases in junctions JJ_1 and JJ_2 is the same, i.e. $\Delta\varphi_1 = \Delta\varphi_2 = \pi(\Phi/\Phi_0)$. When the bias current of the sensor is $I^* \neq 0$, there is an additional phase increase δ across the Josephson junctions. Owing to the structural symmetry, the current is expected to be equally divided between the two branches ($0.5I^*$ in each of them). The additional jump δ of the phase increase across JJ_1 and JJ_2 leads to $\delta\varphi_1 = -(\pi\Phi/\Phi_0) + \delta$; $\Delta\varphi_2 = (\pi\Phi/\Phi_0) + \delta$. According to (2.138), the following expression is valid for the left-hand branch:

$$0.5I^* = -I_0 \sin\left[\left(\frac{\pi\Phi}{\Phi_0}\right) - \delta\right].$$

The expression for the right-hand branch is:

$$0.5I^* = I_0 \sin\left[\left(\frac{\pi\Phi}{\Phi_0}\right) + \delta\right].$$

The summation of these two expressions yields the current I^*:

$$I^* = I_0 \left\{\sin\left[\left(\frac{\pi\Phi}{\Phi_0}\right) + \delta\right] - \sin\left[\left(\frac{\pi\Phi}{\Phi_0}\right) - \delta\right]\right\} =$$

$$= 2I_0 \cos\left(\frac{\pi\Phi}{\Phi_0}\right) \sin\delta.$$

As in (2.138) in this case the maximum possible value of the critical Josephson current is again I_0.

Maximization of I^* with respect to δ in the above equation yields:

$$I_0^* = 2I_0 \left|\cos\left(\frac{\pi\Phi}{\Phi_0}\right)\right|. \tag{2.154}$$

If $I_0 L_0 < \Phi_0$, where L_0 is the self-inductance of the SQUID loop, and if $I_{01} = I_{02}$ (a symmetric SQUID), the expression (2.154) describes the variation of the sensor critical current I_0 due to the magnetic flux variation (Fig. 2.38). When the device is asymmetric ($I_{01} \neq I_{02}$), the relationship $I_0^*(\Phi)$ has a more complex form [89]. The ratio $I_{01}/I_{02} = a$ is termed an asymmetry factor. When $a = 1$ the modulation depth of the SQUID critical current is $\Delta I^* = 2I_0$.

If $I_0 L_0 > \Phi_0$ (for instance, in a symmetric device $a = 1$), the modulation depth Δ_0^* is less than $2I_0$. The modulation of the current I_0^* by the magnetic field leads to the modulation of the entire resistive region of the sensor $I(V)$ curve. As a result, at a fixed value of the current $I^* > I_0^*$, a periodic dependence of the voltage V on

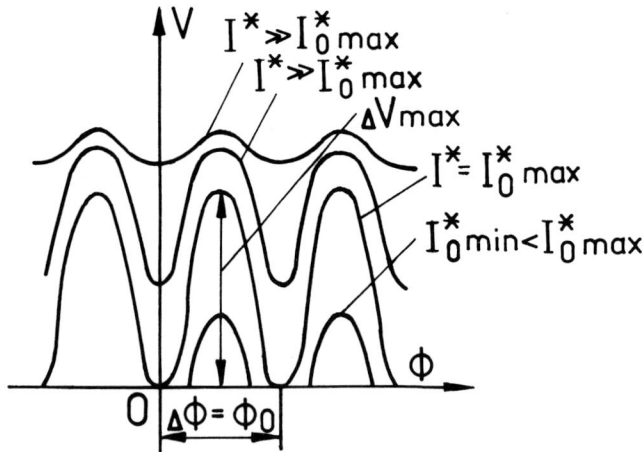

Fig. 2.39. Dependence of the voltage V on the magnetic flux in a DC SQUID.

the magnetic flux is observed. When $I_0 L_0 < \Phi_0$ and $a = 1$, the voltage dependence on the magnetic flux is given by:

$$V(\Phi) = R_N I_0 \left|\sin\left(\frac{\pi \Phi}{\Phi_0}\right)\right|. \tag{2.155}$$

The curve $V(\Phi)$ is shown in Fig. 2.39. The oscillation amplitude of $V(\Phi)$ is $\Delta V_{\max} = R_N I_0$, and for $a \neq 1$ it decreases proportional with $1/a$. If $I_0 L_0 > \Phi_0$ and $a = 1$, the amplitude is $\Delta V_{\max} \sim 2 R_N (I_0 \Phi_0 / L_0)^{1/2}$ at $I^* = I^*_{\max}$.

As a rule the working point of a DC SQUID is chosen in the steepest region of the current, where a maximum value of the transducing coefficient $V_\Phi = (\partial V / \partial \Phi)|_I$ is achieved. In such a way a small variation $\delta \Phi_{\text{ext}} \ll \Phi_0$ of the magnetic flux applied perpendicularly to the ring leads to a voltage V at the output. In fact the DC SQUID converts the magnetic flux into a voltage. Usually a constant current I^* which is more than twice the current I_0 of $JJ_{1,2}$ is forced through the device. Then a voltage $V > 0$ appears between the terminals of $JJ_{1,2}$ whose inductance is L_0. If an external magnetic flux $\Phi_{\text{ext}} < \Phi_0/2$ is applied, a screening current $I_s = \Phi_{\text{ext}}/L_0$ arises in order to maintain a zero magnetic flux inside the material of the ring (Meissner effect).

The screening current increases to reach the value $I_s = \Phi_0/2L_0$ as the flux Φ_{ext} is increased to $\Phi_{\text{ext}} = \Phi_0/2$. With a further increase in Φ_{ext} such a transition of the magnetic flux between its quantum states becomes favorable, concerning energy, that the direction of the screening current is reversed. If $\Phi_{\text{ext}} = +(\Phi_0/2)$, the screening current is $I_s = -\Phi_0/2L_0$, and within the interval $\Phi_0/2 < \Phi_{\text{ext}} \leq 3\Phi_0/2$ the magnitude of I_s is increased until the direction of I_s is reversed again and the whole process is repeated from the beginning. The functional dependence of the screening current on the external magnetic flux has a triangular waveform (Fig.

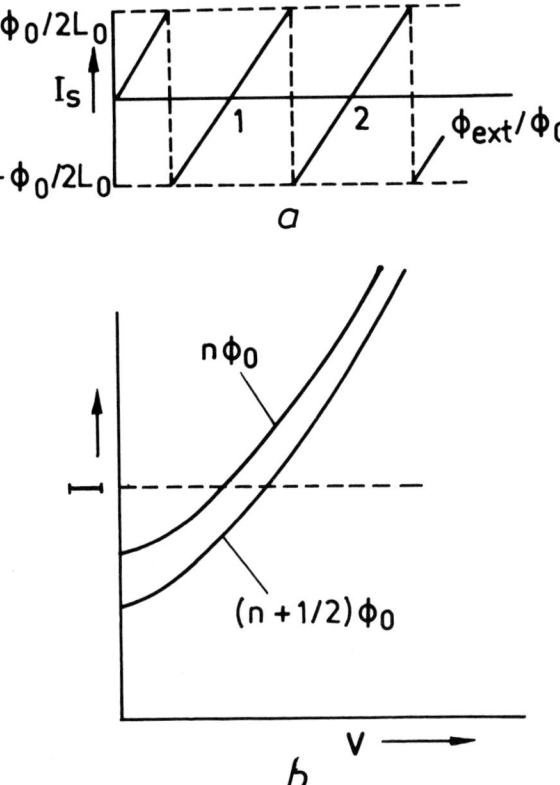

Fig. 2.40. (a) Self-induced current for increasing flux Φ_{ext}. (b) A current–voltage characteristic of a DC SQUID.

2.40a). The jump occurs when $\Phi_{ext} = (n+1/2)\Phi_0$, $n = 0, 1, 2, 3, \ldots$. If the current I^* exceeds $2I_0^*$, the interferometer voltage is a periodic function of the increasing magnetic flux Φ_{ext}. The waveform of the signal is shown in Fig. 2.40a. If a screening current I_s is circulating inside the ring, the resultant currents in the two junctions JJ_1 and JJ_2 differ by $2|I_s|$. This is why if the current across one of the junctions is $(I_0^* - 2I_s)$, the current across the second Josephson contact is I_0^* and is exactly the critical current of this junction. Therefore the critical current I_0^* in a DC SQUID is lower than $2I_0$ by $2|I_s|$. This means that the current I_0^* is an oscillating function of the variation of the applied external magnetic flux. The largest decrease in the critical current magnitude by Φ_0/L_0 occurs at $\Phi_{ext} = (n+1/2)\Phi_0$; the smallest decrease is zero and occurs at $\Phi_{ext} = n\Phi_0$. The current voltage characteristics of the sensor at these extremum points are shown in Fig. 2.40b. Since the critical current of the junction is increased with a decrease in the temperature T, the amplitude of the $V(\Phi)$ relationship is inversely proportional to T. The operation of DC SQUID magnetometers and gradiometers is based on the $V(\Phi)$ dependence.

The current–voltage curve is usually smoothed by normal low-resistance shunts R_1 and R_2 connected in parallel to JJ_1 and JJ_2 (§2.7.3). Two conditions must be fulfilled to achieve maximum sensitivity:

(a) maximum value of $\Delta V_{max} = I_0 R_N$, where at $T \to 0$, $\Delta_{max} \approx \Delta(0)$, and in most superconductors this value is several millivolts;

(b) minimum value of the period $\Delta B = \Phi_0/A_0$ of the $V(B)$ oscillations.

This is why the loop area A_0 must be increased. At a fixed temperature $T_0 < T_c$ and in most practical cases the loop inductance does not exceed 10^{-7} H. The corresponding ring diameter is about 10^{-2} m, the maximum possible value of A_0 being in the 10^{-4} m^2 range. Thus the maximum period of oscillation is about 10^{-11} T [89]. The read-out circuit of a DC SQUID is described in §7.1.3. A comprehensive survey of the operation of SQUIDs is presented in [39,40].

References

[1] W. Shockley, Electrons and Holes in Semiconductors, Van Nostrand Co., Princeton, 1950.
[2] S.M. Sze, Physics of Semiconductor Devices, John Wiley and Sons, New York, 1981.
[3] K. Seeger, Semiconductor Physics, Springer-Verlag, Wien, 1973.
[4] A. Nissbaum, The Theory of Semiconductor Junctions, in: Semiconductors and Semimetals, vol. 15, R.K. Wilardson and A.C. Beer (eds.), Academic Press, New York, 1981.
[5] W. Van Roosbroeck, Bell. Syst. Tech. J., 29 (1950) 560–607.
[6] W. Van Roosbroeck, Phys. Rev., 91-2 (1953) 282–289.
[7] M. Kurata, Numerical Analysis of Semiconductor Devices, Lexington-Books, Lexington, 1982.
[8] S. Selberherr, Analysis and Simulation of Semiconductor Devices, Springer-Verlag, New York, 1984.
[9] A.C. Beer, Galvanomagnetic Effects in Semiconductors, Solid-state Phys., Suppl. 4, Academic Press, New York, 1963.
[10] E.H. Putley, The Hall Effect and Related Phenomena, Butterworths, London, 1960.
[11] P.S. Kireev, Semiconductor Physics, English Translation by MIR Publ., Moscow, 1978.
[12] E.V. Kuchis, Galvanomagnitnie effecti i metodi ih issledovania, Radio i Sviaz, Moskwa, 1990 (in Russian).
[13] C.L. Chien and C.R. Westgate (eds), The Hall Effect and its Applications, Plenum Press, New York, 1980.
[14] H. Weiss, Structure and Applications of Galvanomagnetic Devices, Pergamon Press, Oxford, 1969.
[15] A. Kobus and J. Tuszynski, Hallotrony i Gaussotrony, Wydawnictwa naukowo-techniczne, Warszawa, 1966 (in Polish).
[16] Yu.V. Vorobiov, V.N. Dobrovolski and V.I. Striha, Metody issledovaniya poluprovodnikov, Vishia Skola, Kiev, 1988 (in Russian).
[17] R. Landauer and J. Swanson, Phys. Rev., 91-3 (1953) 555–560.
[18] G.E. Pikus, Sov. Phys. Tech. Phys. 1 (1956) 17–46.
[19] W. Van Roosbroeck, Phys. Rev., 101-6 (1956) 1713–1725.
[20] S.M. Ryvkin, A.A. Grinberg, Y.L. Ivanov, S.R. Novikov and N.D. Potechnica, Sov. Phys. Solid State, 2-4 (1960) 541–553.
[21] A. Chovet, Phys. Status Solidi, A28 (1975) 633–645.
[22] S. Gristoloveanu and J.H. Lee, J. Phys. C, 13 (1980) 5983–5997.
[23] O. Madelung, Z. Naturforsch., 14a (1959) 951–958.
[24] H. Pfleiderer, Solid-state Electron., 15 (1972) 335–353.
[25] S. Gristoloveanu, Phys. Status Solidi, A64 (1981) 683–695.
[26] S. Gristoloveanu, Phys. Status Solidi, A65 (1981) 281–292.

[27] R.S. Popović, Hall Effect Devices, Adam Hilger, Bristol, 1991.
[28] D. Shoenberg, Superconductivity, Cambridge, 1952.
[29] J. Bardeen, Theory of Superconductivity, Handbuch der Physik, Bd. XV, Berlin, 1956.
[30] J. Bardeen and J. Schrieffer, Recent Developments in Superconductivity, in: Progress in Low Temperature Physics, vol.III, C.J. Gorter (ed.), New York, 1961.
[31] A.C. Rose-Innes and E.H. Rhoderick, Introduction to Superconductivity, Pergamon Press, Oxford, 1969.
[32] E.A. Lynton, Superconductivity, Methuen and Co. Ltd., London, 1969.
[33] W. Buckel, Supraleitung, Phys. Verlag, Weinheim, 1972.
[34] B. Josephson, Phys. Lett., 1 (1962) 251–253; Adv. Phys., 14 (1965) 419–451.
[35] R.C. Jaklevic, J. Lambe, A.H. Silver and J.E. Mercereau, Phys. Rev. Lett. 12 (1964) 159–160.
[36] J.E. Zimmerman, P. Thiene and J.T. Harding, J. Appl. Phys. 41 (1970) 1572–1580.
[37] J.E.C. Williams, Superconductivity and its Applications, Pion Limited, London, 1970.
[38] V.N. Alfeev, Poluprovodniky, sverhprovodniky i paraelektriky v krioelektronike, Sovetskoe Radio, Moskwa, 1979 (in Russian).
[39] J. Clarke, Proc. IEEE, 77 (1989) 1208–1223.
[40] H. Koch, in: Sensors, vol. 5, R. Bolk and K.J. Overshott (eds.), VCH, Weinheim, 1989, 381–445.
[41] J.G. Bednorz and K.A. Muller, Z. Phys., B64 (1986) 189–193.
[42] M.K. Wu, J.R. Ashburn, C.J. Torng et al., Phys. Rev. Lett., 58 (9) (1987) 908–910.
[43] A.A. Abrikosov, Osnovi teorii metalov, Nauka, Moskwa, 1987 (in Russian).
[44] A.C. Beer, Hall Effect and the Beauty and Challenges of Science, in: The Hall Effect and its Applications, C.L. Chien and C.R. Westgate (eds.), Plenum Press, New York (1980) 229–338.
[45] E.H. Hall, Am. J. Math., 2 (1879) 287–292.
[46] E.H. Hall, Am. J. Sci., series 3, 20 (1880) 161–186.
[47] J.F. Lin, S.S. Li, L.C. Linares and K.W. Teng, Solid-state Electronics, 24(9) (1981) 827–833.
[48] P. Norton, T. Braggins and Levinstein, Phys. Rev. B, 8(1973) 5632–5653.
[49] D.L. Rode, Phys. Status Solidi, B55 (1973) 687–696.
[50] V.N. Dobrovolskii and V.G. Litovchenko, Perenos elektronov u dirok u poverhnosty poluprovodnikov, Naukova Dumka, Kiev, 1985 (in Russian).
[51] E.V. Kuchis, Metodi Issledovaniya Effekta Holla, Sovetskoe Radio, Moskwa, 1974 (in Russian).
[52] L. Berger and G. Bergman, in: The Hall Effect and its Applications, C.L. Chien and C.R. Westgate (eds.), Plenum Press, New York, 1980, 55–76.
[53] A.C. Beer, Galvanomagnetic Effects in Semiconductors, Academic Press, New York, 1963.
[54] V.N. Dobrovolskii and A.N. Krolevets, Sov. Phys. Semicond., 17(1) (1983) 1–7.
[55] H.H. Jenson and H. Smith, J. Phys. C, 5 (1972) 2867–2880.
[56] A.G. Andreou and C.R. Westgate, Proc. IEEE, 73 (1985) 489–490.
[57] S. Middelhoek and S.A. Audet, Silicon Sensors, Academic Press, London, 1989.
[58] S. Kordić and P.C.M. van der Jagt, Sensors and Actuators, 8 (1985) 197–217.
[59] S. Kordić, Offset reduction and three-dimensional field sensing with magnetotransistors, Ph. D. dissertation, Delft Univ. Techn., The Netherlands, 1987.
[60] B.R. Nag, Electron Transport in Compound Semiconductors, Springer-Verlag, Berlin, 1980.
[61] H.J. Lippman and F. Kuhrt, Z. Naturforsch., 13a (1958) 462–474.
[62] H. Weiss, Physik und Anwendung galvanomagnetischer Bauelemente, Vieweg, Braunschweig, 1969.
[63] J. Haeusler, Solid-state Electronics, 9 (1966) 417–441.
[64] W. Versnel, Solid-state Electronics, 24 (1981) 63–68.
[65] L. Andor, H. Baltes, A. Nathan and H.G. Schmidt-Weinmar, IEEE Trans. Electron Devices, ED-32 (1985) 1224–1230.
[66] G. De Mey, in: Advances in Electronics and Electron Physics, vol.61, P.W. Hawkes, ed., Academic Press, New York, 1983.
[67] J.P. Newsome, Proc. IEE, 110 (1983) 653–659.
[68] T. Mimizuka, Solid-state Electronics, 21 (1978) 1195–1197.
[69] A. Nathan, W. Alegreto, H. Baltes and Y. Sugiyama, IEEE Trans. Electron Devices, ED-34 (1987) 2077–2085.
[70] J. Volger, Phys. Rev., 79 (1950) 1023–1024.

[71] G. De Mey, Appl. Phys. 6 (1975) 189–197.
[72] R.T. Bate and A.C. Beer, J. Appl. Phys., 32-5 (1961) 800–805.
[73] R.T. Bate, J.C. Bell and A.C. Beer, J. Appl. Phys. 32-5 (1961) 806–814.
[74] R.A. Smith, Semiconductors, Cambridge Univ. Press, Cambridge, 1987.
[75] H. Suhl and W. Shockley, Phys. Rev., 75 (1949) 1617–1618.
[76] A. Chovet, Transport, recombinaison et bruit dans les semiconducteurs ambipolaires en presence de champ magnetique, Ph.D. dissertation, I.N.P., Grenoble, France, 1978.
[77] E.I. Karakushan and V.I. Stafeev, Sov. Phys.- Solid State, 3 (1961) 493–498.
[78] V.I. Stafeev and E.I. Karakushan, Magnitodiody, Nauka, Moskwa, 1975 (in Russian).
[79] J. Chretien, Les dispositifs á injection de porteurs en tant que capteurs magnetosensibles, Ph.D. dissertation, I.N.P., Grenoble, France, 1977.
[80] S. Gristoloveanu, l'Onde Electrique, 59 (1979) 68–74.
[81] A. Chovet and S. Gristoloveanu, Rev. Phys. Appl., 19 (1984) 69–76.
[82] S. Gristoloveanu, A. Chovet and G. Kamarinos, Solid-state Electronics, 21 (1978) 1563–1569.
[83] P. Lilienkamp and H. Pfleiderer, Phys. Status Solidi, A43 (1977) 479–486.
[84] Ch.S. Roumenin, C.R. Acad. Bulg. Sci., 38 (1985) 1501–1504.
[85] Ch.S. Roumenin, Sensors and Actuators, A24 (1990) 83–105.
[86] S. Shapiro, Phys. Rev. Lett., 11 (1963) 80.
[87] D.N. Langerberg, D.I. Scalapio and B.N. Taylor, Phys. Rev. Lett. 15 (1965) 294.
[88] I. Giaever, Phys. Rev. Lett., 14 (1965) 904.
[89] S.I. Bondarenko and V.I. Sheremet, Primeneniye sverhprovodimosty v magnitnich izmereniya, Energoatomizdat, St. Petersburg, 1982 (in Russian).
[90] W.C. Steward, Appl. Phys. Lett., 12 (1968) 277–280.
[91] D.E. McCumber, J. Appl. Phys., 39 (1968) 3113–3118.

Chapter 3

Characteristics of solid-state magnetic field sensors

3.1. Introduction

The subject of this chapter is the phenomenological description of the characteristics of solid-state sensors for magnetic fields and the methods for their determination. For this purpose the sensor is regarded as a black box with an input and an output, the particular transducing mechanism of its operation being ignored. The basic input of this class of devices is an external magnetic field B. In most transducers the maximum response to magnetic excitation is achieved when the magnetic field B is normal to the active surface of the device. 2-D and 3-D magnetic-field microsensors have recently been invented. By means of the same magnetosensitive volume they register simultaneously, or in succession, two or three components of the vector $B = (B_x, B_y, B_z)$, and orthogonality is not a limiting factor. Modulating magnetosensors are a subject of consideration in this book (§1.2). They necessarily have an additional input where the supply power (voltage V_s or current I_s) is applied. Depending on the device's construction, the biasing circuitry and the mode of operation, the output of the magnetic-field transducer, generates a voltage V, a current I or a frequency f. The output signal may be analog or digital (carrier-domain magnetometers). Figure 3.1 shows the block-diagram of a magnetosensor used for establishing its characteristics.

The great variety of applications, the heightened requirements for sensor performance and reliability, the research activity and the need to be able to compare the results obtained, the role of the materials used, and the trend towards unification call for the creation of relevant criteria for the evaluation of sensors, magnetosensors included. More than 30 figures of merit are used for the complete characterization of solid-state magnetic-field transducers. In the author's opinion, the consideration of all these characteristics can be made more clear and comprehensible if they are classified into three groups as follows:

(a) Variation of the output OUT as a result of the variation of the input parameter B, all other parameters being constant (e.g. the temperature T, pressure

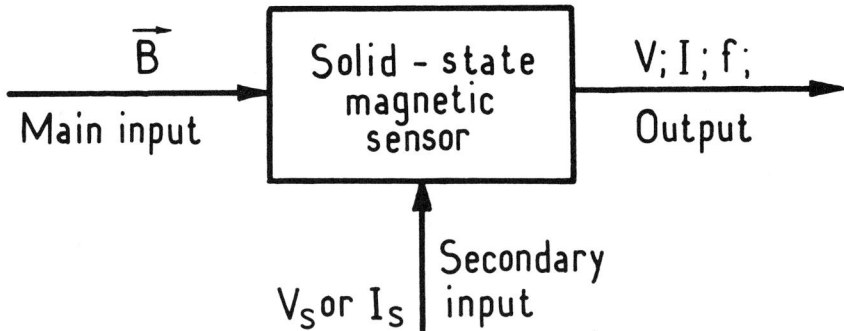

Fig. 3.1. A block-diagram of a solid-state magnetosensor used for the establishing of its characteristics.

P, radiation, vibrations, duration t, etc.). The constancy of all these parameters is denoted by the index C: $OUT(B)_C$;

(b) The behavior of the output signal OUT as a function of all possible types of external influence, such as T, P, t, supply power W_s, etc. in a constant external magnetic field B: $OUT(C)_B$;

(c) Characteristics which describe the requirements relevant to the additional (electric) input and to the magnetosensor as a circuit element, denoted SD.

Table 3.1 presents a classification of the principal figures of merit of magnetosensitive transducers in accordance with the approach discussed above. In addition to Table 3.1, there are other characteristics, which either duplicate those in the table, or are related to fabrication technology, production, market, etc. [1,2]. Unfortunately, in the literature on sensors some figures of merit are quite often arbitrarily defined, magnetosensitivity for instance, or there is no information on the conditions under which they have been determined; others are declared "the best", "the most accurate", with "excellent reliability", "extremely rugged", and for some

Table 3.1

Classification of the principal figures of merit of solid-state magnetic sensors

$OUT(B)_{C=const}$	$OUT(C)_{B=const}$	SD
Sensitivity	Noise	Excitation
Nonlinearity	Offset	Input impedance
Range	Cross-sensitivity	Output impedance
Frequency reponse	Drift	Size
Directivity	Creep	Weight
Resolution	Temperature error	Packaging
Accuracy	Operating life	Sensor material
Hysteresis	Reliability	Environmental condition
Error	Stability	
Output form	Response time	
Repeatability		

magnetosensitive devices the list of characteristics is far from complete. All these factors affect the accuracy of sensor investigations and applications and lead to a certain amount of confusion.

The sensor characteristics are defined below as accurately as possible and whenever necessary the methods for their determination are described.

3.2. Magnetosensor characteristics related to OUT(B)$_C$

3.2.1. Magnetosensitivity

The sensitivity S or the transduction efficiency is the most important figure of merit of solid-state magnetosensitive devices and all other types of sensors. By definition, this parameter S is the ratio of the variation in the output signal (voltage, current or frequency) to the variation in the external magnetic field at a constant temperature, pressure, radiation, etc. Both an absolute sensitivity and a relative sensitivity of the modulating type of magnetosensor can be defined [3]:

$$S_A^{(V)} \equiv \left| \frac{\partial V}{\partial B} \right|_C, \quad [\text{V T}^{-1}];$$

$$S_A^{(I)} \equiv \left| \frac{\partial I}{\partial B} \right|_C, \quad [\text{A T}^{-1}]; \quad (3.1)$$

$$S_A^{(f)} \equiv \left| \frac{\partial f}{\partial B} \right|_C, \quad [\text{Hz T}^{-1}].$$

The expressions (3.1) define the absolute sensitivity. The output is a voltage V, current I or frequency f. The output signal of Hall sensors, for instance, is $V_{\text{out}} \equiv V_H$, and in accordance with (3.1) $V_H = S_A^{(V)} B$. If the Hall terminals are shunted ($V_H = 0$) or the sample is short, $I_{\text{out}} \equiv I_H$ and $I_H = S_A^{(I)} B$.

The relative magnetosensitivity is determined by the ratio of the absolute sensitivity to the supply current or voltage applied to the additional input. The figure of merit is the current-related sensitivity S_{RI} when the additional input is fed by current I_s, and the voltage-related sensitivity S_{RV} when the additional input is fed by a voltage V_s:

$$S_{RI}^{(V)} = \frac{S_A^{(V)}}{I_s} \equiv \left| \frac{1}{I_s} \frac{\partial V}{\partial B} \right|_C, \quad [\text{V A}^{-1} \text{T}^{-1}];$$

$$S_{RI}^{(I)} = \frac{S_A^{(I)}}{I_s} \equiv \left| \frac{1}{I_s} \frac{\partial I}{\partial B} \right|_C, \quad [\text{T}^{-1}]; \quad (3.2)$$

$$S_{RI}^{(f)} = \frac{S_A^{(f)}}{I_s} \equiv \left| \frac{1}{I_s} \frac{\partial f}{\partial B} \right|_C, \quad [\text{Hz A}^{-1} \text{T}^{-1}];$$

$$S_{RV}^{(V)} = \frac{S_A^{(V)}}{V_s} \equiv \left|\frac{1}{V_s}\frac{\partial V}{\partial B}\right|_C, \quad [\text{T}^{-1}];$$

$$S_{RV}^{(I)} = \frac{S_A^{(I)}}{V_s} \equiv \left|\frac{1}{V_s}\frac{\partial I}{\partial B}\right|_C, \quad [\text{A V}^{-1}\,\text{T}^{-1}]; \tag{3.3}$$

$$S_{RV}^{(f)} = \frac{S_A^{(f)}}{V_s} \equiv \left|\frac{1}{V_s}\frac{\partial f}{\partial B}\right|_C, \quad [\text{Hz V}^{-1}\,\text{T}^{-1}].$$

The relative sensitivity is to be preferred in the comparative analysis of different magnetosensors as it "measures" the useful output signal with respect to the energy consumed for its generation. Unfortunately, only the absolute sensitivity S_A is given in many articles and handbooks, which makes an adequate evaluation of the transducing mechanism or the sensor structure more difficult.

3.2.2. Nonlinearities

From a metrological point of view, magnetosensor nonlinearities are errors associated with the hypothesis that the output relationship OUT(B) is represented by a straight line, i.e. OUT(B) = SB. The transverse characteristic of an ideal transducer is a straight line whose origin is the point OUT($B = 0$) and whose slope is $\tan\alpha$, i.e. the origin of the characteristic coincides with that of the x–y coordinate system, where $x \equiv B$ and $y \equiv$ OUT(B). If for certain reasons the position or the shape of the output relationship OUT(B) = SB is changed, the distortion leads to errors in the magnetic-field measurements, which are of three main types [4]:

(i) additive or zero errors;
(ii) multiplicative or sensitivity errors;
(iii) errors due to nonlinearity of the sensor output characteristic.

Additive (zero) errors are independent of the magnetic field. They are parallel shifts (translations) of the relationship OUT(B) = SB, i.e. OUT(B) = $SB \pm$ OUT'($B = 0$), where OUT'($B = 0$) is an additional parasitic signal, called an offset, when it refers to a differential sensor. This error can with equal probability modify the real output relationship in either the positive or negative direction. The additive error value must be added or subtracted from the appropriate value of the ideal characteristic to obtain the correct value of the field **B**.

If the ideal sensor characteristic is OUT(B) = SB, for instance a Hall element or a MAGFET, the multiplicative or sensitivity error is a linear function of the field B. Such an error is typically manifested by the change of the slope $\tan\alpha$ of the output characteristic. In this case the change can also be positive or negative, therefore increasing or decreasing the slope $\tan(\alpha \pm \alpha')$ of the real characteristic. Quite often additive and multiplicative errors coexist. This is the fact illustrated by Fig. 3.2. The total error consists of the two error components.

Depending on the deviation of the real output characteristic from a straight

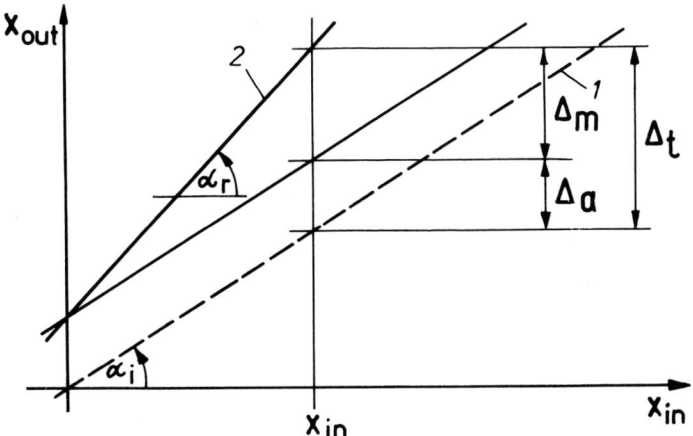

Fig. 3.2. The combined effect of additive and multiplicative errors. 1 = ideal characteristics; 2 = real charcteristics; Δ_α = additive error, Δ_m = multiplicative error; $X_{\text{out}} \equiv \text{OUT}(B)$ and $x_{\text{in}} \equiv B$.

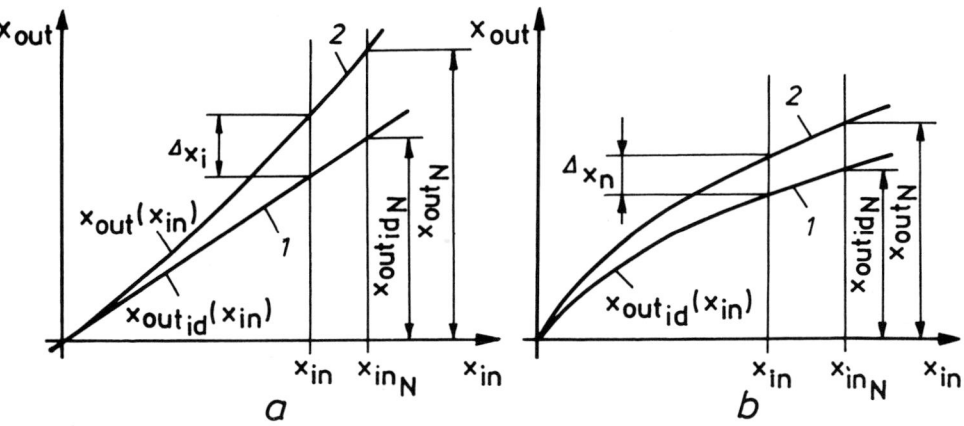

Fig. 3.3. Errors caused by the nonlinearity of the magnetosensor in the case of (a) linear, (b) nonlinear ideal characteristics. 1 = ideal, 2 = real characteristics. The subscript N refers to the nominal value, Δ_{xl} is the linear absolute error and Δ_{xn} denotes the nonlinear absolute error. $x_{\text{out}} \equiv \text{OUT}(B)$ and $x_{\text{in}} \equiv B$.

line, there can be various errors caused by the nonlinearity of the sensor itself. These errors are the most often discussed in sensor literature, although the two other types of errors can be found in almost all cases considered. The ideal output characteristic of certain devices, like magnetoresistors, magnetodiodes, some versions of bipolar magnetotransistors, etc., can be nonlinear, and for practical purposes is approximated by an appropriate nonlinear function. The errors that occur in the cases of a linear and nonlinear ideal output characteristic are presented in principle in Fig. 3.3. It can be concluded from the figure that in both cases the

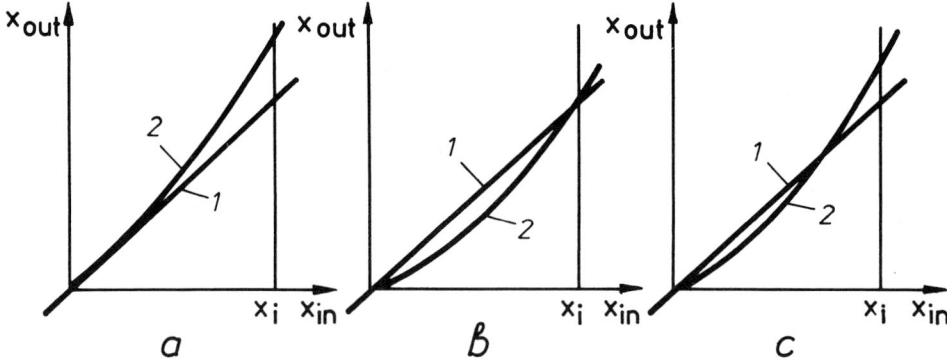

Fig. 3.4. Linearity errors; derivation of: (a) first- , (b) second- , and (c) third-order linearity errors. 1 = ideal, linear, 2 = real magnetosensor measurement characteristics. $x_{\text{out}} \equiv \text{OUT}(B)$ and $x_{\text{in}} \equiv B$.

absolute error is:

$$\Delta\text{OUT} = \text{OUT}_2 - \text{OUT}_1. \tag{3.4}$$

The relative error is:

$$\delta(\text{OUT}) = \frac{|\Delta\text{OUT}_2 - \Delta\text{OUT}_1|}{(\Delta\text{OUT})_1} 100\%. \tag{3.5}$$

When the ideal characteristic is a straight line, the relative error is termed a nonlinearity (NL). According to [5,6], nonlinearities in data acquisition systems can be classified as first-, second-, third-, ..., n-th order nonlinearities. Linearity errors are also first-, second-, third-, ..., n-th order errors, respectively. Figure 3.4 shows typical ideal and real output characteristics of magnetosensors. The first-order linearity error as seen from Fig. 3.4a monotonically increases, whereas the second-order error reaches a maximum and then decreases (Fig. 3.4b). After a region of increase, the third-order linearity error decreases and its sign is inverted (Fig. 3.4c). One more approach to the determination of linearity errors is known from the sensor literature:

$$\text{NL} \equiv 1 - \left\{ \frac{[\text{OUT}(B)/(B)], \text{ kG}}{[\text{OUT}(1)/(1)], \text{ kG}} \right\} 100\%. \tag{3.6}$$

Calibration is necessary in order to determine the real characteristic $\text{OUT}(B)$ of any magnetosensor. A set of monotonically increasing or decreasing values $B_1, B_2, B_3, \ldots, B_n$ of the magnetic field, known in advance, is applied and the corresponding responses $\text{OUT}(B_1), \text{OUT}(B_2), \text{OUT}(B_3), \ldots, \text{OUT}(B_n)$ at the output are registered under the specified conditions.

The data acquired can be presented in a table or in graphical form. Complete information concerning the output characteristic is obtained by means of a calibration cycle. This includes the application of specified values of the magnetic field $B_1, B_2, B_3, \ldots, B_n$ and the registration of the appropriate output responses $\text{OUT}(B_1)$,

OUT(B_2), OUT(B_3), ..., OUT(B_n) in the entire range of increasing or decreasing magnetic fields (or within a part of it) for which the sensor has been designed.

3.2.3. Range of magnetic fields

Each solid-state magnetosensor is designed for a certain optimum range of magnetic-field magnitudes [B_{lower}, B_{upper}], depending on the mechanism of transduction, the mode of operation, etc. This information should necessarily be available when using the device. For instance, if the construction of the device is optimized, Hall elements and bipolar magnetotransistors can detect magnetic fields between $B_{lower} \approx 1\mu T$ and $B_{upper} \approx (1 \text{ to } 2)T$ at $T = 300$ K. The range of the measured magnetic field is usually a result of a trade-off with the real output characteristic OUT(B). The limits B_{lower} and B_{upper} determine the range of magnetic-field magnitudes within which the output signal is approximated as accurately as possible by a straight line, i.e. $S = $ const.

3.2.4. Directivity and sensor excitation by magnetic fields

The magnetosensitivity of many solid-state sensors depends on the angle between the magnetic field B and the active surface of the sensor. If the vectors B and v_{dr} are orthogonal, i.e. $B \perp v_{dr}$, the maximum magnitude of the Lorentz force F_L is observed together with the maximum transducer efficiency (§2.2). Any deviation from the angle $\alpha = \pi/2$ reduces the sensitivity. If the vectors B and v_{dr} are parallel (collinear), in the ideal case the output signal is zero. In fact, owing to shortcomings in the construction of the device, the flow of supply current cannot be perfectly confined and made absolutely parallel to the active surface. Therefore the vector B, being parallel to that surface, generates a certain output signal. If the ratio of the appropriate magnetosensitivities when the vector B is orthogonal or parallel to the active surface of the device is $S_\perp/S_\parallel \approx 50$, the sensor is called a uniaxial transducer [7]. There are magnetosensitive devices, thin-film magnetoresistors for instance, with two sensitive axes. Nevertheless, uniaxial sensors are preferable for practical applications. The optimum angle α of maximum sensitivity is determined experimentally by a 360° rotation of the structure in a plane perpendicular to the active surface.

An important feature of magnetosensitive devices is the manner of their excitation by the magnetic field, i.e. whether the field vector B is perpendicular or parallel to the surface of the structure. In both cases, however, $B \perp v_{dr}$. Depending on the orientation of B with respect to the surface of the device, the solid-state magnetosensors known so far are divided into orthogonal and parallel-field devices.

3.2.5. Frequency response

The dynamic behavior of magnetosensors is very important when rapid variations of the magnetic field are to be measured. The frequency response is by definition the dependence of the amplitude ratio of the output signal to the of the a.c. magnetic field B (or of the phase difference between the output signal and the input signal B) on the frequency of the external sine wave of the a.c. magnetic field B within a specified frequency interval. This figure of merit is usually specified as "within $\pm \ldots \%$ (or $\pm \ldots \% db$) from ... to ... Hz". The dynamic degradation of a magnetosensor in an a.c. magnetic field B is assumed to begin with a drop in the output signal by a factor of $1/\sqrt{2}$, ($3\,db$). Unfortunately, almost no data can be found in the literature about this characteristic, which is of particular practical importance. This is probably due to several effects that perturb the measurements: the skin effect, eddy currents in the samples, a parasitic a.c. voltage induced in the output terminals, the strong influence of the high-frequency magnetic field B upon the capacitance and the inductance of the registration circuit, etc. [8,9]. A coreless coil is to preferred for the excitation of the a.c. magnetic field, since an iron or ferrite core contributes to the hysteresis and to the phase shift, thus introducing a registration error.

Particular set-ups which allow the correct measurement of the frequency response of the magnetosensor, provided the dominating sensor mechanism is known, are presented in [7,10,11]. The parasitic effects that perturb the measurement of the frequency response can be best suppressed by the integration into a single chip of the magnetosensor under investigation, the registration circuitry and the excitation coil.

3.2.6. Resolution

This figure of merit is determined by the smallest variation of the magnetic field that can be detected at the output of the magnetosensor. Quite often the resolution is represented by its average and by its maximum value and is expressed as a certain percentage of the full scale of the output signal. The full output signal range is the difference between the extreme points of the output, e.g. ± 5 V, ± 500 μA, $\pm 5 \times 10^4$ Hz, etc. This characteristic is always measured in the dynamic mode. A low-frequency a.c. magnetic field with $f \leq 10$ kHz is commonly used for that purpose. The minimum output signal variation that can be caused by the field B is the one that generates at the sensor's output terminals the same output power as that of the noise at $B = 0$ within a frequency range of $\Delta f = 200$ Hz at the main frequency $f_0 = 10$ kHz. Figure 3.5 shows a version of a resolution-measurement channel [12]. The homogeneous magnetic field is generated by Helmholtz coils. When $B = 0$, in the absence of a LF signal, the noise power is measured by means of a square voltmeter. Then a growing, calibrating magnetic field is excited by a

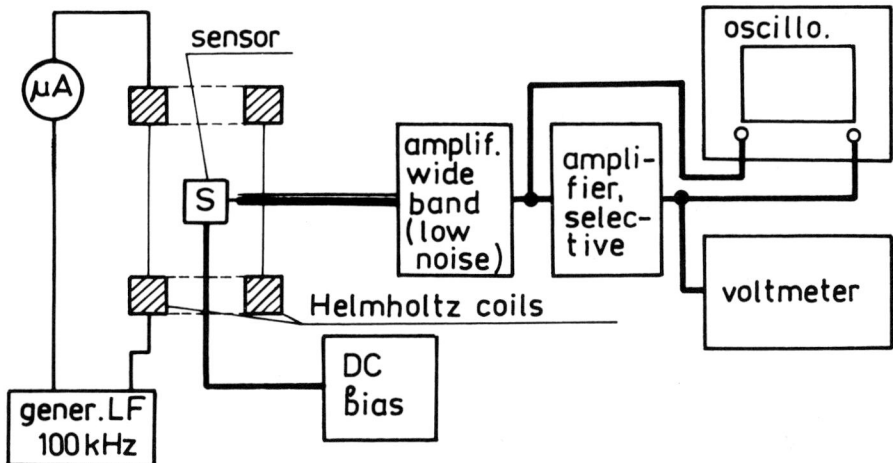

Fig. 3.5. Channel for the measurement of the minimum detected magnetic field $B_{\min}$.

LF generator until the voltmeter indicates double the value of the noise voltage at $B = 0$. This particular magnitude, $B_{\min}$, of the magnetic field is the lowest value detectable by a given magnetosensor when the signal to noise ratio is unity. Accumulators whose noise is negligible are always used as the power supply of the sensor. The electrical and magnetic screening of all circuit components including the sensor and the low-noise amplifier deserve special attention. The Helmholtz magnetic system is also screened in order to suppress the emission of parasitic signals. Bifilar winding of the sensor input and output cables is employed in order to prevent the appearance of a parasitic magnetic field.

3.2.7. Error

The error is the algebraic difference between the magnetic field indicated at the output and its exact value. This figure of merit is expressed as a percentage of the full scale of the magnetosensor output. If the device investigated allows a theoretical determination of the $OUT(B)$ dependence, the theoretical values are assumed to be exact. The error band is defined as the band of maximum deviations of the output values from a specified reference line or curve due to those causes attributable to the sensor. The range of allowable deviations is usually expressed as "$\pm \ldots \%$" of the full scale output, whereas the maximum deviations observed during device testing and calibration are expressed as "$+\ldots\%$" and "$-\ldots\%$" of the full scale output. The error band is determined by at least two calibration cycles in order to ensure repeatability [2]. The error curve is a graphical representation of the errors resulting from a certain number of calibration cycles.

3.2.8. Accuracy

This figure of merit is defined as the ratio of the error to the full output scale or as the ratio of the error to the output expressed as a percentage. The accuracy can be given also in the units used for the measurement of the magnetic field or as "within $\pm \ldots \%$" of the full scale output. The term "error" is to be preferred in magnetosensor specifications.

3.2.9. Hysteresis

This characteristic determines the maximum variation in the output signals for any fixed value B_0 of the magnetic-field magnitude within a certain range of magnetic fields. The value B_0 is reached first by increasing and then by decreasing the external magnetic field. The hysteresis is expressed as a percentage of the full scale output during any calibration cycle.

3.2.10. Output form

Most solid-state magnetic-field sensors, for instance, Hall devices, magnetoresistors, magnetodiodes, bipolar magnetotransistors, generate an analog output signal. There are such transducers, however, whose output is digital, because of their device construction and mode of operation. These are the different types of thyristor-like magnetometers with current filaments: circular carrier-domain magnetometers, unijunction magnetotransistors, etc. The digital output is preferable from the standpoint of the smart approach to magnetosensor solutions.

3.2.11. Repeatability

There is a strong dependence of the performance of magnetosensors and the data acquisition systems containing them on the ability of the transducer to reproduce its output readings when the same direction and magnitude of the magnetic fields **B** are repeatedly applied under the same conditions. The repeatability is defined as the maximum variation in the output readings when $B = \text{const}$, and is expressed as a percentage of the full scale output. At least two calibration cycles are needed in order to determine the repeatability [2].

3.3. Magnetosensor characteristics related to OUT(C)$_B$

3.3.1. Noise

The output noise is a fundamental physical parameter which determines the lowest magnetic-field magnitude $B_{\min}$ that can be detected by the sensor. The noise is a random amplitude and frequency component in the d.c. output signal, measured in terms of the power spectral density, i.e. by the frequency spectrum of the mean square, the effective or the root-mean-square (RMS) value of the fluctuating (noisy) quantity [13]. In fact random voltage or current fluctuations are observed at sensor output terminals when $B = 0$. This noise is not caused by defects in the contacts or by any other effect that can be purposefully eliminated. It is generated on a microscopic level as a result of the random behavior of charge carriers [14]. A more detailed analysis of noise will be presented, because of the importance of this sensor parameter and the scarcity of the information about it in the literature.

In the general case the noise voltage averaged over time, $\overline{V}_N(t)$, is zero, but the noise power averaged over time differs from zero. The mean square noise voltage is determined from the expression [13–16]:

$$\overline{V}_N^2 = \lim_{t \to \infty} \frac{1}{t} \int_0^t [\overline{V}_N(t)]^2 dt. \tag{3.7}$$

Hence the RMS voltage is $\overline{V}_N = (\overline{V}_N^2)^{1/2}$. The noise voltage and its RMS value depend on the frequency range Δf. The harmonic component of the noise voltage is expressed by the spectral noise density function $S_{NV}(f)$. The voltage spectral density is proportional to the power per unit frequency interval. The noise voltage spectral density and the RMS noise are determined from the expression:

$$V_N = \left(\int_{f_1}^{f_2} S_{NV}(f) df \right)^{1/2}, \tag{3.8}$$

where f_1 and f_2 are the limits of the frequency range within which the noise voltage is considered.

If, for instance, the output voltage in (3.1) is replaced by the mean-square noise voltage from (3.8), the following relationship is obtained:

$$\langle B_N^2 \rangle = \int_{f_1}^{f_2} \frac{S_{NV}(f) df}{S_A^2}, \tag{3.9}$$

The mean-square noise-equivalent magnetic induction within the frequency interval (f_1, f_2) is denoted by $\langle B_N^2 \rangle$ and $S_{NV}(f)$ is the noise voltage spectral density at the output of the magnetosensitive device, which is measured in V^2 Hz^{-1} [13,15,16]. The minimum detectable value $B_{\min}$ of the magnetic induction predetermined by the sensor noise in a narrow frequency interval Δf which includes a fixed frequency f_0, with a signal to noise ratio that equals unity, is given by the expression:

$$B_{\min} = \frac{[S_{NV}(f) \Delta f]^{1/2}}{S_A}, \quad [\text{T}]. \tag{3.10}$$

There are two methods for measuring the noise: the analog method, which employs the average noise power, and the digital method, based on the computation of the spectral power by means of fast Fourier transformation algorithms. The

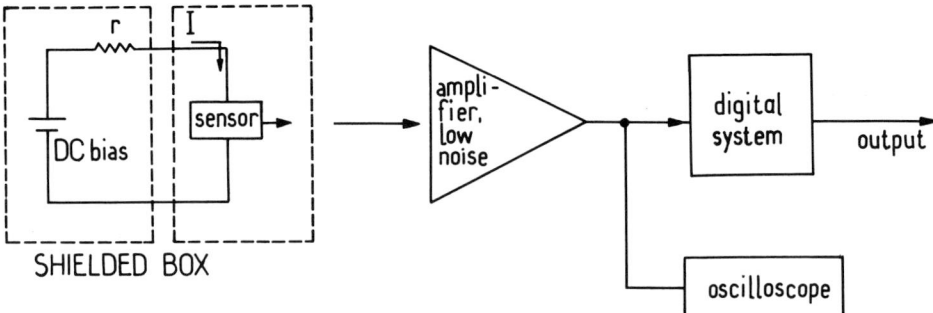

Fig. 3.6. Scheme of the digital noise-evaluation technique.

detailed theory of both methods and their adequate device implementations can be found in [16,17]. The analog technique is time consuming; therefore the fast digital method is to be preferred. In Fig. 3.6 the principle of digital noise registration is presented schematically. The scheme includes: a magnetosensor and a current generator whose power supply is an accumulator; a wide-band low-noise amplifier with a gain in the 10^4 range, whose power supply is also an accumulator; an oscilloscope connected to the amplifier output in order to observe the waveform of the output signal; a digital spectral analyzer which employs a fast Fourier transformation as well as an algorithm for the computation of the spectrum of the sensor output signal at $B = 0$. Special care must be taken of the electromagnetic screening of the sensor, the amplifier and the accumulator power supply.

3.3.2. Offset

The offset is the output signal of the modulating type of sensor (in most cases with a differential output) in the absence of an external magnetic field ($B = 0$). Essentially, this kind of error is a static value or a very slow variation with time of a voltage, current or frequency.

If there is no additional information concerning the magnetic field or the sensor itself, the offset cannot be distinguished from the useful output signal:

$$\text{OUT}(B) = SB + \text{OUT}'(B = 0), \tag{3.11}$$

where $\text{OUT}'(B = 0)$ is the offset; it results from a structural or electrical asymmetry of the sensor.

If the offset is time invariant, it causes a parallel shift of the whole calibration curve (cf. §3.2.2). This error is often expressed as "within $\pm \ldots \%$ of the full scale output". A more relevant definition of the offset is its expression as a signal of an ideal sensor generated by an equivalent magnetic field $B_{\text{off,eq}} \equiv B_{\text{off}}$:

$$B_{\text{off,eq}} \equiv \frac{V_{\text{off}}}{S_A^{(V)}}; \quad B_{\text{off,eq}} \equiv \frac{I_{\text{off}}}{S_A^{(I)}}; \quad B_{\text{off,eq}} \equiv \frac{f_{\text{off}}}{S_A^{(f)}}. \quad (3.12)$$

The offset depends on different external influences: pressure, light and particularly on variations in temperature. In the last case the offset may equally well increase or decrease with temperature. Sensor specifications usually offered by manufacturers include data about the temperature drift of the offset.

3.3.3. Cross-sensitivity and temperature errors

The magnetic sensor cross-sensitivity, C.S., is an undesirable output signal which, instead of being generated by the externally applied magnetic field, is associated with some other environmental parameter like temperature, light, pressure, etc. The figure of merit C.S. is expressed as:

$$\text{C.S.} \equiv \frac{1}{S} \frac{\partial S}{\partial C}, \quad (3.13)$$

where S is the sensor magnetosensitivity (cf. §3.1.–3.3) and C is the source of perturbation. Most often this is the temperature; therefore:

$$\text{T.C.} \equiv \frac{1}{S} \frac{\partial S}{\partial T} 100\%. \quad (3.14)$$

In this case the cross-sensitivity is called the temperature coefficient, T.C., of magnetosensitivity. It is measured in % K^{-1} or % °C^{-1}. The temperature interval ($T_{\min}$, $T_{\max}$) within which the value of the temperature coefficient is constant should necessarily be specified.

The thermal inertness of the magnetosensor, of the magnet exciting the field, of the temperature sensor and of the thermostat used for the experiment must be taken into consideration when the T.C. is determined experimentally. Accurate results can be obtained if the calibration-cycle measurements are performed while the thermostat is being slowly cooled down. In this case the temperature of all the devices inside it is the same. From the author's experience an optimum cooling rate of about 0.3°C min^{-1} can be recommended for modern magnetic-field microsensors, temperature sensors, etc. The experimental results are influenced by the thermal behavior of the material of the permanent magnet used for field excitation. This is why a precalibrated solenoid fed by a current generator is to be preferred for the excitation of a homogeneous field B. The temperature drift of the offset can, moreover, be evaluated by switching off the current in the coil.

A great number of solid-state magnetosensors manifest cross-sensitivity associated with the existence of a component of the magnetic field B which does not coincide with the normal, u_n, to the active surface of the device. Hall elements, bipolar magnetotransistors and other devices serve as examples. This field component usually generates a parasitic output signal which is treated as an error. This type of cross-sensitivity is commonly represented by the ratio of the output signals or magnetosensitivity values for the same magnitude of the magnetic field, which is successively directed along one of the three orthogonal coordinate axes, x, y and z: $S_x : S_y : S_z$ [7].

There are semiconductor magnetosensors which demonstrate a remarkable asymmetry of their galvanomagnetic properties when the direction of the magnetic field is reversed from $+\boldsymbol{B}$ to $-\boldsymbol{B}$. For instance, InSb and GaAs three-terminal orthogonal Hall elements [18], and silicon three-terminal parallel-field Hall devices [19], possess such asymmetry. It is characterized by the dimensionless parameter called the directional figure of merit:

$$\mathrm{DFM} \equiv \left| \frac{\mathrm{OUT}(+\boldsymbol{B})}{\mathrm{OUT}(-\boldsymbol{B})} \right|. \tag{3.15}$$

3.3.4. Drift and creep

With respect to magnetosensors, drift is the undesirably slow variation with time of the output signal when the magnetic field remains constant. Therefore this variation is not a function of the magnetic field. If the drift is temperature induced, it is determined as $\mathrm{OUT}(B)/\Delta T$ and is measured in V °C^{-1}, A °C^{-1}, Hz °C^{-1}. If a sophisticated technology is used in the implementation of the microsensor, the temperature drift is reversible when T is repeatedly increased and decreased within the same temperature interval. Generally, when $B = 0$ or $B = $ const, the drift is a nonlinear function of varying environmental parameters (temperature, pressure, light, etc.). Owing to certain technological shortcomings the drift may be irreversible.

Creep is also a parasitic magnetosensor signal. It is a small and slow variation of the output within a certain period of time when all environmental parameters, including the field $\boldsymbol{B}$, are kept constant. The creep may equally well be in the positive or in the negative direction. It often fluctuates around a certain value which may be time dependent. In the general case the creep is an irreversible effect and efforts to eliminate it are concentrated mainly on the technology and on optimizing the mode of sensor operation.

3.3.5. Stability, operating life and reliability

By stability is meant the ability of the solid-state magnetosensor to preserve its performance characteristics unchanged for a relatively long time. The main criterion for stability is the repeatability within a certain time interval of the output readings originally obtained upon calibration of the device in a normal room environment. Stability is typically expressed as "within ... % of the full scale output" over a time interval of ... weeks, months or years. This figure of merit is of particular importance for smart magnetosensor solutions.

The operating life is determined as the minimum period of time during which continuous and intermittent rating applied to the magnetosensitive device does not bring its transducing performance beyond certain limits [2].

Reliability is characterized by the probability that the magnetosensor performance is maintained during a certain time interval without changing its characteristics.

3.3.6. Response time

This magnetosensor characteristic is defined as the time needed by the output signal to reach a certain percentage of its final value as a result of a step change of the magnetic field. This figure of merit of modern solid-state devices is in the range 10^{-14} to 10^{-8} s [1,7–9]. The response time may be indicated in catalogs, for instance, as "98% response time ... μs".

3.4. Characteristics of the magnetosensor as a circuit element

3.4.1. Electrical excitation, input and output impedance

Electrical excitation is a characteristic which is associated with the modulating nature of solid-state magnetosensors and with their additional input in particular.

Complete information about the voltage and/or the supply current necessary for operation of the device should be available. The excitation is usually expressed as a range of allowable voltage and/or current values. The maximum value of the excitation voltage or current, which does not cause damage of the device or performance degradation under room conditions should also be specified.

The additional input impedance presented to the power supply is measured between the excitation terminals. Unless otherwise specified, the input impedance is measured in a room environment in the absence of an external magnetic field and with open-circuited output terminals.

The output impedance is the impedance presented by the transducer to the associated external circuitry. It is measured between the output terminals of the magnetosensor under room conditions with open-circuited additional input terminals.

3.4.2. Room conditions

Magnetosensors most often operate under the following environmental conditions [2]:
Temperature: $(25 \pm 10)°C/(77 \pm 18)°F$;
Relative humidity: 90% or less;

Barometric pressure: 26 to 32 inches Hg.

The other characteristics in the last SD column of Table 3.1 are clear and do not need special comment. They are included in the specifications of sensor producers.

References

[1] S. Middelhoek and S.A. Audet, Silicon Sensors, Academic Press, London, 1989.
[2] H.N. Norton, Sensor and Transducer — Selection Guide, Elsevier Sci. Publ., Oxford, 1990.
[3] S. Middelhoek and D.J.W. Noorlag, J. Phys. E: Sci. Instrum., 14 (1981) 1342–1352.
[4] A. Boros, Measurement Evaluation, Elsevier, Amsterdam, 1989.
[5] A. Lenk, Electromechanische Systeme, in: Systems with Auxiliary Power, vol. 3, Verlag Techn., Berlin, 1975.
[6] V.S. Gutnikov, A. Lenk and U. Mende, Sensorelectronik, Verlag Techn., Berlin, 1984.
[7] A.W. Vinal, IBM J. Res. Develop., 25(3) (1981) 196–201.
[8] A. Kobus and J. Tuszynski, Hallotrony i Gaussotrony, Wydawnictwa naukowo — techniczne, Warszawa, 1966 (in Polish).
[9] H. Weiss, Physik and Auwendung galvanomagnetischer Bauelemente, Vieweg, Braunschweig, 1969.
[10] V. Zieren, Integrated silicon multicollector magnetotransistors, Ph.D. dissertation, Delft Univ. Techn., The Netherlands (1983).
[11] J.L. Lopez and J.C. Licini, J. Appl. Phys., 53 (1982) 8389–8391.
[12] J. Chretien, Les dispositifs à injection de porteurs en tant que capteurs magnetosensibles, Ph.D. dissertation, I.N.P., Grenoble, France, 1977.
[13] A. Chovet, Ch.S. Roumenin, G. Dimopoulos and N. Mathieu, Sensors and Actuators, A, 21–23 (1990) 790–794.
[14] M.J. Buckingham, Noise in Electronic Devices and Systems, Ellis Horwood and John Wiley, New York, 1983.
[15] R.S. Popović, Hall Effect Devices, Adam Hilger, Bristol, 1991.
[16] A. Chovet and P. Victorovitch, l'Onde Electrique, 57 (1977) 699–707; 57 (1977) 773–783; 58 (1978) 69–80.
[17] A. Chovet, Transport, recombinaison et bruit dans les semiconducteurs ambipolaires en présence de champ magnétique, Ph.D. dissertation, I.N.P., Grenoble, France, 1978.
[18] Y. Sugiyama, Fundamental research on Hall effect in inhomogeneous magnetic fields, Res. of the Electrotechnical Laboratory, No. 838, Tokyo, Japan, 1983.
[19] Ch. S. Roumenin, C.R. Acad. Bulg. Sci., 39(11) (1986) 65–68.

Chapter 4

Hall sensors and magnetoresistors

4.1. Introduction

An analysis of the physical aspects of the Hall effect and the magnetoresistance effect has been presented in Chapter 2. The most important results can be summarized as follows: the maximum Hall voltage and minimum magnetoresistance are observed when the lines of current flow in a magnetic field $B \neq 0$ and at $B = 0$ do not change their position and the orientation of the equipotential lines is changed by the Hall angle due to the magnetic field.

If the equipotential lines remain unchanged and the lines of current flow are tilted at the Hall angle, the maximum magnetoresistance and minimum Hall voltage are observed.

The Hall effect is manifested by the well-known device of classical construction: a parallelepiped of a conducting material, very thin in the direction of the field B, is equipped with two input (supply) contacts on two opposite sides and two output terminals located on the other two opposite sides of the sample (Fig. 2.2). This is exactly the geometry used by E. Hall in his experiments more than 100 years ago. The adaptation of the theoretical results presented in Chapter 2 to the design of the Hall sensors includes:

(a) geometry optimization;

(b) the correct choice of materials with adequate galvanomagnetic and transport parameters;

(c) suitable fabrication which must first of all be compatible with IC technology. The application of the Hall element as an electron device, on the other hand, is determined by its properties. They are characterized by the corresponding figures of merit as described in Chapter 3.

The magnetoresistance effect is manifested by a two-terminal semiconductor sample with such a shape as to achieve the maximum geometrical magnetoresistance. In this respect the Corbino disk, where no Hall effect occurs, has the most suitable structure. Carrier mobility is rather important for the performance of this type of sensor, since the variation $\Delta R(B)$ is proportional to $\mu^2 B^2$. In spite of the maturity of silicon technology, silicon cannot be used in devices fabricated for practical applications, owing to the low magnetoresistance observed. The n-type InSb and n-type GaAs are very promising materials for such sensors because of the high carrier mobility. For instance, the bulk mobility of electrons in n-type InSb is about 8×10^4 cm^2 V^{-1} s^{-1} at $T = 300$ K, whereas in n-type Si it is approximately 1500 cm^2 V^{-1} s^{-1}.

It has been shown in Chapter 2 that in any piece of a conducting material with finite or semifinite dimensions the Hall effect occurs, provided the sample is equipped with two contacts for the supply current and an external transverse magnetic field is applied. This phenomenon can be detected if the sample has at least one output terminal (the other output terminal can be replaced by a potentiometer [1]). The above facts justify the assumption that the parallelepiped shape used so far should not be preferred to other possible constructions of Hall sensors with different geometries. The heuristic value of this idea is in the fact that the Hall effect can be combined with particular device designs which have advantages concerning metrology, technology of fabrication, etc. For instance, the location of all contacts upon one and the same surface makes fabrication of the device compatible with conventional planar IC technology. The presence of only three terminals allows the combination of the magnetoresistance effect with the Hall effect. If a solid-state sensor can be excited by an external magnetic field B parallel to the chip surface, there is a promising opportunity for the construction of functional magnetometers which can measure simultaneously or successively the two (B_x, B_y) or three (B_x, B_y, B_z) components of the vector B. Overcoming of the routine approach to the rectangular Hall structure has given rise to the invention of a number of nontrivial Hall sensors, termed parallel-field sensors, whose characteristics are very attractive from the viewpoint of their practical application (cf. §4.3).

There are two principal methods of applying an external magnetic field B to a solid-state sensor while, at the same time, preserving the orthogonality $I \perp B$. The first method is to apply a field B which is perpendicular to the plane of the structure. Sensors activated in such a manner will be referred to subsequently as orthogonal. The second method is the so-called parallel-field excitation and the devices activated in this way are referred to as parallel-field sensors.

So far, solid-state magnetosensors, including Hall elements and magnetoresistors, have usually been classified in the literature as vertical or lateral, depending on the direction of the current flow (perpendicular to the plane of the structure or parallel to it). In the general case, however, the carrier trajectories are not straight lines and include both vertical and parallel components of velocity. This is why it is the author's opinion that the manner of sensor activation by the magnetic field is a more universal criterion for classification it is this that has been adopted in this book.

4.2. Orthogonal Hall sensors

4.2.1. Hall-sensor shapes

Shown in Fig. 4.1 are the shapes of Hall structures known so far. From the standpoint of symmetry, which makes the fabrication technology more simple and improves the electrical parameters (minimum offset, linearity, etc.), the sensors presented can be divided into two groups. The first group (b–k) includes those structures symmetrical only with respect to the supply contacts and/or the output contacts. The second group (l–r) includes sensors with a higher degree of symmetry: they are invariant under 90° rotation, and the input and output terminals are interchangeable. An arbitrarily shaped structure (a) has been added to complete the classification. Although far from the optimum, this shape is used for the characterization of wafers in the semiconductor industry. Shapes (b) and (c) are analogous to the Hall structure described in §2.3, Fig. 2.2. If the positions of the sensor terminals H_1 and H_2 are absolutely symmetrical, with a supply voltage V the two contacts are equipotential and in the ideal case there is a zero offset $V_{H1,2} = 0$, when $B = 0$. In a magnetic field $\boldsymbol{B} \neq 0$, the total Hall voltage generated in the sensor appears between terminals H_1 and H_2. The plates must meet the requirement $l \gg w$, and the contacts H_1 and H_2 must have minimum dimensions in order to increase the geometrical correction factor G. A specific feature of the three-terminal version is the asymmetry of the output characteristic $V_H(\boldsymbol{B})$ with respect to the direction of the magnetic field $\boldsymbol{B}$. In high-mobility materials like n-type InSb and n-type GaAs, when $l \leq w$, the even magnetoresistance effect and the odd Hall effect coexist, thus determining the strong asymmetry of the output signal. The voltage at terminal H (Fig. 4.1c) is half the effective voltage V_H. The bridge-shaped structures (d) and (e) are also analogous to those in Fig. 4.1b, c. It is worth noting that their low resistivity metal terminals C_1, C_2 and H_1, H_2 have no direct contact with the active region and their shunting effect upon the Hall voltage V_H is negligible. This is the case presented in Fig. 2.5a. In fact the contacts of sensors (d) and (e) are replaced by the virtually same material as that of the plates. These geometrical configurations are the best approximations to an infinitely long sample with a factor $G \approx 1$.

Although the contacts C_1 and C_2 are far from the active volume, in the butterfly shaped device (f) the geometrical correction factor satisfies the condition $G < 1$, mainly because of the shunting influence of H_1 and H_2 upon V_H. This sensor has been designed to register nonhomogeneous magnetic fields and is convenient for application together with ferrite or μ-metal flux concentrators [3].

Design (g) is analogous to structures (b)–(f) and has planar insulating dams close to the Hall terminals H_1 and H_2 to reduce the offset. The component of the electric field parallel to longitudinal axis of the sensor is reduced and the inevitable misalignment of H_1 and H_2 has a smaller effect. This device shape is often used in

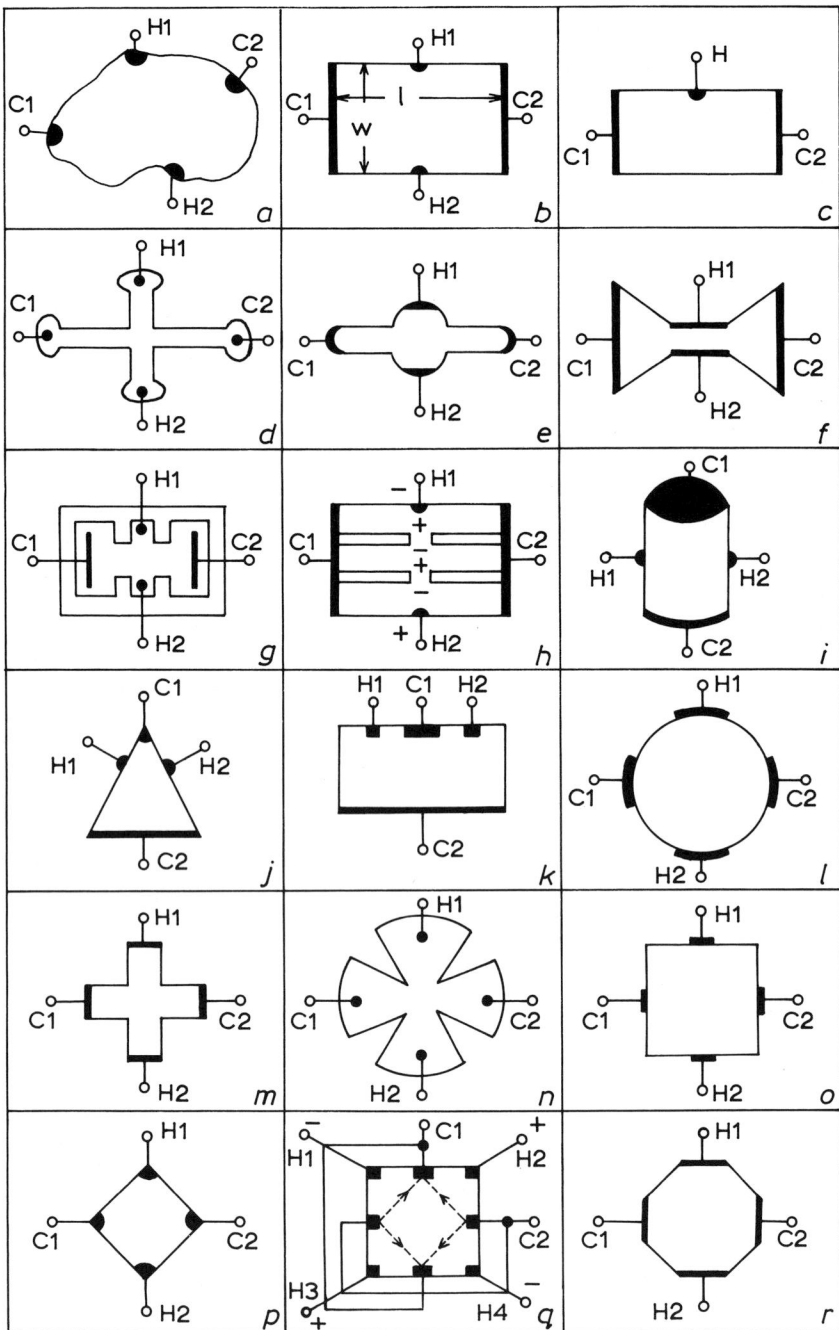

Fig. 4.1. Various shapes of orthogonal Hall sensors. Current contacts are denoted by C_i, Hall terminals by H_i.

integrated sensors. Such transducers are fabricated in n-type silicon epilayers and all contacts are located on the same side of the structure.

If the generated Hall voltage is low, structure (h) can be used for its enhancement. Narrow isolating slits divide the device into three identical rectangularly shaped parts, thus yielding three separate Hall sensors. They are supplied in parallel and the Hall voltages are summed in series. To a first approximation V_H is increased by a factor of 3 [4]. The increased active area is the main disadvantage of this discrete Hall device.

An analysis of the cylindrical structure (i) can be found in [5]. It has been proved that at $B =$ const the Hall voltage increases linearly with the ratio a=height/radius, and when $a \geq 3$ saturation occurs. This device is convenient for quality evaluation in the production of semiconductor materials.

In the triangular Hall device (j), the shift of sensor terminals H_1, H_2 towards the contact C_1 leads to an increase in the voltage V_H [6]. The optimum distance between H_1 or H_2 and the contact C_1 is about 1/3 of the length of the insulated side. Such specific behavior of the sensor is explained by the increase in the carrier drift velocity, and hence in the Hall field, as a result of the narrowing of the structure, and by the increase in the effective factor G. Another modification of the triangular device is (k) [6].

The Hall sensors (l)–(r) have equivalent contacts and therefore some compensation methods can be used to suppress the offset, cf. §9.5.1. The Maltese-cross and four-leaf clover shaped structures (m) and (n), like (d) and (e), are approximations of an infinitely long sample as their metal contacts are not connected directly to the active region, and thus the correction factor is $G \approx 1$ [7].

Some Hall structures included in integrated sensors are combinations of identical elements, like structure (q) [8]. It has four supply contacts and four Hall contacts and can be regarded as two identical structures of type (k), connected in parallel. As a result the input and output resistances are reduced, but the magnetosensitive volume remains the same. The insulating dams around all terminals confine the supply current and reduce considerably the offset caused by the piezo-resistance effect. The directions of the current have been represented by dashed lines in Fig. 4.1q.

4.2.2. Functional relations of the different geometrical shapes of Hall sensors

The methods for the determination of the Hall voltage in structures with different shapes have been listed in §2.3.4. Conformal mapping is the most illustrative method for an intuitive understanding of the influence of geometry upon Hall sensor characteristics. Its application for the above purpose was first described in [9], and later in [5,10–14]. It will be used to determine which one of all the shapes shown in Fig. 4.1 is the most suitable for Hall sensors, i.e. whether their magnetosensitivity can be improved by a proper choice of the geometry.

The possibility of illustrating the properties of the analytical functions by a particular geometrical image in the complex half-plane has motivated the application of conformal mapping to the solution of various problems of electrical engineering, heat distribution, astronomy, etc. Ptolemy's stereographical projection, for instance, used for the flat representation of the firmament is a conformal transformation. According to Riemann's theorem, from the theory of conformal mapping, each simply connected region R from the complex plane (z) can be conformally mapped onto the inner part of the unit circle $|w| < 1$ on the plane (w). There is a reversible unambiguous correspondence between the points in the original region and their images in the mapped region. Moreover, the angle between two intersecting lines upon the original plane is equal to the angle between their images in the (w) plane [15]. This approach can be applied to Hall structures with different shapes if they meet the following requirements [5]:

(i) There must be no space charges in the active sensor region, i.e. electroneutrality must be established;

(ii) The devices must be two-dimensional, i.e. they must be planar or quasi-planar (their thickness t must be much less than the two other dimensions l and w, $t \ll l, w$). This condition, in the general case, sets limitations on the application of conformal mapping to intrinsic samples and samples in which magnetoconcentration occurs;

(iii) Uniformity of both the external magnetic field and the galvanomagnetic properties of the material all over the device plane considered.

The current density j and the electric-field vector E determine the intersecting lines of interest in Hall sensors. According to (iii) the Hall angle Θ_H between them is preserved by the conformal mapping. The next step is the transformation of the unit circle from the (z) plane into the upper complex half-plane $\text{Im}(w) > 0$. For this purpose the Möbius relation is used:

$$w = \frac{a + bz}{c + dz}, \tag{4.1}$$

where the complex constants a, b, c and d fulfill the condition $a/c \neq b/d$.

As a result of the above procedure, the boundary of the unit circle passes into the pertinent "circumference", a boundary of the upper half-plane $\text{Im}(w) > 0$. Therefore if the initial simply connected region R has the shape of some of the Hall devices, presented in Fig. 4.1 (triangle, square, circle, hexagon, octagon, etc.), the potentials and the current contacts will be transformed into the respective arcs of the unit circle, and subsequently into respective sections of the real axis u of the upper half-plane $\text{Im}(w) > 0$, $(w = u + iv)$.

Schwarz–Christoffel's transformation is determined by the expression:

$$U = f(z) = \int_0^z (X - a_k)^{\alpha_k - 1} dX. \tag{4.2}$$

It is used in the conformal mapping of the upper complex half-plane onto a polygon whose vertices are A_k and the internal angles are $\pi\alpha_k$, where α_k is a parameter within the interval [0;2] and a_k are the points from the real axis which correspond to A_k [10]. In the general case, the approximation of (4.2) is rather complicated but in the case of a rectangle the transformation is reduced to:

$$U = \frac{b}{2K(k)} \int_0^z \frac{dx}{[(1-x^2)(1-k^2x^2)]^{1/2}}, \tag{4.3}$$

where the ratio of the width a to length b is given by the relation:

$$\frac{a}{b} = \frac{K(1-k^2)^{1/2}}{2K(k)}. \tag{4.4}$$

The ratio a/b defines the parameter k of the transformation, and K denotes a complete elliptic integral of the first kind.

The validity of the reverse transformation of geometrical forms (rectangle → upper complex half-plane → unit circle → arbitrary simply connected region) like those in Fig. 4.1 follows from the theory of conformal mapping. The above analysis leads to the conclusion that there is an unambiguous reversible correspondence between the rectangularly shaped (classical) Hall sensors and the arbitrarily shaped ones. Figure 4.2 illustrates the connection, via the upper complex half plane, of the

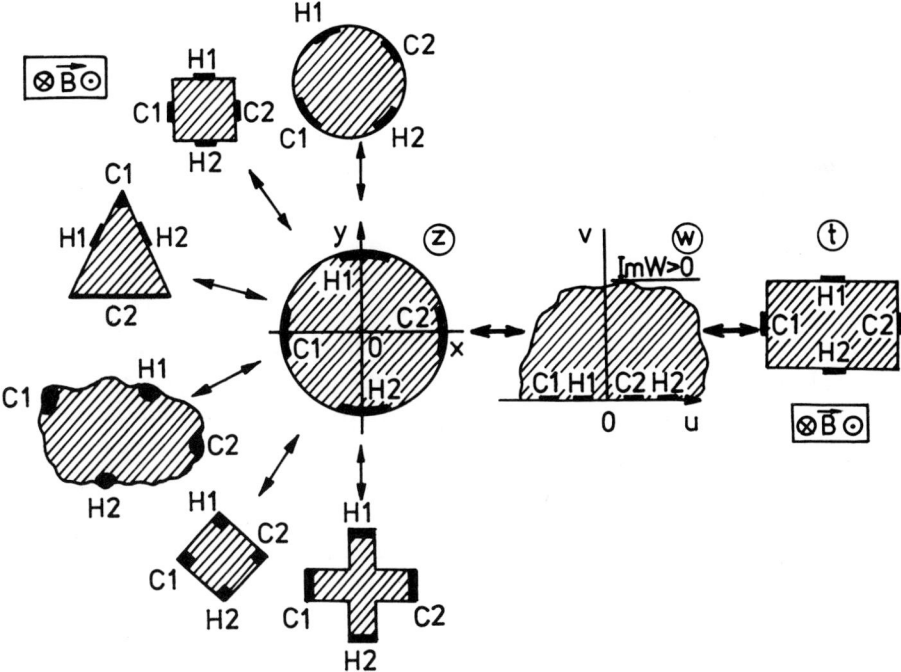

Fig. 4.2. The connection of selected orthogonal Hall sensors with the conventional rectangular Hall device via conformal mapping.

selected geometrical shapes of the Hall sensor to its classical rectangular version [16]. It can be concluded that the electrical efficiency of the Hall plates is invariant with respect to their shape, i.e. there are no such properties of the rectangular Hall device that cannot be observed in a structure with another simple shape (a circle, a square, etc.), provided the contact dimensions and positions are properly chosen [9]. This is why all geometrical forms of Hall sensor are equivalent as far as the geometrical factor $G \approx 1$ is concerned (the shunting effect of ohmic contacts has been neglected). This important conclusion explains why it is impossible to improve the magnetosensitivity of the Hall sensor by changing the geometry. In particular, the long Hall device ($l \gg w$) is equivalent to an element with equal length and width ($l = w$) and reduced contact dimensions. In the extreme case of an infinitely long Hall structure ($l/w \to \infty$) and finite contact dimensions the equivalent Hall sensor has point electrodes.

4.2.3. The geometrical correction factor in Hall sensors

In §2.3.4. the geometrical correction factor G was introduced in order to take into account the influence of the shape of the Hall device on the generated output voltage V_H:

$$G = \frac{V_H}{V_{H\infty}}. \tag{4.5}$$

The voltage V_H is a characteristic of the actual device, and $V_{H\infty}$ refers to an infinitely long sample with point contacts without shunting effects. In the general case the factor G is a function of the device geometry and the Hall angle. It is used to describe the decrease in the output signal due to the imperfect spatial confinement of the current forced through structures with different shapes. One of the main causes of the reduction in the Hall voltage is the shunting action of the supply contacts This behavior can be inferred from the distribution of the equipotential lines in a semiconductor sample with a finite length and low resistivity contacts for $B \neq 0$, Fig. 2.5b. The shorter the sample, the better pronounced is the effect. Figure 4.3 shows the electric field, the equipotential lines and the lines of current flow in a square sample ($l = w$) [17,18]. The equipotential lines are distorted throughout the whole structure. The perturbation is caused by the current contacts 1 and 2 and by the finite dimensions of the Hall terminals 3 and 4.

The dependence of the Hall voltage on the ratio l/w in a low magnetic field $|\mu B \ll 1|$ is shown in Fig. 4.4 in compliance with the theoretical analysis in [19]. According to [20], the geometrical correction factor G for a rectangular Hall sample with point output terminals can be approximated by the expression:

$$G \simeq 1 - \frac{16}{\pi^2} \exp\left(-\frac{\pi l}{2w}\right) \left[1 - \frac{8}{9} \exp\left(-\pi \frac{l}{w}\right)\right] \left(1 - \frac{\Theta_H^2}{3}\right). \tag{4.6}$$

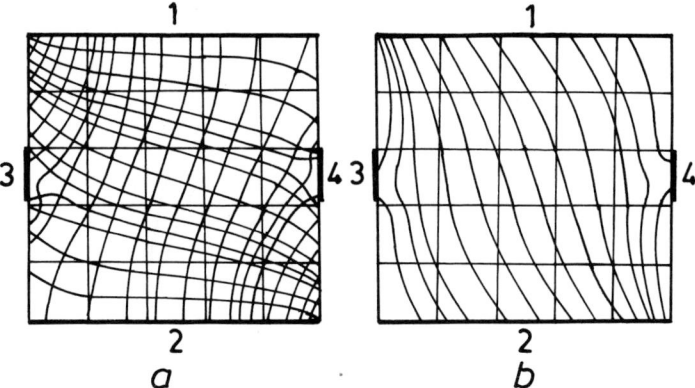

Fig. 4.3. (a) Distribution of the equipotential lines and the electric field in a square sample. (b) Distribution of the lines of current flow in a square sample. The Hall angle is $\Theta_H = 36°$ [17].

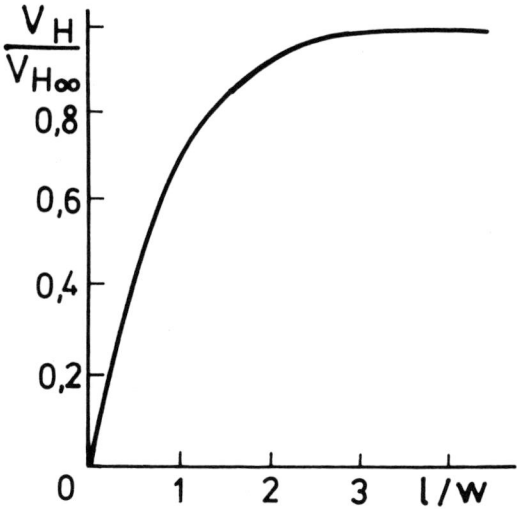

Fig. 4.4. Dependence of the normalized Hall voltage $V_H/V_{H\infty}$ on the ratio l/w; $V_{H\infty}$ is the Hall voltage at $l/w \to \infty$.

The above expression is valid if $0.85 \leq l/w < \infty$ and $0 \leq \Theta_H \leq 0.45$ [21]. The following expression is used for high values of the Hall angles:

$$G \simeq 1 - \frac{8}{\pi} \exp\left(-\frac{\pi}{2}\frac{l}{w}\right) \operatorname{ctg} \Theta_H. \tag{4.7}$$

Table 4.1 presents the values of the factor G as computed from (4.6) for several values of the ratio l/w as well as the corresponding errors δ, in %, when the value of G is not taken into account [17].

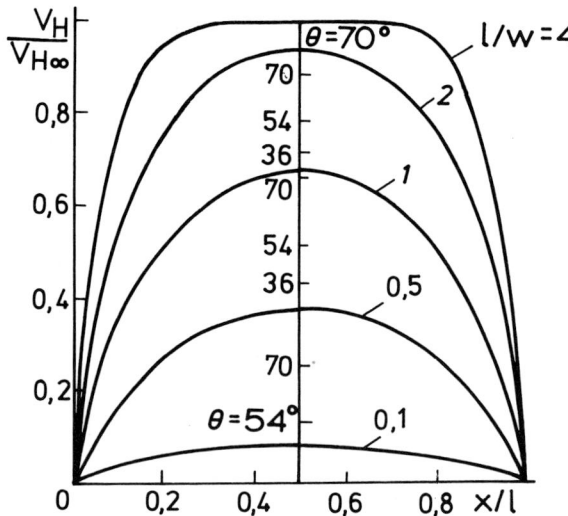

Fig. 4.5. Longitudinal distribution of the normalized Hall voltage $V_H/V_{H\infty}$ in rectangular samples with different values of the l/w ratio. The values of the Hall voltage with different Hall angles Θ_H have been marked upon the vertical axis through a point $x/l = 0.5$ [22].

If $l/w \geq 4$, the factor G can be eliminated in the considerations and the sample can be regarded as a long plate with good accuracy, provided the Hall contacts are located in the middle of the structure where the equipotential lines are not perturbed (Fig. 2.5b). For small values of Θ_H and short samples ($l/w < 1$), the geometrical correction factor is determined by the formula $G \approx 0.75 l/w$. It can be seen from Fig 4.5 that a 20% displacement of the Hall terminals from the middle of a rectangular device with $l/w = 4$ is large enough to cause shunting of the respective supply current contact by the Hall voltage [22]. An expression for the geometrical correction factor of a relatively long Hall sample with $l/w > 1.5$ and small output contacts ($s/w < 0.18$) is derived in [11]:

$$G \simeq \left[1 - \frac{16}{\pi^2} \exp\left(-\frac{\pi}{2}\frac{l}{w}\right) \frac{\Theta_H}{\tan \Theta_H}\right]\left(1 - \frac{2}{\pi}\frac{s}{w}\frac{\Theta_H}{\tan \Theta_H}\right), \qquad (4.8)$$

Table 4.1

Values of the geometrical factor G as computed from (4.6) for several values of the ratio l/w, together with the corresponding errors δ, when the G factor is not taken into account

l/w	0.5	1	2	3	4	5
G	0.394	0.683	0.930	0.985	0.997	0.999
δ (%)	60.6	31.7	6.96	1.46	0.3	0.06

where s is the length of the Hall terminals.

The term in the square brackets of (4.8) represents the influence of the supply current contacts, cf. (4.6), and the second term in (4.8) is associated with the Hall terminals. According to (4.8), when $l/w > 3$ and $s/w < 1/20$, the correction factor is $G \approx 1$.

The factor G of a short Hall sample with point Hall contacts is approximated by the expression [20]:

$$G \simeq 0.742 \frac{l}{w} \left[1 + \frac{\Theta_H^2}{6} \left(3.625 - 3.257 \frac{l}{w} \right) \right]. \tag{4.9}$$

The accuracy of (4.9) is about 1%, if $l/w < 0.35$ and $\Theta_H < 0.45$. At small Hall angles $\Theta_H \approx 0$, the expression (4.9) reduces to:

$$G \approx 0.742 \frac{l}{w}. \tag{4.10}$$

Increasing the magnetic induction B, i.e. the Hall angle Θ_H, effectively "lengthens" the plate. This is explained by the geometrical magnetoresistance effect which arises near the supply current contacts. The sample resistance in the neighborhood of the supply current contacts increases together with B. It can be seen from Fig. 4.5 that the increase in Θ_H leads to an increase in the voltage V_H measured in the middle of a short structure. This is equivalent to a "lengthening" of the sample.

The geometrical correction factor G of Hall sensors whose geometry is invariant under a 90° rotation (Fig. 4.1l–r) can be approximated by the general expression [23]:

$$G(P, \Theta_H) \simeq 1 - g(P) \frac{\Theta_H}{\tan \Theta_H}, \tag{4.11}$$

where P is the ratio of the sum of the contact lengths to the total length of the sample boundary and $g(P)$ is a sample geometry function which is specific for each particular shape [(l)–(r)].

If $(\Theta_H/\tan \Theta_H)$ in (4.11) is developed in a series and only the first two terms are considered, the following expression is derived in compliance with [24,25]:

$$G(P, \Theta_H) \simeq 1 - g(P) \left(1 - \frac{1}{3} \Theta_H^2 \right). \tag{4.12}$$

If in (4.12) the coefficients G_0 for $B = 0$ and β are introduced, an approximation similar to (4.9) is obtained:

$$G = G_0 (1 + \beta \Theta_H^2). \tag{4.13}$$

The comparison of (4.12) and (4.13) yields for G_0 and β of Hall plates with relatively small contacts:

$$G_0 \approx 1 - g(P); \tag{4.14}$$

$$\beta \approx \frac{1 - G_0}{3G_0}. \tag{4.15}$$

The above expressions are valid for an octagon if $P \leq 0.73$ and $G \geq 0.4$ (Fig. 4.1r). For long contacts the expressions for these introduced coefficients are:

$$G \approx 0.742 \frac{l}{w}; \tag{4.16}$$

$$\beta \approx 0.604 - 0.543 \frac{l}{w}. \tag{4.17}$$

Expressions (4.16) and (4.17) are good approximations, when $l/w < 0.35$ and $G_0 < 0.26$ [24,25]. Therefore, at small magnetic fields ($\mu B \ll 1$), the geometrical correction factor is expressed by means of a square function of the Hall angle Θ_H. At $\Theta_H \to \pi/2$ the G-factor should tend to unity irrespective of the shape of the device. Therefore, as the Hall angle is increased, the portion of the current lines which is parallel to the isolating boundaries grows longer. This is why in a high magnetic field the Hall voltage is less affected by the shunting effects, i.e. the Hall sensor is effectively longer than in the case of small values of Θ_H.

According to (4.8) the G-factor is influenced by the shape of the Hall terminals. From the standpoint of minimum disturbance of the electric-field distribution, the optimum output contacts are those shown in Fig. 4.1d, e, m, and n. They are not directly connected to the active transducing region. The point idealization of the Hall contacts is suitable for a theoretical analysis but the practical implementation of such contacts is not possible. In the general case, the Hall terminals have finite dimensions and have a shunting effect upon the Hall field E_H in the sample [26] (Fig. 4.3). The finite dimensions of the output contacts do not only reduce the effective length l^* of the sensor; the small current across them worsens the noise characteristics of the device [18].

The conclusions concerning the geometrical factor G as a measure of the amount of short circuiting effects, are valid for a Hall plate of arbitrary geometry, i.e. for any shape of Hall device.

4.2.4. Modes of operation of Hall sensors

As discussed in Chapter 2, there are two complementary methods of Hall-effect registration (output voltage or output current), irrespective of the manner of activation of the sensor by the magnetic field. In both cases the output is a linear function of the supply current or voltage and the magnetic field B.

4.2.4.1. Hall-voltage mode of operation

The Hall voltage in regular shaped structures in the absence of parasitic signals has been considered in Chapter 2. In this section the Hall voltage will be discussed in a homogeneous, arbitrarily shaped Hall plate, Fig. 4.6, with extrinsic conductivity under isothermal conditions. The results obtained will be valid for all the Hall configurations in Fig. 4.1. The supply current flows via contacts 1 and 3 located on two opposite sides of the sample, while contacts 2 and 4 are used for registration of the output voltage. The current in the device is represented by solid lines and the equipotential lines are dashed. The point contacts 2 and 4 have, in the general case, different potentials; therefore a certain voltage V_{off}, termed the offset, arises across the differential output 2–4 in the absence of an external magnetic field $\boldsymbol{B}$. Moreover, at the output a noise voltage $V_N(f)$ is generated which depends on the frequency band considered. When an external magnetic field $\boldsymbol{B}$ is applied, a Hall voltage $V_H \sim IB$ appears between terminals 2 and 4. The geometrical factor of the structure in Fig. 4.6 is $G < 1$ owing to the imperfect spatial confinement of the current I. This is why an additional signal V_{MR}^* is generated across terminals 2 and 4 by the physical and geometrical magnetoresistance. If the magnetic field is a function of the time t, an additional signal component $V_{\text{ind}}(B, t)$ appears at the output. Therefore the total output voltage between terminals 2 and 4 under isothermal conditions when $B \neq 0$ will be:

$$V_{2-4} = V_{\text{off}} + V_N(f) + V_H + V_{MR}^* + V_{\text{ind}}. \tag{4.18}$$

Only the voltage V_H is regarded as a useful signal; all other components are referred to as disturbances. The voltage V_{ind} can be virtually eliminated by a proper choice of the device's construction and the position of the supply and output wires. The output voltage is the integral of the electric field over an arbitrary line that connects terminals 2 and 4. For that purpose the area of the Hall plate is

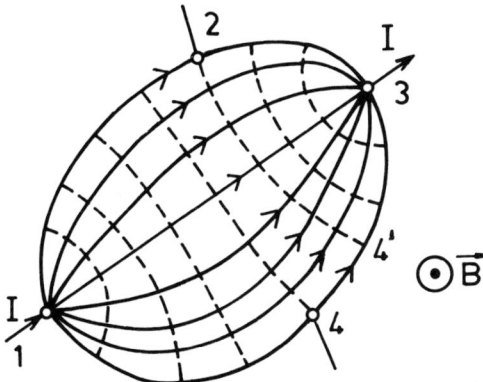

Fig. 4.6. An arbitrary shaped Hall plate with current contacts (*1* and *3*) and Hall contacts (*2* and *4*).

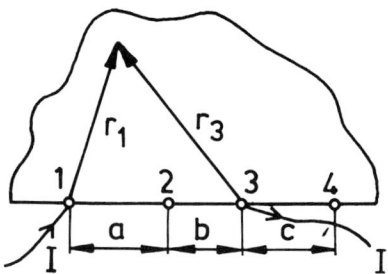

Fig. 4.7. Mapping of the plate from Fig. 4.6 upon the complex half-plane.

conformally mapped onto a half-plane (Fig. 4.7). The potential φ of a point at a distance r_1 from contact 1 and distance r_3 from contact 3 is given by the expression [27]:

$$\varphi = \varphi_1(r_1) + \varphi_3(r_3) = \frac{I}{\pi t} \int_{r_1}^{r_3} \frac{dr}{r} = \frac{I\rho}{\pi t} \ln \frac{r_3}{r_1}, \qquad (4.19)$$

where t is the thickness of the semiconductor plate and $I \equiv I_{1-3}$.

At $B = 0$, (4.19) yields:

$$V(B=0) = \frac{I(0)}{\pi a \sigma(0)} \ln \frac{b(a+b+c)}{ac}. \qquad (4.20)$$

The distances a, b, and c have been specified in Fig. 4.7.

According to (4.20) the input resistance of the Hall device in Fig. 4.6 is defined as $R(0) = V(0)/I(0)$. When an external magnetic field is applied, the Hall field is determined from the expression:

$$\boldsymbol{E}_H = \frac{\mu_H}{\sigma(B)} \boldsymbol{j} \times \boldsymbol{B}. \qquad (4.21)$$

The replacement of rot $\boldsymbol{E}_H$ in (2.40) by the corresponding expression from (4.21) yields:

$$\operatorname{rot} \boldsymbol{j} = 0. \qquad (4.22)$$

The current density at a constant current via contacts 1 and 3 is determined by solving the equations (2.39) and (4.22) under the boundary condition $j_n = 0$, (2.43). Therefore it can be concluded that the field $\boldsymbol{B}$ does not change the positions of the lines of current flow. According to (4.21) the Hall field $\boldsymbol{E}_H$ is perpendicular to the vector $\boldsymbol{j}$, while the externally applied field $\boldsymbol{E}$ and the current $\boldsymbol{I}$ are collinear with the current density $\boldsymbol{j}$. This is why for $\boldsymbol{B} \neq 0$ the dashed lines in Fig. 4.6 are the lines of the Hall field $\boldsymbol{E}_H$, and the solid lines are the lines of the applied external field $\boldsymbol{E}$ and the supply current $\boldsymbol{j}$. If we neglect the noise, the potential difference between contacts 2 and 4 is found from the general expression:

$$V(B) = \int_{L_{2-4'}} (\boldsymbol{E}_H \cdot \boldsymbol{u})dl + \int_{L_{4'-4}} (\boldsymbol{E} \cdot \boldsymbol{u})dl =$$
$$= V_H + V_{\text{off}}, \tag{4.23}$$

where the integration path $L_{2-4'}$ has been chosen in such a manner that anywhere between points 2 and $4'$ its tangent is collinear with the vector of the Hall field $\boldsymbol{E}_H$. This curve actually determines the Hall voltage V_H. The line $L_{4'-4}$ goes along the insulating boundary and coincides with the external field $\boldsymbol{E}$. Therefore the second integral in (4.23) determines the offset V_{off}; $\boldsymbol{u}$ is a unit vector, which is a tangent to both $L_{2-4'}$ and $L_{4'-4}$ and dl is an element of length [28].

Two additional unit vectors are introduced: $\boldsymbol{n}$ is the normal to the integration curve and $\boldsymbol{h}$ is collinear with the magnetic field $\boldsymbol{B}$, i.e. $\boldsymbol{B} = \boldsymbol{h}B$, where $\boldsymbol{n} = \boldsymbol{u} \times \boldsymbol{h}$. The expression (4.21) can be used to obtain:

$$\int_{L_{2-4'}} (\boldsymbol{E}_H \cdot \boldsymbol{u})dl = \frac{\mu_H B}{\sigma(B)} \int_{L_{2-4'}} (\boldsymbol{j} \cdot \boldsymbol{n})dl =$$
$$= \frac{\mu_H B}{\sigma(B)t} I(B), \tag{4.24}$$

since $t \int_{L_{2-4'}} (\boldsymbol{j} \cdot \boldsymbol{n})dl = I(B)$.

The problem of finding the electric field $\boldsymbol{E}$ when $\boldsymbol{B} \neq 0$ is identical with that of determining the field $\boldsymbol{E}$ when $\boldsymbol{B} = 0$, the only difference being that $\sigma(0)$ and $I(0)$ must be replaced by $\sigma(B)$ and $I(B)$, respectively. If these substitutions are taken into account, the second term on the right-hand side of (4.23) becomes equal to the right-hand side of (4.20). The resistance of the plate in a magnetic field will be $R(B) = V(B)/I(B)$. From (4.20), (4.23) and (4.24) is found the variation $\Delta R(B) = R(B) - R(0)$ of the resistance due to the external magnetic field:

$$\Delta R(B) = \frac{\mu_H B}{\sigma(B)t} + \frac{1}{\pi t \sigma(0)} \left[1 - \frac{\sigma(0)}{\sigma(B)} \right] \ln \frac{b(a+b+c)}{ac}. \tag{4.25}$$

The first term on the right-hand side of (4.25) is associated with the Hall voltage between contacts 2 and 4. The second term results from the change in the offset signal between contacts 2 and 4 when a magnetic field $\boldsymbol{B} \neq 0$ is applied. This result is of fundamental importance for the operation of Hall sensors in high magnetic fields, and if it is not taken into account a registration error occurs. The offset equals zero at arbitrary values of the magnetic field B if at $B = 0$ terminals 2 and 4 are equipotential. In a low magnetic field, to a first approximation, $\sigma(B) \approx \sigma(0)$ and the second term in (4.25) can be neglected even if the positions of the output contacts are not symmetrical. If the contacts of the structure in Fig. 4.6, instead of being on the boundary surface, are located at a certain distance from it (this case is often met with when the Hall effect is used for semiconductor wafer characterization), the registered signal includes both the Hall signal V_H and the geometrical magnetoresistance voltage V_{MR}^*. When the distance from the wafer boundary to the contact is large enough, only physical and geometrical magnetoresistance occur. This result is explained by the fact that the Hall field is the consequence of the accumulation of charge upon the boundary surfaces, and in the configuration discussed the registration probes are far from them. To complete the discussion it must be mentioned that the parasitic effects arising at the Hall terminals when the Hall voltage V_H is used for material characterization, to a first approximation, can be eliminated by the measurement of $V_H(+I, +B)$, $V_H(+I, -B)$, $V_H(-I, +B)$ and $V_H(-I, -B)$, with a subsequent averaging of the results obtained.

4.2.4.2. Basic characteristics of Hall sensors

(a) Hall voltage. According to (4.23) and Fig. 4.6, the first term is the Hall voltage between 2 and 4'. Equation (2.27) in its vector form yields for V_H:

$$V_H = \int_{L_{2-4'}} R_H [\boldsymbol{j} \times \boldsymbol{B}] dl. \qquad (4.26)$$

For a homogeneous semiconductor plate and an orthogonal field $\boldsymbol{B}$, equation (4.26) can be written in the form:

$$V_H = \frac{R_H}{t} B \int_{L_{2-4'}} j t \, dl. \qquad (4.27)$$

The integral in (4.27) is the total supply current in the sensor. Therefore:

$$V_H = \frac{R_H}{t} I B. \qquad (4.28)$$

Expression (4.28) is the same as (2.34). In the real Hall devices presented in Fig. 4.1, where all contacts have finite dimensions, the voltage V_H is corrected by the geometrical factor G, cf. §4.2.3:

$$V_H = G \frac{R_H}{t} I B. \qquad (4.29)$$

When the extrinsic conductivity is well pronounced, in a n-type semiconductor for instance, the Hall voltage, in compliance with Chapter 2, is determined from the formula:

$$V_H = G \frac{r_H}{qnt} I B. \qquad (4.30)$$

Hall sensors are often supplied by a constant voltage V. By making use of (2.34) and assuming a low field ($\mu_H B \ll 1$), one can obtain the following formula for V_H:

$$V_H = \mu_H \left(\frac{w}{l}\right) G V B. \qquad (4.31)$$

It can be seen from (4.31) that, at $V = $ const, the Hall voltage increases with an increase in the ratio w/l. For a very short sample the G-factor is determined from (4.9). The maximum value of V_H with a fixed product VB can be obtained from (4.31) as well:

$$V_H \approx 0.742 \mu_H V B. \qquad (4.32)$$

For instance, for $V = 1$ V and $B = 1$ T at $T = 300$ K in a n-type silicon sample ($\mu_{H_n} \approx 0.15$ m^2 V^{-1} s^{-1}), the Hall voltage determined from (4.32) is $V_H \approx 0.126$ V. For n-type GaAs ($\mu_{H_n} \approx 0.9$ m^2 V^{-1} s^{-1}) the corresponding value is $V_H \approx 0.67$ V. In fact, the above results show that the maximum Hall voltage which can be generated by a n-type Si sensor is about 12% of the supply voltage V at $B = 1$ T and for an n-type GaAs sensor the corresponding value under the same conditions is about 67% of the supply voltage [24,25].

The efficiency η is of considerable importance for the operation of Hall sensors:

$$\eta = \frac{W_{\text{out}}}{W_{\text{in}}}, \qquad (4.33)$$

where W_{out} is the power dissipated by the load R_L connected to the sensor output, and W_{in} is the power transferred to the additional second input. The corresponding expressions for W_{out} and W_{in} are the following:

$$W_{\text{out}} = I_{\text{out}}^2 R_L; \quad W_{\text{in}} = I_{\text{in}}^2 R_{\text{in}}, \qquad (4.34)$$

where R_{in} is the internal resistance, measured between the supply contacts of the sensor at $B = 0$. The output current is $I_{\text{out}} \equiv I_H = V_H/(R_s + R_L)$, where R_s is the output resistance, measured between the two sensor contacts. As a result the relationship $W_{\text{out}} = V_H^2 R_L/(R_s + R_L)^2$ is obtained. When the output resistance R_s and the load R_L are equal ($R_s = R_L$), by making use of (4.28), is obtained the power W_{out}:

$$W_{\text{out}} = \frac{I_{\text{in}}^2 B^2 R_H^2}{4t^2 R_s}. \qquad (4.35)$$

In accordance with (4.33)–(4.35) the parameter η is determined from the formula:

$$\eta = \frac{R_H^2 B^2}{4t R_{\text{in}} R_s}. \qquad (4.36)$$

Since R_{in} and R_s are proportional to $\rho = 1/q\mu_n n$ (for an n-type semiconductor), and the Hall coefficient R_H according to (2.33) is proportional to $1/qn$, the following equation is valid for η:

$$\eta = m \left(\frac{\mu_n B}{t}\right)^2, \qquad (4.37)$$

where the coefficient m varies in the range $2 \leq m \leq 5$ in agreement with [29]. Therefore, the efficiency η is a square function of carrier mobility.

The maximum value of the Hall voltage $V_{H,\text{max}}$ is determined by the maximum allowable supply current I_{max} via the Hall transducer. This current determines the maximum temperature of operation. In the stationary mode the supplied power must be equal to the power dissipated by the sensor:

$$I_{\text{max}}^2 R_{\text{in}} = \alpha A \Delta T, \qquad (4.38)$$

where α is the heat dissipation constant, A is the area of the device involved in heat dissipation, and ΔT is the difference between the maximum allowed temperature and the temperature of the environment. The current $I_{\max}$ can be determined from (4.38) by assuming a parallelepiped-shaped device, for instance that in Fig. 4.1b, with a resistance $R_{\text{in}} = \rho l/wt = l/wtq\mu_n n$; the voltage $V_{H,\max}$ is obtained in compliance with (4.28):

$$V_{H,\max} \simeq wAB\left(\frac{2\alpha\mu_{Hn}\Delta T}{qnt}\right)^{1/2}. \tag{4.39}$$

It can be inferred from (4.30) and (4.31) that the functional relationship between the Hall voltage V_H and the dissipated input power $W_{\text{in}} = IV$ is:

$$V_{H\max} = G\left(\frac{w}{l}\right)^{1/2} r_H \left(\frac{\mu_{Hn}}{qnt}\right)^{1/2} W_{\text{in}}^{1/2} B = G\left(\frac{w}{l}\right)^{1/2} \mu_{Hn} \left(\frac{W_{\text{in}}\rho}{t}\right)^{1/2} B. \tag{4.40}$$

The factor $\mu_H \rho^{1/2}$ in (4.40) is an important parameter associated with the technology. If the geometrical correction factor $G.(w/l)^{1/2}$ is optimized by taking into account the corrections for small Hall angles and ideal point contacts $s \to 0$ [20], equation (4.40) yields the optimum geometry of rectangular Hall structures: $l/w \approx 1.35$, where $G = 0.81$. These results are presented in Fig. 4.8 [20]. It can be seen from the figure that the Hall voltage has a maximum when the supply current has a fixed value and the sample is long ($l/w > 3$), or the input voltage has a fixed value and the Hall sensor is short ($l/w < 0.1$), or when the dissipated power has a fixed value and the length to width ratio of the sensor is $l/w \approx 1.3$ to 1.4.

(b) Offset voltage. It can be seen from equation (4.23) and Fig. 4.6 that the second term is the offset voltage and in its nature it is always a d.c. signal. In a parallelepipedal sample (Fig. 4.9) with $B = 0$, the voltage V_{off} is determined from the equation:

$$V_{\text{off}} = \rho \frac{I}{wt}\Delta l, \tag{4.41}$$

where Δl is the geometrical misalignment of the sensor contacts with respect to the equipotential plane.

The origin of the offset presented in Figs. 4.6 and 4.9 is not the only factor responsible for the appearance of this imperfection. The reasons that cause it can be divided into external, associated with external influences on the Hall structure, and internal, associated with the properties of the device itself.

The external reasons are:

(1) Crystal damage due to the fabrication process as a result of which regions of local resistivity variation appear in the active sensor volume.

(2) Mechanical stress and strain in the integrated sensor chip after packaging [30,31]. The substrate resistivity becomes nonuniformly distributed because of the piezo-resistance effect. On the other hand, a high-temperature sensitivity is

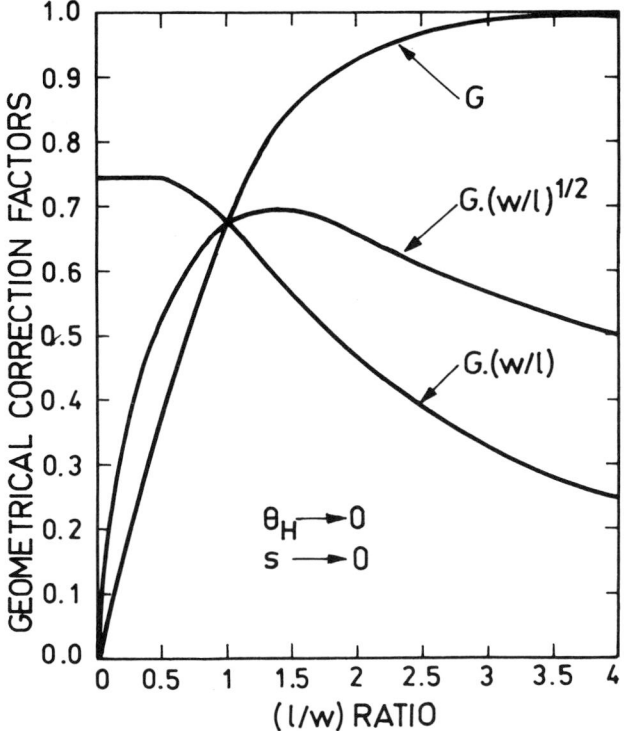

Fig. 4.8. Low-field geometrical correction factor for rectangular silicon Hall plates with point-contact Hall terminals as a function of l/w [20].

characteristic of piezo-resistance thus leading to an additional offset.

(3) Tolerance of geometry, misalignment of Hall terminals in particular, which generates an extra voltage at the output (Figs. 4.6, 4.9). Only this reason for the offset is commonly discussed in published works, thus putting a limit to a deeper understanding of the problem.

The most important internal reasons for the offset are:

(4) Nonuniform temperature distribution in the Hall structure due to the supply power. The temperature gradient across the sensor results in a resistivity imbalance and an additional offset. The temperature difference is proportional to I^2 and will introduce between the Hall terminals a thermoelectric potential proportional to I^2.

– Nonuniform heat dissipation as a result of the variation in sheet resistivity across the structure due to uneven thickness or a nonuniform distribution of doping impurities.

– Nonuniformities of the temperature coefficient of the resistivity of the Hall device, which depends on dopant density. Therefore even homogeneous external heating can change the offset. The reason mentioned in point 4 can influence the

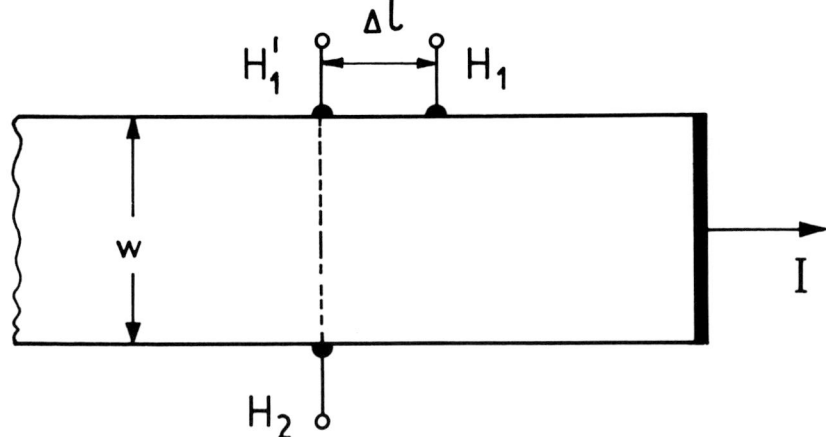

Fig. 4.9. Arising of geometrical offset Δl of the sensor contacts with respect to the equipotential line $H_1' H_2$.

offset only in the temperature gradient; the thickness gradient or the density of doping impurities gradient has a component parallel to the length of the device. At high supply voltages these effects are the dominating reason for the additional offset.

(5) Generation of a thermoelectric voltage across the Hall terminals when they have different temperatures (Seebeck effect). The generated voltage is $V_{th} = \alpha_s \Delta T$, where ΔT is proportional to the heat obtained. The heat in turn is proportional to V^2, and α_s is the Seebeck coefficient. In silicon, for instance, this parameter is $\alpha_s \approx 0.7$ mV K^{-1} and, when ΔT is several degrees, the thermoelectric voltage may reach about 2 mV. With a standard magnetosensitivity of 0.7 V T^{-1} this value corresponds to a parasitic magnetic induction $B \approx 3$ mT.

(6) If heat is generated at one of the supply contacts and absorbed at the other one (Peltier effect), a longitudinal heat flux arises in the sensor and a thermoelectric voltage proportional to the supply current I appears across the Hall terminals.

(7) The supply current in the Hall sensor can generate a magnetic field of its own. The distribution of its magnitude along the width of the device is investigated in [32]. The self-induced Hall voltage is:

$$V_H = \int_{-w/2}^{w/2} R_H j B_z dy. \tag{4.42}$$

According to (4.42) this "offset" disappears if the bias current density is evenly distributed. But if there is a concentration gradient of doping impurities along the y-axis, the parasitic signal arises. In Hall measurements this effect can be neglected, since at $I = 10$ mA, for instance, when $w = 250$ μm and $t = 10$ μm, the doping density varies by 50% across the whole wafer (5 cm), and the self-induced magnetic field is about 20 nT.

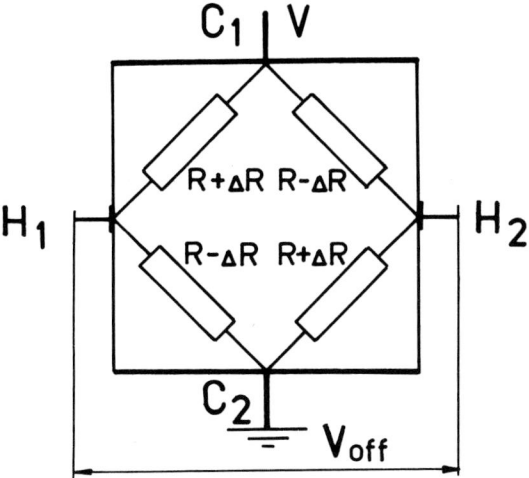

Fig. 4.10. Equivalent bridge circuit model of a Hall sensor.

(8) When alternating magnetic fields are measured, an a.c. voltage $V_{\text{ind}} = L\partial B/\partial t$ is induced in the Hall plate between the output terminals and the connecting wires, where L is the circuit inductance. Although this is not an offset in the exact meaning of the word, since at $B = 0$ the induced voltage is $V_{\text{ind}} = 0$, this parasitic signal, nevertheless, is a source of registration errors. For instance, an alternating magnetic field whose amplitude is $100mT$ at a frequency of 100 Hz will induce a voltage of about $440\mu V$ in a section with a 1 mm^2 surface [32].

All the reasons for the offset listed above can be illustrated by the simple bridge circuit (the offset model) of the Hall structure shown in Fig. 4.10 [33]. There is a relationship between the offset and the asymmetry $\pm \Delta R$ of the bridge:

$$V_{\text{off}} = \frac{\Delta R}{R} V_{\text{in}}. \tag{4.43}$$

In the ideal case, when the values of all resistors are equal, there is no offset, i.e. $V_{\text{off}} = 0$.

(c) Noise voltage. It has been pointed out in §3.3.1 that random voltage or current fluctuations are generated across the output terminals of magnetosensors in the absence of a magnetic field, cf. (3.7) and (3.8). The noise voltage depends on the frequency band f_1 to f_2. It has both a generation origin and a modulation origin. The generation noise is associated with the internal fluctuations of the electromotive force, and to a first approximation, is independent of the biasing mode. Thermal noise belongs to this type of noise. The modulation noise voltage is determined by the fluctuations of the device parameters, mainly the conductance, i.e. by the current flow in the sensor. The shot noise and the noise due to carrier generation and recombination are modulation noises.

(1) The thermal noise in Hall sensors is determined by the absolute temperature according to the well-known Nyquist relation: $S_{NV}(f) = 4k_B T R(f)$. A semiempirical formula can be found in [26] for S_{NV} in a rectangular sample:

$$S_{NV}(f) \simeq 8k_B T (\pi \mu_n q N t)^{-1} \ln\left(\frac{w}{s}\right), \tag{4.44}$$

where N is the total number of carriers, the other notations having already been defined.

Generally, the impedance of a Hall sensor does not manifest reactance at low frequencies and the dynamic output resistance is frequency independent. The thermal noise comes from a white-noise source and prevails at high frequencies. At the same time the thermal noise is directly dependent on the shape and geometry of the device. Thermal noise can also depend on the supply voltage if the internal resistance of the device is controlled by this voltage. This is the case, for instance, with MOSFET, JFET or MESFET Hall sensors, where the gate voltage together with the processing parameters determine the channel resistance. Thermal noise is often dominates in MAGFETs.

(2) The shot noise is associated with the discrete nature of electricity. It can be observed when the bias current is controlled by charge transfer across a potential barrier. This is usually the collector current in bipolar magnetotransistors. According to Schottky's relationship, the spectral density of the shot noise is given by the formula:

$$S_{NI}(f) = 2q I_0, \tag{4.45}$$

where I_0 is the bias current.

It can be seen from (4.45) that this kind of noise originates from a source of white noise whose magnitude directly depends on the mean biasing current I_0 in any sensor.

(3) The voltage of the generation–recombination (G–R) noise is due to carrier generation or recombination. In fact it is related to the number of quasi-free carriers in the transducer. Since generation and recombination are independent random processes, the instantaneous carrier concentrations fluctuate about their equilibrium values with a "Lorentzian" spectrum:

$$S_{NI}(f) \sim \frac{I_0^2 \tau}{1 + (2\pi f \tau)^2}, \tag{4.46}$$

where τ is the minority-carrier lifetime.

The fluctuating voltage which corresponds to the G–R noise plays a substantial role at relatively low frequencies. The trapping phenomena due to impurities and defects in GaAs Hall sensors serve as an example.

(4) The $1/f$ noise has a conductivity-modulation origin and prevails in magnetosensors at low frequencies. The following relationship between the voltage and the spectral density of the $1/f$ noise has been established experimentally:

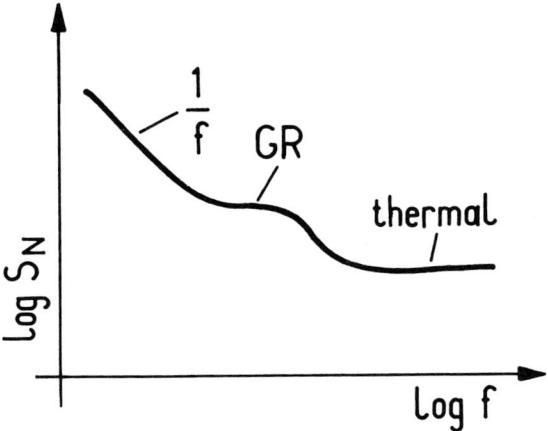

Fig. 4.11. A typical current noise spectrum of a Hall sensor or magnetoresistor.

$$S_{NV}(f) \simeq \alpha \left(\frac{V_0}{l}\right)^2 (2\pi N t f)^{-1} \ln\left(\frac{w}{s}\right), \tag{4.47}$$

where V_0 is the supply voltage, N is the total number of carriers in the sensor, and α is a dimensionless coefficient, the so-called Hooge parameter.

In fact, α is a characteristic rather of specific processes in the device than of the material itself. It varies through a wide range of values $10^{-9} \leq \alpha \leq 2 \times 10^{-3}$ [35,36]. As yet there is no generally accepted explanation of the origin of the $1/f$ noise. Discussions as to whether it results from surface trapping or from fluctuations of carrier mobility in the bulk still continue.

In Hall sensors as well as in the other types of magnetotransducers the different kinds of noise briefly discussed above coexist. The spectral density of the resultant noise is a sum of the individual contributions, but only the thermal noise, the G–R noise and the $1/f$ noise, in particular, are of practical importance. Detailed information concerning noise can be found in [24,34–41]. A typical noise spectrum of a Hall sensor is presented as an illustration in Fig. 4.11.

(d) Signal-to-noise ratio. It was shown in Chapter 3 that the signal-to-noise ratio (SNR) is a parameter of principal importance for the practical application of solid-state magnetosensors, including Hall devices. In a constant magnetic field the offset and the noise voltage cannot be distinguished from the useful Hall signal. For convenience a "signal-to-noise ratio" can be introduced for the offset: $\text{SNR}(f = 0) = V_H/V_{\text{off}}$, and at high frequencies $\text{SNR}(f) = V_H(f)/[S_{NV}(f)\Delta f]^{1/2}$, where Δf is a narrow frequency interval which includes a fixed frequency f_0; $S_{NV}(f)$ is the voltage of the noise spectral density, (4.44)–(4.47). The expressions for the Hall voltage (4.28), for the offset (4.41) and the offset SNR can be used to obtain:

$$\text{SNR}(f=0) = \mu_H B \frac{w}{\Delta l}. \tag{4.48}$$

The expression (4.48) is correct for very long Hall devices with very small output contacts and $\Delta l \neq 0$. It has been pointed out in section (c) that at low frequencies the $1/f$ noise dominates in Hall sensors. The expression (4.47) can be used to obtain the SNR [24,26,42]:

$$\text{SNR}(f) \simeq \frac{\mu_H}{\alpha^{1/2}} \left[\frac{2\pi Ntlw}{\ln(w/s)}\right]^{1/2} G \left(\frac{w}{l}\right)^{1/2} \left(\frac{f}{\Delta f}\right)^{1/2} B. \tag{4.49}$$

At high frequencies the thermal noise dominates and the corresponding SNR according to (4.44) will be:

$$\text{SNR}(f) = \left(\frac{\pi \mu}{8 k_B T q N t \ln(w/s) \Delta f}\right)^{1/2} r_H G I B, \tag{4.50}$$

where I_0 is the bias current.

From (4.48)–(4.50) follows the important conclusion that the SNR of Hall devices can be increased by the use of semiconductor materials with high carrier mobility.

(e) Sensitivity. The absolute and relative magnetosensitivities of modulating sensors such as Hall devices, for instance, have been defined according to (3.1)–(3.3). The expression (4.29) for the Hall voltage can be used to obtain the supply-current-related sensitivity S_I:

$$S_I = G \frac{r_H}{qnt}. \tag{4.51}$$

If the Hall plate is nonuniformly doped over its thickness t, the product nt in (4.51) must be replaced by the charge density N_s of the carriers upon the surface, hence:

$$S_I = G \frac{r_H}{q N_s}. \tag{4.51'}$$

The substitution made is correct only if the carrier mobility and the Hall factor do not vary across the active layer of the device [24]. Such an assumption is approximately valid in low-doped layers with a small gradient of impurity concentration. The typical values of S_I are within the $80 \leq S_I \leq 400$ V A^{-1} T^{-1}-interval [24]. There is a strong dependence of the figure of merit S_I on the material of the plate, since $r_H \approx 1$, irrespective of the material.

The supply voltage-related sensitivity S_V is defined similar to (3.3). In the case of highly doped materials the following formula is obtained according to (4.31):

$$S_V = \mu_H \frac{w}{l} G. \tag{4.52}$$

In silicon Hall devices typically $S_V \approx 0.07$ T^{-1}, and in GaAs sensors $S_V \approx 0.2$ T^{-1} [24]. The voltage-related sensitivity increases as the effective length of the structure decreases, yet there is an upper physical limit. The following is obtained with the aid of (4.32):

$$S_{V,\max} = 0.742\mu_H. \tag{4.52'}$$

In n-type Si Hall devices this value is 0.126 T^{-1} at $T = 300$ K and in n-type GaAs sensors the corresponding value is 0.67 T^{-1}.

(f) Hall-voltage nonlinearities. The nonlinearity definition in §3.2.2, when adapted to the output Hall characteristic, acquires the form:

$$\text{NL} = \frac{V_H - V_{H_0}}{V_{H_0}}, \tag{4.53}$$

where V_H is the real output voltage measured experimentally and V_{H_0} is the appropriate value taken from the straight line which best approximates the experimental $V_H(B)$ relationship.

When $B \to 0$ and $I \to 0$, in compliance with [26] the characteristic V_{H_0} is given by the formula:

$$V_{H_0} = \left[\frac{\partial^2 V_H(B,I)}{\partial B \partial I}\right] BI. \tag{4.54}$$

The meaning of the second derivative in (4.54) is the initial sensitivity S_0 of the Hall device. In the general case $V_H = S_0 BI$, where the coefficient S_0 may depend on B and I, and (4.53) can be written in the form:

$$\text{NL} = \frac{S(B,I) - S_0}{S_0} = \frac{\Delta S}{S_0}. \tag{4.55}$$

Equations (4.29) and (4.55) can be used to obtain the magnetosensitivity:

$$S = \frac{R_H}{t} G. \tag{4.56}$$

If some of the factors in (4.56) depend on the magnetic field or on some supply current, a nonlinearity occurs at the sensor output. Depending on whether the reason for the nonlinearity lies in R_H, G or t, the following types of nonlinearity can be distinguished: nonlinearity due to the material, nonlinearity due to the geometry of the device, and in the case of an integrated sensor, nonlinearity due to the influence of the electric field across the p–n junction which isolates the Hall sensor from the substrate [43]. The influence of the geometrical correction factor G, i.e. of the shunting effect of the contacts, has been the subject of a great proportion of the investigations on nonlinearity. Structures shaped like those in Fig. 4.1d and m have a geometrical factor $G \approx 1$ and its effect upon the nonlinearity NL can be neglected. The effect of the field B and the supply current I on the Hall factor, and hence on NL, has not been studied in detail. From the analysis in [43] it is known that in a low magnetic field ($\mu B \ll 1$), the influence of the scattering factor r_H upon R_H (according to (2.28) $R_H \sim r_H$) and upon the nonlinearity is expressed by NL $= -\gamma \mu_H^2 B^2$, where the value of the coefficient γ is $\gamma \approx 1$.

The main source of nonlinearity in integrated Hall sensors is the electric field across the p–n junction. It is associated with the modulation of the Hall microsensor thickness t by the Hall voltage itself: the voltage V_H modulates the potential difference across the isolating junction. The variation in this potential determines the width of the p–n junction space-charge region, hence the effective thickness t^* of the Hall device. This is why the Hall voltage and the magnetosensitivity depend on t^*. This type of nonlinearity is proportional both to the magnetic induction and to the supply current. For n-type materials NL $\simeq (S_0^2/Gr_n)(\varepsilon q n/2(V_{bi} - V_H))^{1/2} BI$, where ε is the permittivity of the material and V_{bi} is the built-in voltage [43].

The nonlinearities described above refer to the static output characteristic $V_H(B)$ of the Hall sensor when the active resistance R_L of the load between terminals H_1 and H_2 is much higher than the output resistance of the transducer. The nonlinearity when the Hall device is operating in the dynamic mode, i.e. when the Hall current I_H flows across the load R_L, is discussed in §9.5.3.

(g) Temperature coefficient of magnetosensitivity. According to the definition in §3.3.3 and (3.13), (3.14), the temperature coefficient of the magnetosensitivity of Hall sensors is T.C. $\equiv (1/S)(\partial S/\partial T)$, where S is the sensitivity according to one of the definitions in (3.1)–(3.3). Depending on the supply mode (constant current or constant voltage), (4.52) can be used to define two temperature coefficients:

$$\text{T.C.}_I = \frac{1}{r_H}\frac{\partial r_H}{\partial T}; \quad \text{T.C.}_V = \frac{1}{\mu_H}\frac{\partial \mu_H}{\partial T}. \quad (4.57)$$

The formulae for T.C.$_I$ and T.C.$_V$ are valid at a constant carrier concentration ($n = $ const), i.e. the semiconductor material must be in the temperature saturation range where $n = N_D$ is independent of temperature. Moreover, a small Hall angle and a negligible junction field effect are assumed. The coefficient T.C.$_I$ in (4.57) has the same behavior with temperature as the Hall scattering factor r_H. Its value is $(1/r_H)(\partial r_H/\partial T) \approx 10^{-3}$ K^{-1} in n-type Si with a donor concentration of 1.75×10^{14} to 2.1×10^{15} cm^{-3} in the temperature range 200 K $\leq T \leq$ 400 K. In n-type GaAs with n in the 3×10^{15}- to 1.6×10^{17}-cm^{-3} interval, the corresponding value of the temperature coefficient is $(1/r_H)(\partial r_H/\partial T) \leq 0.3 \times 10^{-3}$ K^{-1} in the temperature range 200 K $\leq T \leq$ 320 K [26].

The temperature coefficient T.C.$_V$ (4.57) has the same behavior as the Hall mobility μ_H, where $\mu_H \sim T^{-2.4}$ for phonon scattering. The corresponding temperature coefficient at $T \approx 300$ K is $(1/\mu_H)(\partial \mu_H/\partial T) \approx -8 \times 10^{-3}$ K^{-1}.

(h) Frequency response. The Hall coefficient R_H is practically independent of the frequency; therefore the phase and amplitude of the Hall voltage do not change if the supply current, the magnetic field or both parameters change at high frequencies [32]. The following effects are important in the analysis of the frequency dependence of the Hall voltage:

(a) the skin effect;
(b) excitation of eddy currents in the plate;

(c) the inducing of an additional a.c. voltage across the Hall terminals.

Owing to the skin effect, the effective thickness of the semiconductor plate depends on the frequency f. Therefore the Hall voltage V_H becomes dependent on the frequency. The skin effect is minimized by reducing the ratio of the thickness t to the depth of penetration $\delta = (\rho/2f\mu)^{1/2}$ of the normal component of the electric field; μ is the permeability of the material. It has been established that in the frequency interval $0 \leq f \leq 50$ MHz, the skin effect does not influence V_H in GaAs samples, but if the frequency of operation of the device is several GHz, the thickness must be about 10 μm [32].

Alternating magnetic fields can induce eddy currents in the semiconductor plate even at relatively low frequencies. The induced current is closed via the current contacts and generates a component which is added to the alternating field B and has the same frequency, but its phase differs by $\pi/2$. This additional magnetic field in the presence of a constant bias current creates an extra voltage at the output which is added to V_H. Owing to their circulation origin, the eddy currents do not directly influence the Hall signal. The corresponding Hall fields on the two floating surfaces are equal in magnitude but with opposite signs, thus cancelling each other. The width w is the sensor dimension that most strongly influences the eddy currents.

In an alternating magnetic field, parasitic voltages of the type $V_{HB} = -d\Phi/dt = -AdB/dt$ may be generated across the current and Hall contacts, where Φ is the magnetic flux inside the corresponding loop, with an effective area A, which is determined by the layout of the connection leads. This negative effect can be compensated for by avoiding loops or twisted leads. A specific feature of Hall sensors operating at ultra high frequencies is their ability to play the role of multipliers of the electric component E and the magnetic component B of the electromagnetic field (self-generating behavior) without an external power supply.

4.2.4.3. Hall-current mode of operation

It has been shown in §2.3.2 that the Hall current I_H in short samples ($l \ll w$) is an equivalent alternative to the Hall voltage and can also be used as a sensor signal to measure the magnetic field. In the Hall-current mode of operation, the Lorentz deflection manifests itself the most prominently. In compliance with Van der Pauw's dual principle [44], the samples in Fig. 2.9b and Fig. 2.9c are equivalent. As the theory of this mode of operation has been discussed in detail in §2.3.2, this paragraph will describe devices for practical applications. The output signal I_H must be transferred from the Hall structure to an appropriate data-acquisition system in order to be used. Otherwise the current I_H flows via the ohmic contacts and the power supply. To prevent this effect one of the current contacts is divided into two parts by an isolation slit whose width is δ (Fig. 4.12a). In this case half of the total current, $I_H/2$, in the structure plays the role of a differential signal,

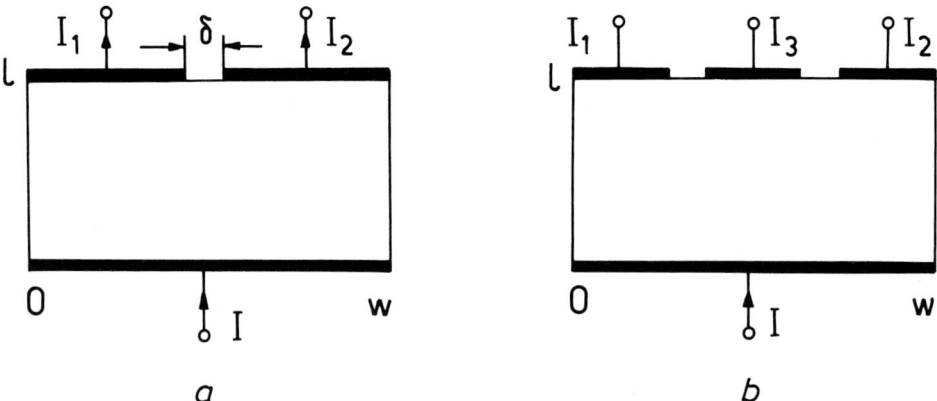

Fig. 4.12. Hall-current sensor: (a) with dual terminals and (b) with triple terminals; the output is I_1 and I_2, respectively.

$\Delta I_H \equiv I_H/2$, between the two parts of the split contact. The other half of the total Hall current flows via the lower terminal which has not been split. Different versions of a circuit for the registration of the current I_H are shown in Figs. 9.4 and 9.5. If the semiconductor plate is homogeneous, if there is uniform contact upon the surface and isothermal conditions for the measurement of the signal are provided, and if the magnetoresistance and the surface leakage currents can be neglected, the following relationship holds good for the Hall current:

$$\Delta I_H = \left(\frac{l}{w}\right) \mu_H B I, \tag{4.58}$$

where l and w are the short-sample length and width respectively, μ_H is the Hall mobility, I is the total bias current of the sensor $I = I_1 + I_2$, Fig. 4.12a, and B is the magnetic induction [45].

The values of ΔI_H can be used to obtain the mobility from (4.58).

The analysis of the most importance figures of merit in the Hall-current mode of operation is analogous to that for the Hall-voltage operation mode (cf. §4.2.4.2b–h). The relative magnetosensitivity S_{RI} of the sensor in Fig. 4.12 is determined from the expression:

$$S_{RI} = \left|\frac{I_1(B) - I_2(B)}{I_1(B) + I_2(B)}\right|, \tag{4.59}$$

since $I_1(B) + I_2(B) = I_1(0) + I_2(0) = I(0)$.

Another device structure designed for operation in the Hall-current mode is presented in Fig. 4.12b. Unlike the device in Fig. 4.12a, the upper metal contact has been partitioned into three parts, the external contacts being the output terminals of the sensor [46]. Such a device has a higher magnetosensitivity S_{RI}. If the magnitude of ΔI_H is approximately that of the current of the transducer in Fig. 4.12a, in the device in Fig. 4.12b, the middle contact leads to a considerable reduction of the initial current component $I(0)$ in (4.59). As a result, the relative variation of the useful output signal is increased. The output terminals of the sensors in Fig. 4.12, in the absence of a magnetic field and when $B \neq 0$, must be equipotential in order to eliminate the leakage currents. For this purpose the circuits in Figs 9.4 and 9.5 can be used. These two orthogonally activated devices are compatible with IC technology and are suitable for the implementation of integrated sensors, by means of MOS technology for instance.

Long samples ($l \gg 0$) can operate in the Hall-current mode too. For this purpose the voltage V_H must be shunted by a measuring instrument with a low resistance. In this case the generated current is $I_H = 0.742 \mu_H B I$ [27].

4.2.5. Requirements as to Hall sensor materials, fabrication technology and packaging

The analysis in §2.3.1 and §4.2.4.2 shows that the Hall structure is a good magnetosensor if it is prepared from a material with a low concentration of doping impurities and high carrier mobility. Semiconductors are the only materials that meet these requirements and because of this fact only semiconductor Hall devices are used in magnetometry for the practical purposes. The grounds for such a conclusion are the following:

(i) According to equation (4.28), the intensity of the Hall effect is proportional to the Hall coefficient $R_H \sim 1/qn$, cf. (2.33). This coefficient, in its turn, is inversely proportional to the carrier concentration n. This is why materials with a low concentration of doping impurities are to be preferred.

(ii) According to (4.31), the Hall voltage explicitly depends on the carrier mobility μ ($V_H \sim \mu$). Therefore the sensor material used should necessarily have high carrier mobility.

(iii) On the other hand, the carrier concentration n and mobility μ are important parameters related to power dissipation, cf. (4.40), i.e. $V_H \sim W^{1/2}(\mu/n)^{1/2}$.

Table 4.2 presents a summary of the main physical parameters of the semiconductor materials widely used for fabrication of Hall sensors. It can be seen from the table that the mobility μ_n of electrons is much higher than the mobility μ_p of holes. That is why n-type semiconductors are most often used in Hall devices. The corresponding values for the case of intrinsic conductivity are also given in Table 4.2.

In this respect the bandgap width has a considerable negative influence upon sensor materials, and on their behavior with temperature in particular. $A^{III}B^V$ compounds are preferred in order to achieve a high value of the Hall angle $\Theta_H \sim \mu B$. But the high temperature sensitivity of the Hall coefficient in these narrow-bandgap materials makes them applicable only within a narrow temperature interval where they manifest extrinsic conductivity. The absence of an IC technology for the fabrication of $A^{III}B^V$ sensors (except for GaAs) entails the need for discrete

Table 4.2

Important material parameters of selected semiconductors for galvanomagnetic sensors at $T = 300$ K

Material	E_g (eV)	μ_n (m^2 V^{-1} s^{-1})	μ_p (m^2 V^{-1} s^{-1})	ρ ($\Omega\cdot$cm)	$-R_H$ (cm^3 C^{-1})	n (cm^{-3})	$\mu_H\sqrt{\rho}$	$\alpha(RT)$ (K^{-1})	Remarks	Refs.
1. Ge	0.67	0.39	0.19	45	9×10^4	2.4×10^{13}			intrinsic	[47]
2. Si	1.12	0.14	0.05	2.3×10^5	4×10^8	1.5×10^{10}		$+5 \times 10^{-3}$	intrinsic	[47]
3. Si	1.12	0.13		2	2.5×10^3	2.5×10^{15}	1840		n-type	[48]
4. GaAs	1.42	0.85	0.045	7.8×10^7	6×10^{11}	9.2×10^6		$+8 \times 10^{-4}$	intrinsic	[49]
5. GaAs	1.42	0.64		0.32	2.1×10^3	3×10^{15}	3680		n-type	[8]
6. InAs	0.36	2.5	0.046	0.08	2.1×10^3	3×10^{15}		$+1 \times 10^{-3}$	nearly intrinsic	[8]
7. InAs		2.2		0.006	125	5×10^{16}	1700		n-type	[8]
8. InSb	0.17	7.5	0.075	0.005	380	2×10^{16}		-2×10^{-2}	nearly intrinsic	[50,51]
9. InSb		5.5		0.0013	70	9×10^{16}	1990		n-type	[8]

off-chip temperature sensors to compensate for the influence of temperature. Therefore a low concentration of doping impurities and high carrier mobility are not the only criteria for the choice of a semiconductor material. The bandgap width E_g must be taken into account as well [52]. In this respect Si and GaAs are unique materials. Silicon can be used up to 425 K and GaAs up to 475 K.

Another serious disadvantage of InSb and InAs in spite of the high carrier mobility is that batch fabrication of devices is impossible. Hence most Hall sensors made from such materials are expensive. Moreover, these materials exhibit a low resistivity ρ, thus raising the problem of compatibility of the Hall sensor with electronic circuits. On the other hand, according to (4.40), the dissipated power depends on the product $\mu_H \rho^{1/2}$. Table 4.2 leads to the conclusion that despite the lower carrier mobility Si has the same $\mu \rho^{1/2}$ product as InSb and InAs, whereas for open-circuit Hall device applications the "bad low-mobility" Si is the best material [8]. The rapid advance of the GaAs IC technology and especially the focused ion-beam technique for the formation of submicron patterns directly in the semiconductor substrate leads to the optimistic expectation that the performance to price ratio of such integrated devices, in particular of sensors, will approach that of silicon transducers.

When the optimum choice of the semiconductor material has been made, the fabrication technology becomes a factor of primary importance for the quality, reproducibility, stability of parameters, reliability and economic effectiveness. Microelectronic technologies meet all these requirements. This is why Si or GaAs IC technology is commonly used nowadays for the production of modern Hall sensors. Magnetic-field microsensors are the new wave that determines the present and future development in this field [52–55]. The main advantages of the planar technology for magnetosensor fabrication are the following:

– Batch production, device miniaturization, low cost, parameter reproducibility, stability of the characteristics and reliability;

– The sensor, the signal processing circuit and, if necessary, a microprocessor can be integrated onto a single chip. The application of integrated Hall sensors is greatly facilitated by the favorable combination of desired characteristics (level of the output signal, impedance, nonlinearity compensation, offset, temperature degradation of magnetosensitivity, multisensing, etc.). Furthermore, the smart approach allows a considerable suppression of the disturbances introduced by long connecting wires, especially when low magnetic fields are measured. Thus a high reliability and accuracy are achieved, while the overall noise level is minimized.

The advanced technological process for fabricating bipolar integrated circuits and especially the formation of n-type silicon epilayers and GaAs collector layers are the best methods of achieving the required high mobility and low concentration of carriers. The insulating $p-n$ junction is reverse biased and serves as a boundary that separates the active Hall region from the other parts of the chip. The heavily doped

n^+-collector regions are used as input and output terminals. The surface merged plate fabricated in such a way is passivated by a SiO_2 layer.

There is only a slight dependence of Hall mobility on n in 5- to 10-μm thick epitaxial Si layers with carrier concentration in the range $10^{15} \leq n \leq 10^{16}$ cm^{-3}. Because of this fact, carrier mobility, being the only process-dependent parameter, predetermines a highly reproducible magnetosensitivity at a constant supply voltage [8]. Unfortunately, in this mode of operation mobility is dependent on temperature; therefore magnetosensitivity depends on temperature as well (cf. §4.2.4.2g). The constant-current bias, on the other hand, shows a more process-dependent magnetosensitivity, because of its dependence on the epitaxial-layer doping and thickness. Processing can facilitate the minimization of the offset. The dependence of the offset on mechanical stress for different crystallographic orientations and directions of the current in the Hall sensor has been determined in ref. [56]. When the current flows in the $\langle 100 \rangle$ direction parallel to the (110) plane, there is a minimum offset. Just the opposite effect is observed when the current flow is in the $\langle 110 \rangle$ direction parallel to the (100) plane. In this case the offset is highly sensitive to mechanical stress and can be used for its registration. The maximum value of the offset voltage is $V_0 = (\pi_l - \pi_t) X V / 2(3 - 2w/l)$, where π_l and π_t are the appropriate longitudinal and transversal piezo-resistance coefficients, and X is the mechanical stress applied. This parasitic effect in Hall sensors can be reduced by the l/w ratio whose optimum value is in the range $1.5 \leq w/l \leq 2$ [52].

The recent development of IC technology has led to the fabrication of superlattices and to the possibility of manufacturing Al(As)/GaAs, Al(GaAs)/GaAs and other types of heterostructures on their basis. Such modulation-doped semiconductor layers can be used for the formation of quantum wells with a two-dimensional electron gas at room temperature and low electric fields. The heterostructures include modulation-doped layers whose thickness limits the thickness of the device's active region to less than 10 nm. High electron mobility is observed in such Hall sensors based on superlattice technology. They can operate both in the Hall-voltage and in the Hall-current mode and manifest high linearity and magnetosensitivity, a low level of $1/f$ noise, a slight temperature dependence of parameters, etc. The possibility of manufacturing such monolithic Hall microsensors is rather advantageous from the standpoint of integrated smart magnetometers. Detailed information on the physical mechanisms, performance and fabrication of superlattice-based devices can be found in [57–59].

In addition to the processing compatibility of Hall microsensors with integrated circuits, the compatibility of electrical parameters is rather important for the quality and characteristics of the device. There are two parasitic electrical effects that need particular discussion: (a) the variation of the effective thickness of the epilayer which is isolated from the semiconductor substrate by a p–n junction; (b) leakage currents. A longitudinal potential gradient arises as the bias current flows along the active region of the sensor. This gradient is superimposed on the reverse bias of the

p–n junction, thus causing a positional modulation of the width of the space-charge region and degradation of the homogeneity of the Hall sensor. This disadvantage can be minimized achieved by the doping of the epilayer up to an optimum concentration $nt \approx 10^{12}$ cm^{-2} [26]. The effect of leakage currents is still a subject of research, but most probably they contribute to the generation–recombination noise. Additional information on the problems of the fabrication of Hall devices is given in [55].

Like all other semiconductor devices, Hall sensors must be encapsulated in order to protect them from light, humidity, dust, chemical corrosion, etc. The package is responsible also for the electrical connections of the sensor chip with the external circuits. The principal requirements of magnetic sensors as to the package are the same as those of conventional integrated circuits. Furthermore, the frames of Hall devices must be nonmagnetic, since standard frames have ferromagnetic properties that cause magnetic-field screening and saturation. Hysteresis is observed as well [60,61]. Bronze frames give satisfactory results. The thermal expansion of the chip and the package must also be taken into account, since an inadequate combination of their thermal coefficients may lead to an additional offset due to mechanical stress. A new approach aimed at the virtual elimination of the mechanical stress caused by packaging is described in [24,25]. The stress-sensitive section of the chip is mechanically isolated from the substrate. Some applications require a specific shape and very small dimensions of the package, especially when ferromagnetic concentrators are used to enhance the magnetic field.

4.2.6. Review of orthogonal Hall sensors

4.2.6.1. Device structure and characteristics of MOSFET Hall sensors

In the past solid-state Hall transducers used to be fabricated as a discrete plate similar to that in Fig. 2.2 and that in Fig. 4.1b. The materials employed were Ge and Si, and after the $A^{III}B^{V}$ compounds were synthesized, InSb, InAs, InP and GaAs found a wide application owing to the high mobility of carriers. Since the transducing efficiency of Hall sensors increases with a decrease in the dimension of the device which is parallel to the external magnetic field B, cf. (2.34), there is a trend towards the minimization of that dimension, i.e. towards the fabrication of "two-dimensional" parallelepipeds. The creation of MOSFET and MAGFET Hall sensors has, perhaps, been the most refined solution to that problem thanks to the development of planar IC technology and its methods for thin-film growth and deposition [62]. A n-channel version of a MOS Hall sensor is shown in Fig. 4.13.

If a potential, V_G, that exceeds the threshold voltage of the device, V_T, is applied to the gate, i.e., $V_G > V_T$, a thin n-type inversion channel is formed on the surface of the p-type silicon substrate. This channel serves as the body of the Hall plate. The supply contacts of the sensor are the n^+-source (S) and the n^+-drain (D). The additional n^+-terminals H_1 and H_2 are the output contacts to the inversion

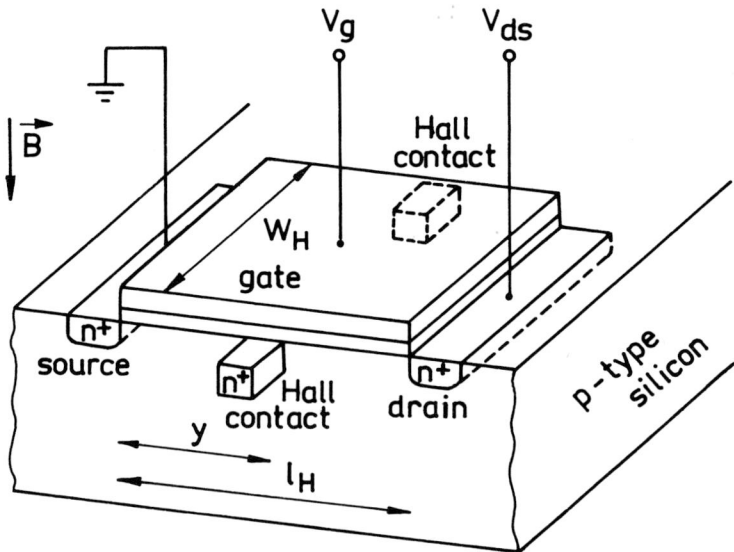

Fig. 4.13. An *n*-channel MOS Hall sensor.

layer. The advantage of the thin Hall layer (typically $\sim 10^2$ Å) when generating a high voltage V_H, cf. (2.34), is limited by the relatively low surface mobility μ_{ch} of carriers in the channel in comparison with the bulk mobility. The experimental determination of the surface mobility of holes in a *p*-channel Hall sensor yields $\mu_{ch} \approx 150$ cm^2 V^{-1} s^{-1} [62]. When the MOS device is operating in the linear region, its Hall signal V_H is a linear function of the drain–source voltage V_{DS}, and once saturation has set in, the further variation of V_H can be neglected. The Hall voltage depends also on the gate–source bias, and the maximum voltage V_H is generated at a gate potential which is close to the MAGFET threshold voltage. A theoretical analysis of the operation of the device under pinch-off conditions together with *p*-channel experimental results is presented in [63]. The conclusion drawn in [13] concerning the invariance of the Hall voltage with respect to gate geometry is of fundamental importance. The application of relaxation methods to obtain the optimum position (y/l) of Hall terminals along the channel length in a rectangular shaped MOS structure is presented in [64]. The y/l ratio is rather important in achieving a high magnetosensitivity, since in the pinch-off mode the drift velocity of the channel carriers gains its maximum value near the drain D. Hence the magnitude of the Lorentz force is a maximum in that region. It has been proved that the optimum condition of the Hall contacts H_1, H_2 is at $0.7 \leq y/l \leq 0.8$ A detailed analysis of the MAGFET in the triode and in the saturation region can be found in [65]. The Hall voltage is given by the expression:

$$V_H = \frac{Gw\mu_{Hn}BI_{DS}}{\beta l[(V_G - V_T)^2 - 2yI_{DS}/\beta l]^{1/2}}, \qquad (4.60)$$

where the drain current is $I_{DS} = \beta[(V_G - V_T)V_{DS} - V_{DS}^2/2]$; $\beta = \mu_n C_{ox}(w/l)$; $\mu_{Hn} = r_n\mu_n$; C_{ox} is the gate oxide capacitance per unit area; $\mu_n \equiv \mu_{ch}$.

N-MOSFETs manifest a higher magnetosensitivity as a result of higher electron mobility in the inversion layers compared with that of holes. For instance, the surface mobility of electrons in the n-channel, is about 450 cm^2 V^{-1} s^{-1} [66]. According to (3.1)–(3.3) and §4.2.4.2e, the expressions for the MAGFET magnetosensitivities are:

$$S_A \equiv \left|\frac{\partial V_H}{\partial B}\right|_{I=\text{const}}, \quad [\text{V T}^{-1}];$$

$$S_{RI} \equiv \left|\frac{1}{I}\frac{\partial V_H}{\partial B}\right| = \frac{r_n G}{qnt}, \quad [\text{V A}^{-1}\text{ T}^{-1}]; \qquad (4.61)$$

$$S_{RV} \equiv \left|\frac{1}{V}\frac{\partial V_H}{\partial B}\right| = \mu_H\left(\frac{w}{l}\right)G, \quad [T^{-1}].$$

In n-channel MOSFET Hall sensors, for instance, at $I_{DS} = 0.1$ mA and $V_{GS} = V_{DS} = 2.7$ V, a magnetosensitivity of 1000 V A^{-1} T^{-1} has been achieved [66]. In the respective MAGFETs at $V_{GS} = V_{DS} = 5$ V and $I_{DS} = 0.5$ mA with a position $y/l = 0.7$ of the Hall terminals H_1 and H_2, the transducer efficiency is 640 V A^{-1} T^{-1} [67]. The orthogonal MAGFET figures of merit found in the sensor literature are summarized in Table 4.3. Unfortunately, the data concerning such important sensor characteristics as nonlinearity, noise, etc., are rather scarce or cannot be found. A MOSFET Hall device with a virtual elimination of the shunting effects irrespective of the w/l ratio is described in [68]. It has been manufactured by means of a double polysilicon MOS process. The device dimensions are $w = 10$ μm and $l = 10$ μm.

The conventional low-resistivity source and drain regions are replaced by the so-called distributed current sources that directly inject an uniformly distributed current into the active sensor region. As a result of the extremely high output resistance of the current sources, this Hall MAGFET has the behavior of an infinitely long sample with $G \approx 1$. With a supply voltage 5 V and a 0.1 mA bias current, the following values of magnetosensitivity are obtained from (4.56): $S_A = 0.7$ V T^{-1}, $S_{RI} = 4 \times 10^3$ V A^{-1}T^{-1} and $S_{RV} = 14\%$ T^{-1}.

A theoretical analysis of the influence of a Gaussian distribution of doping impurities and a homogeneous doping upon the galvanomagnetic properties of n-channel MOSFETs has been made in [69]. It has been shown that in the case of a nonuniform doping profile, the magnetosensitivity should be higher in the triode regime of operation. The optimum positioning of Hall contacts is at a distance of $0.6l$ from the source.

In addition to the classical MOS Hall plates, there are MAGFET devices with

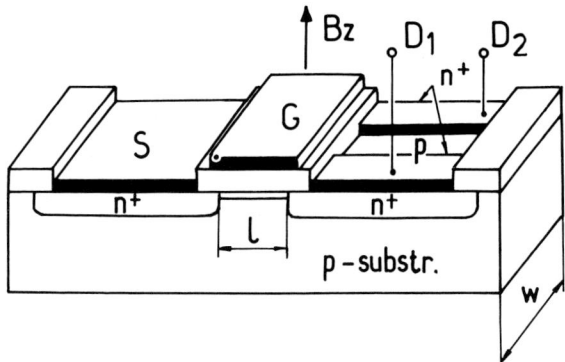

Fig. 4.14. Split-drain magnetosensitive MOSFET

split-drain regions D_1 and D_2 [62] (Fig. 4.14). They are activated by an orthogonal magnetic field, and their shape is similar to those described in §4.2.4.2. The magnetic field B, whose direction is normal to the plane of the structure, causes a deflection of the lines of the current flow in the channel region. The results of a numerical modeling of this effect are illustrated in Fig. 4.15 [63,70,71]. The resulting asymmetry $\Delta I_D(B)$ of the two drain currents ($I_{D1} = I_{D2}$ at $B = 0$) is a measure of the magnetic induction. The relatively low channel mobility leads to limited magnetosensitivity, cf. (4.59). Nevertheless, the values of the absolute sensitivity may be rather high, since this type of sensor usually operates in the saturation region, and the load resistors R_1 and R_2 are connected to the drains as shown in Fig. 4.16. The expression for the output signal is:

$$\Delta V_D = 2\Delta I_D R = \mu_H \left(\frac{l}{w}\right) G R I_{DS} B. \tag{4.62}$$

The voltage drop across the load resistors is $RI_{DS} = V_R$ and the supply voltage of the classical MOS Hall sensor is $V_{DS} = I_{DS} R_{ch}$, where R_{ch} is the effective channel resistance. From the expression (4.61) for S_{RV} the important conclusion can be drawn that the transducing efficiency of the split-drain device is equivalent to that of the conventional MOS Hall sensor.

A Si MAGFET with complementary split-drain pairs, fabricated by means of CMOS technology, is described in [24,72]. Each complementary pair includes a n-channel and a p-channel device integrated into the chip, similar to a CMOS amplifier. High sensitivities $S_A = 1.2$ V T^{-1} and $S_{RI} = 1.2 \times 10^4$ V A^{-1} T^{-1} have been achieved with a supply voltage of 10 V and a bias current of 100 μA.

The magnetosensitivity and noise characteristics of n-channel silicon and GaAs MAGFETs as a function of their dimensions and of the width and position y/l of the lateral drain (sensor) contacts have been investigated experimentally [73,74]. These structures include a source, a main drain and two additional reverse-biased drains located symmetrically on both sides of the channel (similar to the Hall

Table 4.3

Parameters of selected orthogonal MAGFET Hall sensors

Type	l/w	Sensitivity, S	Equiv. noise (μT Hz$^{-1/2}$)	Equiv. offset (mT)	T.C. of S (% K^{-1})	Refs.
p-channel, Si	5	0.4% T^{-1}				[62]
p-channel, Si	3	0.6% T^{-1}				[63]
n-channel, Si	0.4	3.1% T^{-1}				[13]
n-channel, Si	1	3.5% T^{-1}		20		[66]
n-channel, Si	1.2	6.4% T^{-1}		14	-0.4	[67]
n-channel, Si	0.1	14% T^{-1}				[68]
CMOS dif. ampl., split-drain, Si	0.3	1.2 V T^{-1} 1.2×10^4 V A^{-1} T^{-1}				[24,72]
n-channel, Si, split-drain	1	9.6% T^{-1}	7			[73]
n-channel, Si, open split-drain	1	172 V A^{-1} T^{-1} 109 mV T^{-1}	0.15			[73]
n-channel, Si, split-drain	1	10% T^{-1}	10			[73]
n-channel, Si, open split-drain	1	110 mV T^{-1}	0.2			[73]
n-channel, GaAs, split-drain	4	70% T^{-1}	25			[73]
n-channel, GaAs, open split-drain	4	140 mV T^{-1}	4			[73]
n-channel, GaAs, open split-drain	4	180 mV T^{-1}	8			[74]
n-channel, Si, split-drain	3	7 mV T^{-1}	10			[74]
n-channel, Si, split-drain	1	2.5 V T^{-1}			-1	[76]
p-channel, Si, split-drain	1	2.2 V T^{-1}				[77]
n-channel, Si, 3-drain	1	6.2% T^{-1}	0.76			[46,78]
n-channel, Si, 3-drain	3	20 nA T^{-1}	22			[74]
n-channel, GaAs, 3-drain	4	24 μA T^{-1}	10			[74]
n-channel, GaAs, 3-drain	4	27 μA T^{-1}	2.3			[74]

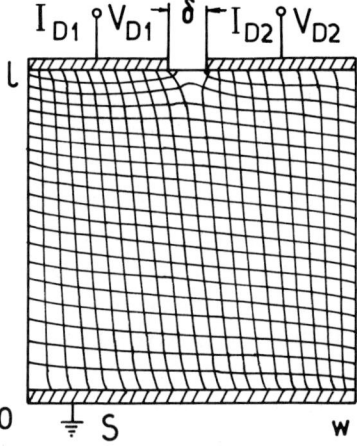

Fig. 4.15. Numerical modeling of a split-drain MAGFET. Equipotential and current lines for $w/l = 4$ μm/4 μm, $\delta = 1.4$ μm and $B = 1$ T [70,71].

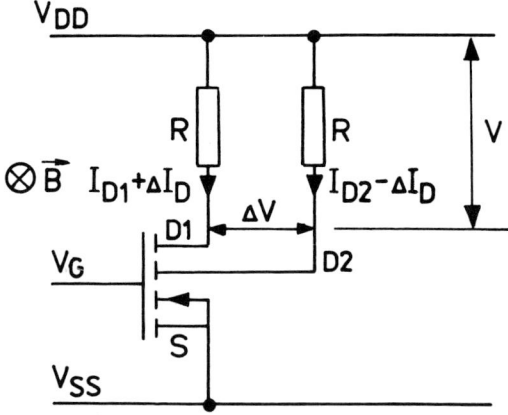

Fig. 4.16. A circuit with a split-drain MOSFET sensor.

MOSFET). It has been found that the best magnetosensitivity S_{RI} is achieved when $w = l$, the drain terminals are narrow and are located near the main drain at $y/l = 0.9$. For instance, in a n-channel silicon version of the split-drain device at $V_{GS} = V_{DS} = 5$ V and $I_{DS} = 24$ μA, the value of the parameter S_{RI} is 9.6% T^{-1}, and at frequencies $f \leq 10$ kHz the noise varies as $1/f$ with a noise-equivalent magnetic induction $B_{\min} \approx 7$ μT Hz$^{-1/2}$. However, if the two additional drains are used as Hall contacts with $y/l = 0.5$, operating in the open circuit mode at a current $I_{DS} = 630$ μA, the noise-equivalent magnetic induction will be $B_{\min} = 150$ nT Hz$^{-1/2}$ with $S_A = 109$ mV T^{-1}. In a GaAs n-channel split-drain MAGFET with $w = 10$ μm and $l = 40$ μm at $V_{GS} = -1.5$ V, $V_{DS} = 5$ V and $I_{DS} = 4$ μA, the relative current magnetosensitivity is about 70% T^{-1}, i.e. about 7 times as high as that of the respective silicon devices. If in the same GaAs MAGFET the two lateral drains are used as Hall contacts ($y/l = 0.5$) with $V_{GS} = 0$ V, $V_{DS} = 5$ V and $I_{DS} = 200$ μA, the maximum measured sensitivity is $S_A = 180$ mV T^{-1}, where $B_{\min} \approx 8$ μT Hz$^{-1/2}$. Table 4.3 presents data from [73,74] concerning different split-drain Si and GaAs MAGFETs. According to [73], the achievement of a maximum resolution $B_{\min}$ is not correlated to a particular operational mode with a maximum magnetosensitivity. An attempt has been made in [73,74] to identify and compare the dominating sources of noise ($1/f$, G–R, thermal and shot-noise) in different types of integrated magnetosensors, including Si and GaAs MAGFETs. It has been shown that in split-drain Si MAGFETs with different values of $w = l$, i.e. $w/l = 1$, at a supply voltage $V_{GS} = V_{DS} = 5$ V and drain current in the interval 20 μA $\leq I_{DS} \leq$ 150 μA the dominating noise sources are the trapping $1/f$ noise and the thermal noise. $B_{\min}$ is (5 to 20) μT Hz$^{-1/2}$, when the sensitivity is about 10% T^{-1}. The same noise sources have been identified in n-channel silicon Hall MOSFETs with different values of w/l, when $y/l = 0.5$, $V_{GS} = V_{DS} = 5$ V, and the drain current $I_{DS} = 0.65$ mA. The sensitivity achieved is 110 mV T^{-1} and B_{min} is $(0.15 \sim 0.25)$ μT Hz$^{-1/2}$. In nonoptimized split-drain GaAs MAGFETs with $l = 40$ μm and $w = 10$ μm, the dominating sources of noise are trapping $1/f$ noise, diffusion noise, G–R noise and thermal noise. When the operational mode is determined by $V_{GS} = 0$, $3V \leq V_{DS} \leq 5$ V and $I_{DS} \approx 0.2$ mA, the sensitivity achieved is within the interval 40 to 80% T^{-1} and B_{min} is between 10 and 40 μT Hz$^{-1/2}$.

It has been shown in ref. [75] that the asymmetry in split-drain MAGFETs due to the bias conditions and the parasitic lateral conductivity can be eliminated if both drains operate at the same bias $V_{D1}(B) = V_{D2}(B)$ in a magnetic field and when $B = 0$. Only, in this case, the output current signal $\Delta I_{D1,2}(B)$ is a function of the real (intrinsic) magnetosensitive characteristics of the device.

A new geometry for an n-channel split-drain MAGFET is suggested in [76]. The two drains are separated by the minimum space of about 10 μm that can be achieved by presented technology. Each drain is L-shaped in order to increase the collecting area for the electrons deflected by the Lorentz force. A minimum channel width is achieved by means of notches on both sides of the gate. An absolute sensitivity of 2.5 V T^{-1} is reached at a current $I_{DS} = 25$ μA. A circuit that interconnects several MAGFETs into a tree in such a way that the total current enters its root and then is successively split by the branches is used to attain the maximum effect of the magnetic field. This approach leads to a 1.5-fold increase in the transducing constant in comparison with a single sensor, provided all other conditions are the same.

Integrated high-resolution 32 × 32 or 64 × 64 sensor arrays consisting of identical p-channel split-drain MAGFETs fabricated by a 3 μm CMOS process are described in [77]. Each one of the identical elements occupies an area of 100 μm^2 and the chip dimensions of the 64 × 64 array are 9.2 × 7.8 mm^2. The dissipated power with a 5-V supply voltage is 12.5 mW and the sensitivity of an individual sensor is 2.2 V T^{-1}. The many applications of this scanning MAGFET device include field mapping, registration of fringing fields in electric motors, and the determination of various nonmagnetic variables, for instance its use as a tactile sensor.

A triple-drain Hall sensor configuration (D_1, D_2, D_3), with an n-channel MOSFET, whose shape is similar to that in Fig. 4.12b, is simulated and investigated in [46,78]. The two outer drains D_1 and D_3 serve as output terminals. The relative current imbalance due to an orthogonal magnetic field applied to a square triple-drain structure is about 50% higher than that in a split-drain device. According to the data in [78], when the dimensions of the Si device are 100 × 100 μm^2 ($w/l = 1$), the sensitivity of the triple-drain MAGFET is 6.2% T^{-1}, while the sensitivity of a double-drain version is 4.2% T^{-1} in the same operational mode. The noise and the magnetosensitivity of two GaAs triple-drain n-channel sensors and a triple-drain Si MOSFET are investigated in [74]. At frequencies $f \leq 1kHz$, the $1/f$ noise dominates. In spite of the greater magnetosensitivity of the GaAs sensor compared with the silicon device, the noise level in the GaAs devices is higher, perhaps because of the worse characteristics of the metal–GaAs interface.

The double-drain MOSFET can be used not only as a sensor, but also as a tool for fundamental investigations on Hall mobility and for the simulation of kinetic phenomena in inversion channels. This approach has been successfully applied to an n-channel GaAs MAGFET [45], and an n- channel Si MAGFET [79].

Some concluding remarks on MAGFET Hall sensors. The orthogonal MOS Hall devices are the earliest integrated versions of the classical Hall elements. They definitely prove the compatibility of Hall devices and ICs concerning materials, technology and electrical parameters and, hence, the possibility of integrating Hall sensors with the appropriate circuitry on a single chip without auxiliary processing steps. It was the investigation of MOS Hall devices at cryogenic temperatures in high magnetic fields that in 1980 led to the discovery of the quantum Hall effect in two-dimensional systems. This phenomenon has become a new and extremely accurate, and easily reproducible, metrological standard of resistance and fundamental constants. Detailed information on this unique effect and its applications can be found, for instance, in [80].

The idea of planar input and output contacts upon one of the surfaces of the silicon or GaAs substrate is the basis for the future improvement of Hall microsensors. MAGFETs are particularly suited to smart solutions when circuits for signal processing are integrated together with the sensor on one and the same chip. Another important advantage of MOS Hall structures is the possibility of using them for the in situ characterization of the electronic properties of different FET modifications used in VLSI and ULSI circuits. From the data in Table 4.3 it can be concluded that, although MAGFETs are rather promising as Hall sensors, there are lots of problems to be solved concerning their figures of merit. The noise, nonlinearity, temperature behavior and other characteristics have not been studied in detail. The nonuniformity induced by the Hall voltage across the channel and its effect upon the sensor characteristics and, especially, on magnetoresistivity and linearity is a problem which has not been thoroughly investigated and, in the author's opinion, deserves special attention. In the absence of a magnetic field ($B = 0$), at constant V_{GS}, V_{DS} and I_{DS}, the properties of the channel in a direction perpendicular to the current I_{DS} are homogeneous. The variations in the carrier concentration in the channel and in the width of the depletion layer are longitudinal (i.e. along the source–drain direction). In an orthogonal magnetic field the Lorentz force $\boldsymbol{F}_L$ causes a redistribution of the channel carriers in a direction perpendicular to the current $\boldsymbol{I}_{DS}$ and to the field $\boldsymbol{B}$. As a result a transverse potential difference ($+0.5V_H$ and $-0.5V_H$) is generated in the Hall plane of the MOS structure. This potential difference is linearly distributed across the channel. Hence at a constant gate voltage the modulation of both the carrier concentration in the channel and the depletion layer width by V_H can be expected owing to the additional electric charge

$\sigma_n(E_H)$. Therefore a variation in the effective thickness t^* of the active sensor region occurs in the direction of the field $\boldsymbol{B}$ and across the channel. According to (4.28) and (4.61), the Hall voltage itself and the Hall sensor magnetosensitivity should be influenced by the modulation of t^*. In the general case it is not obvious that both a decrease and an increase in the thickness t^* relative to the central source–drain axis ($t(+0.5V_H)$; $t(-0.5V_H)$) will compensate the characteristics of the device. Furthermore, the Hall voltage increases in the source to drain direction, i.e. an inhomogeneous Hall effect occurs, and thus in compliance with §2.3.5 a circulation current j^* arises around the active sensor region. The overlapping of the output n^+-terminals with the gate electrode also influences the characteristics of the device. In a certain sense the problems considered above correspond to the variation of the thickness t with the Hall voltage V_H in bulk bipolar devices [43]. In MOS Hall sensors, the effect discussed should be more pronounced in comparison with bipolar integrated versions, since the origin of the inversion layer is electrostatic, and because of this fact the channel is more sensitive to varying electric potentials. In bulk bipolar Hall structures the active region is confined mainly by its metallurgical boundaries.

The improvement of GaAs devices is the way to avoid the relatively low magnetosensitivity of Si MAGFETs due to the low surface mobility of electrons in the channel (less than half the bulk value). Recent results from experiments with GaAs devices indicate that MAGFETs are not just an intermediate stage in the development of Hall sensors. They are a modern trend in the evolution of GaAs technology.

4.2.6.2. Bulk bipolar Hall sensors — review of device structures and sensor characteristics

Bipolar IC technology is used for the production of a great proportion of the Hall sensors commercially available. A top view and a cross-section of such a Hall plate with n-type conductivity are shown in Fig. 4.17.

The basic idea of this sensor was first described in [81]. The sequence of processing steps is the same as that of integrated n–p–n transistors. Reverse biased p–n junctions surrounding the Hall device are responsible for its isolation from the p-substrate. Shallow n^+ diffusion regions in the n^- layer serve as the input and output contacts of the microsensor. The n-Si epilayer used as an active transducing region must meet the following requirements in order to guarantee a high magnetosensitivity: low carrier concentration and high electron mobility in the bulk. The typical carrier concentrations and epilayer thicknesses are $10^{15} \leq n \leq 10^{16}$ cm^{-3} and $5 \leq t \leq 10$ μm, respectively. Table 4.4 summarizes the device characteristics of selected bulk bipolar Hall structures.

The high performance of a sensor depends on the homogeneity of the Si epilayer and on the accuracy of the positions of the Hall contacts. This is why the main reason for the large offset $B_{0,eq} \simeq 80$ mT in [82] is the epilayer nonuniformity.

Table 4.4

Parameters of selected orthogonal bulk Hall sensors, whose device structures have been shown in Fig. 4.1

No.	Device struct. mater.	l/w	n^- Hall region	Epi-doping N_D (cm^{-3})	Epi-thickn. t (μm)	Sensitivity S	Equiv. magn.-field noise (mT)	Equiv. offset (mT)	T.C. of offset (mT K^{-1})	T.C. of sensitivity (% K^{-1})	NL (%)	Frequency resp. (kHz)	Refs.
1	(b), Si	2	epi	10^{15}	10	0.6 V T^{-1}	7×10^{-4}	17				15	[81]
2	(b), Si	1	epi	10^{15}	8	0.5 V T^{-1}		80					[82]
3	(b), Si	1	epi			0.4 V T^{-1}		25				10	[83]
4	(b), Si	1.2	epi	10^{15}	10	$B_{switch} = 50$ mT							[84]
5	(q), Si	1	epi	2.5×10^{15}	12	7.2% T^{-1}		10	0.025	-0.6			[8]
6	(b), Si	1.2	epi + impl.	10^{16}	15	7.6% T^{-1}		10		-0.45			[30]
7	(b), Si Hall IC TL 173	1	epi			15 V T^{-1}		7	0.2	-0.18		10	[85]
8	(b), Si Hall IC SAS 231		epi			100 V T^{-1}		35	0.4		±2		[86]
9	(p), Si	1	epi			300 V A^{-1} T^{-1}							[87]
10	(m), Si	1	epi + impl.	4×10^{15}	14	290 V A^{-1} T^{-1}					comp		[89]
11	(m), Si	1	epi + impl.	1.5×10^{15}	14	3300 V A^{-1} T^{-1}					comp		[89]
12	(b), Si		epi	5×10^{14}	4	1 V T^{-1}, 7% T^{-1}							[90]
13	(b), GaAs	2	epi	5×10^{14}		22 V T^{-1}				-0.05			[91]
14	(b), GaAs	1	epi + impl.	5×10^{17}	0.13	0.17 V T^{-1}		8		-0.04			[93]
15	(m), GaAs	1	epi + impl.	10^{17}		500 V A^{-1} T^{-1}			0.5	-0.05	±0.1		[94]
16	(m), GaAs	1	epi			15 V T^{-1}				-0.03	±0.2		[95]
17	(m), GaAs	1	epi + impl.	10^{17}	0.4	0.2 V T^{-1}, $T = 300$ K; 0.13 V T^{-1}, $T = 77$ K; 0.19 V T^{-1}, $T = 4.2$ K			0.03	-0.03	±0.03, $T = 300$ K; ±2, $T = 77$ K; ±0.6, $T = 4.2$ K		[96]
18	(o), GaAs	1	epi + impl.	$(1.1$–$3.8)$ 10^{17}	0.25	65 V A^{-1} T^{-1}	4×10^{-3}	comp.					[98]
19	(m), GaAs with concentr.	1	epi + impl.	10^{17}		(350–1100) V A^{-1} T^{-1}	7×10^{-5}			-0.08		10^4	[99]
20	(b), GaAs	4	epi + MOCVD + impl.	10^{17}	0.35	220 V A^{-1} T^{-1}		3.4			±2		[101]

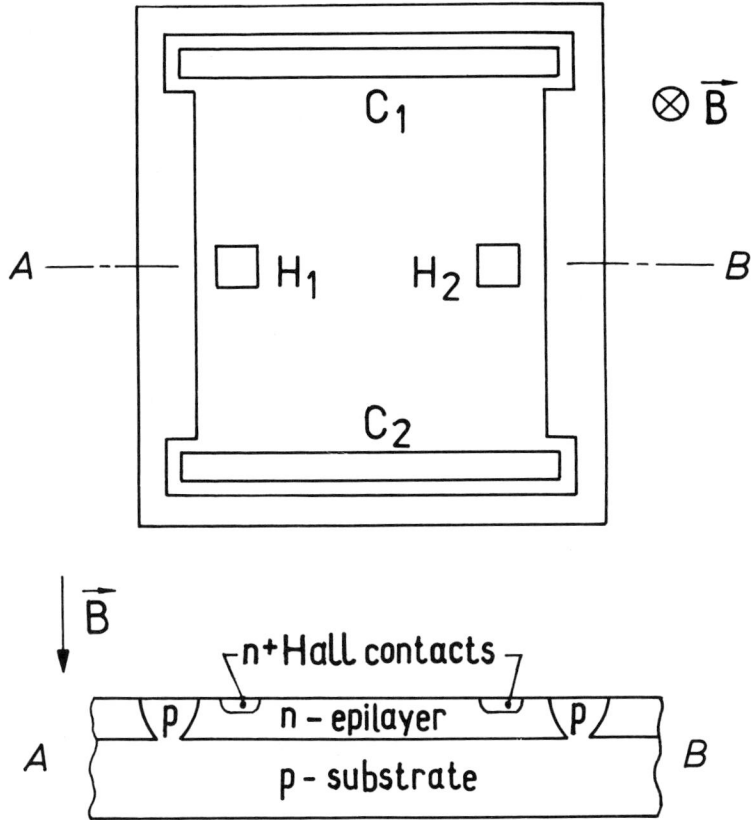

Fig. 4.17. Top-view and a cross-section of an n-type bulk bipolar Hall device.

An attempt has been made in [83] to include in the processing sequence the formation of four additional diffusion contacts on the active region in order to eliminate the initial offset of about 25 mT. Furthermore, they can be used to reduce the temperature dependence of the magnetosensitivity. A theoretical analysis is presented in [84] aimed at the minimization of errors due to mask misalignment and etching. The experimental results obtained from a Hall sensor in a triggered action agree with theoretical predictions. A sophisticated multi-electrode device which consists of four symmetric silicon Hall plates surrounded by a common isolation ring to confine the current and to compensate for the stress-induced offset is described in [8]. Ion implantation is used in order to achieve a uniform doping of the Hall region [30]. The measured mean magnetosensitivity is near the optimum value that corresponds to an electron mobility $\mu_H \approx 1300$ cm^2 V^{-1} s^{-1} (i.e. 0.13 m^2 V^{-1} s^{-1}) in the bulk of the epilayer. An isolation ring can be used for offset minimization in rectangular-shaped Hall structures like that in Fig. 4.1g.

Commercial general-purpose Hall-effect devices are available from various com-

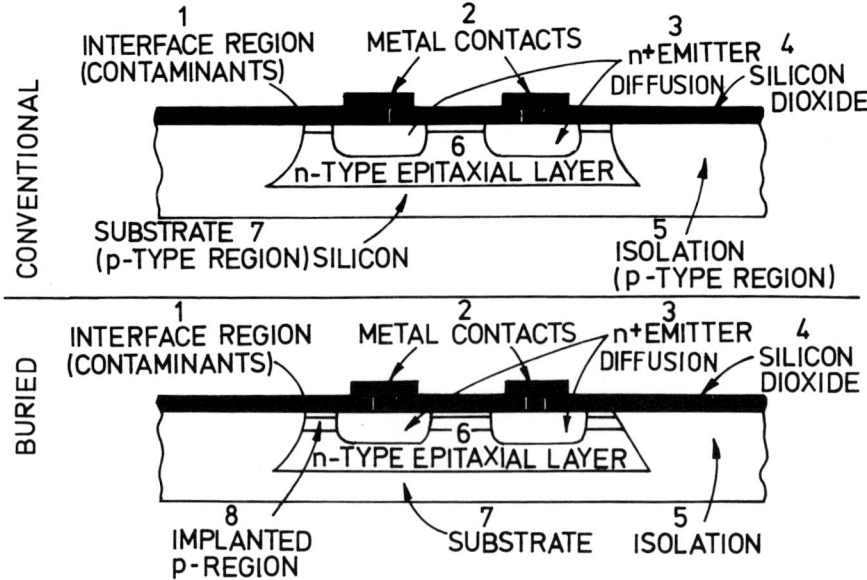

Fig. 4.18. A cross-section of a buried Hall sensor realized using IC technology [88].

panies including Siemens, Texas Instruments, Honeywell, and Sprague. In addition to the Hall epilayer sensor, such silicon analog ICs include on-chip amplification and stabilization circuitry. For instance, the magnetosensitivity of the LOHET (Honeywell) device is 75 V T^{-1} and that of the UGN 3501 (Sprague) is 7 V T^{-1}. The specifications of two other commercially available Hall ICs with an analog output can be seen in Table 4.4 [85,86]. A modification of the epilayer of a Hall microsensor with interchangeable input and output contacts is described in [87].

The device from Fig. 4.17 can be improved as shown in Fig. 4.18. This is the so-called buried Hall sensor [88]. The structure is similar to the pinched collector layer resistor in bipolar integrated circuits. A shallow p-type layer is used to separate the active sensor region from the Si–SiO$_2$ interface. The shallow p-layer serves also as an electrostatic screen between the Si–SiO$_2$ interface and the Hall sensor.

The improvements in technology described above contribute to the stabilization of the characteristics and to the minimization of the influence of surface contamination.

A detailed analysis of output nonlinearity in silicon Hall microsensors has been made in [25,43,89]. The method and the proper circuitry suggested for nonlinearity compensation are considered in §9.5.3.

Buried n-Si uniformly doped Hall sensors with different donor concentrations jacketed by a heavily doped p-type material have been investigated. The active device region was formed by phosphorous-ion implant and diffusion, followed by a heavier implant and shallow diffusion of boron. Nonlinearity is compensated for by changing the sensor's magnetosensitivity. For this purpose the effective thickness of the sensor is modulated by varying the reverse bias of the jacketing p–n junction.

A Hall microsensor implemented in the collector region of a silicon lateral n–p–n transistor is suggested in [90]. A specific feature of this device is the formation of the Hall contacts in the depleted layer of the reverse biased collector p–n junction. The minority carriers injected by the emitter move in the base towards the collector at an increasing drift velocity v_{dr}. At sufficiently large values of the collector–base voltage V_{CB}, the drift velocity is saturated, i.e. $v_{dr} \to v_{dr,sat}$ Therefore, the Lorentz force and the magnetosensitivity reach their maximum values in the depletion region where the Hall terminals are located, cf. §2.3.3.

The higher electron mobility in GaAs (about 5.5 times as high as that in Si) and the wider temperature interval of operation ($\sim-70°C$ to $250°C$) motivate the research activity aimed at creating epitaxial GaAs Hall microsensors [91,92]. Better homogeneity of the n-epilayer is achieved by a Se-ion implant [93]. A typical Hall sensor structure with a GaAs epilayer is shown in Fig. 4.19 [94].

This cross-shaped device is most often used for industrial production. Detailed information on the processing steps and the dependence of the sensor characteristics on the ion implantation doses and on the dimensions l and w can be found in [94]. A high-precision current meter and a wattmeter, which include a GaAs Hall sensor similar to those in [92] and [93], are described in [95]. High quality GaAs Hall devices fabricated by means of Si implantation into a semiinsulating GaAs substrate designed for operation in a wide temperature range, including cryogenic temperatures $4.2 \leq T \leq 77$ K, are suggested in [96]. Owing to their performance, these Hall devices are among the best GaAs sensors reported so far. An investiga-

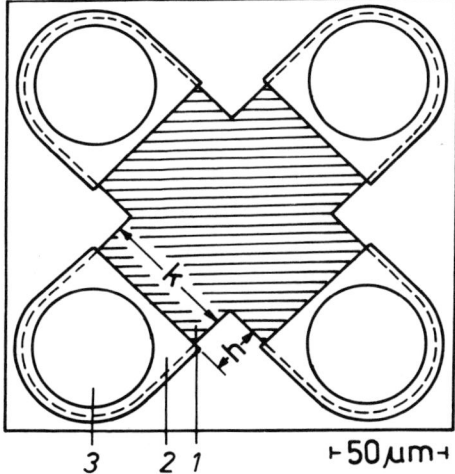

Fig. 4.19. Device structure of a GaAs Hall sensor; 1 = active area; 2 = metallization; 3 = contact region [94].

tion of the temperature dependence of the offset of GaAs sensors is presented in [97] and an accurate method for the offset compensation has been proposed that uses the device itself as a temperature sensor. An epitaxially grown GaAs symmetrical transducer with high stability and a magnetosensitivity of approximately 65 V A^{-1} T^{-1} has been proposed for applications in low-field magnetometry [98]. A $4\mu T$ minimum magnitude of the magnetic field has been achieved and an offset compensating circuit has been described.

A whole family of GaAs Hall sensors implanted with different doses of silicon are presented in [99]. Their parameters at a 5 mA bias current are given in Table 4.4. Furthermore, a new type of butterfly-shaped CoMoSiBFe-based magnetic-field concentrator has been suggested that allow the registration of extremely low magnetic fields in the 70×10^{-9} T range.

An integrated magnetosensitive circuit has been created that includes amplifiers, a comparator, a TTL compatible output buffer and Si-implanted rectangular shaped GaAs Hall element used as a sensor [100].

A promising implementation of GaAs Hall devices has been achieved by growing the GaAs layer by the MOCVD (Metal Organic Chemical Vapour Deposition) technique. MOCVD enables a high quality to be achieved for the device and precise control of sensor parameters in the -190 to $100°C$ temperature interval [101].

The problems of creating integrated InSb Hall devices, although being far from their solution, are not obstacles to the utilization of this material in magnetometry. As a result of the high electron mobility μ_n, which reaches values as high as 73×10^3 cm^2 V^{-1} s^{-1}, the Hall effect and the magnetoresistance effect are very well pronounced. This fact has motivated the combination of the two effects in the three-terminal Hall element [102]. A highly asymmetric output signal depending on the direction of the field B is characteristic of this sensor. The evolution of the classical two-terminal magnetoresistor and the four-terminal Hall device into a three-terminal Hall sensor with the shape shown in Fig. 4.1c is presented in Fig. 4.20a. The middle contact is used for the summation of the even magnetoresistance signal $V_{MR} \sim B^2$ and the odd Hall signal $V_H \sim B$: $V_D(B) = 1/2[V_{MR}(B) + V_H(B)]$. Illustrated in Fig. 4.20b are the respective output characteristics. The three-terminal device manifests a highly asymmetric output signal depending on the sign of the magnetic field B in the current drive mode. As shown in [103], the asymmetry increases with increasing Hall angle $\Theta_H \approx \mu_H B$. The output voltage at positive values of Θ_H is a linear function of the induction B and at negative Θ_H is almost constant (Fig. 4.20b). A directional figure of merit (DFM) has been introduced in order to evaluate this remarkable asymmetry, cf. §3.3.3. A DFM ≈ 20 in a magnetic field $B = 1T$ has been achieved by three-terminal n-type single-crystal InSb sensors with $\mu_n \approx 73 \times 10^3$ cm^2 V^{-1} s^{-1} and $R_H \approx 330$ cm^3 C^{-1}, fabricated by a mechanical polishing and chemical etching of InSb plates whose dimensions are $l = w = 4.2$ mm and $t = 60$ μm [103]. Three-terminal Hall element arrays with

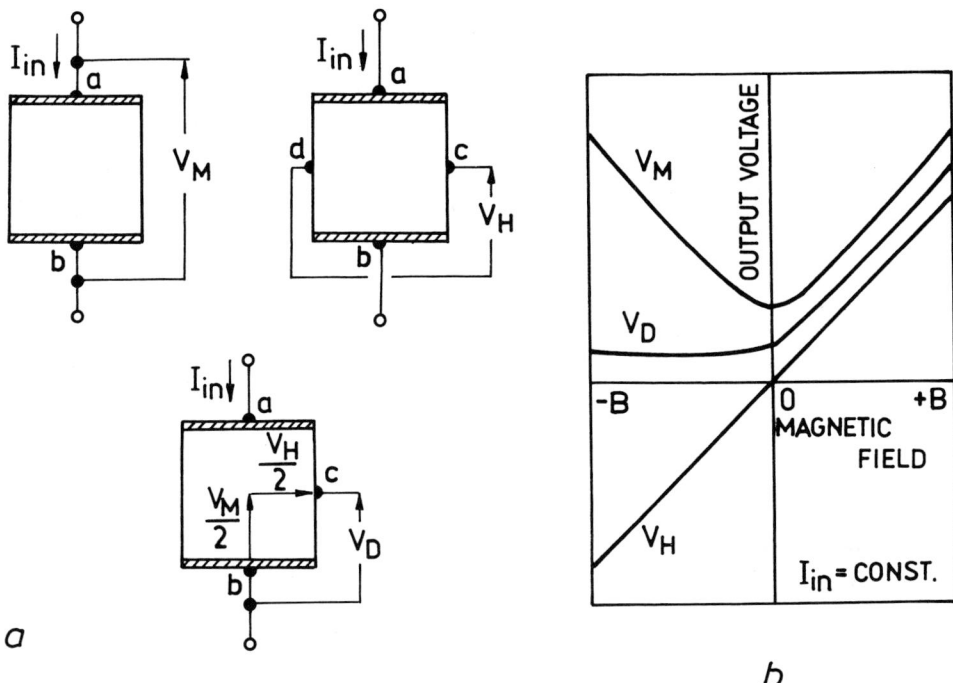

Fig. 4.20. Evolution of the two-terminal magnetoresistive element and the four-terminal Hall device into a three-terminal Hall sensor (a) and a characteristic of the three-terminal Hall element in comparison with those of conventional galvanomagnetic sensors (b).

an individual device size of 4 μm have been developed to detect nonuniform and locally inverted magnetic fields generated from stripe domains. There are only a few published works [104,105], concerning the two-dimensional numerical simulation of Hall sensors subjected to locally inverted magnetic fields that pierce only a small part of the active region. It has been shown that, depending on the device geometry and domain dimensions, the two fundamental manifestations of the Hall effect (voltage and current) coexist in a complicated manner which is determined by the nonuniformity of the local Hall angle. Irrespective of the complex galvanomagnetic picture, the displacement of the bubble relative to the axis of symmetry of the device generates a symmetric response at the output, $\gamma \sim 1.5$ mV μm^{-1}. Increasing the sensor's supply current or voltage does not change the spatial resolution γ.

Concluding remarks on bulk bipolar Hall sensors. The data in Table 4.4 give reasons for a relatively complete characterization of orthogonal bulk bipolar Hall devices by contrast with MAGFETs. A rapid advance in epitaxial GaAs Hall sensors is observed, as well. There is a remarkable fact: the typical concentrations of doping impurities $N_D \simeq 10^{15}$ to 10^{16} cm^{-3} and thicknesses $t \sim 5$ to 10 μm of Si epilayers in bipolar integrated circuits are almost the optimum values for Hall sensors. This is a serious prerequisite for the on-chip integration of the Hall transducer itself, the signal conditioning circuitry and, if necessary, a microprocessor with the properly chosen software for processing of magnetic information. Moreover, this smart approach allows most of the typical shortcomings of sensors (offset, nonlinearity, drift, etc.) to be successfully overcome at the stage of fabrication. Although the comparison of the magnetosensitivity of Hall devices is hampered by the absence of a generally accepted standard definition of this most important figure of merit, in all cases GaAs modifications have a higher transducing efficiency.

The stability of bulk bipolar Hall sensors as well as that of the other types of Hall sensor is an important problem which unfortunately has not been studied in detail. Hall sensor are input converters of magnetic energy and their temporal stability determines the overall stability of data acquisition systems that employ magnetosensitive devices. It can be concluded from the results in [104,106] that the Hall coefficient can vary by about 2.5% owing to the mechanical stress which results from encapsulation. This piezo-Hall effect in lightly doped n-type Si samples is highly sensitive to the crystallographic orientation. The crystal surface is another factor that determines the stability of magnetosensitivity. The thermal growth of SiO_2 may be accompanied by mobile-ion contamination of the oxide layer. The migration of mobile ions cause instabilities on the Si–SiO_2 interface. The offset is perturbed by the random fluctuations of the distribution of mobile charges. It has been found experimentally that the effect of surface instability upon the magnetosensitivity of the sensor is proportional to the transducing efficiency itself. More sensitive devices manifest a greater instability. The technology of buried Hall transducers allows the minimization of these adverse effects [24].

The investigations into noise in bipolar Hall microsensors are incomplete. Although this kind of magnetosensor should necessarily be the subject of future detailed research, such sensors are still the best available when the practical applications require small-offset, high-frequency modes of operation and a small temperature coefficient.

4.2.6.3. New trends in the development of orthogonal Hall sensors
Progress in the field of Hall devices is in the direction of submicron Hall sensors on the basis of GaAs, InAs or other materials of this type as well as two-dimensional electron-gas devices on the basis of modulation-doped superlattices.

A. Submicron Hall sensors. The detection of submicron displacements, vibrations, bubble domains, velocity, magnetic patterns in bank-notes and credit cards, reading of magnetic disks and tapes, memories, angular displacements and other sophisticated applications require magnetic sensors capable of measuring a nonuniform, locally inverted magnetic induction $B_z(x, y)$ [103]. High spatial resolution can be achieved by means of the miniaturization of Hall sensors. Figure 4.21 is a schematic illustration of a typical arrangement of a strongly inverted magnetic field and the

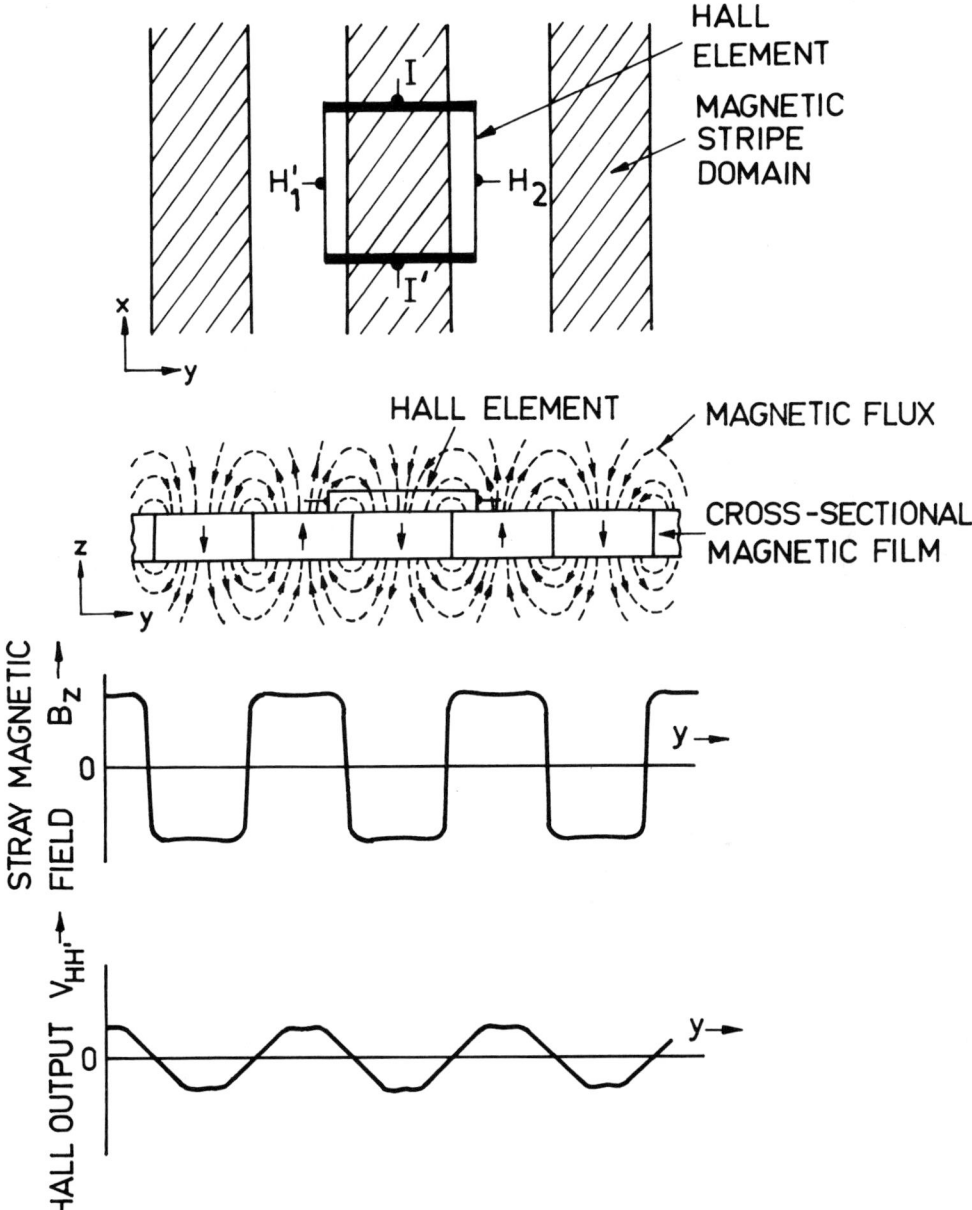

Fig. 4.21. Schematic diagram of typical arrangement of locally inverted magnetic fields generated from stripe domains and Hall output illustrated as a function of this arrangement.

dependence of the Hall output on the displacement with respect to the y-axis of the sensor. Submicron Hall devices are designed in such a manner that power dissipation, the presence of a high electric field and the effects of small geometry are taken into account. The output signal V_H of a submicron Hall sensor is given by the expression:

$$V_H = R_s I B, \qquad (4.63)$$

where

$$R_s = 1/n_s q; \qquad (4.64)$$

R_s is the sheet Hall coefficient, n_s is the sheet concentration of carriers.

In the general case, the input current I cannot exceed a certain maximum value I_{max}. Therefore, the magnetosensitivity S_{max}, being the most important parameter, is determined as:

$$S_{max} = R_s I_{max}. \qquad (4.65)$$

In conventional Hall devices with large dimensions, I_{max} is limited by the maximum power that can be dissipated. Just the opposite situation is observed in submicron Hall sensors, where dissipated power is not so important since the reduced dimensions make heat conductivity more efficient. Instead, the influence of the high field becomes the dominating factor. In this case, the drift velocity of carriers reaches its maximum value $v_{max} \equiv v_{sat}$, and this is the evidence for the high electric field arising as a result of the forcing of an input current I through the sample. In real Hall microsensors the current I is, to a certain extent, limited by the drift velocity v_{max}. The maximum current is:

$$I_{max} = n_s v_{max} w, \qquad (4.66)$$

where w is the width of the sensor's active region.

Thus, the maximum magnetosensitivity is obtained from equations (4.64)–(4.66):

$$S_{max} = v_{max} w. \qquad (4.67)$$

Two important results follow from (4.67): (a) the parameter S_{max} does not depend on the concentration or mobility of carriers; (b) S_{max} decreases with a reduction in the sensor dimensions. There are two device parameters of principal importance: the input resistance R_{in} and the offset. Their values increase as the sensor dimension w shrinks [107]. The processing steps and related problems that arise when scaling down single-crystal GaAs or InSb Hall sensors to an area of several μm^2 are described in [103]. The galvanomagnetic properties and noise in cross-shaped Hall magnetosensors are described in [108]. A sensitivity of 1 V T^{-1} at 10^{-3} W input power has been achieved in InSb samples. The magnetosensitivity of GaAs samples is 2.5 V T^{-1} at an input power of 3×10^{-3} W. At frequencies $f \leq 1$ kHz, the $1/f$ noise dominates and in InSb Hall structures $B_{min} \approx 1.3$ nT. This noise level is lower than the noise in GaAs sensors with $n \sim 10^{16}$ cm^{-3}. A Hall microsensor with a sensitive region whose size is 0.5 μm, fabricated by maskless Si-ion implantation into a semiinsulating GaAs crystal, is described in [107]. It has been shown that an accurate pattern definition and high performance can be achieved by a minimum number of processing steps. The maximum sensitivity registered experimentally is 48 mV T^{-1} and, in agreement with the model discussed above, it does not depend on the sheet concentration. In GaAs the maximum drift velocity is $v_{\max} \approx 2 \times 10^7$ cm s^{-1} at $T = 300$ K. This value corresponds to the sensitivity $S_{\max} \approx 60$ mV T^{-1}. It can therefore be concluded that a high electric field arises in the active region. A quarter-micron-width Greek-cross Hall sensor with dimensions 0.25×0.25 μm^2 fabricated by focused ion beam implantation and suitable for inclusion in ULSI circuits is described in [109]. When the applied electric field is 10 kV cm^{-1}, the measured magnetosensitivity is 85 V A^{-1} T^{-1}, $S_A = 34$ mV T^{-1} at a supply current of 0.4 mA. At frequencies below 10 Hz the $1/f$ noise dominates in this submicron structure and at $f \approx 500$ Hz the noise is of the G–R type. Hooge's noise parameter in the particular GaAs devices considered is $\alpha \approx 4.5 \times 10^{-6}$.

B. Two-dimensional electron-gas Hall devices. The necessary conditions (high mobility and low sheet concentration of carriers) for achieving high magnetosensitivity in Hall microsensors ($R_H \sim 1/n_s$) are most completely satisfied by heterojunction structures with a two-dimensional electron gas (2-DEG). The molecular-beam epitaxial growth technique (MBE) is used for the fabrication of the modulation-doped (AlAs/GaAs)$_{SL}$GaAs superlattice structures.

A cross-section of the semiconductor (AlAs/GaAs)$_{SL}$GaAs superlattice structure is shown in Fig. 4.22 [57]. The buffer and the modulation-doped 1.5-nm-AlAs/2.5-nm-GaAs superlattice layers were grown on a (100) GaAs substrate. Only the central region of the GaAs superlattice has been selectively doped by Si. The 2-DEG appears in a two-dimensional quantum well at the superlattice–buffer layer interfaces. The thickness of the 2-DEG active layer is about 10 nm. Shown schematically in Fig. 4.23 is the energy-band diagram of a superlattice structure.

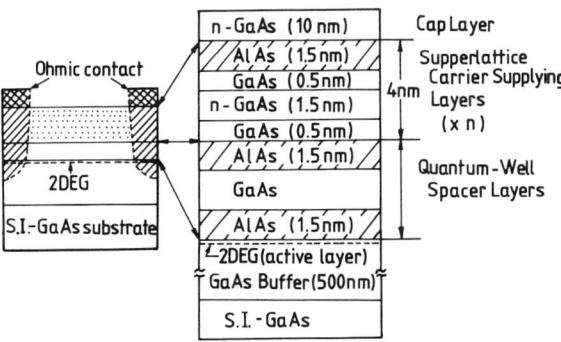

Fig. 4.22. Cross-sectional view of the superlattice structure with an undoped GaAs quantum well inserted into the AlAs barrier adjacent to the GaAs buffer. The 2-DEG is used as the active layer for the Hall sensor [57].

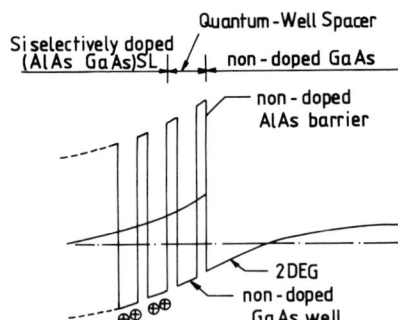

Fig. 4.23. Energy-band diagram of the superlattice structure [57].

A Greek-cross-shaped 2-DEG Hall sensor with $l = 364$ μm and $w = 200$ μm whose geometrical correction factor is $G \approx 0.8$ is described in [57]. The offset is 0.1% of the input voltage. The magnetosensitivities achieved are 1200 V A^{-1} T^{-1} and $S_A = 18$ V^{-1} T^{-1} at a supply current $I = 9.5$ mA. A small temperature dependence of about -0.11% K^{-1} of the magnetosensitivity is observed in the interval $200 \leq T \leq 400$ K. The thermal-noise source dominates in this type of Hall microsensor. The minimum detectable magnetic field is $B_{min} \approx 10$ pT. This result justifies the expectation that 2-DEG Hall sensors could be used for the registration of biomagnetic fields. A 2-DEG split-drain Hall sensor whose maximum output is 30% T^{-1} when $w = l$ is described in [57] as well. The magnetosensitivity is about twice as high as that of Si dual-drain MOS devices. A variation in impedance of 3.8% when $B = 0.25$ T and 25% at $B = 1$ T has been observed experimentally in a 2-DEG superlattice magnetoresistor with $l/w = 1.7$ [57].

The noise spectrum across the current and the Hall terminals of a 2DEG (AlGa)As/GaAs Van der Pauw element is investigated in [110]. The spectrum definitely indicates a $1/f$ dependence of the noise level without any hump due to Lorentzian generation–recombination noise. The maximum magnetosensitivity of this heterostructure is 5.7 V T^{-1}, i.e. approximately double the magnetosensitivity of conventional GaAs sensors. The minimum detectable field has been found to be 2 nT at $f = 1$ kHz from the signal-to-noise ratio of the Hall voltage at maximum sensitivity to the corresponding noise voltage.

Split-contact magnetoresistors have been implemented on the basis of modulation-doped AlAs/GaAs superlattice structures grown on a semiinsulating GaAs substrate by means of molecular-beam epitaxy. The 2-DEG at the hetero-interface serves as an active layer with high carrier mobility. Owing to thermal noise, the value of B_{min} is about 2.4 nT [59]. With $l/w = 2$, the registered magnetosensitivity is 46% T^{-1}, which is about ten times as high as that of the corresponding Si MAGFET sensor.

The general conclusion is that superlattice 2-DEG Hall devices achieve a sensitivity which is extremely high for this class of transducers and which approaches the physical limits even without a high electric field. These sophisticated magnetosensors can be integrated together with signal processing FETs onto a single chip in order to construct high-performance Hall ICs. In the author's opinion the synthesis of high-technology, superlattice 2-DEG semiconductors, for instance, with the classical Hall effect can very soon be expected to lead to new extraordinary results and applications.

4.3. Parallel-field Hall sensors

E. Hall could hardly have expected that the galvanomagnetic phenomenon discovered by him could occur when the vector B of the magnetic field is parallel to the plane of the solid-state structure. Special attention should be paid to the fact that the effect considered is not the so-called planar Hall effect, as a result of which a transverse magnetoresistance $V_{MR} \sim B^2$ arises and a nonlinear component appears at the Hall output. This effect has had no practical applications so far and is not discussed in the present book. More detailed information concerning this problem can be found in [111]. Transducers activated by a parallel magnetic field with preserved orthogonality of the current and the field B are the subject of discussion in this section. Only when $I \perp B$, is the Hall signal a linear function of the induction B. It is mere chance, in the author's opinion, that for more than one hundred years the Hall effect has been thoroughly investigated and applied in sensor electronics based only on orthogonally activated devices. The evolution of the idea of parallel-field Hall elements begins with overcoming the routine approach to the shape of Hall devices discussed in §4.2.2. The theory of conformal mapping can be used to prove the invariance of the galvanomagnetic efficiency of a Hall plate with respect to its shape. There are applications, however, in vector magnetometry for instance, when high resolution is required along with the registration of two or three components of the vector B simultaneously, or successively, by means of one and the same structure with fixed orientation. The Hall effect can be used to solve this problem by the implementation of microsensors that can be activated both by orthogonal and by parallel magnetic fields. IC technology and the possibility of using the Hall effect for the registration of a magnetic field which is parallel to the surface of the device were the basis for the creation, in 1983-1984, of a new generation of microsensors termed parallel-field or conformally transformed [112,113]. In [113] they are called vertical Hall-effect devices, owing to the direction of the current which in the structure considered is vertical with respect to the surface of the chip. In accordance with the author's viewpoint discussed in Chapter 3, the manner of sensor magnetic-field excitation is a more universal classification parameter. Therefore the term parallel-field transducers is preferred for this class of Hall devices.

4.3.1. Conformal mapping and parallel-field Hall devices

The invariance of a Hall device's electrical and galvanomagnetic properties with its geometry has been proved in compliance with §4.2.2 and Fig. 4.2. The upper complex half-plane is the main connection in conformal mapping. Until recently this procedure was referred to only as a formal subsidiary operation. The innovative approach to conformal mapping lies in the attempts to achieve sensor implementations of the geometric images from the upper complex half-plane $\text{Im}(w) > 0$. Figures 4.24 and 4.25 are the respective graphical illustrations of the transformations of a classical rectangular Hall plate into a half-plane device, and of a circular Hall plate which is invariant under a 90° rotation into a half-plane device with interchangeable current and sensor contacts. All plates are assumed to be two-dimensional and homogeneous, and have constant thickness. An important feature of their "unfolding" into planar images is the preservation of the sequence of current and Hall-contact locations and the constancy of terminal potentials. A more detailed derivation of the transformations presented in Figs. 4.24 and 4.25 can be found, for instance, in [9,11,14,15,23,24].

The practical implementation of such uncommon devices is based on the formation of rectangular ohmic strips upon one and the same side of a semiconductor plate. They serve as planar supply and output terminals. Since the trajectories of the majority carriers in the bulk of such structures are curved lines that begin and end at the metal contacts on the top surface of the chip, the Lorentz force affects, in this

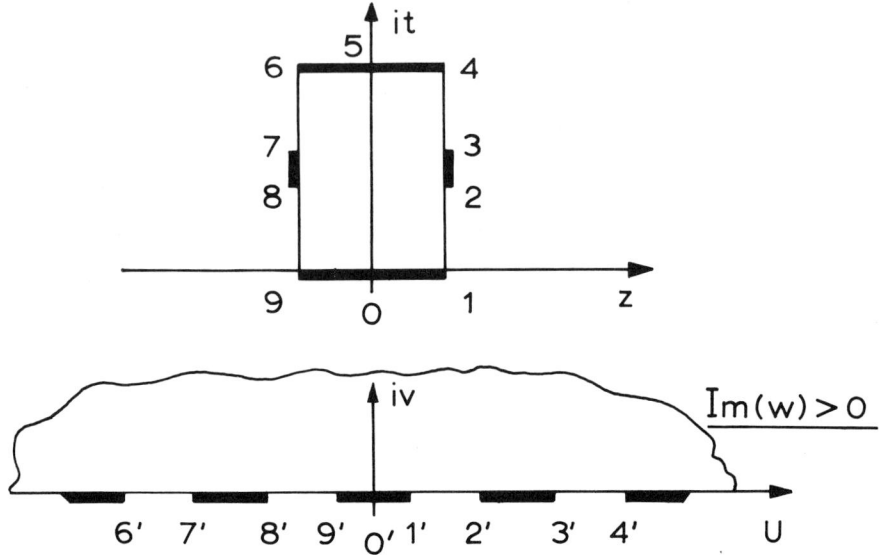

Fig. 4.24. Conformal mapping of a rectangular Hall sensor into the half-plane (parallel-field) device.

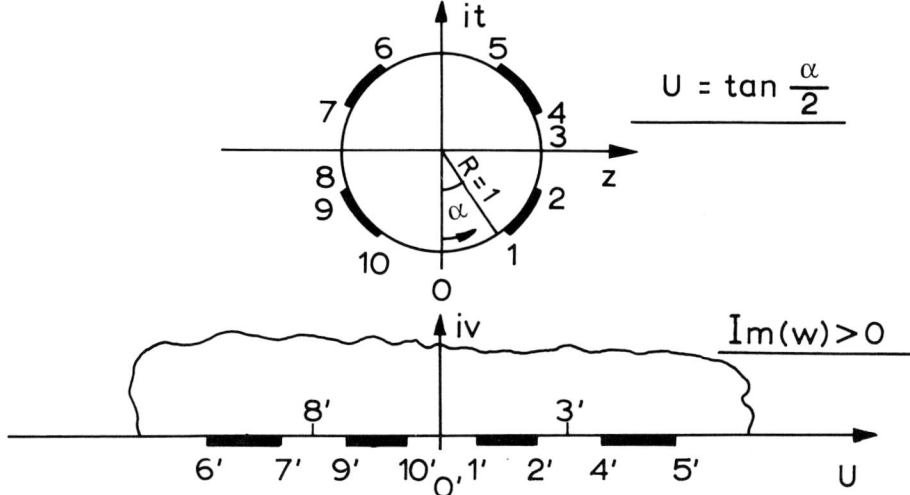

Fig. 4.25. Galvanomagnetically equivalent circular Hall plate and half-plane (parallel-field) Hall sensor, obtained by conformal mapping.

particular case, both the lateral and the vertical components of the drift velocity, v_{dr}, if the field B is parallel to the plane of the semiconductor substrate. If these sensors were to be a physical reality, they would have the following advantages from the standpoint of IC technology:

(a) all or almost all contacts are located on one and the same side of the silicon plate;

(b) the magnetosensitivity is expected to be high, since the carrier mobility attains the bulk value characteristic typical of a given semiconductor material (the active device volume is part of the substrate material);

(c) the possibility of in-plane registration of the components of the vector B.

Originally the idea of parallel-field or conformally transformed silicon Hall sensors was realized by a planar three-terminal micro-device with a differential output [112], and independently by a semicircular six-terminal vertical CMOS Hall element [113]. It has been established by experiment that the operation of novel parallel-field transducers is based on the Hall effect, i.e. the output signal (current or voltage) depends linearly on the induction B and on the supply current and is an odd function of the sign of the field B and the current I. The devices described in [112,113] are not the only ways to realize conformally transformed Hall sensors, but they have been used to test a really heuristic idea that has led to the appearance of a new class of transducing elements. Other types of parallel-field Hall microsensors made using standard IC technology have subsequently been developed. All designs of parallel-field Hall devices known so far are shown in Fig. 4.26 [6,16,24,43,112–129]. In principle, the transition from a classical rectangular Hall element to the

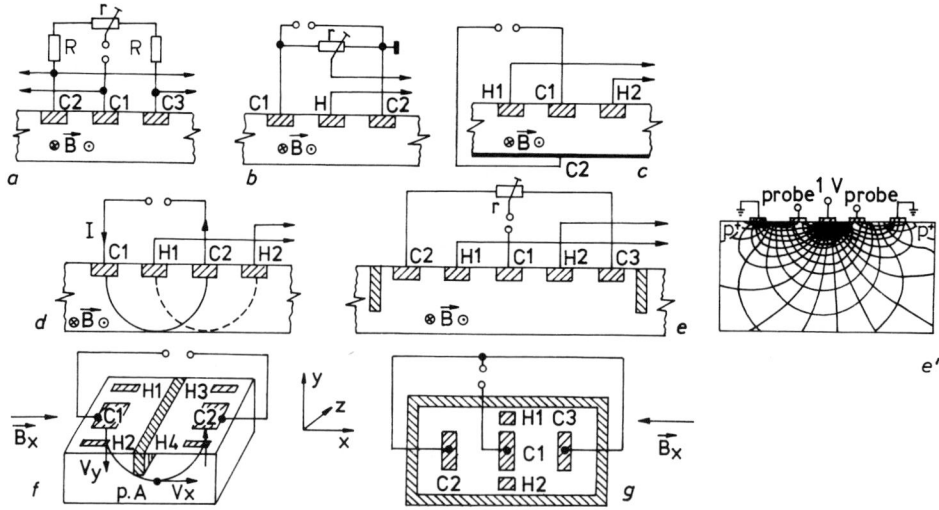

Fig. 4.26. Device structures of parallel-field Hall microsensors known so far [16].

modifications in Fig. 4.26 should not change the magnetosensitivity (c.f. §4.2.2), provided the shunting effect of contacts and the offset are negligible, the material is homogeneous and the devices are spatially confined in an adequate manner. In fact, the Hall voltage V_H generated across terminals H_1 and H_2 along the line of integration H_1–H_2 is found from the expression:

$$V_H = \int_{H1}^{H2} \mathbf{E}_H dl, \qquad (4.68)$$

where $\mathbf{E}_H = R_H \mathbf{j} \times \mathbf{B}$ is the local Hall field, R_H is the Hall coefficient and $\mathbf{j}$ is the current density.

Since the current $\mathbf{j}$ and the field $\mathbf{B}$ are mutually perpendicular, the resulting Hall field is $E_H = R_H jB$ and

$$V_H = R_H B \left(\frac{1}{t}\right) \int_{H1}^{H2} jt dl, \qquad (4.69)$$

where t is the thickness of the conformally transformed sensor in the direction of the magnetic field.

Therefore:

$$V_H = R_H B I \frac{1}{t}, \qquad (4.70)$$

where I is the total supply current in the device.

It can be concluded, from a comparison of (4.70) with the well-known expression (4.28) for the Hall voltage in a rectangular sample, that the transducing efficiency is invariant in spite of the unusual geometry. The isolation of the active sensor region of parallel-field devices in the semiconductor plate itself (usually n-type) is achieved by the conventional methods of IC technology — a deep p^+-isolating ring, for instance. The author believes that the existence of this class of Hall sensors is related also to the fact that the Hall effect is an odd galvanomagnetic phenomenon. If an even magnetoresistance effect is the principle of operation of the original orthogonal device which has been conformally transformed, the magnetic field will induce a synphase variation of the properties in the respective symmetric parts of the parallel-field modification, irrespective of the direction of the current. Thus no differential output signal will appear at the output. Detailed information about all types of parallel-field Hall microsensors known so far can be found in [16].

4.3.2. Device structures, operating principle and characterization of parallel-field Hall microsensors

According to the physical nature of the Hall effect, the conditions necessary for its manifestation are that the current I and the external magnetic field B should be perpendicular and that a boundary surface should exist to which the Lorentz force $F_L \sim v_{dr} \times B$ would redistribute the majority carriers with a drift velocity v_{dr} (the absence of such a boundary excludes the occurrence of that galvanomagnetic phenomenon). These conditions are fulfilled in the integrated sensors with strip input and output contacts (along the z axis) presented in Fig. 4.26. In fact, the magnetic field $B \| z$ applied perpendicularly to the cross-section, i.e. parallel to the chip surface, by the F_L force simultaneously disrupts the symmetry of the current and potential pictures in the structure (the electrical symmetry at a field $B_z = 0$ derives from the geometry). As a result of this magnetic deformation, a Hall signal is generated on the plane of the chip, detectable by the respective output terminals Hk ($k = 1, 2, 3, 4$). Unlike the classical long Hall plates, the characteristic feature of the galvanomagnetic properties of most of the devices shown in Fig. 4.26, in which the substrate itself serves as an active sensor region, is the coexistence of the Hall effect and the geometrical magnetoresistance effect (the Hall elements are neither short nor long).

To a smaller degree, this property should be manifest in sensors confined by a p-ring and manufactured by bulk CMOS technology [113], and in those embedded in an n-epilayer and made with standard bipolar processes [120,122]. Within these structures, the supply current has to flow within a region relatively well confined by the technology. The occurrence of the magnetoresistance is associated with the incomplete compensation of the Lorentz deflection by the Hall field E_H owing to the nonlinear shape of the current paths. As a result, the magnetic field B induces an imbalance in both the potential and current distributions. Beside their general behavior as Hall sensors, the devices in Fig. 4.26 have some features that need to be discussed separately.

Figure 4.26a presents a cross-section of the structure and the biasing circuit of a triple parallel-field Hall microsensor. The three ohmic contacts $C1$, $C2$ and $C3$

Table 4.5
Parameters of silicon parallel-field Hall microsensors whose device structures are shown in Fig. 4.26

No.	Device structure	Output	Sensitivity S	Equivalent magnetic-field noise density ($T^2\,Hz^{-1}$)	Equivalent offset (mT)	Linear range (T)	NL (%)	Temperature coefficient of S (% K^{-1})	Refs.
1	(a), n-substrate as active reg.	differ.	$70\,mV\,T^{-1}$, $I = 10\,mA$, $l_{C1,2} = 300\,\mu m$		5	±1	1	0.1	[112,115]
2	(b), n-substrate as active reg.	single, after comp. MR	$75\,mV\,T^{-1}$, $I = 10\,mA$, $l_{C1,2} = 200\,\mu m$			±1	1	0.1	[116]
3	(b), n-substrate as active reg.	single	DFM ~ 4, $|B| = 1\,T$						[116]
4	(c), n-substrate as active reg.	differ.	$100\,mV\,T^{-1}$, $I = 10\,mA$, $l_{C1,2} = 200\,\mu m$		6	±1	1	0.1	[6]
5	(d), n-substrate as active reg.	differ.	$120\,mV\,T^{-1}$, $I = 9\,mA$, $l_{C1,2} = 300\,\mu m$		10	±1	1		[118]
6	(d), n-substrate, confin. by p-ring	differ.	$300\,V\,A^{-1}\,T^{-1}$						[119]
7	(e), n-substrate, confin. by p-ring	differ.	$450\,V\,A^{-1}\,T^{-1}$	3×10^{-13} ($f = 100\,Hz$), 10^{-15} ($f = 100\,kHz$)			1		[113,114]
8	(e), n-substrate as active reg.	differ.	$190\,mV\,T^{-1}$, $I = 10\,mA$		7	±1	1	0.1	[117]
9	(e), n-substrate as active reg.	differ.	$24.7\,mV\,T^{-1}$, $I = 1.0\,mA$, $l_{C1,2} = 80\,\mu m$			±1	2		[78]
10	(e), n-substrate as active reg.	differ.	$66.6\,mV\,T^{-1}$, $I = 1.0\,mA$, $l_{C1,2} = 80\,\mu m$			±1	2		[78]
11	(e), n-epilayer	differ.	$101\,mV\,T^{-1}$			±0.4			[125]
12	(g), n-epilayer	differ.	$75\,V\,A^{-1}\,T^{-1}$					0.1	[120,127]
13	(g), n-epilayer	differ.	$1243\,V\,A^{-1}\,T^{-1}$						[122]
14	(g), n-epilayer	differ.	$41\,V\,A^{-1}\,T^{-1}$	10^{-10} ($f = 40\,Hz$)					[128]

have strip geometry and are equidistant from one another. The differential output voltage $V_{C2,C3}$ is taken directly from the terminals $C2$ and $C3$. The magnetic field B is applied perpendicularly to the cross-section [112,115].

Experiments were carried out on samples manufactured by standard planar technology on n-type silicon wafers. The substrate serves as an active sensor region. The inevitable offset $V_\text{off} \equiv V_{C2,C3}$ ($B = 0$) is annulled in the field $B = 0$ by means of a trimmer r (Fig. 4.26a). The voltage $V_{C2,C3}(B)$ is a linear function of the induction B and the supply current I_{C1}, changing its sign with the polarity of the field B and of the current I_{C1}. The output signal increases parallel to a rise in the load resistor R to a certain value $R \geq 4$ kΩ, after which saturation is observed. With a decrease in the distance $l_{C1,C2} = l_{C1,C3}$, the voltage $V_{C2,C3}(B)$ increases substantially. Depending on the working point, geometrical distances and the load resistor R, the magnetosensitivity may vary within a wide range. Table 4.5 contains typical values of the basic sensor parameters of the devices in Fig. 4.26. An interesting effect in the device structure of Fig. 4.26a is observed when terminals $C1$ and $C2$ are used as outputs. A section of negative magnetoresistance appears in the curve $V_{C1,C2}(B)$, i.e. $V_{C1,C2}$ decreases with increasing of B, after which the signal increases as a square function, $V_{C1,C2} \sim B^2$. If, however, contact $C3$ is disconnected, the negative magnetoresistance disappears and a typical dependence $V_{C1,C2} \sim B^2$ is obtained.

The performance of this sensor structure can be explained by the occurrence of the Hall effect and the geometrical magnetoresistance. In Fig. 4.26a the lines of flow of the currents $I_{C1,C2}$ and $I_{C1,C3}$ are curved, beginning and ending on one side of the substrate. The influences of the external magnetic field B_z, via the Lorentz force F_L, in individual sections of the trajectory are different. For instance, under the ohmic contacts $C1$, $C2$ and $C3$, to a first approximation, the velocity of the dominant part of the carriers v_dr is parallel to the axis y, and the carrier deflection prevails, i.e. $\Theta_{Hn} \simeq -\mu_n B$ (for an n-type substrate). Owing to the deflection θ_H of the current I_{C1} under the central contact $C1$, the output Hall signal $V_{C2,C3} \sim I_{C1} B$ appears on output terminals $C2$ and $C3$. The magnetoresistance voltage $V_{MR} \equiv V_{C1,C2}(\pm B) = V_{C1,C3}(\pm B) \sim B^2$ represents a synphase (additive) component of the output $V_{C2,C3}(B)$. Due to differential sensor circuit, this MR component is compensated. On the other hand, the Hall field develops only if the current is forced to flow parallel to an isolating boundary. Owing to the nonlinear trajectory of currents $I_{C1,C2}$ and $I_{C1,C3}$, the velocity v_dr has two components v_x and v_y. Therefore mainly the velocity v_x is responsible for the generation of the Hall field at $B_z \neq 0$. This is the reason why Hall voltages are generated on the upper surface between contacts $C1$ and $C2$, and $C1$ and $C3$. They have a nonuniform distribution on the chip surface. The maximum values of the Hall voltages are reached in the middle between $C1$ and $C2$ and $C1$ and $C3$. This conclusion has been verified by investigation of the magnetoresistance (MR) with respect to the orientation of the field B in the structure shown in Fig. 4.26a. Only two contacts have been connected to the biasing circuit, namely $C1$ and $C2$, or $C1$ and $C3$. In our case it has been established that at $I_s =$ const., when the field $\boldsymbol{B}$ is oriented along the x axis $(\boldsymbol{B}\|x)$, the change in MR is about 30% smaller than in the case $(\boldsymbol{B}\|y)$ and $(\boldsymbol{B}\|z)$. In the field $(\boldsymbol{B}\|x)$, the Lorentz force $\boldsymbol{F}_L$ acts only upon the velocity v_y, which determines the essential role played by the deflection near the ohmic contacts $C1$, $C2$ and $C3$. In the field $(\boldsymbol{B}\|y)$ or $(\boldsymbol{B}\|z)$ the increase in MR is not as significant as in the case $(\boldsymbol{B}\|x)$, because a Hall field is generated that is a counterbalance to the lateral magnetic "pressure" on the carriers. The negative magnetoresistance at terminals $C1$ and $C2$, or $C1$ and $C3$, (Fig. 4.26a) is related to the summing up of the even magnetoresistance effect and the odd Hall effect [16].

It should be pointed out that the short-circuit effects inevitably reduce the output Hall signal in the device in Fig. 4.26a. A successful attempt has been made in the introduction of two-dimensional numerical modeling of complex kinetic processes in the presence of a magnetic field for the sensor shown in Fig. 4.26a [70].

The parallel-field microsensor shown in Fig. 4.26b is characterized by the possibility of the voltage $V_{H,C2}(\boldsymbol{B})$ being asymmetrical with respect to the sign of the field $\boldsymbol{B}$ [116]. These directional galvanomagnetic properties are likewise associated with the simultaneous generation of the magnetoresistance MR and the Hall effect. Since the voltage $V_{C1,C2}(\boldsymbol{B})$ is an even function of $\boldsymbol{B}$ and V_H is an odd one, depending on the polarity of the field $\boldsymbol{B}$, these two galvanomagnetic signals are summed or subtracted, thus determining the asymmetry of the curve $V_{H,C2}(\boldsymbol{B})$. In the experiments carried out with the device structure in Fig. 4.26b manufactured in n-type silicon with resistance $\rho \approx (6-9)\Omega\cdot\text{cm}$ have displayed a directional figure of merit $\gamma_{\text{DFM}} \equiv V_{H,C2}(+\boldsymbol{B})/V_{H,C2}(-\boldsymbol{B}) \sim 4$, when $B = 1T$. Analogous to the results and ideas presented in ref. [103] concerning the Hall effect in nonhomogeneous magnetic fields in InSb and GaAs, similar solutions could also be sought for the sensor of Fig. 4.26b. Linear applications of the structure in Fig. 4.26b are of interest to magnetometry. One of the possible neutralization circuits for the MR component and the obtaining of a zero offset has also been shown. In this case, the linear output signal is taken between the H contact and the r trimmer and is half the effective Hall voltage developed in the sample.

In the device structure in Fig. 4.26c, the bottom surface of the chip is used as contact $C2$, and the other contact $C1$ and the sensor terminals $H1$ and $H2$ are located on the top surface. The silicon substrate serves as a sensor region [6]. It has been established that on shortening the distance $l_{C1,H1} = l_{C1,H2}$ the Hall voltage increases substantially. This result can be explained by the considerable difference in the area of the current contacts $C1$ and $C2$. Close to $C1$, the density of the current lines at $I_{C1,C2} =$ const increases, as well as the velocity v_{dr} of the carriers. Since the field $\boldsymbol{E}_{H1,H2} \sim v_{\text{dr}} \times \boldsymbol{B}$ it also increases. However, the shifting of terminals $H1$ and $H2$ closer to contact $C1$ is limited by their shunting influence on the voltage V_H.

The parallel-field microtransducer in Fig. 4.26d with strip geometry for the contacts was suggested for the first time and examined in ref. [118]; the substrate serves as the active sensor region. After computer simulations of that device structure, a version was manufactured by a standard bulk CMOS-compatible process, whereby the offset has practically been eliminated [119]. The operation of the device is similar to that of the sensor shown in Fig. 4.26a. The current $I_{C1,C2} \sim n v_{dr}$ generates the Hall voltage on the chip surface close to the contacts $C1$ and $C2$ ($v_{dr} = v_x + v_y$). When one of the Hall terminals H_1 is obligatorily located between the contacts $C1$ and $C2$, and the other terminal $H2$ is outside the area restricted by $C1$ and $C2$, the structure in Fig. 4.26d can register the Hall signal $V_H(B_z)$. What is more, it is of no importance on which side of the two contacts $C1$ or $C2$ the Hall terminal $H2$ is located. In this device the current contacts are interchangeable [16].

The parallel-field Hall microsensor shown in Fig. 4.26e exists in two basic versions: the first concerns the structures in which the active region is an n-substrate surrounded by a p-type ring as shown in the figure (a reversed biased p–n junction ring substrate), manufactured by bulk CMOS technology [113,114]; in the second version, the silicon substrate itself is used as the active sensor region [117] (where a p-type ring is omitted). Table 4.5 also includes the data from [78] for two microsensors similar to the device from [117]. In the first microsensor the dimensions of the n^+ regions are 150×10 μm^2 and in the other one, which is more sensitive, the respective dimensions are 30×10 μm^2.

The purpose of the ring, [113], is to modulate the effective thickness of the device. The operating principle of the transducer is the same as that of the sensors already described in Fig. 4.26a–d. The annulling of the offset in the field $B_z = 0$ is achieved by the trimmer r, i.e., $V_{H1,H2}(B = 0) = 0$. In the CMOS version [43], an original way of eliminating the offset has been suggested: additional offset-correction contacts OL and OR are formed between the terminals $C2$ and $C3$ and the respective sensor terminals $H1$ and $H2$, and the shallow p-layer is used to obtain the buried structure. A comparison of the magnetosensitivity of the structures in Fig. 4.26e without a deep p-type ring shows that the transducing efficiency of the sensor in Fig. 4.26a is about three times lower than that of the device in Fig. 4.26e, whereas the sensitivities of the transducers in Fig. 4.26d and e practically coincide [117,118]. In general terms, these results corroborate the invariance of the sensitivity of the Hall elements with respect to geometry. Numerical 2-D modeling of current and potential lines (Fig. 4.26e$'$) using the Green-function approach is obligatory for ensuring optimum parameters of the sensor in Fig. 4.26e [70,114]. The initial results along these lines are encouraging. It has been pointed out [26] that the equivalent input magnetic-field noise spectral density for the microsensor in Fig. 4.26e, for $I = 0.5$ mA and $W \simeq 1$ mW, at 100 Hz is 3×10^{-13} T^2 Hz^{-1} (1/f noise) and at 100 kHz it is $\sim 10^{-15}$ T^2 Hz^{-1} (thermal noise). A recent report describes a silicon magnetic-field device with a frequency output [125]. A structure similar to the one in Fig. 4.26e was used as a transducer cell. The frequency output is based on two I^2L ring oscillators. The Hall cell is embedded in an n-type epitaxial layer on a

p-type substrate. This sensor structure has no buried layer. A magnetosensitivity of about 101 mV T^{-1} has been obtained at a power consumption $W \simeq 3.1$ mW.

The operation of the microsensor in Fig. 4.26f [121], is based on the nonlinear trajectories of the carriers, shown by a line starting and ending on the upper surface containing $C1$ and $C2$. The ohmic contacts $C1$ and $C2$ are equipotential planes with current lines perpendicular to them at $B = 0$. This is the reason why the effective trajectory in a first approximation is symmetrical with respect to point A, where the vertical velocity $v_y = 0$, and $v_x = v_{dr}$. Right under the contacts $C1$ and $C2$ the velocity $v_y = v_{dr}$, and the lateral component $v_x \sim 0$; the directions of v_y under $C1$ and $C2$ have opposite signs, i.e., $|-v_y(C1)| = |+v_y(C2)|$. When an external magnetic field B_x parallel to the substrate and the Hall contacts $H1$, $H2$, $H3$ and $H4$ is applied, the Lorentz force F_L acts on the vertical velocity v_y. The deflections of the carriers to the left and right of point A have opposite signs, $\pm F_L \sim \pm v_y \times B_x$. As a result two Hall voltages with opposite signs are generated: $-V_{H1,2}$ and $V_{H3,4}$ on the upper surface near contacts $C1$ and $C2$, which are detected by terminals $H1$–$H2$ and $H3$–$H4$. Essentially, the current $I_{C1,2}$ creates two individual Hall signals in the structure. A further increase of the magnetosensitivity of the Hall sensor can be obtained by forming a deep p^+-ring in the n-type Si substrate surrounding the contacts $C1$ and $C2$ and the Hall contacts $H1$–$H2$ and $H3$–$H4$, and dividing the area between electrodes $C1$ and $C2$ into two equal parts. This p^+-ring restricts the effective dimensions of the sensor region and optimizes the influence of the field B_x on the vertical component v_y. The basic difference between the parallel-field Hall microsensor in Fig. 4.26f and that described in Fig. 4.26g [120,122], is that in the latter another buried n^+-layer is formed under the current contacts in an n-type epitaxial layer. Owing to the low ohmic n^+-layer, a vertical current I_y generates the Hall voltage V_H. The source current I has a significant vertical component I_y, which is the result of the low-resistance n^+ buried layer. This current I_y is responsible for the sensitivity to B_x. The product sensitivity S[V A^{-1} T^{-1}] is calculated as $S = 0.74 R_H K (l/w)/t$.

Reference [122] presents descriptions of the sensors in Fig. 4.26g, in which reduction of the width of the structures and, hence, enhanced magnetosensitivity up to 1243 V A^{-1} T^{-1} in the so-called trenched version have been achieved by the use of sophisticated processing steps.

A CMOS magnetosensitive IC consisting of an on-chip parallel-field Hall sensor like that i Fig. 4.26e and a BJT amplifier is described in [129]. An overall sensitivity of 40 V T^{-1} at a supply current $I = 1$ mA and a noise spectral density less than 1 μT Hz$^{-1/2}$ at $f = 100$ Hz have been achieved by this integrated device. Because of its low power consumption, this Hall IC is a promising candidate for practical application.

There are parallel-field Hall sensors, Fig. 4.27, based on junction field-effect transistors (JFETs). The semiconductor material used is germanium [130], or silicon [131]. Unlike other conformally transformed versions in which the output

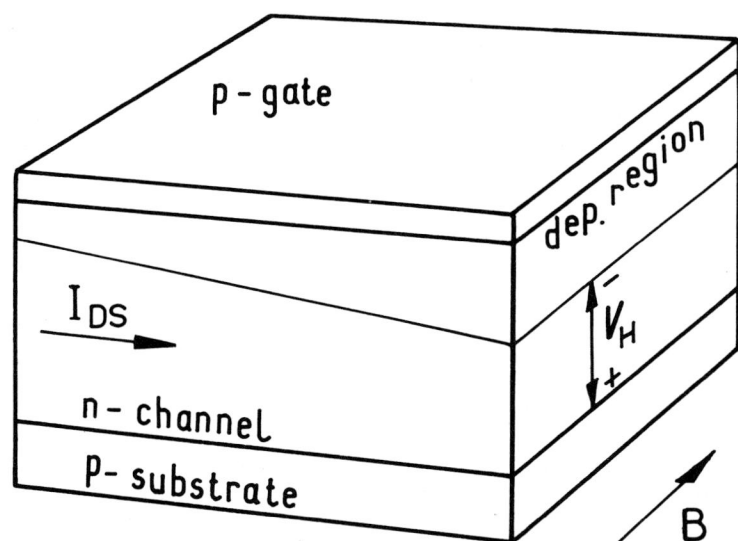

Fig. 4.27. A JFET Hall sensor with channel thickness modulated by the Hall voltage V_H [130].

is a differential signal across the Hall terminals, in JFET devices the magnetic field B modulates the junction bias $V_{DS}(B)$ and the current $I_{DS}(B)$. The variation of the drain current is a measure of the magnetic induction. The results have been obtained at a temperature $T = 77$ K [130], while at $I_{DS} = 30$ mA in a 0.6 T magnetic field the relative variation $\Delta I_{DS}(B)/I_{DS}(0)$ is about 0.035%. These sensors, however, possess high cross-sensitivity to the other two components of the magnetic field vector.

4.3.3. Outlook and conclusions

A class of semiconductor magnetosensitive devices with planar input and output contacts has been created. These sensors can measure an external magnetic field which is parallel to the plane of the device, and their properties are determined by the Hall effect. A connection has been found between these transducers and the classical Hall element, based on conformal mapping. The analysis of the results of the Hall effect in orthogonal and parallel-field devices leads the author to assume that solid-state structures with planar contacts offer a more general and fundamental approach to observing the consequences of the deflection of majority carriers by the Lorentz force. In principle, the classical arrangement of the orthogonal Hall device can be transformed into a parallel-field version without affecting its electrical and galvanomagnetic properties. This conclusion follows from the identical origin of sensor operation irrespective of a modification of the geometry of the device.

Which is the best amidst the variety of parallel-field Hall microsensors? Unfortunately, for the time being, there is no unambiguous answer to this question. The underlying cause is above all in the blank spaces in Table 4.5. There are as yet few data about the various device structures. A comparative analysis is difficult even regarding the basic sensor parameter — the magnetosensitivity S. Information on this matter is presented in three different ways: $S_A \equiv [\text{V T}^{-1}]$; $S_{RI} \equiv [\text{V A}^{-1} \text{T}^{-1}]$ and $S_{RV} \equiv [\text{T}^{-1}]$ (§3.2.1). The lack of a generally accepted adequate method of defining sensitivity results in certain misunderstandings in the comparison of similar parallel-field transducers. At this stage, results have shown that the devices whose sensor region is suitably limited by a p^+-ring or is confined by the epilayer have higher magnetosensitivity. The optimization of conformally transformed Hall transducers requires the development of adequate numerical methods that take into account the "semiconfined" nature of sensor structures, especially when the geometrical correction factor of a particular device modification is to be determined.

Although novel Hall sensors are still at the beginning of their evolution, future priority research on parallel-field microdevices should be along the following lines:

(a) The introduction of GaAs as an alternative material for the fabrication of these sensors;

(b) Vector magnetometry, compass and smart solutions.

4.4. Magnetoresistance sensors

Generally magnetoresistors are two-terminal semiconductor sensors whose operation is based on the geometrical magnetoresistance effect. In a magnetic field B their resistance is increased irrespective of the direction of the field. The principle of operation of these transducers, the influence of geometry, carrier mobility, etc. have been considered in §2.4.

4.4.1. Magnetoresistor design and materials

It can be seen in §2.4.2, and from equations (2.74) and (2.74′) that the magnetoresistance effect is most strongly manifested in the Corbino disc (Fig. 2.15b). Interesting results concerning the influence of the shape upon the magnetoresistance are illustrated in Fig. 4.28 [132]. The samples shown in Fig. 4.28b–e, f and j manifest the same magnitude of the magnetoresistance effect as the Corbino disc. The bold lines in Fig. 4.28 are the metal electrodes and the thin lines denote the isolating boundaries of the semiconductor plates. Almost the same magnetoresistance is observed in the structures in Fig. 4.28g–i as that of the Corbino disc. The following causes, as seen from Fig. 4.28, eliminate the Hall effect in the device structures shown:

(a) one of the electrodes encompasses the other one;

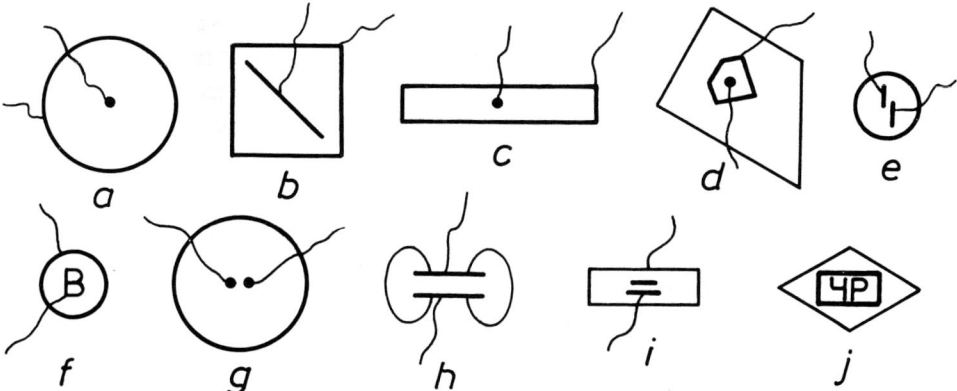

Fig. 4.28. Sample shapes equivalent to the Corbino disk with respect to the magnetoresistance effect.

(b) both electrodes are encompassed by an external, closed conducting loop;

(c) the sample periphery is at a considerable distance from the two terminals, whose shape is not important.

The greater the variation of the active resistance per unit variation of the magnetic induction, the larger the signal from the magnetoresistive element. The initial resistance of the sample should have a considerable value to make the measurement of the variation $R(B) - R(0)$ easier and more accurate. In this respect, all the device structures presented in Fig. 4.28 have the serious disadvantage that their initial resistance $R(0)$ is low, especially when the material is InSb or InAs, whose resistivity is in the range 10^{-3} to $10^{-2} \Omega \cdot$cm. Because of this fact the InSb Corbino configuration has resistance $R(0) \approx 0.05$ to $0.5\ \Omega$, and the resistance $R(0)$ of a disc whose thickness is 10 μm is only about 5 Ω [133].

The influence of the material of the sensor upon the magnetoresistance effect is most strongly manifested by the carrier mobility. In most practical cases it approaches the Hall mobility μ_H, i.e. $\mu_M \approx \mu_H$. It has been shown in Table 4.2 that InSb has the highest carrier mobility. This is why it is preferred as the material for magnetoresistive components. The narrow bandgap, however, leads to a high temperature coefficient of the resistance. Its value is about -2% K^{-1} at $T = 300$ K. This problem can be partly overcome by increasing the density of doping impurities, and therefore the extrinsic conductivity, in order to widen the temperature interval of operation. The temperature dependence of the normalized internal resistance $R(T)/R\ (T = 20°C)$ for selected semiconductor materials is shown in Fig. 4.29.

The main requirement in the design of a magnetoresistance sensor is to minimize the l/w ratio as seen from eqs. (2.74), (2.74′), (2.75), (2.75′). High conductivity, $\sigma \approx 200$ to $800\ (\Omega \cdot$cm$)^{-1}$, is typical of extrinsic (doped) InSb. Because of this, the ideal single Corbino disc and the short structure ($l \ll w$) have no practical importance. Several magnetoresistors can be connected in series to increase the output resistance $R(0)$. Corbino discs connected in series are shown in Fig. 4.30a.

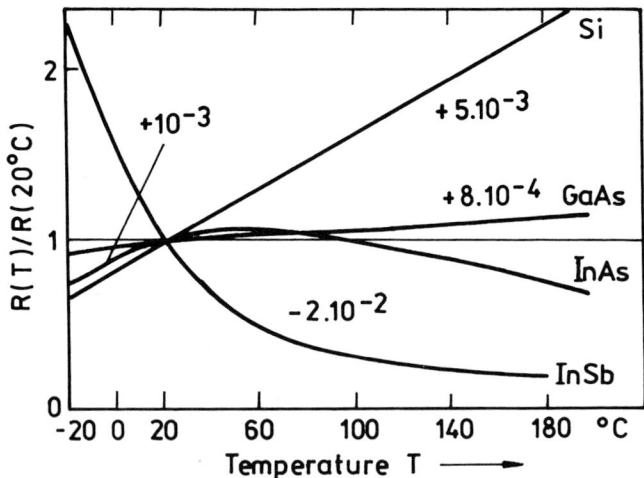

Fig. 4.29. Temperature dependence of the resistance $R(T)/R(T=20°C)$ of selected semiconductors.

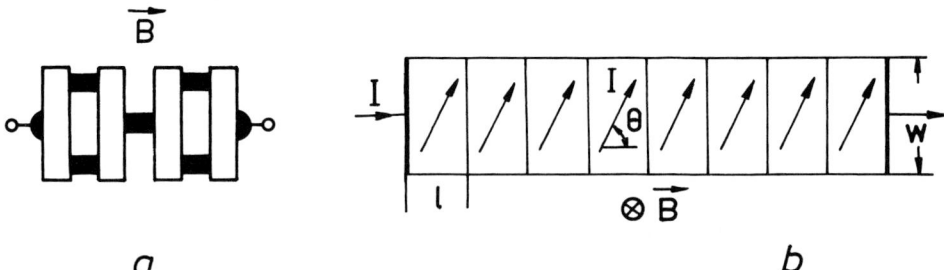

Fig. 4.30. Connection of Corbino discs in series (a), and a semiconductor plate with short-circuit straps and current paths (b).

The shortcoming of such an arrangement is the increased length of the device in the direction of the field **B**, which leads to a greater distance between the poles of the magnet, thus reducing the excitation field.

This disadvantage has been avoided by the structure in Fig. 4.30b, where the magnetoresistors connected in series have been replaced by a long semiconductor plate. Metal strips are deposited upon its surface to divide it into regions whose lengths are smaller than their widths. In such a way each region serves as an individual magnetoresistor. This approach leads to the minimization of the Hall effect and makes for the achievement of an output resistance $R(0)$ in the range 100 to 1000 Ω. The gridded InSb magnetoresistors fabricated by means of evaporation techniques are a similar solution to the problem. The shunting strips of high conductivity metals are deposited perpendicularly to the direction of the supply current I [111,134]. The initial resistance is controlled by the cross-section and by the length of the magnetoresistance system. The magnetoresistors covered by metal strips (Fig. 4.30b) can be replaced by a material with built-in high conductivity regions formed during the crystal growth. Sensors of this type have been implemented on the basis of the eutectic InSb + 1.8% NiSb alloy [111,135].

The compound NiSb whose resistivity is lower by two orders of magnitude than that of the semiconductor material forms crystal needles in the InSb whose diameters are about 1 μm; the length of each needle is approximately 50 μm and the distance between two separate needles is ≤ 10 μm. The needles should necessarily be perpendicular to the crystal axis and to the magnetic field. This requirement is illustrated by Fig. 4.31 [136]. The resultant length-to-width ratio in this structure is $l/w \leq 0.2$, and the current I flows in a zigzag manner. The plates fabricated in such a way are then sliced parallel to the NiSb needles and their thickness is reduced to about 25 μm. Photolithography and chemical etching techniques are used to prepare meander-shaped magnetoresistors with high resistance $R(0)$. Such magnetoresistors are shown in Fig. 4.32. Indium, In, is deposited upon the contact areas by a galvanic process. The resistance meander is glued onto a soft-iron or ceramic carrier and the wire connections are soft soldered to the pads. The completed sensor is shown in Fig. 4.32. The whole structure is protected by a film of lacquer.

Novel microelectronic processing techniques are also employed for the fabrication of magnetoresistors. The magnetic field is concentrated onto the device's active region by means of ferrite carriers upon which the magnetoresistor itself is mounted. Detailed information about some perfections of the fabrication technology of semiconductor magnetoresistors is included in [136].

The semiconductor compound $(HgTe)_{1-x} + (CdTe)_x$ has been lately used for the manufacture of magnetoresistors. The bandgap width and carrier mobility

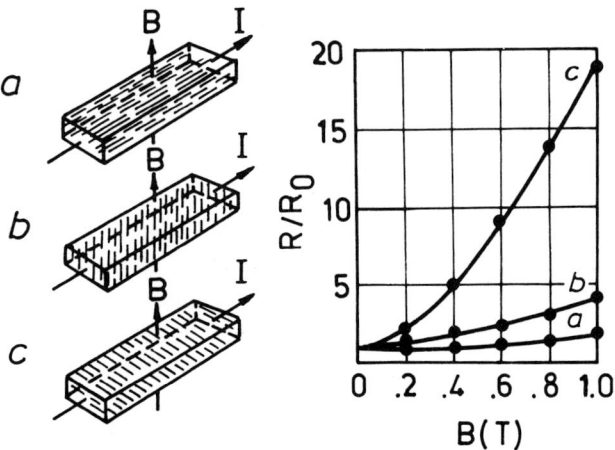

Fig. 4.31. Increase in resistance with a magnetic field.

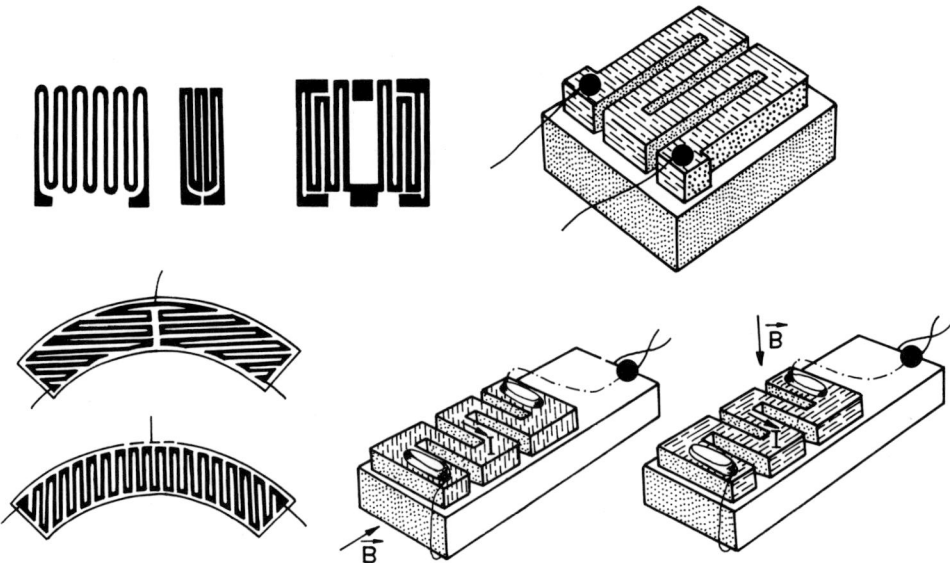

Fig. 4.32. Layout and examples of contact pads of meander-type magnetoresistors.

in such materials depend on the composition and temperature. The compound with $E_g = 0$ exhibits the maximum mobility; hence, its magnetosensitivity at any temperature exceeds that of InSb devices. Low initial resistance $R(0)$, being the typical shortcoming of magnetoresistors, is also manifested by the (HgTe)$_{1-x}$ + (CdTe)$_x$ devices. The maximum transducing efficiency in the cryogenic temperature interval $20 \leq T \leq 100$ K is observed at $x = 0.1$ [137].

4.4.2. Properties and characterization of magnetoresistors

The main metrological characteristics of two-terminal magnetoresistance sensors are the initial resistance $R(0)$, which can vary between less than 1 Ω and tens of kΩ, and the magnetosensitivity $S_{MR} = \partial R(B)/\partial B$, which is measured in Ω G^{-1} or Ω T^{-1}. The relationship

$$\frac{\Delta R(B)}{R(0)} = F(B),$$

where $\Delta R(B) = R(B) - R(0)$, is often used for the characterization of the transducing efficiency. Usually the supply current is in the range $1 \leq I \leq 100$ mA. Magnetoresistors operate in the constant current mode ($I = $ const) and the absolute magnetosensitivity $S_A = \partial V/\partial B|_I$ [mV T^{-1}] can be used for their characterization.

The resistance $R(0)$ is determined by the geometry of the device and by the conductivity of the sensor material. With the typical magnetoresistor transverse dimensions $w = 80$ μm and $t = 25$ μm, the desired value of the resistance $R(0)$ is obtained by varying the length l. The conductivity of InSb, for instance, is "tuned" by the concentration of doping impurities. The three concentrations used by SIEMENS have been widely accepted as standard [138]:

$\sigma = 200$ (Ω·cm)$^{-1}$ undoped $n = 2 \times 10^{22}$ m^{-3} D-type;
$\sigma = 550$ (Ω·cm)$^{-1}$ Te-doped $n = 6 \times 10^{22}$ m^{-3} L-type;
$\sigma = 800$ (Ω·cm)$^{-1}$ Te-doped $n = 2 \times 10^{23}$ m^{-3} N-type.

In a low magnetic field at a current $I = $ const, the output magnetoresistance signal according to equation (2.74) is a square function of B, i.e. $V_{MR}(B) \sim R(B) \sim B^2$. In high fields the relationship is almost linear: $V_{MR}(B) \sim R(B) \sim B$. The magnetoresistor output is an even function of the polarity of $\boldsymbol{B}$; that is why this type of magnetosensors cannot be used to determine the direction of the vector $\boldsymbol{B}$. The resistivity of the material and carrier mobility as well as the magnetosensitivity of the device decrease with increasing doping concentration (Fig. 4.33). The dependence of the relative magnetoresistance $R(B)/R(0)$ on the angle φ between the direction of the current $\boldsymbol{I}$ and the vector $\boldsymbol{B}$, the axis of the NiSb needles being perpendicular to the field $\boldsymbol{B}$, has been investigated and is described in [111].

A proportionality between $R(B)/R(0)$ and $\sin^2 \varphi$ is found to be the resultant relationship. An analogy with this case is observed when the current $\boldsymbol{I}$ is perpendicular to $\boldsymbol{B}$ and the angle ψ between the vector $\boldsymbol{B}$ and the NiSb needle axis is varied: $R(B)/R(0) \sim \sin^2 \psi$.

Square-shaped magnetoresistors ($l = w$) exhibit an interesting and useful property. In a relatively high field, the dependence of the resistance on B is given by the relationship $R(B) = (R_H/t)B$, where R_H is the Hall coefficient and t is the thickness of the MR element. Thus the increase in resistance is linear with the

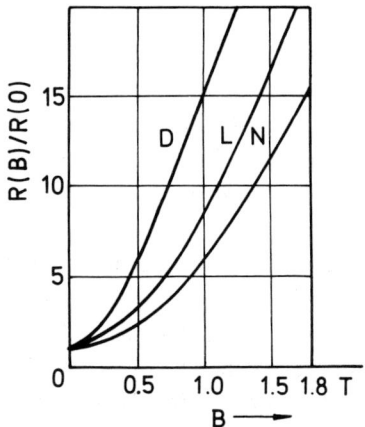

Fig. 4.33. The magnetic-field dependence of the normalized resistance $R(B)/R(0)$ of magnetoresistors in the InSb/NiSb eutectic for three different doping grades [138].

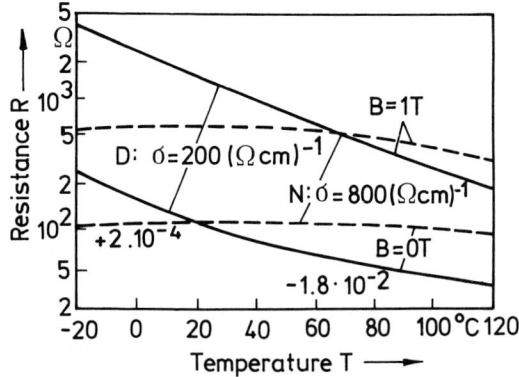

Fig. 4.34. The temperature dependence of the resistance of a magnetoresistor in InSb/NiSb for D and N doping grades, for $B = 0$ T and $B = 1$ T [138].

Table 4.6

Values of the ratio $R(75°C)/R(25°C)$ at magnetic fields $B = 0$ T and $B = 1$ T for various materials [138]

Material	$R(75)/R(25)$ (%) for $B = 0$ T			$R(75)/R(25)$ (%) for $B = 1$ T		
	min.	main value	max.	min.	main value	max.
D	45	47	55	28	28	35
L	74	84	94	53	63	75
N	90	95	99	76	82	89

applied magnetic field B, and the sensitivity is just equal to that of a Hall sensor of the same material and with the same thickness [136].

The temperature coefficient in the case of a doped D-type InSb with extrinsic conductivity at $T = 300$ K is high (about -2% K^{-1}). With increasing doping level, the transition to extrinsic conductivity is shifted toward higher temperatures T, although the magnetosensitivity is reduced at the same time owing to the decrease of mobility (Fig. 4.34) [138]. According to the data in [138], the temperature coefficient of magnetoresistors depends on the composition of the material, the magnetic induction and the temperature interval. These relationships can be seen from Table 4.6 as well.

Magnetoresistors can operate in a wide frequency range between a d.c. signal and microwave frequencies. The limiting frequency f_L is determined by the relaxation time, which is a function of the conductivity and the dielectric constant of the material. In InSb these parameters can be neglected. It has been established by accurate measurements that the output signal of magnetoresistance sensors, to a first approximation, has the same magnitude at $f = 34$ GHz as under d.c. conditions [136].

4.4.3. Multiterminal magnetoresistance devices

A three-terminal magnetoresistive element which is a functional combination of two identical magnetoresistors is shown in Fig. 4.35 [136].

The two outer terminals are the electrical input and the middle terminal is the output. This sensor has the following principle of operation: the magnetic field is applied only to a part of the three-terminal structure by means of an excitation magnet that can be moved along the length of the device. The resistance of the upper half $R_a(B)$ is much higher than the resistance $R_b(0)$ of the lower part. The output potential of the middle electrode with respect to the lower terminal is $V_{out} = V_{in} R_b/(R_a + R_b) \ll V_{in}$. As the magnet is moved towards the lower part of the structure, V_{out} is increased. If the magnet has a symmetrical position between R_a and R_b, i.e. when $R_a = R_b$, the output signal is $V_{out} = 1/2 V_{in}$, and, finally, when the magnetic field is applied only to the lower magnetoresistor R_b, the following relationships are valid: $R_a \ll R_b$ and $V_{out} \approx V_{in}$. The sensor described is actually a contactless potentiometer. The sensor shown in Fig. 4.35 can be realized also by means of discrete resistors separated by a sufficiently small distance. There is always a certain offset voltage $V_{out}(B=0)$ in the three-terminal sensor described, but many applications require a zero output, which cannot be achieved by the arrangement presented in Fig. 4.35. There is a possibility of achieving a zero output by connecting the three-terminal device to a Wheatstone bridge. Such a differential magnetoresistor is shown in Fig. 4.36, where $R_a \equiv R_2$ and $R_b \equiv R_1$ are

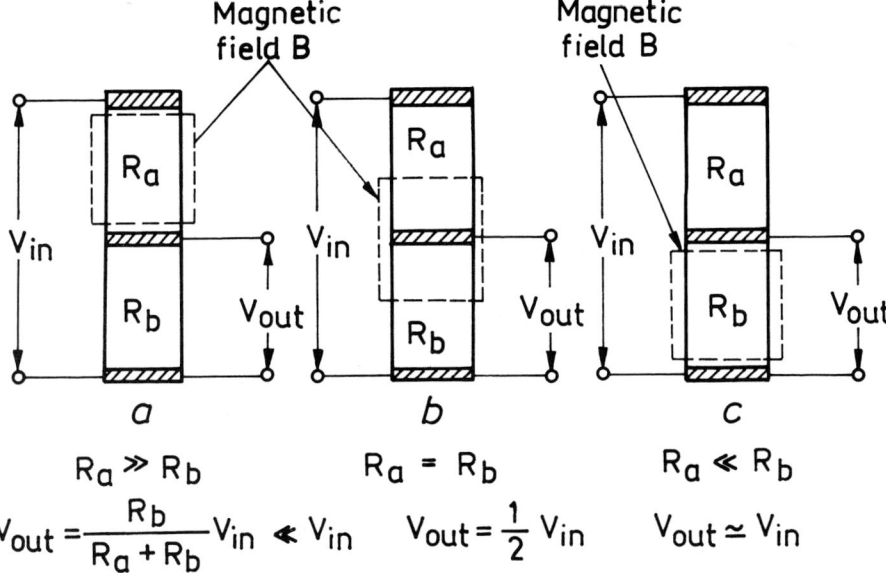

Fig. 4.35. Principle of contactless potentiometer.

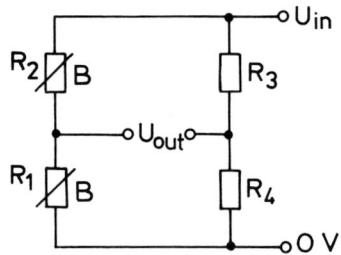

Fig. 4.36. Signal conditioning of a differential magnetoresistor; R_1 and R_2: magnetoresistors; R_3 and R_4: fixed resistors.

the resistances of the three-terminal sensor and R_3 and R_4 are resistors with fixed values. By a proper choice of R_3 and R_4, a zero output $V_{out}(0)$ can be adjusted at $B = 0$. The voltage $V_{out}(B)$ of the differential configuration depends on the effective variation ΔB which results from the movement of the excitation magnet along R_1 and R_2 and is due to the strong nonlinear dependence of the $R(B)$ characteristic on the biasing magnetic field B_0, as well. The following relationships are valid for the differential magnetosensitive circuit in Fig. 4.36 in a low magnetic field [138]:

$R_3 = R_4;$

$B_1 = B_0 + \Delta B;$

$B_2 = B_0 - \Delta B;$

$R_1 = R + \Delta R = R + \dfrac{dR}{dB}\Delta B;$

$R_2 = R - \Delta R = R - \dfrac{dR}{dB}\Delta B.$

The mean voltage is determined from the expression $V_{R1} = V_{in} R_1/(R_1 + R_2)$. Therefore $V_R = V_{in}/2$ and the output differential signal will be:

$$V_{out} = V_{R1} - \frac{V_{in}}{2} = V_{in}\left(\frac{R_1}{R_1 + R_2} - \frac{1}{2}\right) = V_{in}\frac{R_1 - R_2}{2(R_1 + R_2)} =$$

$$= V_{in}\frac{2(dR/dB)\Delta B}{4R} = V_{in}\frac{(dR/dB)\Delta B}{2R}.$$

This is why the following expression is valid for the ratio of the output voltage to the input voltage:

$$\frac{V_{out}}{V_{in}} = \frac{1}{2}\frac{1}{R}\frac{dR}{dB}\Delta B;$$

hence the magnetosensitivity will be:

$$\frac{V_{out}}{\Delta B} = \frac{1}{2}\frac{1}{R}\frac{dR}{dB_0}\frac{V_{in}}{2}. \tag{4.71}$$

The overall sensitivity $V_{\text{out}}/\Delta B$ of the differential magnetoresistive circuit is proportional to the relative variation (dR/RdB), which, in its turn, depends on the temperature and on the level of the magnetic biasing B_0. According to (2.74), when $C_1 = 1$, the relative variation in resistance at a small Hall angle Θ_H is:

$$\frac{1}{R}\frac{dR}{dB} = \frac{2\mu_H^2 B}{1 + \mu_H^2 B^2}. \tag{4.72}$$

Rare earth/cobalt magnets exciting a field $B_0 \simeq 0.4T$ are usually used to achieve a high transducing efficiency for the circuit in Fig. 4.36. This differential circuit has an improved temperature stability and its working point can be chosen in the most sensitive region of the $R(B)$ characteristic.

A four-terminal semiconductor ring can be used for the functional implementation of the differential bridge circuit in Fig. 4.36, which, by analogy with the three-terminal device, consists of four integrated magnetoresistors, each terminal being turned to an angle of 90° with respect to its neighbor (Fig. 4.37). The supply voltage is applied to two opposite terminals, and the other two opposite terminals serve as the output for V_{out}. The magnetic-field excitation system consists of two magnets fixed opposite to each other whose dimensions are the same as those of the individual magnetoresistive sections. The two magnets can be simultaneously rotated with respect to the center of the ring. The resistance of each one of the four identical magnetoresistors is increased or decreased depending on the angular displacement α of the magnetic field. A voltage is thereby generated across the output contacts [136]. This functional magnetoresistance device generates an output signal whose level is high compared with that of other arrangements and can reach 70 to 80% of the supply voltage V_{in}. The theoretical expression for the output voltage is $V_{\text{out}} = (K_M - 1)V_{\text{in}}X/(K_M + 1)$, where K_M is the relative increase of the resistance per unit length of one of the elements from Fig. 4.37 in a magnetic field and X is the normalized angular displacement. A linear output $(V_{\text{out}}/V_{\text{in}})X$ is characteristic of this sensor in the $\alpha = \pm 30°$ range. An InSb ring whose thickness is 10 μm has been used for fabrication of the device and the diameter is 3 mm. About a 10% variation of the resistor output voltage is observed as the temperature is varied in the interval $-10 \leq T \leq 60°C$. There are different device structures whose input resistance is 1 to 5 kΩ and the typical supply voltage is about 8 V, the input power being in the 100 mW range. The variation of the magnetosensitive output is about $3V$ with nonlinearity less than $\pm 0.5\%$ in the range $-20 \leq T \leq 100°C$. The temperature coefficient is less than 0.01% °C^{-1} [136].

Circuits which include an operational amplifier or silicon $n-p-n$ transistor for the compensation of the temperature dependence of the differential magnetoresistive sensors (cf. §9.5.2) are suggested in [138]. Parallel-field, three-terminal sensors with "negative magnetoresistance" (cf. §4.3.2) are of interest in connection with certain applications. The well-defined "zero" that serves as a distinction between

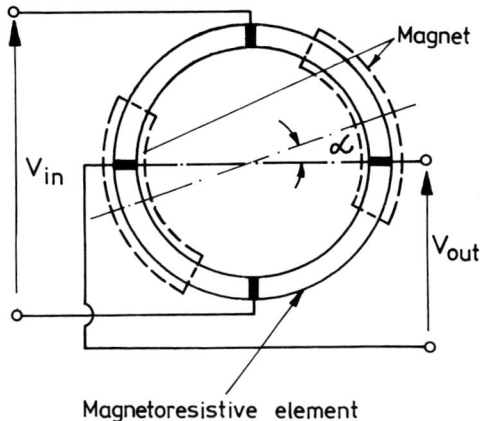

Fig.4.37. Four-terminal contactless device.

the polarity of $+B$ and $-B$ magnetic fields can very well be used for signal detection by the modulating technique.

The semiconductor magnetoresistance sensors described together with magnetoresistors based on the anisotropic magnetoresistance effect [139-142], are among the transducers most widely used when a high signal-to-noise ratio and high performance are required. The on-chip integration of magnetoresistance devices together with the appropriate circuitry may very soon bring new sophisticated solutions and applications.

References

[1] E.H. Putley, The Hall Effect and Semiconductor Physics, Dover Publ., New York, 1960.
[2] S. Kataoka and Y. Sigiyama, Proc. IEE, 126 (1979) 141–146.
[3] J. Chretien, Les dispositifs à injection de porteurs en tant que capteurs magnetosensibles, Ph.D. dissertation, I.N.P., Grenoble, France, 1977.
[4] M. Epstein, H.M. Sachs and L.J. Greenstein, Proc. IRE, 47 (1959), 2014.
[5] G. De Mey, Adv. Electron. and Electron Phys., 61 (1983) 1–62.
[6] Ch.S. Roumenin, C.R. Acad. Bulg. Sci., 40(2) (1987) 51–54.
[7] V. Der Pauw, Phillips Res. Rep., 13 (1958) 1–9.
[8] G.S. Randhava, Microelectronics J., 12(6) (1981) 24–29.
[9] R.F. Wick, J.Appl. Phys., 25 (1954) 741–756.
[10] J. Haeusler, Solid-state Electron., 9 (1966) 417–441
[11] J. Haeusler, Arch. Electrotech., 52 (1968) 11–19.
[12] H.J. Lippmann and F. Kuhrt, Z. Naturforsch., 13(a) (1958) 462–474.
[13] R.S. Hemmert, Solid-state Electron., 17 (1974) 1039–1043.
[14] W. Versnel, Solid-state Electron., 24 (1981) 63–68.
[15] G.A. Korn and T.M. Korn, Mathematical Handbook for Scientists and Engineers, McGraw-Hill, New York, 1961.
[16] Ch.S. Roumenin, Sensors and Actuators, A, 30 (1992) 77–87.

[17] E.V. Kuchis, Galvanomagnitnie effecti i metody ih issledovaniya, Radio i Sviaz, Moskwa, 1990 (in Russian).
[18] E.V. Kuchis, Metody Issledovaniya Effecta Holla, Sovetskoe Radio, Moskwa, 1974 (in Russian).
[19] I. Isenberg, B.R. Russel and R.F. Freene, Rev. Sci. Instr., 19(10) (1948) 685–688.
[20] H.J. Lippmann and F.Kuhrt, Z. Naturforsch., 13(a) (1958) 474–483.
[21] F. Kuhrt and H.J. Lippmann, Hallgeneratoren, Springer-Verlag, Berlin, 1968.
[22] A.C. Beer, Galvanomagnetic Effects in Semiconductors, Academic Press, New York, 1963.
[23] W. Versnel, J. Appl. Phys., 52 (1981) 4659–4666.
[24] R.S. Popović, Hall Effect Devices, Adam Hilger, Bristol, 1991.
[25] R.S. Popović and B. Hälg., Solid-state Electron., 31 (1988) 681–688; Sensors and Actuators, A, 21–23 (1990) 908–910.
[26] H.P. Baltes and R.S. Popović, Proc. IEEE, 74 (1986) 1107–1132.
[27] Yu. V. Vorobiov, V.N. Dobrovolski and V.I. Striha, Metody Issledovaniya Poluprovodnikov, Vishia Skola, Kiev, 1988 (in Russian).
[28] L.J. Van Der Pauw, Philips Techn. Rundsch., 20 (1958/9) 230–234.
[29] A.W. Saker, F.A. Cunnel and J.T. Edmond, J.Sci. Instr., 6 (1955) 217.
[30] Y. Kanda, M. Migitaka, H. Hamamoto, H. Morozumi et al., IEEE Trans. Electr. Devices, ED-29 (1) (1982) 151–154.
[31] Y. Kanda, IEEE Trans. Electr. Devices, ED-29 (1) (1982) 64–70.
[32] H. Weiss, Structure and Applications of Galvanomagnetic Devices, Pergamon Press, Oxford, 1969.
[33] Y. Kanda and M. Migitaka, Phys. Status Solidi, A35 (1976) K115–K118.
[34] T.G.M. Kleinpenning and L.K.J. Vandamme, J. Appl. Phys., 50 (1979) 5547.
[35] F.N. Hooge, Phys. Lett., 29A (1969) 139–140.
[36] L.K.J. Wandamme, in: Noise in Physical Systems and $1/f$ Noise, M.Savelli, G. Lecoy and J.P. Nongier (eds.) Elsevier, Amsterdam (1983) 183–192.
[37] A. Van der Ziel, P.H. Handel, X. Zhu and K.H. Duh, IEEE Trans. Electron Devices, ED-32 (1985) 667–671.
[38] A. Chovet, Ch.S. Roumenin, G. Dimopoulos and N. Mathieu, Sensors and Actuators, A, 21–23 (1990) 790–794.
[39] M.J. Buckingham, Noise in Electronic Devices and Systems, Ellis Horwood and John Wiley, New York, 1983.
[40] A. Chovet and P. Victorovitch, l'Onde Electrique, 57 (1977) 699–707; 57 (1977) 773–783; 58 (1978) 69–80.
[41] H.M. Vaes and T.G.M. Kleinpenning, J. Appl. Phys., 48 (1977) 5131–5134.
[42] T.G.M. Kleinpenning, Sensors and Actuators, 4 (1983) 3–9.
[43] R.S. Popović, Sensors and Actuators, 17 (1989) 39–53.
[44] H.H. Jensen and H. Smith, J.Phys. C, 5 (1972) 2867–2880.
[45] A.G. Andreou and Ch.R. Westgate, Proc. IEEE, 73 (1985) 489–490.
[46] A. Nathan, A.M. Huiser, Can. J. Phys., 63 (1984) 695–698.
[47] A.S. Grove, Physics and Technology of Semiconductor Devices, Wiley, New York, 1967.
[48] P. Norton, T. Braggins and H. Levinstein, Phys. Rev. B, 8 (1973) 5632–5653.
[49] R. Mueller, Grundlagen der Halbeiter-Elektronik, Springer Verlag, Berlin, 1971.
[50] N. Kotera, J. Shigeta, K. Narita, T. Oi, et al., IEEE Trans. Magn., MAG-15 (1979) 1946–1955.
[51] A.C. Beer, Solid-state Electron., 9 (1966) 339–351.
[52] V. Zieren, Integrated silicon multicollector magnetotransistors, Ph. D. dissertation, Delft Univ. of Technology, The Netherlands, 1983.
[53] S. Middelhoek, J.B. Angell and D.J.W. Noorlag, IEEE Spectrum, 17(2) (1980) 45–56.
[54] V. Zieren, D.J.W. Noorlag, S. Middelhoek and E.A. Wolsheimer, Eur. Electron., 1(5) (1981) 10–16.
[55] S. Middelhoek and S.A. Audet, Silicon Sensors, Academic Press, London, 1989.
[56] Y. Kanda and M. Migitaka, Phys. Status Solidi, A38 (1976) K41–K44.
[57] Y. Sugiyama, H. Soga and M. Tacano, J. Crystal Growth, 95 (1989) 394–397; Proceedings of the 4th Conf. on Solid-state Sensors and Actuators — TRANSDUCERS '87, Tpkyo (Japan) (1987)

547–550.
[58] M. Tacano, Y. Sugiyama and H. Soga, Solid-state Electron., 32 (1989) 49–55.
[59] Y. Sugiyama, H. Soga, M. Tacano and H.P. Baltes, IEEE Trans. Electron Devices, ED-36 (1989) 1639–1643.
[60] Ch.S. Roumenin and B. Bojkov, University Annual (Techn. Phys.), 22 (2) (1985) 33–48 (in Bulgarian).
[61] G. Schneider, Microelectron. Reliab., 28 (1988) 75–92.
[62] R.C. Gallagher and W.S. Corac, Solid-state Electron., 9 (1966) 571–580.
[63] P.W. Fry and S.J. Hoey, IEEE Trans. Electron Devices, ED-16 (1969) 35–39.
[64] W.N. Carr and S.T. Hong, SWIEECO Record of Techn. Papers, 22 (1970) 166–170.
[65] G.R. Mohan Rao and W.N. Carr, Solid-state Electron., 14 (1971) 995–1001.
[66] A. Yagi and S. Sato, Jpn. J. Appl. Phys., 15 (1976) 655–661.
[67] M. Hirata and S. Suzuki, Proc. 1st, Sensor Symp. Japan (1981) 305–310.
[68] R.S. Popović, Sensors and Actuators, 5 (1984) 253–262.
[69] R. Agrawal, P.K. Yadava, R. Dwivedi and S.K. Srivastava, Sensors and Actators, A, 28 (1991) 21–28.
[70] A. Nathan, A.M.J. Huiser and H.P. Baltes, IEEE Trans. Electron Devices, ED-32 (1985) 1212–1219.
[71] L. Andor, H.P. Baltes, A. Nathan and H.G. Schmidt-Weinmar, IEEE J. Solid-state Circuits, SC-20 (1985) 819–821; IEEE Trans. Electron Devices, ED-32 (1985) 1224–1230.
[72] R.S. Popović and H.P. Baltes, J. Solid-state Circuits, SC-18 (1983) 426–428.
[73] N. Mathieu, P. Giordano and A. Chovet, Sensors and Actuators, A, 32 (1992) 656–660 and 682–687.
[74] N. Mathieu, A. Chovet, R. Fauquembergue, P. Descherdeer and A. Leroy, Sensors and Actuators, A, 25–27 (1991) 741–745.
[75] X. Zheng and S. Wu, Sensors and Actuators, A, 28 (1991) 1–5.
[76] D. Misra, T.R. Viswanathan and E.L. Heasell, Sensors and Actuators, 9 (1986) 213–221.
[77] J.J. Clark, Sensors and Actuators, A, 24 (1990) 107–116.
[78] D.R. Briglio, Characterization of CMOS magnetic field sensors, M.Sc. thesis, Univ. of Alberta, Edmonton, Canada, 1988.
[79] D.R. Briglio, A. Nathan and H.P. Baltes, Can. J. Phys., 65 (1987) 842–845.
[80] R.E. Prange and S.M. Girvin (eds.), The Quantum Hall Effect, Springer-Verlag, Berlin, 1987.
[81] G. Bosch, Solid-state Electron., 11 (1968) 712–714.
[82] J.E.L. Hollis, Measurement and Control, 6 (1973) 38–40.
[83] R.J. Braun, IBM J. Res. Develop., July (1975) 344–352.
[84] G. Bjorklund, IEEE Trans. Electron Devices, ED-25 (5) (1978) 541–544.
[85] M.J.Thorn Proc. 29th IEEE Vehicular Technology Conf., Arlington Hts. (USA) (1979) 226–229.
[86] H. Lachman, Components Report, Data Sheet Siemens AG, 14(6) (1979) SAS 321.
[87] I. Fujimoto and Y. Imamura, Nat. Techn. Rep., 21 (1975) 681–691.
[88] (No author given), Electron. Weekly, April 29 (1985) 59–61.
[89] R.S. Popović, Proceedings of the 4th Int. Conf. on Solid-state Sensors and Actuators — TRANSDUCERS '87, Tokyo (Japan) (1987) 539–542.
[90] Y. Takahana, K. Miyauachi, K. Tsuruta and K. Tsuboi, ISSCC Dig. Techn. Papers, Philadelphia (USA), 24 (1981) 42–43.
[91] A. Thanailakis and E. Cohen, Solid-state Electron., 12 (1969) 997–1000.
[92] A. Hojo, S. Tanaka and I. Kuru, Proceedings of the 7th Conference on Solid-state Devices, Tokyo (Japan) (1975) 261–266.
[93] T. Inada, T. Ohkubo, M. Kitahara, Y. Kanda and T. Hara, Electron. Lett. 14 (1978) 503–505.
[94] E. Pettenpaul, J. Huber and H. Weidlich, Solid-state Electron., 24(8) (1981) 781–786.
[95] K. Matsui, S. Tanaka and T. Kobayashi, Proceedings of the 1st Sensor Sympos. (Japan) (1981) 37–40.
[96] T. Hara, M. Mihara, N. Toyoda and M. Zama, IEEE Trans. Electron Devices, ED-29 (1982) 78–82.
[97] B. Wilson and B.E. Jones, J. Phys. E: Sci. Instr., 15 (1982) 364–366.

[98] P. Daniil and E. Cohen, J. Appl. Phys., 53(11) (1982) 8257–8259.
[99] P. Extance and G.D. Pitt, Proceedings of the Int. Conf. on Solid-state Sensors and Actuators — TRANSDUCERS '85, Boston (USA) (1985) 304–307.
[100] T.R. Lepkowski, G. Shade, S.P. Kwok, M. Feng et al., IEEE Electron Device Letters, EDL-7(4) (1986) 222–224.
[101] R. Campesato, A. Passaseo, C. Flores and S. Verni, Sensors and Actuators, A, 32 (1992) 651–655.
[102] S. Kataoka, H. Yamada, Y. Sugiyama and H. Fujisada, Proc. IEEE, 59 (1971) 1349.
[103] Y. Sigiyama, Fundamental research on Hall effect in inhomogeneous magnetic fields, Res. of the Electrotechnical Laboratory, No. 838 (Tokyo, Japan), 1983.
[104] A. Nathan, W. Allegretto, H.P. Baltes and Y. Sugiyama, IEEE Electron Device Lett., EDL-8 (1987) 1–3; IEEE Trans. Electron Devices, ED-34 (1987) 2077–2085.
[105] A. Nathan, W. Allegretto, H.P. Baltes and T. Smy, Can. J. Phys., 65 (1987) 956–960.
[106] B. Hälg, J. Appl. Phys., 64 (1988) 276–282.
[107] T. Kanayama, M. Oasa, H. Hiroshima and M. Komuro, Proceedings of the 4th Int. Conf. on Solid-state Sensors and Actuators — TRANSDUCERS '87, Tokyo (Japan) (1987) 532–535; J. Vacuum Sci. and Techn., B, 6(3) (1988) 1010–1013.
[108] Y. Sigiyama and S. Kataoka, Proceedings of the 3th Int. Conf. on Solid-state Sensors and Acuators — TRANSDUCERS '85, Boston (USA) (1985) 308–311.
[109] M. Tacano, T. Kanayama, H. Hiroshima, M. Komuro and Y. Sugiyama, J. Appl. Phys., 62 (1987) 4301–4303.
[110] M. Tacano, Y. Sugiyama and T. Taguchi, IEEE Electron Device Lett., EDL-8 (1987) 22–23.
[111] H. Weiss, Physik und Anwendung Galvanomagnetischer Bauelemente, Vieweg, Braunschweig, 1969.
[112] Ch.S. Roumenin and P.T. Kostov, Planar Hall-effect device, Bulg. Patent N 37208 (Dec. 26, 1983).
[113] R.S. Popović, IEEE Electron Device Lett., EDL-5 (1984) 357–358.
[114] A.M. Huiser and H.P. Baltes, IEEE Electron Device Lett., EDL-5 (1984) 482–484.
[115] Ch.S. Roumenin and P.T. Kostov, C.R. Acad. Bulg. Sci., 38 (1985) 1145–1148.
[116] Ch.S. Roumenin, C.R. Acad. Bulg. Sci., 39(11) (1986) 65–68.
[117] Ch.S. Roumenin and P.T. Kostov, Semiconductor Hall element, Bulg. Patent N 39283 (Jan. 8, 1985); C.R. Acad. Bulg. Sci., 39(5) (1986) 63–66.
[118] Ch.S. Roumenin, Hall effect sensor, Bulg. Patent N 41974 (May 6, 1986); C.R. Acad. Bulg. Sci., 40(11) (1987) 59–62.
[119] U. Falk, Sensors and Actuators, A, 21–23 (1990) 751–753.
[120] K. Maenaka, T. Ohgusu, M. Ishida and T. Nakamura, Electron. Lett., 23 (1987) 1104–1105.
[121] Ch.S. Roumenin, Hall-effect device, Bulg. Patent N 80103 (June 10, 1987).
[122] T. Nakumura and K. Maenaka, Sensors and Actuators, A, 21–23 (1990) 762–769.
[123] R. Popović and W. Heidenreich, in: Sensors, Vol. 5, R. Bolk and K.J. Overshott (eds.), VCH, Weinheim, 1989, Ch. 3.
[124] H.P. Baltes and A. Nathan, in: Sensors, Vol. 1, T. Grandke and W.H. Ko (eds.), VCH, Weinheim, 1989, Ch. 7.
[125] K. Holzlein and J. Larik, Sensors and Actuators A, 25–27 (1991) 349–355.
[126] Ch.S. Roumenin, Sensors and Actuators, A, 24 (1990) 83–105.
[127] K. Maenaka, T. Ohgusu, M. Ishida and T. Nakamura, Proceedings of the 4th Int. Conf. on Solid-state Sensors and Actuators — TRANSDUCERS '87, Tokyo (Japan) (1987) 523–526.
[128] K. Maneaka, M. Tsukahara and T. Nakamura, Sensors and Actuators, A, 21–23 (1990) 747–750.
[129] R.S. Popović, Proceedings of the 12th Yugoslav Conference on Microelectronics, MIEL 84, Niš (Jugoslavia) (1984) 299–307.
[130] G.I. Rekalova, T.V. Persiyanov and G. Shtyubner, Sov. Phys. Semicon., 10 (1976) 213–214.
[131] V.V. Alekseev and I.M. Vikulin, Instrum. Exp. Techn., 27 (1984) 185–187.
[132] M. Green, Solid-State Electron., 3 (1961) 314–316.
[133] A. Kobus and J. Tuszynski, Hallotrony i Gaussotrony, Wydawnictwa naukovo-techniczne, Warszawa, 1966 (in Polish).
[134] H.H. Wieder, Hall Generators and Magnetoresistors, Pion, London, 1971.
[135] H. Weiss and M. Wilhelm, Z. Phys., 176 (1963) 399–408.

[136] S. Kataoka, Recent development of magnetoresistive devices and applications, Circular of the Electrotechnical Laboratory, No. 182 (Tokyo, Japan), 1974.
[137] I.M. Vikulin, L.F. Vikulina and V.I. Stafeev, Galvanomagnitniye Pribori, Radio i Sviaz, Moskwa, 1983 (in Russian).
[138] Sensoren — Magnetfeldhalbleiter, Teil 1, Datenbuch, SIEMENS, 1982/83.
[139] J.H. Fluitman, in: Summer Course 1982: Solid-state Sensors and Transducers, vol. II, W. Sansen and J. van der Spiegel (eds.), Leuven, Belgium, Katholieke Universiteit, 1982, IX-1-23.
[140] G. Chiron, G. Delapierre and D. Rousseau, Sensors and Actuators, 4 (1983) 369–374.
[141] U. Dibbern, Sensors and Actuators, 10 (1986) 127–140.
[142] U. Dibbern, in: Sensors, vol. 5, R. Bolk and K.J. Overshott (eds.), VCH, Weinheim, 1989, 341–380.

Chapter 5

Magnetodiode sensors

The classical magnetodiode sensor (MD) is a two-terminal semiconductor device whose operation is based on the superposition of high carriers injection by one or two (a p^+–n and an n^+–n) junctions, magnetoconcentration and surface recombination. There are multiterminal versions with improved electrical control of the mode of operation and higher magnetosensitivity. Information about the magnetic field is derived from the superlinear current–voltage characteristic. The materials most widely used for the fabrication of this class of magnetosensitive transducers are Ge, Si, GaAs, InSb, etc. The theoretical basis of magnetodiodes has been presented in §2.5.

The heuristic value of the material implementation of geometrical images from the upper complex half-plane into real Hall devices, with considerable manufacturing and metrological advantages, has been discussed in §4.3.1. It was pointed out that the "unfolding" of orthogonal magnetosensors is physically meaningful if their operation is an odd function of the field B. Although orthogonal MD structures are nonuniform owing to the high injection level, and in this case conformal mapping is not a trivial task, the possibility of an odd MD effect occurring at surface recombination rates $s_1 \gg s_2$ or $s_1 \ll s_2$ leads to a novel approach to the development of parallel-field MD versions. In the "unfolding" of such orthogonal MD structures around to a certain axis of symmetry, the p^+–n and n^+–n junctions become planar and the sensors acquire a structurally-symmetric form. Because of this understanding, the class of MD devices is enriched by useful technical solutions, differential magnetodiodes for instance (cf. §5.1). The fundamental importance of the idea of conformally-transformed or parallel-field transducers for the advance of solid-state magnetosensors can be seen in the analysis of bipolar magnetotransistors in Chapter 6.

5.1. Device structures, materials and operation of MD sensors

The MD sensors and Hall devices known so far can be classified as orthogonal and parallel-field transducers. The basic MD device is presented schematically in

Fig. 5.1. The orthogonal versions are (a)–(f) and the structures (g)–(x) are parallel-field devices. Novel proposals for parallel-field MD devices are also shown in Fig. 5.1.

Almost intrinsic Ge, with $\rho \approx 50$ Ω·cm, and Si, with $\rho \approx 15$ to 20 kΩ·cm, were used for the implementation of the earliest discrete MD versions [1–6]. They have a parallelepiped shape (Fig. 5.1a), and include an injection junction and an ohmic contact. The dominating contribution of magnetoconcentration to the magnetodiode effect had not been completely explained at the beginning of the evolution of MD sensors; therefore, regions with different rates of surface recombination had not been formed in those long magnetodiodes. Planar MD devices (Fig. 5.1b), prepared without using the standard methods of IC technology have been tested with the purpose of making the current path longer and increasing the effective base length l^*. For instance, in the Ge magnetodiode In has been employed for the formation of the injecting p^+–n junction and Pb has been used to prepare the ohmic contact. Wide bandgap materials like Si and GaAs are to be preferred in order to increase the temperature interval of operation. The optimum l/L_p ratio for this type of MD device is $3 \leq l/L_p \leq 7$, since in shorter structures the diode curve is hardly controlled by the magnetic field and at $l \gg L_p$, the resistance of the base region is crucial. An important feature of this kind of sensor is that, irrespective of its polarity, the magnetic field always leads to reduced conductivity.

The action of a sensor can be considerably improved by the formation of a region with a high surface recombination rate (Fig. 5.1c) achieved by mechanical abrasion and a second n^+–n contact [7]. In this case the magnetodiode effect is completely utilized and the sensitivity of such versions is polar and superior to that of long MDs. The transducing efficiency also can be increased by the series connection of two p^+–n–n^+ MD sensors with such a mutual orientation that one of them increases its resistance while the resistance of the other decreases when exposed to a magnetic field (Fig. 5.1c). If an identical pair of MD sensors is added in parallel to the first pair, the result is a bridge circuit (Fig. 5.1c) with a differential output, and improved linearity and temperature stability [8].

If only a small part of the MD area is occupied by the p^+–n junction, the injected carriers move in a relatively narrow channel (filament). When an orthogonal field B is applied, the filament is distorted and changes its position in the structure owing to the Lorentz force. Such an effect can be observed in MD sensors with negative resistance and an S-type region of the $I(V)$ curve. The negative differential resistance is a phenomenon of interest, since it occurs in other types of bipolar magnetosensitive device as well. Therefore it will be briefly discussed. The physical mechanism of the negative differential resistance is associated with the redistribution of the externally applied voltage between the p–n junction and the base region of the diode. The resistance of the base region drops when the injection of minority carriers is enhanced. As a result, the supply voltage is redistributed in such a manner that the voltage drop in the base region is decreased and the

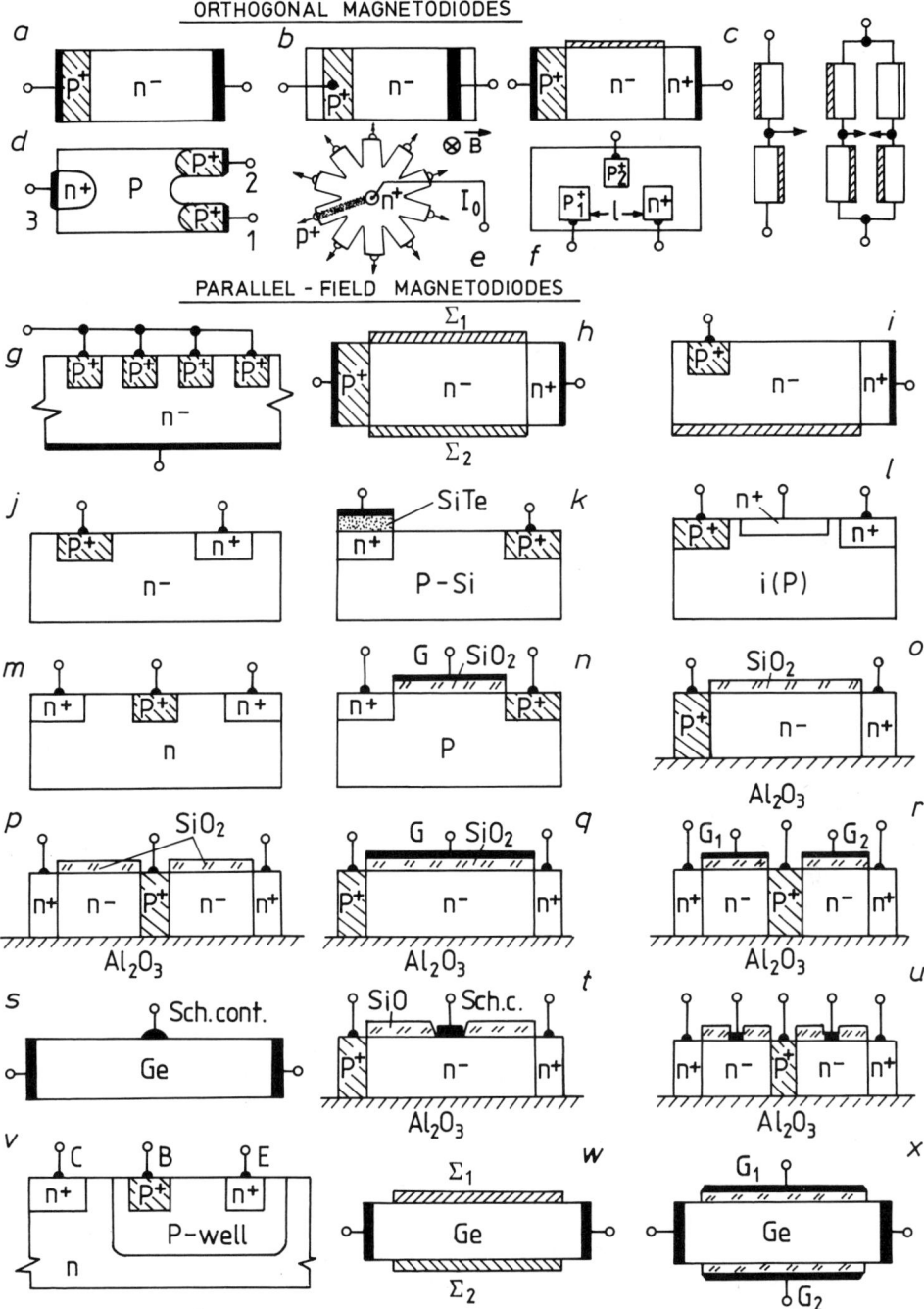

Fig. 5.1. Various shapes of orthogonal and parallel-field magnetodiodes.

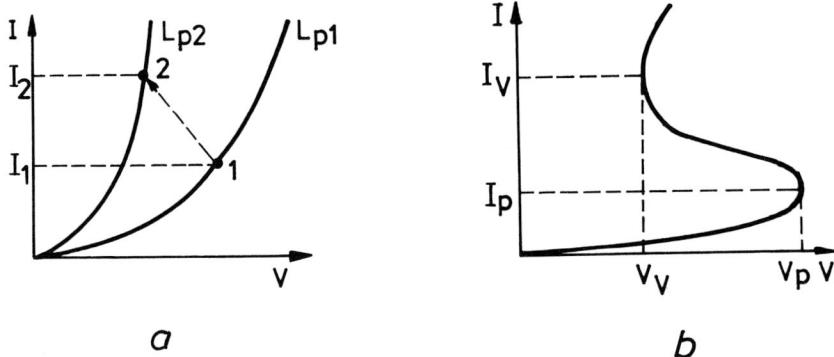

Fig. 5.2. (a) Formation of an S-type negative-resistance region of a diode $I(V)$ characteristic due to the increase of length L_p. (b) A typical S-type $I(V)$ curve.

voltage across the injection junction is increased. The concentration of both types of carrier in the base region is further increased and this leads again to a decrease of its resistance and to a new redistribution of the supply voltage. This process is repeated and a mechanism of positive feedback appears, which is responsible for the avalanche increase of the carrier concentration, i.e. it satisfies the necessary condition for the existence of a region of negative differential resistance of the $I(V)$ curve when $B = 0$. But the resistance of the p–n junction is reduced with an increase of I. This is why the appearance of an S-type region of the $I(V)$ characteristic is possible if the decrease in the resistance of the base region with increasing I exceeds that of the p–n junction resistance. Such a superlinear increase in the base conductivity requires an additional mechanism. The main cause for the occurrence of an S-type region of the $I(V)$ characteristic is the increase in the diffusion length $L_p \sim (\mu_p \tau_p)^{1/2}$ of injected carriers when their concentration in the base is increased. The $I(V)$ characteristics of a forward biased p^+–n–n^+ diode with two values L_{p1} and L_{p2} of the diffusion length ($L_{p1} < L_{p2}$) is shown in Fig. 5.2a. It can be seen from the figure that the increase of L_p in the interval I_1 to I_2 leads to a transformation of the first $I(V)$ curve into the second one. There is an important fact: the increase of L_p must take place in an interval [I_1 to I_2] which does not include the point ($I = 0, V = 0$). If the increase of L_p begins immediately from the point $I = 0$, no negative resistance occurs. The increase of L_p can be associated with an increase of τ_p. Such a mechanism can be observed in diodes whose base region is compensated as a result of doping by both donor and acceptor impurities. Sb or Bi, for instance, can serve as donor impurities in Ge, whereas Au can be used as an acceptor [8]. A typical S-shaped current–voltage curve can be seen in Fig. 5.2b. Based on the analysis of this dependence at low currents, a slow increase of the injected carrier concentration in the base region can be predicted until the peak-point (I_p, V_p) is reached and a superlinear increase is observed after that point. Therefore an S-type $I(V)$ characteristic can be obtained by a decrease of the injection level at small magnitudes of the forward current I_f.

The formation of an additional n^+-contact to the base region can be used to shunt the p^+–n junction. In real p–n junctions, the recombination of carriers in the p–n junction space-charge region can play the role of such a "shunting" resistor, provided there are mechanisms of surface leakage, etc. If the diode is biased by a constant voltage V, the concentration of carriers and, therefore, the increase in the current I will be abrupt, i.e. "vertical", until the whole bulk of the device is filled with nonequilibrium plasma and the recombination process compensates the injection. There is no S-shaped region of the $I(V)$ curve in this case. But if the p^+–n–n^+ structure is biased by a constant current I, the level of the avalanche injection is controlled by the external circuit and an S region appears in the $I(V)$ curve. Moreover, in the region of negative resistance a longitudinal current filament (a domain) of rather high conductivity is formed. The filament spreads from the respective point of the p^+–n junction to the n^+–n contact in a conic angle Ψ, which is determined by the current I, i.e. by the injection level. Outside the domain, when the base region is homogeneous, the density of minority carriers decreases exponentially:

$$\Delta p_i(x) = p_i(y_i) \exp\left[(y_1 - y)/L_{\text{eff}}\right], \tag{5.1}$$

where y is the distance from the filament axis x; y_i is the radius of a given ith cross-section, perpendicular to the spreading direction of the filament; p_i is the concentration of holes on the boundary of the filament for the i-th cross-section; and L_{eff} is a parameter related to the diffusion of minority carriers outside the filament at $y \geq y_i$.

The existence of a filament with high carrier density is common for all cases of an S-type $I(V)$ curve. Owing to different external reasons, including the magnetic field $\boldsymbol{B}$, the current domain can be shifted from the place of its formation. This fact is very important for the operation of certain types of MD sensor. This property of the filament is used in the MD devices presented in Fig. 5.1d–e. In the first case (d), depending on the direction of the magnetic field $\boldsymbol{B}$, the Lorentz force can "switch" the conductivity channel from the first $p^+ - n$ junction to the second one or from the second junction to the first one. Therefore, this MD sensor can serve as a magnetic switch.

Rotation magnetometers (e) are another promising field of application of MDs with an S-type $I(V)$ characteristic. The Lorentz force causes a rotation of the current filament, and the frequency of the output signal at one of the p^+-terminals is used to determine the induction B. This idea has been successfully implemented in p-type InSb MDs designed for operation at $T = 77$ K (Fig. 5.1d and e) [9]. The discrete rotation sensor (e) has 10 terminals [9]. The increase of the current leads to a filament self-stabilisation and to an increase of the density of the nonequilibrium plasma outside the filament. Therefore the conduction channel can cover two adjacent p^+–n junctions together. Output terminal sectioning as shown in Fig. 5.1d and e is used to reduce this effect.

The device structure in Fig. 5.1f is a magnetosensor which includes two forward biased p^+–i–n^+ diodes and an n^+-contact on one and the same substrate [10]. Au diffusion has been used for the fabrication of a high resistivity silicon region with $\rho \approx 55$ Ω·cm. The action of this MD is based on the Lorentz-force controlled injection in the long p_1^+–i–n^+ diode after the voltage V has exceeded the threshold value V_1 determined by the electrostatic interaction of the p_1^+–i junction with the auxiliary p_2^+–i–n^+ diode. This combined sensor can operate both in the standard $I(V)$ characteristic mode at $V_2(p_2^+) \neq 0$ and in the mode of S-type $I_1(V_1)$ and $I_2(V_2)$ curves with a distance l_0 between the p_1^+ and the n^+ regions ($l \geq l_0$).

In the parallel-field discrete magnetodiode in Fig. 5.1g the p^+–n contact has been divided into sections to increase the length of the lines of current flow [4]. The device actually consists of four MDs connected in parallel with a common base region. Higher magnetosensitivity has been determined from experiments with a Ge version of the device in Fig. 5.1g [4], in comparison with the case of a p^+–n junction covering the entire top surface. The parallel-field discrete MD in Fig. 5.1h has two injecting junctions and has been designed in accordance with the requirements associated with the magnetoconcentration effect. The two opposite surfaces (top and bottom) exhibit different rates of recombination, for instance, $s_1(\Sigma_1) \gg s_2(\Sigma_2)$. The recombination rate s_1 has been achieved by mechanical abrasive processing (grinding) of the surface Σ_1, while s_2 is the result of a very smooth mechanical polishing of Σ_2 and subsequent chemical treatment. The impurity concentration in the base region is low. Ge and Si versions of the device have been manufactured, [1–7,11–16].

The device geometry of the discrete MD in Fig. 5.1i with a planar p^+–n junction leads to a higher magnetosensitivity [13]. This modification of the conventional device structure enhances the MD effect by means of two additional effects. The first is a decrease or increase in the length of the current path depending on the polarity of the field B. The second effect leads to a better injection–magnetoconcentration coupling. In a magnetic field, the space-charge region is directly modulated by the increase in the carrier concentration.

The parallel-field device structure in Fig. 5.1j can be fabricated by an IC process since the two injecting junctions are planar. This type of MD sensor requires a low recombination rate on the top surface with a high recombination rate on the bottom of the substrate. The top surface of silicon MDs fabricated by an IC process is passivated by a thermally grown SiO_2 layer — therefore the recombination rate s is low. These MDs exhibit polar magnetosensitivity enhanced by magnetic-field modulation of the length (increase or decrease) of the current path [5].

High resistivity silicon with $\rho = 20$ to 30 kΩ·cm can be used for the fabrication of a MD sensor with a memory (Fig. 5.1k) [17]. A specific feature of the structure is the thin insulating layer of SiTe inserted between the n-region and the respective

Al contact. This layer is responsible for the existence of a region with a negative differential resistance of the $I(V)$ curve with "low" and "high" conductivity deeply modulated by the MD effect in a magnetic field $B = \pm 0.3$ T. After the removal of the field B, the device preserves its conductivity originally induced by the magnetic field. The erasure of "magnetic information" is achieved by a high-current pulse whose duration is 10 μs.

A planar $p^+-i(p)-n^+$ sensor has been suggested [18], in order to eliminate or minimize the effect of the surface recombination rates, whose values in MDs can hardly be controlled (Fig. 5.1l). High resistivity silicon with $\rho \geq 2 \times 10^4$ $\Omega\cdot$cm and $L_p \approx 7 \times 10^{-2}$cm has been used for fabrication of the device. An additional, elongated n^+-strip is located between the n^+-region and the p^+-region. The additional n^+-strip is kept at a positive potential with respect to the p^+-region in order to reverse the bias of the n^+-i junction. In this case the recombination upon the top surface has been replaced by the extraction of carriers (i.e. $s \to \infty$). In such a way the transducing efficiency is considerably increased.

The parallel-field device in Fig. 5.1m is a planar differential magnetodiode which combines a common p^+-n junction and two n^+-n contacts on one and the same substrate [19]. The differential output (the two outer terminals) provides for a better temperature stability and improved linearity. The magnetosensitivity of this type of sensor can be further enhanced by a proper treatment of the bottom surface to achieve a high recombination rate in comparison with the top surface which has been passivated by a thermally grown SiO_2 layer.

The sensor presented in Fig. 5.1n resembles a MOS transistor except for the source and drain regions whose doping is p^+ and n^+ respectively. The task of the gate is to optimize, by means of the field effect, the effective length of the current path, which is then modified by magnetoconcentration. Regions of majority-carrier accumulation, depletion and inversion are successively arranged along the channel at $V_G \neq 0$ as a result of the double injection, [20,21].

The evolution of IC technology towards VLSI and ULSI offers new opportunities for the creation of MD sensors of improved performance. Silicon on sapphire (Al_2O_3) magnetodiodes, denoted as SOS MDs are a typical example in this respect. The minority-carrier lifetime in these structures is in the range 1 to 10 ns and the corresponding diffusion length is $L \sim 1$ μm, which is comparable with the thickness of conventional silicon epilayers in the SOS system. The crystallographic and thermal properties of Si and Al_2O_3 produce a high density of defects and generation–recombination centers at the Si–Al_2O_3 interface, thus leading to a natural asymmetry of the surface recombination velocities: low upon the Si–SiO_2 interface and high on the Si–Al_2O_3 interface. Because of this asymmetry SOS is the "ideal" system for the miniaturization of MDs. High magnetosensitivity in low magnetic fields is achieved in such a way. The recombination rate on the Si–SiO_2 interface is in the range 1 to 10 m s^{-1} and has the value 100 m s^{-1} on the Si–Al_2O_3 interface [13,15,20,22–25], i.e. the recombination rates on both surfaces differ by a factor of 10^2. Moreover, these MDs are parallel-field devices (Fig. 5.1o). The idea underlying the operation of the differential version in Fig. 5.1m has been further developed in the device in Fig. 5.1p by making use of SOS technology. The author believes this is a promising solution that can ensure high performance of the transducer. The version of an integrated MD fabricated by a SOS polysilicon gate process (Fig. 5.1q), was proposed in [26,27]. Ion implantation for the formation of the n^+- and p^+-regions and self-aligned gates are characteristic of this process. The task of the gate voltage is the same as in the case presented in Fig. 5.1n. A differential SOS MD with two gates similar to the devices in Fig. 5.1m and p is shown in Fig. 5.1r. The gate voltages V_{G1} and V_{G2} modulate the internal parameters of the two functionally connected sensors. As a result, it becomes possible to achieve a zero offset and improved linearity and temperature stability of the output.

Schottky MDs are three-terminal devices whose operation is based on the magnetoconcentration or MD-effect controlled current across a reverse biased metal–semiconductor junction. A discrete version of a magnetoconcentration-based Schottky MD (Fig. 5.1s), is described in [28]. The device structure includes a parallelepiped of quasi-intrinsic Ge with $\rho \approx 60$ $\Omega \cdot$cm and two ohmic supply contacts. The two surfaces towards which electrons n_0 and holes p_0 are deflected by the magnetic field B have been subjected to special treatment to achieve highly asymmetric rates of surface recombination. Au has been deposited upon one of these surfaces by evaporation to prepare a Schottky junction. According to [28], the reverse current I_s of the Schottky diode has a diffusion and a thermo-ionic component and can be described by the formula:

$$I_s = \frac{q(|\mu_n| + \mu_p)}{|\mu_n|} \left[\frac{D}{L} n_s + A^* T^2 \exp\left(\frac{-q\varphi_n}{k_B T} \right) \right], \tag{5.2}$$

where D, L, A^*, T and φ are the diffusion constant, the diffusion length, the Richardson constant, the temperature and the height of the junction potential barrier for electrons, respectively. The concentration of electrons under the Schottky junction is n_s.

The reverse current I_s is modulated by the density δn of carriers additionally concentrated by the Lorentz force in the region between the depletion layer of the Schottky junction and the neutral bulk of the semiconductor plate. Depending on the direction of the field B, the excess of carriers δn may be positive [$\delta n(+B) > \delta n(0)$] or negative [$\delta n(-B) < \delta n(0)$]. It is the excess $\delta n(B)$ that strongly modulates the current I_s and is responsible for the Schottky MD magnetosensitivity $\delta I_s(B) \sim \delta n(B)$. The operation of the integrated parallel-field Schottky MD shown in Fig. 5.1t is based on the MD effect in SOS structures at high injection across the p^+–n and n^+–n junctions [29–31]. It follows from §2.5.3.2 that the carrier distribution in the magnetodiode is two-dimensional. It has two independent components: longitudinal (along the x-axis) and transverse (along the y-axis). In the case of a Schottky MD, the carrier distribution can be presented in the form:

$$n_s \sim n_0 m_L \left(x = \frac{l}{2}\right) m_T \left(y = -\frac{t}{2}\right), \qquad (5.3)$$

where m_L and m_T are the longitudinal and transverse carrier distributions respectively. The equilibrium concentration of carriers in the base region is n_0. The Schottky junction is on the surface with a low recombination rate in the middle ($x = l/2$) of the sample.

In the semiconductor regime of operation of the MD in a low magnetic field (§2.5.2.2) the following expressions can be used to describe the two components m_L and m_T [30]:

$$m_L\left(x = \frac{l}{2}\right) = \frac{\tau_V s}{n_0} \sim \left[\frac{1}{|\mu_n|} + \frac{1}{\mu_p}\right]^{-1} \tau_V \frac{V}{l^2} \qquad (5.4)$$

and

$$m_T\left(y = -\frac{t}{2}\right) \sim 1 + \frac{\delta n_s(\mathbf{B}, s)}{n_0}. \qquad (5.5)$$

The transverse density m_T depends on the field $\mathbf{B}$ and on the recombination rate s upon the Si–SiO$_2$ interface. The dependence of the transverse distribution of carriers on the Schottky-diode reverse current I_s is determined by the longitudinal density m_L (5.4). On the other hand m_L is a linear function of the applied longitudinal voltage V, i.e. increasing V leads to a stronger influence of m_T upon the current I_s, cf. (5.3). It follows from the analysis of the Schottky MD presented in [28] that the magnetosensitivity increases as the recombination rate in the metal–semiconductor junction is reduced, i.e. an increase of the potential barrier φ_n is equivalent to an increase of the curvature of band bending. Figure 5.3 illustrates the fundamental dependence of the magnetic control of the reverse current I_s of a SOS Schottky MD on the induction and polarity of the field $\mathbf{B}$, and also shows the external circuitry [6]. The second term in (5.1), which corresponds to the thermo-ionic current, diminishes the relative contribution of the diffusion current and reduces the MD effect. This is why the replacement of the Schottky junction with a p–n junction [28], or with a metal–insulator–semiconductor (MIS) structure [29], has been suggested.

The differential Schottky MD in Fig. 5.1u combines the advantages of the devices already shown in Fig. 5.1m, p and r.

The silicon sensor in Fig. 5.1v is based on the magnetodiode principle and has been manufactured by a standard bulk CMOS process [32]. The reversely biased collector junction C plays the role of the Si–Al$_2$O$_3$ interface with high velocity of carrier recombination in SOS MDs (Fig. 5.1o). By analogy with the device in Fig. 5.1l, the recombination of carriers has been replaced by collection with $s \to \infty$. The active base region is a p-well whose thickness is about 12 μm, the substrate is a collector, the n^+ and p^+ regions serve as an emitter and as a base contact, respectively. The output signal is taken from terminals E and B. For instance, if the injected electrons are deflected downwards by the field $\mathbf{B}$, the current I_C will

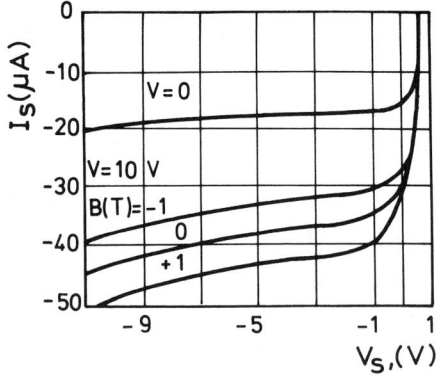
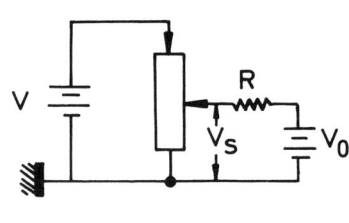

Fig. 5.3. Influence of magnetic field at $T = 300$ K on the reverse characteristic $I_s(V_s)$ in an SOS Schottky micromagnetodiode, ($n_0 = 5 \times 10^{15}$ cm^{-3}, $l = 20$ μm, $w = 200$ μm, $t = 0.65$ μm) and external circuitry [15].

increase, and the concentration of minority carriers in the base region (the p-well) will be reduced, thus leading to an increase in its resistance and in the output signal $V_{EB}(\boldsymbol{B})$.

Two discrete parallel-field sensors whose operation is based on the classical magnetoconcentration effect are presented in Fig. 5.1w and x to make the discussion complete [33]. They are made of almost intrinsic Ge with $\rho \approx 40$ $\Omega\cdot$cm. A region of high surface recombination rate has been formed upon one of the surfaces (Fig. 5.1w). Additional gate electrodes G_1 and G_2 upon the surfaces Σ_1 and Σ_2 are used in the device in Fig. 5.1x in order to achieve a reproducible control of the recombination rates s_1 and s_2 by making use of the field effect. According to the data in ref. [33], the devices in Fig. 5.1w and x exhibit high magnetosensitivity.

5.2. Effect of device parameters and temperature on the current–voltage characteristics of integrated MDs

Unlike a Hall sensor whose output signal increases or decreases, starting from zero, provided there is no offset, i.e. $V_H(B = 0) = 0$, generally, the magnetosensitivity of MD devices is determined by the comparison of their current–voltage characteristics at $B = 0$ and $B \neq 0$. Therefore detailed information concerning the MD $I(V)$ curves at $B = 0$ for different device parameters and temperatures is necessary for the optimization of the transducer's performance. At the same time, modern trends in the development of MDs require their miniaturization and the application of adequate IC technology — the SOS process, first of all. This is why special attention is paid in this section to the $I(V)$ characteristics of SOS MD sensors at $B = 0$.

It follows from the analysis in §2.5.3.2 and from Fig. 2.24 that at $B = 0$ different conductivity regimes in MDs can be distinguished: exponential, ohmic ($I \sim V$), semiconductor ($I \sim V^2$), insulating ($I \sim V^3$) and diffusion. According to the experimental results in [30,31], in SOS MDs for instance, the insulating and the diffusion modes are not observed. This follows from the fact that the minority-carrier lifetime is rather short and with typical majority-carrier concentrations $n_0 \sim 10^{15}$ cm^{-3} in the base region, the electric field needed to achieve a very high injection level is in the 10^4 V cm^{-1} range. At such electric fields a breakdown in the structure takes place before any of the two regimes of operation mentioned above is able to set in. In fact, a potential difference of about 100 V is necessary for the transition from the semiconductor (2.118) into the insulating mode (2.120). An impurity concentration of approximately 5×10^{14} cm^{-3} is required to achieve the insulating mode in SOSs with a base length $l \sim 10$ μm at a reasonable voltage $V \sim 10$ V [30]. On the other hand, the transition from the ohmic into the semiconductor regime corresponds to much lower electric fields. The experimental $I(V)$ curves of different SOS MDs at $B = 0$ are presented in Fig. 5.4 [30,31]. It is evident from the figure that the square conductivity ($I \sim V^2$) is well pronounced in accordance with the semiconductor double injection, while other curves correspond to ohmic conductivity ($I \sim V$) or to an intermediate mode $I \sim V^\alpha$, where $1 < \alpha < 2$. A

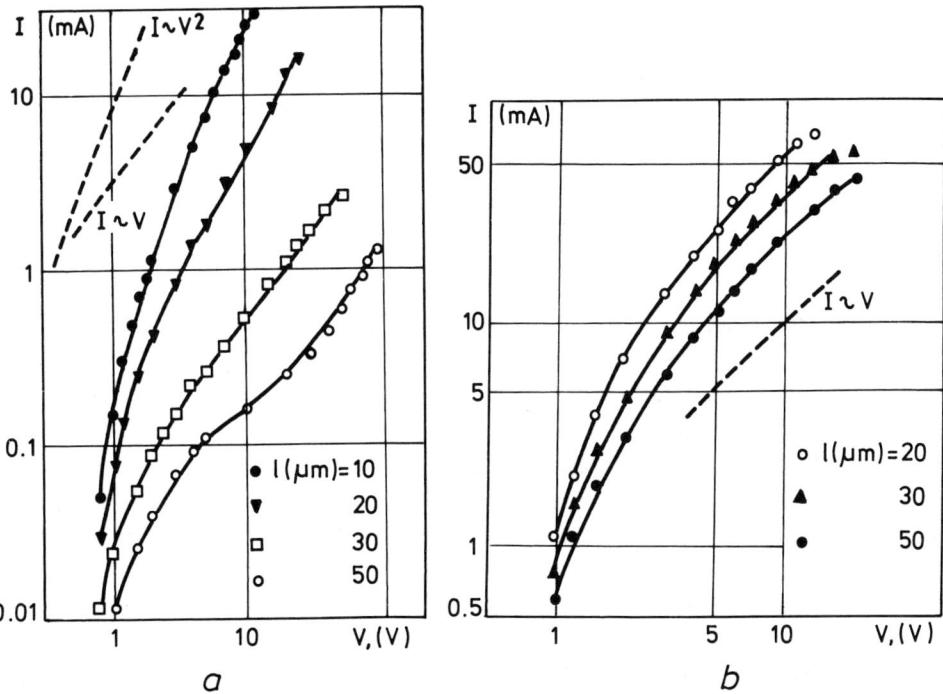

Fig. 5.4. SOS p^+-n-n^+ magnetodiode $I(V)$ characteristics at different values of the length l; $T = 300$ K, $w = 50$ μm; (a) $n_0 = 7 \times 10^{15}$ cm^{-3}; (b) $n_0 = 10^{18}$ cm^{-3} [30].

high injection level is achieved (cf. §2.5.3.2) when $\delta n \approx \delta p \gg n_0 > p_0$. At a low equilibrium concentration (n_0, p_0) of carriers, the semiconductor regime is achieved irrespective of the length of the device (Fig. 5.4a) [30]. At high concentrations of doping impurities in the MD base region ($n_0 \geq 10^{17}$ cm^{-3}) the injection level is not sufficient to entail the semiconductor mode, and hence the conductivity remains ohmic (Fig. 5.4b) [30]. At moderate doping ($n_0 \sim 10^{16}$ cm^{-3}), the conductivity may be square or ohmic, depending on the diode length and on the applied voltage.

The parameter l (length) is even more important for the conductivity of MD devices than the applied voltage. The dependence of the current on the voltage in the ohmic, semiconductor and insulating modes is V (2.117), V^2 (2.118), and V^3 (2.120), respectively, whereas the corresponding dependencies on l are : l^{-1}, l^{-3} and l^{-5}. For instance, in a SOS MD with $l \leq 20$ μm the injection level needed to establish the semiconductor mode is easily achieved. This regime sets in immediately after the mode of exponential conductivity $I \sim \exp(qV/Ck_BT)$ with a coefficient $C = 1.7$. Just the opposite behavior is manifested by long p^+-n-n^+ MDs: the semiconductor regime sets in at a higher bias V after the region of ohmic conductivity. The voltage V_{tr}, which corresponds to the transition from the ohmic to the semiconductor mode, is obtained by a comparison of equations (2.117) and (2.118):

$$V_{tr} \simeq \frac{l^2}{\mu_p \tau_{sc}}. \tag{5.6}$$

If the minority-carrier transit time τ_{tr} is assumed to be approximately equal to the effective lifetime τ_{sc} of the carriers, the following relationship is valid:

$$\tau_{tr} \simeq \frac{l}{\mu_p(V/l)} \simeq \tau_{sc}. \tag{5.7}$$

This is why, in compliance with (5.6) and (5.7), the transit time τ_{tr} must be shorter than τ_{sc}, i.e. $\tau_{tr} < \tau_{sc}$ or $V > V_{tr}$, in order to establish the semiconductor mode. These conditions are more difficult to satisfy in long MDs, since $V_{tr} \sim l^2$.

The effective lifetime τ_{eff}, determined experimentally from the $I(V)$ curves of SOS MDs whose only difference is the length l, decreases with increasing l [30,31]. This inversely proportional relationship, $\tau_{eff} \sim 1/l$, indicates that the parameter l in the expression (2.118) for the MD semiconductor regime must be replaced by an effective value l_{eff}. The discrepancy between the experimental results and (2.118) is removed if $l_{eff} = l - 3$ μm is used, and approximately equal values of τ_{eff} are obtained [30]. This approach is correct if the density of recombination centers is assumed to be independent of the MD length l.

S-type negative-resistance regions of the $I(V)$ curves are sometimes observed in long SOS MDs with $l \geq 30$ μm when the electric field applied is $E \geq 10^4$ V cm^{-1}. This behavior can be explained by the existence of recombination centers and traps whose density is comparable with the impurity concentration in the base region. Under such conditions the excess of electrons differs from that of holes, their lifetimes differ, and complicated $I(V)$ relationships with regions of negative differential resistance and a current filament are observed [30,31].

In principle, the width w of SOS MD sensors does not influence the current density and hence the mode of operation when $B = 0$. The parameter w is important when the maximum current via the structure is to be determined.

There is a substantial dependence of MD operation on temperature when $B = 0$. It has been found as a result of detailed investigations on this problem that in the interval $77 \leq T \leq 350$ K the relatively long SOS MDs with $l \sim 50$ μm exhibit an ohmic behavior at low temperatures, [30,31]. Moreover, breakdown takes place before the semiconductor regime has set in. Double injection occurs as the temperature is increased to $T \geq 240$ K and the transition from the ohmic to the semiconductor mode takes place at a bias whose value decreases with increasing temperature. In short SOS MDs with $l \sim 10$ μm the semiconductor regime is manifested in the whole temperature interval $77 \leq T \leq 350$ K and at low temperatures double injection occurs (after the ohmic mode) at a relatively low voltage $3 \leq V \leq 5V$. At temperatures $T \geq 200$ K, the semiconductor regime

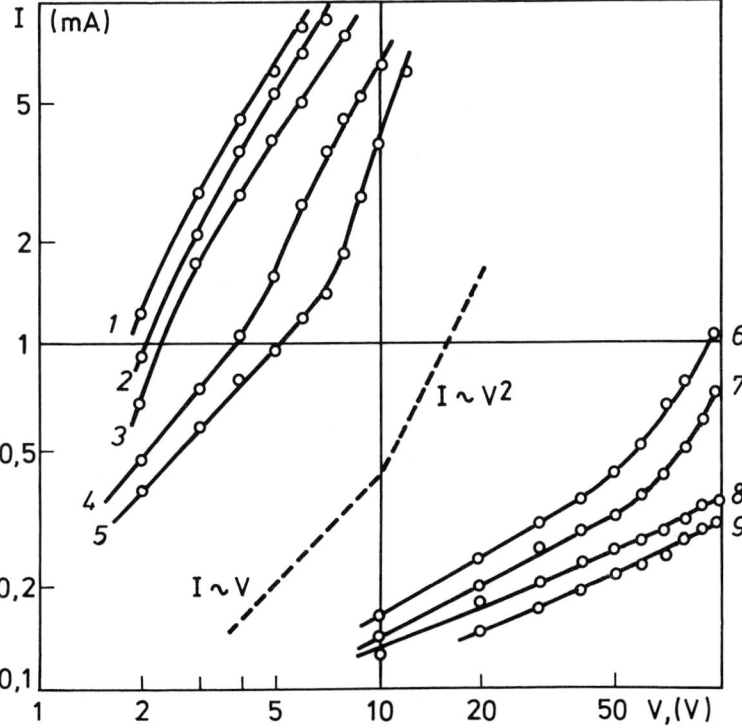

Fig. 5.5. $I(V)$ characteristics of SOS p^+-n-n^+ MDs with two different lengths l_1 and l_2 at different temperatures T ($n_0 = 5 \times 10^{15}$ cm^{-3}); $w = 50$ μm and $t = 0.65$ μm. $l_1 = 10$ μm: $T = 353$ K (1), 295 K (2), 240 K (3), 190 K (4), 166 K (5); $l_2 = 50$ μm: $T = 333$ K (6); 295 K (7); 203 K (8); 155 K (9).

(2.118) sets in immediately after the exponential mode and no ohmic conductivity is observed. Figure 5.5 presents a summary of the principal conclusions concerning the influence of device parameters and temperature on SOS MD behavior when $B = 0$ [30].

5.3. Determination of the MD sensor magnetosensitivity

A magnetodiode is a nonlinear electronic component whose output signal, depending on the mode of operation, device structure, temperature, characteristics of the surface, etc., is in most cases an exponential, square or other type of power function of the magnitude and direction of the magnetic field (cf. §2.5.3.1 and §2.5.3.2). Owing to these facts, special comments are necessary concerning the criteria for magnetosensitivity described in §3.2.1, (3.1)–(3.3). The conception of MD sensor transducing efficiency is discussed in detail in [6]. By the introduction of criteria for MD magnetosensitivity, a successful attempt has been made to eliminate the vague definitions that stand in the way of the comparison of different devices. As a result of the nonlinearity of the output characteristic, the transducing efficiency of these bipolar sensors is a function of the field B, and the introduction of four basic criteria for magnetosensitivity is advisable:

The first criterion refers to the absolute slope of the $I(B)$ or $V(B)$ curve at a given induction B or within an interval ΔB of field magnitudes. The absolute slope is measured under the following conditions:

– the maximum allowable power which does not entail irreversible changes in the MD device due to its heating is applied;

– the electric excitation is accomplished at constant current in order to measure the voltage slope, at constant voltage to measure the current slope, and at constant voltage and current to measure the resistance slope;

– the variation of the magnetic field around the selected value B_0 is small, i.e. $\Delta B \approx 0$.

The measured average slopes $(\Delta V/\Delta B)$ [V T^{-1}]; $(\Delta I/\Delta B)$ [A T^{-1}] and $\Delta R/\Delta B)$ [Ω T^{-1}] of the MD are most representative in the neighborhood of $B_0 \approx 0$ or at $B_0 \geq 0.1T$ [6]. The average slope $(\Delta I/\Delta B)$ at $B_0 \leq 0.1T$ and at $B_0 \approx 0$ is illustrated as an example in Fig. 5.6a. The dynamic slope is defined as (dV/dB), (dI/dB) or (dR/dB). Small values of the magnetic induction in the neighborhood of the selected point B_0 are usually used to determine the dynamic slope. The additional a.c. magnetic field is $B_0 \approx 10^{-4}$ T (Fig. 5.6b).

The maximum magnetosensitivity at $B_0 \approx 0$ is achieved by such a treatment of the two active surfaces of the MD that leads to highly asymmetric recombination rates $s_1 \to 0$ and $s_2 \to \infty$. In the semiconductor regime, which is the most favorable, the valid relationships for the current magnetosensitivity S_I and for the voltage magnetosensitivity S_V are as follows [14]:

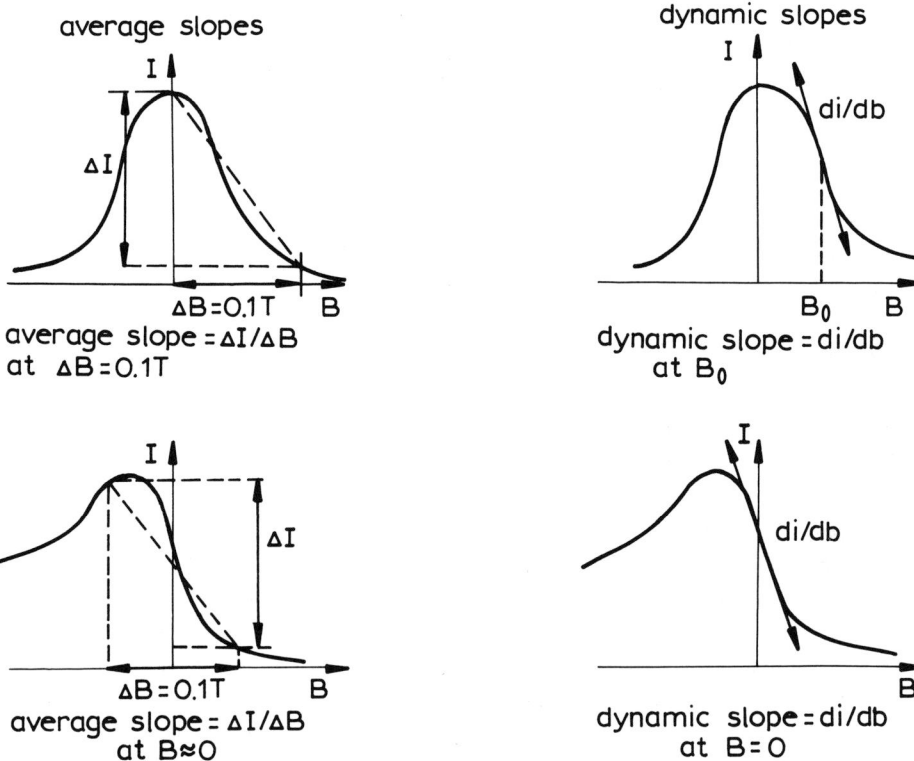

Fig. 5.6. Average and dynamic slopes at $B = B_0$ and $B \approx 0$.

$$S_I = \left|\frac{\partial I}{\partial B}\right|_V = a_1 a_2 \frac{V^3}{l^4}; \tag{5.8}$$

$$S_V = \left|\frac{\partial V}{\partial B}\right|_I = -\frac{a_1}{2a_2}\left(\frac{\tau_V}{\tau_{\text{eff}}}\right)^2 Il^2; \tag{5.9}$$

where

$$a_1 = \tfrac{9}{8} q |\mu_n| \mu_p \tau_V (n_0 - p_0) \tag{5.10}$$

and

$$a_2 = -\frac{1}{t}\left(|\mu_n|\mu_p \tau_V \operatorname{th}\frac{t}{l^*} \operatorname{th}\frac{t}{2l^*}\right). \tag{5.11}$$

The carrier lifetime in the bulk is denoted by τ_V, $l^* = (D^*\tau_V)^{1/2}$ is the diffusion length in the bulk, and D^* is given by the expression:

$$D^* = \left(\frac{2k_B T}{q}\right)\left(\frac{|\mu_n|\mu_p}{|\mu_n| + \mu_p}\right).$$

According to (5.8), the current magnetosensitivity S_I is proportional to V^3 and inversely proportional to l^4. The voltage sensitivity S_V is proportional to Il^2, equation (5.9). Therefore, the sensitivities S_I and S_V increase with increasing supply voltage. High values of the current magnetosensitivity S_I are characteristic for short magnetodiodes, whereas long MDs exhibit high values of the voltage magnetosensitivity S_V.

The following expression for the parameter S_V is deduced in [25] at very low values of the magnetic induction ($B_0 \leq 1$ mT), provided the distribution of carriers between the Si–SiO$_2$ and the Si–Al$_2$O$_3$ interface of SOS structures is exponential:

$$S_V \equiv \left|\frac{\partial V}{\partial B}\right|_I = \frac{q(|\mu_n| + \mu_p)\tau_{\text{eff}}(s_2 - s_1)V^2}{8k_B T l}. \tag{5.12}$$

The conclusion that the current I and the sensitivity S_V are square functions of the operating voltage V can be drawn from the comparison of (5.12) and (2.118).

The first criterion allows the comparison of the absolute sensitivity S_A of different MDs. It can be used, on the other hand, to choose the preferable mode of operation (I = const or V = const) of a particular MD sensor. The higher the sensitivity S_A, the easier the processing of the MD output signal. The MD absolute magnetosensitivity, similar to the case of Hall sensors and magnetoresistors, does not give any information concerning the minimum detectable magnetic field, $B_{\min}$, determined by the noise in the device.

The second criterion is associated with the relative slope whose analog is the relative magnetosensitivity. The relative slope is expressed in % T^{-1} and is measured under the same conditions as S_A: $(1/\langle V \rangle)(\Delta V/\Delta B)$; $(1/\langle I \rangle)(\Delta I/\Delta B)$; or $(1/\langle R \rangle)(\Delta R/\Delta B)$. $\langle \ldots \rangle$ denotes the "mean" value of the respective electrical parameter. The relative slope makes it possible to compare MDs based on the same physical mechanism, but made of different materials, Ge, Si, or $GaAs$, for instance, or fabricated by different technologies. This criterion can be used to select the preferable semiconductor material, the impurity concentration in the base region, etc. The relative sensitivity makes it possible to compare long magnetodiodes with a single p^+–n junction and double-injection versions.

The third criterion refers to the power slope ($\Delta W/\Delta B$). It characterizes the ability of the device to operate as a modulating transducer which transforms the external supply energy dissipated in it into a magnetic-field-dependent quantity. Only the mean slope measured ($\Delta W/\Delta B$) is physically meaningful. This parameter is determined at fixed values of the current, the voltage, the resistance or the energy in an external magnetic field. Generally speaking, the maximum energy transfer from the additional second input to the output is observed at the maximum value of the applied magnetic field. The slope $(\Delta W/\Delta B) = (W_B - W_{B+\Delta B})/\Delta B$ of dipole MD devices, in particular, is determined at a constant voltage V, current I, or resistance R. The positive sign of the slope $(\Delta W/\Delta B)$ means that the energy has been transferred to the magnetosensitive terminals or the power consumed by the

MD has decreased when a magnetic field has been applied. Usually, the variation of ΔB in the experiments does not exceed 0.1 T.

The fourth criterion is related to the minimum value of the magnetic field detectable by the MD sensor. This is a dynamic parameter. The minimum detectable value B_{min} is the value which transfers to the output of the MD device a power equal to the power of the sensor's intrinsic noise within a certain frequency interval Δf around the fixed frequency f_0, i.e. the signal-to-noise ratio equals unity. The MD should not necessarily operate in the maximum power dissipation mode in order to achieve the lowest possible value of B_{min}. It has been found that at high injection levels the MD intrinsic noise is proportional to the second power of the supply current, and the output signal is a linear function of the current I [6]. Therefore, the supply current I should be decreased while still preserving the high injection in order to achieve a lower value of B_{min} at $S/N = 1$. The minimum signal detectable by MDs, as in the case of other galvanomagnetic sensors, can be further improved by a factor of 10^2 by means of ferrite or μ-metal conic concentrators of the magnetic field. The noise measuring equipment has been described in §3.3.1.

5.4. Influence of the magnetic field, device parameters, temperature and noise on MD sensor magnetosensitivity

When an external magnetic field B is applied perpendicularly to the current I in an MD device, depending on their polarity, the carriers are deflected towards one of the two active surfaces Σ_1 or Σ_2. In the general case the $I(V)$ characteristic is modified dramatically by the field B. Typical $I(V)$ dependencies of MD sensors, obtained experimentally regardless of their discrete or IC implementation, are shown in Fig. 5.7.

The supply voltage has been chosen as a parameter (curves 1,2 and 3). The three cases presented in Fig. 5.7a–c are typical manifestations of the semiconductor double-injection regime at different ratios of the recombination velocity s_1 and s_2 (cf. §2.5.3.2). The polarity which causes a Lorentz deflection of the electron-hole plasma towards the surface Σ_1 with the low recombination velocity $s_1 \to 0$ has been adopted as the positive polarity ($+B$) of the magnetic field, whereas the negative polarity ($-B$) causes a carrier deflection towards the surface Σ_2 with $s_2 \to \infty$. In a magnetic field ($+B$), the current I increases to reach a maximum as the induction B is increased, then a drop of I is observed with a further increase in B (Fig. 5.7a). This behavior is explained by the increased carrier concentration on Σ_1 (in SOS MDs this is the Si–SiO$_2$ interface) and by the fact that in real devices $s_1 \neq 0$. When the maximum value of the current I has been reached, a further increase of B leads to a reduced mean density of the plasma, and hence the current I decreases. If the applied voltage is low (curve 1), the $I(+B)$ characteristic does

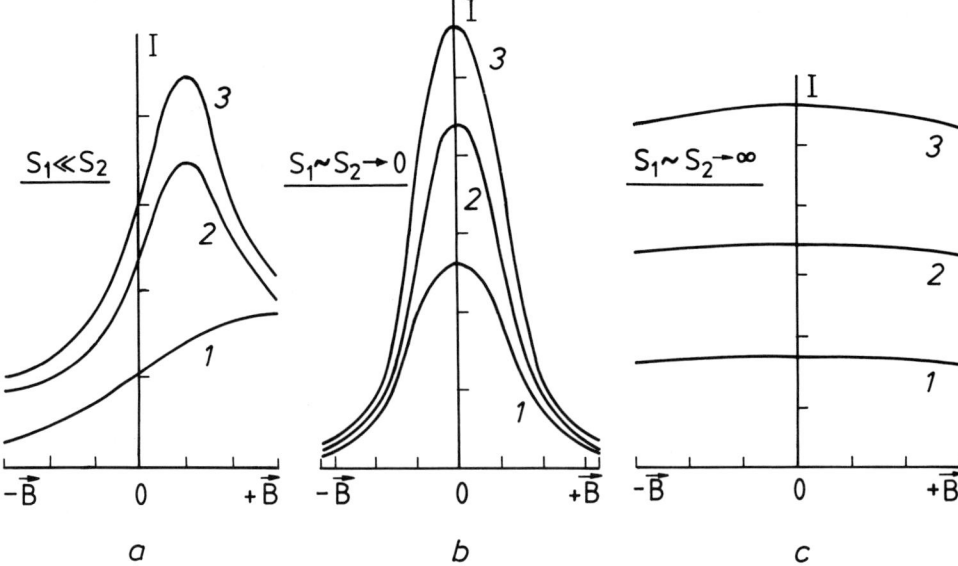

Fig. 5.7. Typical experimental $I(\boldsymbol{B})$ dependences of p^+–n–n^+ MD sensors: (a) $s_1 \ll s_2$; (b) $s_1 \approx s_2 \to 0$; (c) $s_1 \approx s_2 \to \infty$.

not exhibit a maximum, since the amount of the injected carriers is low and the carrier concentration on Σ_1 does not reach saturation. When the polarity of the magnetic field is reversed, i.e. in a field $(-\boldsymbol{B})$, and when the magnetic induction is increased, more and more carriers are deflected towards the surface Σ_2 with the high recombination rate s_2. In SOS MDs this is the Si–Al$_2$O$_3$ interface. The current I decreases as a result. At certain values of B, the longitudinal injection can be compensated for by the rapid surface recombination, and the square-function semiconductor mode ($I \sim V^2$) is replaced by ohmic conductivity ($I \sim V$). The comparison of the two characteristics $I(+\boldsymbol{B})$ and $I(-\boldsymbol{B})$ presented in Fig. 5.7a leads to the conclusion that conductivity is highly nonlinear when a magnetic field is applied. Nevertheless, an approximately linear relation $I \sim \boldsymbol{B}$ can be obtained in the neighborhood of point $B \approx 0$. No maximum is reached by the characteristics in Fig. 5.7a and the variation of the current I, caused by the field $\boldsymbol{B}$, is reduced as the doping impurity concentration in the base region of the MD is increased. This result is associated with the low injection level, which is insufficient for the magnetodiode effect to occur.

If the recombination rate on both surfaces Σ_1 and Σ_2 of the MD is low, i.e. $s_1 \approx s_2 \to 0$, even $I(\boldsymbol{B})$ characteristics with a maximum at $B = 0$ are observed (Fig. 5.7b). No magnetosensitivity is manifested by the MD when the recombination velocity on both surfaces is high i.e. $s_1 \approx s_2 \to \infty$ (Fig. 5.7c). It should be noted that the effect of the injection is partially compensated for by the fast surface

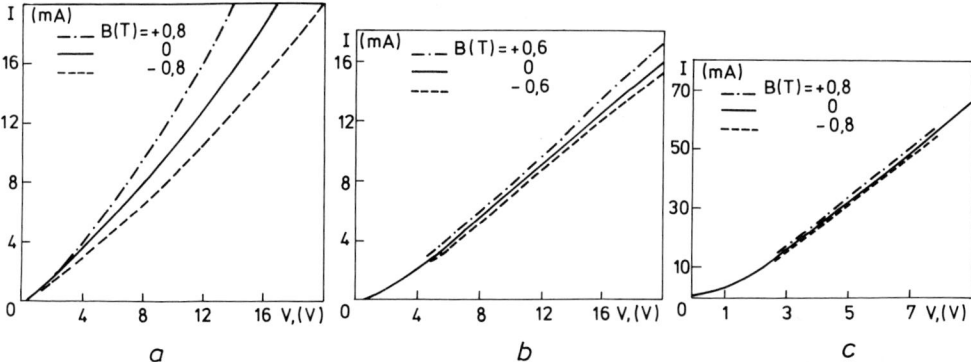

Fig. 5.8. Effect of the magnetic field upon the $I(V)$ characteristics of SOS MDs with different doping concentrations in the base epilayer; $l = 20$ μm, $w = 50$ μm, $t = 0.65$ μm, $T = 300$ K: (a) $n_0 = 5 \times 10^{15}$ cm^{-3}; (b) $n_0 = 2 \times 10^{16}$ cm^{-3}; (c) $n_0 = 10^{17}$ cm^{-3} [30].

recombination upon Σ_1 and Σ_2, and the square-function law of the semiconductor regime is violated.

It has already been discussed in §2.5.3.2 and §5.2 that the semiconductor mode of operation is necessarily used to achieve sufficient magnetosensitivity of integrated SOS MDs. It is easier to obtain the desired $I \sim V^2$ double-injection relationship by reducing the doping level of the diode base region. An increased transducing efficiency of the samples with a low concentration of doping impurities is expected as a result. This conclusion is confirmed by Fig. 5.8a–c, where the characteristics of different MDs are compared [30]. The effect of impurity concentration on the $I(V)$ curves in a magnetic field $B \neq 0$ is quite well pronounced in the sample with $n_0 = 5 \times 10^{15}$ cm^{-3} (Fig. 5.8a). The magnetic control of the $I(V)$ characteristic is slightly loosened when the doping level is 2×10^{16} cm^{-3} (Fig. 5.8b), and in the third case (Fig. 5.8c), with $n_0 = 1 \times 10^{17}$ cm^{-3}, almost no influence of the magnetic field is observed. From the viewpoint of fabrication technology, it is hard to achieve a doping level below 1×10^{15} cm^{-3} in SOS structures. A poor reproducibility of sensor characteristics is exhibited, moreover, by devices with moderate doping levels ($n_0 \sim 5 \times 10^{15}$ cm^{-3}) owing to the high density of defects in the epilayer.

The length l of the base region is the next parameter that influences the magnetosensitivity of SOS MDs. The magnetic control of the $I(V)$ characteristics of SOS devices which differ only in the length of the base region is illustrated by Fig. 5.9a and b [30].

At a constant voltage V, short samples ($l \leq 20$ μm) exhibit a higher sensitivity S_I. The variation in voltage caused by the magnetic field at a constant current ($I = $ const) is larger in long devices. This result is in a good agreement with (5.7) and (5.8) and can be explained by an increase in the resistance of the device instead of enhancement of the MD effect. In fact, the variation of $\tau_{\text{eff}}(B)$ plays

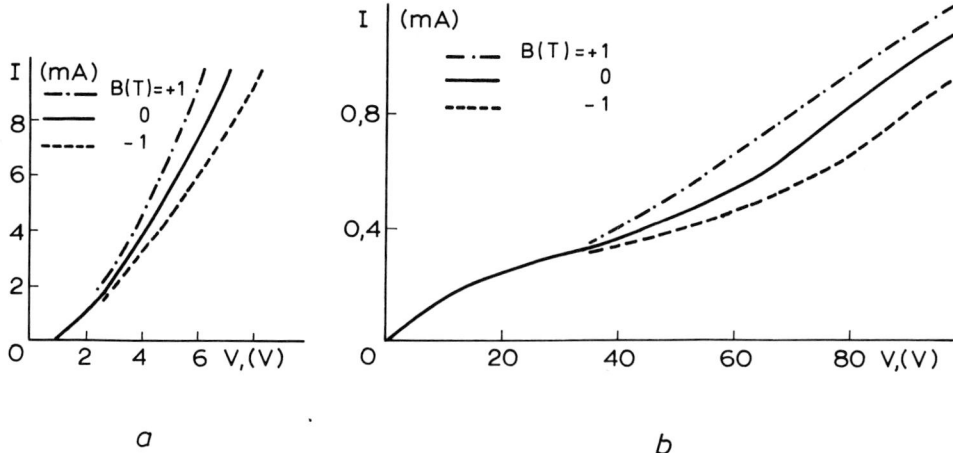

Fig. 5.9. Effect of the magnetic field on the $I(V)$ characteristics of SOS p^+-n-n^+ MDs with different lengths l_1 and l_2; $w = 50$ μm, $t = 0.65$ μm, $T = 300$ K, $n_0 = 7 \times 10^{15}$ cm^{-3}: (a) $l_1 = 10$ μm; (b) $l_2 = 50$ μm [30].

a more important role in short MDs because of the higher injection level in them. The applications of long SOS MDs are limited to a certain extent since the double-injection semiconductor mode sets in at higher values of the supply voltage.

It can be seen from the example in Fig. 5.10, illustrating the effect of w on magnetosensitivity, that at $V = $ const the magnetic control of the current $I(B)$ is better pronounced in wide MD devices. Conversely, narrow SOS MD sensors manifest high sensitivity S_V at $I = $ const.

The sample thickness t must be comparable with the ambipolar diffusion length L_{amb} of carriers to achieve a pronounced MD effect. Usually the epilayer thickness of SOS MDs is $t \sim 1$ μm. This is the optimum value which satisfies the condition $t \approx L_{amb}$.

In addition to controlling the injection and the MD effect, the temperature T also influences the performance of this class of bipolar devices. It can be seen from the experimental results in [15,30,31], for instance, that in Ge and Si MDs the sensitivity S_V decreases with increasing temperature. The smaller the supply current I, the more rapid is the sensitivity decrease. The sensitivity may be reduced by a factor of 3 to 5 in the temperature interval $200 \leq T \leq 350$ K. This result is explained by the rapid decrease of the supply voltage at $I = $ const. Unlike S_V, the parameter S_I increases with temperature.

The frequency response of MD devices in an a.c. magnetic field is, generally speaking, a function of the effective lifetime of the minority carriers, i.e. $f \sim 1/\tau_{eff}$. The cut-off frequency f_0 is increased when the magnetic field causes a carrier deflection towards the surface with the high recombination rate, whereas the Lorentz deflection towards the surface with $s \approx 0$ leads to a decrease of f_0. The

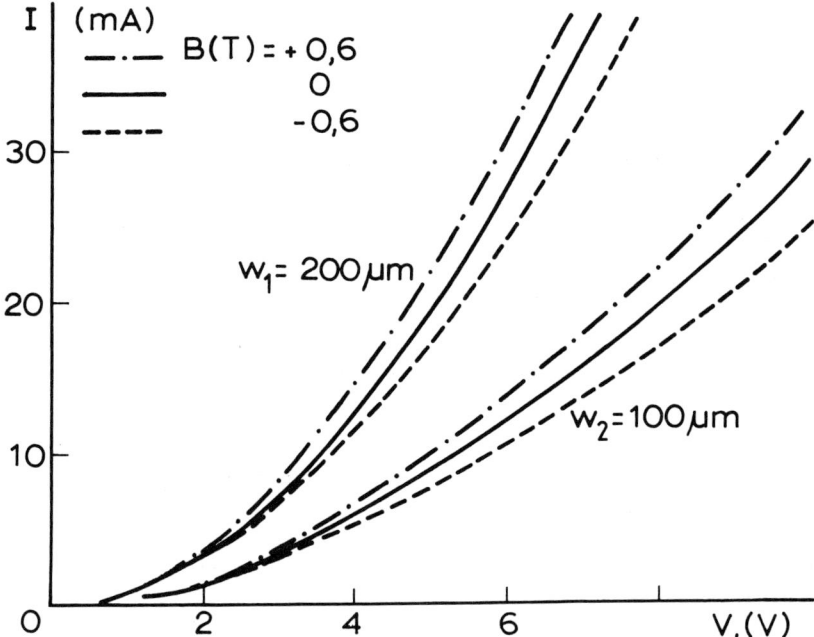

Fig. 5.10. Effect of the magnetic field upon the $I(V)$ characteristics of SOS p^+–n–n^+ MDs with different widths w_1 and w_2; $l = 10$ μm; $t = 0.65$ μm, $T = 300$ K, $n_0 = 7 \times 10^{15}$ cm^{-3} [30].

experimentally determined cut-off frequency of Ge p^+–n–n^+ MDs is $f_0 \sim 10$ KHz, [4,34], but a smooth frequency response is manifested by discrete Si planar p^+–n–n^+ MD devices up to a frequency $f_0 \approx 1$ MHz [34]. According to the data in [35], the cut-off frequency of SOS and Schottky MD sensors is $f_0 \sim 10$ to 100 MHz with $\tau_{\text{eff}} \sim 10^{-9}$ to 10^{-8}s.

Noise is the main factor that determines the lowest value, $B_{\min}$, of the magnetic field which can be detected by MD sensors. The level of intrinsic noise in magnetosensitive devices based on magnetoconcentration is rather low. For instance, in Ge MC sensors, the noise power spectral density is proportional to I^2 and its value is less than 10^{-14} V^2 Hz^{-1} at a frequency $f = 1$ kHz. The source of this noise is $1/f$ at $f \leq 1$ kHz, but at $f > 1$ kHz generation–recombination and diffusion noises are observed. The $1/f$ noise is rather sensitive to the magnetic field [36]. If this noise is associated with surface phenomena, a magnetic field $B = 0.5$ T can increase by a factor of 10^3 the fluctuation spectral density in MC devices at low frequencies. The noise level is higher by an order of magnitude in MDs than in MC sensors, provided the other conditions are the same. In Ge and SOS MD devices the noise is $1/f$ at frequencies $f < 1$ kHz [15]. Conductivity fluctuations in the base region of the structures are the source of that noise. The noise in the Schottky MDs has a similar character [31]. According to the data in [37], the noise in SOS MDs

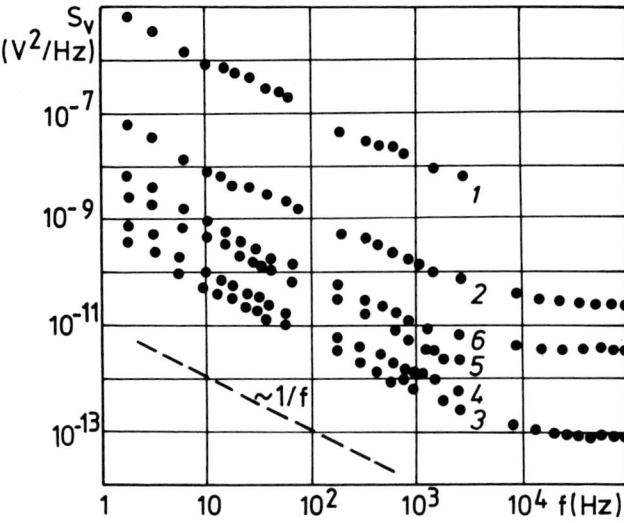

Fig. 5.11. Voltage noise power spectral density $S_V(f)$ for several p^+–n–n^+ micromagnetodiodes at $T = 300$ K. 1: $n_0 = 5 \times 10^{15}$ cm^{-3}, $l = w = 50$ μm, $I = 0.8$ mA, $V = 70$ V; 2: $n_0 = 5 \times 10^{15}$ cm^{-3}, $l = 10$ μm, $w = 50$ μm, $I = 9$ mA, $V = 8.9$ V; 3: $n_0 = 7 \times 10^{15}$ cm^{-3}, $l = 10$ μm, $w = 50$ μm, $I = 1$ mA, $V = 2$ V; 4: $n_0 = 7 \times 10^{15}$ cm^{-3}, $l = 10$ μm, $w = 200$ μm, $I = 9.9$ mA, $V = 3.3$ V; 5: $n_0 = 7 \times 10^{15}$ cm^{-3}, $l = 20$ μm, $w = 200$ μm, $I = 5.5$ mA, $V = 7.2$ V; 6: $n_0 = 7 \times 10^{15}$ cm^{-3}, $l = 10$ μm, $w = 50$ μm, $I = 10.2$ mA, $V = 6$ V [37]

increases with increasing length l and voltage V. Furthermore, the noise decreases with an increase in the doping concentration in the base region or the width of the structure. When the double-injection mode sets in, the noise levels become proportional to I^2, and the noise voltage spectral density is $S_V(f) \sim lV/wtn_0 f$ [15]. Figure 5.11 shows the power spectral density of the voltage fluctuations S_V [V^2 Hz^{-1}] for several SOS devices, biased with current generators [37]. For instance, in an MD with $l = 10$ μm, $w = 50$ μm and $I = 1$ mA, the noise spectral density is 10^{-12} V^2 Hz^{-1} near $f = 1$ kHz; its sensitivity being around 10 V T^{-1}, it follows that, with a selective circuit such as $\Delta f/f = 10^{-3}$, a magnetic field $B = 10^{-7}$ T can be detected for $S/N = 1$ [37]. Under similar operating conditions the noise level in Ge p^+–n–n^+ MDs is of the same order of magnitude as that of the noise in SOS MDs. Magnetosensitivity is not the only parameter to increase as the temperature decreases. The noise level and the diode resistance increase as well. In Ge MDs the noise power spectral density is increased 15 times when the temperature is lowered from +20°C to −50°C. The magnetic field leads also to an increase in noise level in MDs because of the increase in the resistance.

Some general conclusions can be drawn concerning the influence of noise on the performance of MDs. There is a substantial dependence of the noise level on the static resistance of the device. Hooge's model most adequately describes the $1/f$ noise in MD sensors [38]:

$$S_V(f) = V^2 \frac{\alpha}{Nf}, \qquad (5.13)$$

where N is the number of carriers, V is the supply voltage and α is the Hooge parameter, whose value is about 2×10^{-3}.

The expression (5.13) gives a qualitative explanation of the existing data concerning the $1/f$ noise in MDs. The noise level, for instance, is higher in long diodes than in short MDs, since at $I = $ const the square of the voltage, V^2, is proportional to l^3. Besides, at the same current, the voltage V in an MD of width w_1 is lower than in a structure of width w_2, when $w_1 > w_2$. Therefore, the noise level is lower in the device whose width is w_1. The influence of the magnetic field B upon the $1/f$ noise can be explained in a similar way by the variation of the static resistance.

5.5. MD sensor review

5.5.1. Orthogonal MD devices

The earliest discrete orthogonally activated MDs were fabricated from almost intrinsic Ge and Si. No special regions with predetermined recombination rates were formed. The devices are long structures with one p^+-n junction (Fig. 5.1a and b) [1–4,39]. As for the even geometrical magnetoresistance effect, the resistance of the MD samples increases when a magnetic field B is applied. Table 5.1 is a summary of the published sensor characteristics of MDs whose device structures are shown in Fig. 5.1. Magnetosensitivity related information is the main problem when characterizing MDs. The parameter S is often designed as the current variation ΔI, voltage variation ΔV, or resistance variation ΔR of the device in a wide range of values ΔB of the magnetic induction. Usually $\Delta B = 0.5$ T and $\Delta B = 1$ T, although the $I(B)$, $V(B)$ and $R(B)$ relationships are highly nonlinear. On the other hand, the data concerning the device dimensions, the operating mode, the noise, and other characteristics are not always available. This is a serious hindrance to the comparative analysis of this class of sensors. The criteria suggested in §5.3 can be applied mainly to modern SOS MDs because of the availability of the necessary data. According to [1–4,39] the magnetosensitivity of discrete Ge versions is about 0.5×10^3 V A^{-1} T^{-1}. The formation of a region with a high surface recombination rate in Ge p^+-n-n^+ structures (Fig. 5.1c) makes the $V(B)$ and $I(B)$ characteristics odd. At a supply voltage $V = 5V$, the current magnetosensitivity is 12 mA T^{-1} in the $-50 \leq B \leq 50$ mT interval provided the $I(B)$ characteristic is linear [7,34]. Owing to the quasi-intrinsic conductivity of the Ge material used, there is a considerable influence of temperature on the single MD sensor characteristics. For instance, the temperature variation of the sensitivity is nonlinear in the range $-20 \leq T \leq 80°$C, and at $B = 0.1$ T the output signal at 80°C is about half the value at $-20°$C. A pair of magnetodiodes can be used (Fig. 5.1c) to minimize the influence of temperature, to

Table 5.1

Parameters of selected magnetodiode sensors, whose device structures are shown in Fig. 5.1.

Device struct. mater.	Doping n_0 (cm^{-3}) resist.	Dimensions $l \times w \times t$	Sensitivity S	Equival. magn.-field noise	T.C. of S (% K^{-1})	Freq. resp. (kHz)	Refs.
(a,b) Ge	50 Ω·cm	4×1×0.2 mm	0.5×10^3 V A^{-1} T^{-1}			10	[1–4]
(c) Ge	50 Ω·cm	3×0.6×0.4 mm	12 mA^{-1} T^{-1}		−0.11	10	[7,34]
(c) Ge	50 Ω·cm	3×0.6×0.4 mm	10 V T^{-1}			10	[7,34]
(a,b) Ge S-type	50 Ω·cm	4×1×0.2 mm	0.5 V mT^{-1}				[4]
(c) Si	20 kΩ·cm	1×0.5×0.5 mm	30 V T^{-1}			1	[39]
(a) Si S-type	2 kΩ·cm	0.15×0.55×– mm	50 V T^{-1}				[39]
(a) Si S-type	1 kΩ·cm	1×1×1 mm	500 V T^{-1}				[39]
(a) GaAs S-type		0.08×–×– mm	10^6 V A^{-1} T^{-1}				[39]
(a) GaAs S-type			10^8 V A^{-1} T^{-1}				[39]
(c) InSb S-type			5×10^5 V A^{-1} T^{-1}				[4]
(d) InSb S-type	10^2 Ω·cm	1×0.05×– mm	12 mA mT^{-1}				[9]
(e) InSb S-type	10^2 Ω·cm		2.5 kHz mT^{-1}				[9]
(f) Si	55 Ω·cm	0.1×–×– mm	2 V T^{-1}				[10]
(g) Ge	50 Ω·cm		10 V T^{-1}			15	[4]
(g) Ge	50 Ω·cm	20×4×3 mm	20 mA mT^{-1}			15	[4]
(h) Ge	50 Ω·cm	7×1×0.25 mm	60 V T^{-1}	10^{-13} V^2 Hz^{-1}, $f \leq 1$ kHz		100	[13,15,37,41]

Type	Resistivity	Dimensions	Sensitivity		Noise	Frequency	Refs
(i) Ge	50 Ω·cm	5×2×1 mm	300 V T^{-1}		10^{-13} V^2 Hz^{-1}	100	[13,15,37,41]
(h) Si	20 kΩ·cm	400×–×– μm	40 V T^{-1}			10	[39]
(h) Si S-type	30 kΩ·cm	300×–×– μm	60 V T^{-1}				[42]
(j) Si S-type	20 kΩ·cm	100×–×30 μm	10 V T^{-1}			10^3	[16]
(j) Si	20 kΩ·cm	3×1×0.6 mm	80 V T^{-1}				[5]
(j) Si S-type	10 kΩ·cm	0.15×–×– mm	10^5 V A^{-1}T^{-1}				[43]
(m) Si differ.	10^{15} cm^{-3}	0.4×0.15×0.3 mm	10^2 V A^{-1}T^{-1}	−0.2		100	[19]
(n) Si	10 kΩ·cm	1×–×0.3 mm	20 V mA^{-1}T^{-1}				[20]
(o) Si SOS	10^{15} cm^{-3}	10×50×7 μm	10 V T^{-1}, 150 mA T^{-1}		$B_{min} = 10^{-7}$ T	100 MHz	[22,37]
(o) Si SOS	7×10^{15} cm^{-3}	31×20×4 μm	4.9 V T^{-1}	−0.1			[25]
(o) Si SOS	5×10^{15} cm^{-3}	50×50×0.65 μm	14 V T^{-1}, 2.2 mA T^{-1}	−0.1	1 μV^2 Hz^{-1}	100 MHz	[15,30,31]
(q) Si SOS	10^{15} cm^{-3}	15×100×0.65 μm	10% T^{-1}	−0.1	$B_{min} = 5$ μT	100 MHz	[26,27,44]
(s) Ge Sch. MD	60 Ω·cm	2×3×1 mm	250 V T^{-1}		$B_{min} = 10^{-7}$ T	100	[28]
(t) Si SOS, Sch.	5×10^{15} cm^{-3}	20×50×0.65 μm	20 V T^{-1}	−0.1	$B_{min} = 10^{-9}$ T	100 MHz	[29–31,41]
(v) Si CMOS	4×10^{15} cm^{-3}	126×48×10 μm	25 V T^{-1}				[32]
(w) Ge	40 Ω·cm	8×0.6×0.2 mm	50 V T^{-1}	−0.5		1	[33]
(x) Ge	40 Ω·cm	8×0.6×0.2 mm	15 V T^{-1}	−0.5		1	[33]

enhance the output signal and widen the magnetic induction interval of linear sensitivity. This arrangement provides for a constant voltage $V_{out}(B=0)$ in the interval $-50 \leq T \leq 80°C$ and the mean variation of the sensitivity S_V is 15%. At $T = 20°C$, S_V is 10 V T^{-1} in the interval $-50 \leq B \leq +50$ mT. Germanium p^+–n MDs (Fig. 5.1a, b) which exhibit an S-type negative resistance at $T = 77$ are described in [4]. The mechanisms responsible for the occurrence of the negative resistance are the increase of τ_{eff}, the existence of ohmic leakage channels that shunt the p^+–n junction and the change in the injection coefficient. The effect of the magnetic field can be observed in the region of negative resistance. At $B = 5$ mT, the sensitivity S_V is 0.5 V mT^{-1}. According to [4], this value exceeds by about 100 times the sensitivity of conventional Ge MDs at room temperature.

By analogy with Ge devices, the first Si MD sensors (Fig. 5.1a) did not include regions with high velocity of surface recombination, and hence their conductivity can only be reduced by the magnetic field. The material used for their fabrication is high resistivity compensated p-Si with $\rho > 20$ kΩ·cm. The p-type material is preferred owing to the higher mobility of injected electrons. Figure 5.12 presents the dependence of the sensitivity S_V on the magnetic field of Si MD devices of the type shown in Fig. 5.1a with two lengths l_1 and l_2 of the base region ($l_1 < l_2$) at two current magnitudes. Longer structures manifest higher transducing efficiency at high values of the current I. A silicon p^+–p–n^+ MD (Fig. 5.1c) without a region of high surface recombination rate, whose magnetosensitivity is 30 V T^{-1} at a current $I = 3$ mA, is described in [39]. Silicon has a wider bandgap than Ge, therefore there is a choice of deep level impurities which can assist the mechanism of the S-type differential resistance [40]. The current–voltage characteristics of Si MDs with an S-type region are like those shown in Fig. 5.2b. There is a strong dependence of point $V_V(B)$ on the magnetic field, and the dependence of $V_p(B)$ is negligible. The voltage magnetosensitivity defined as $\Delta V_V / \Delta B$ of a p^+–n MD with

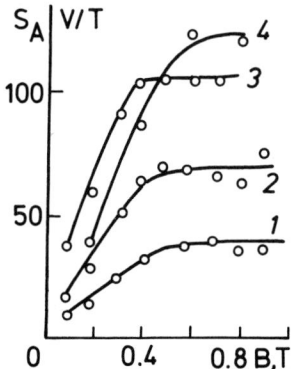

Fig. 5.12. $S_V(B)$ dependences of two silicon MDs: $I = 3$ mA (curves 1 and 3), $I = 1$ mA (curves 2 and 4) [4].

the device structure shown in Fig. 5.1a is 50 V T^{-1} at the temperature $T = 300$ K and current $I = 10$ mA. The semiconductor material has been compensated with Ni and its resistivity is $\rho \sim 2$ kΩ·cm [4]. The negative resistance disappears as the magnetic induction is increased to values of $B \geq 0.4$ T.

An almost linear dependence of $V_V(B)$ and a high transducing efficiency is manifested by a silicon MD of the type shown in Fig. 5.1a. Its base region, with a resistivity $\rho \sim 10^5$ Ω·cm, is compensated by Ir. At the current $I = 10$ mA the sensitivity defined as $S_V = \Delta V_V / \Delta B$ is 500 V T^{-1} and at $B \geq 0.5$ T the resistance is negative.

Discrete MDs with the device structure shown in Fig. 5.1a have also been fabricated out of $A^{III}B^V$ compounds like GaAs and InSb. The electron mobility in these materials is well above that in Ge and Si. P-type GaAs compensated with Ni, Fe, Ti or Cr is commonly used. The current–voltage characteristics of many MDs are S-type and the magnetic field, irrespective of its polarity, leads to reduced conductivity. Besides, the magnetic field B has a considerable effect on the valley-point V_V, whose displacement can be used as a criteria for magnetosensitivity. GaAs MDs are high-voltage devices, unlike other MDs. At $B = 0$, the value of V_p is in the range 30 to 150 V and V_V is in the range 10 to 50 V. The sensitivity of an S-type GaAs MD compensated with Ni is presented in Fig. 5.13. The parameter V_V has been used to determine the sensitivity at $T = 300$ K [39]. It can reach 10^6 V A^{-1} T^{-1} in low magnetic fields. In Cr-doped GaAs MDs the transducing efficiency reaches 10^8 V A^{-1}T^{-1}. Ion implantation can be employed to improve the parameters of S-type GaAs MDs by reducing V_V, V_p and l, for instance, since in this case the diffusion length L_n is 1 to 2 μm.

Because of the narrow bandgap of InSb ($E_g \approx 0.17eV$), S-type MDs made of this material can operate at low temperatures only. The extremely high electron mobility determines the magnetosensitivity $\Delta V_V / \Delta B$ of the n^+-p-p^+ device structures in Fig. 5.1c which do not include a surface with a high recombination rate. Their sensitivities can reach 5×10^5 V A^{-1} T^{-1} at $T = 77$ K. A bistable flip-flop sensor

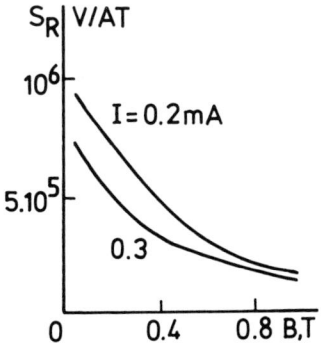

Fig. 5.13. $S_R(B)$ dependence of an S-type GaAs MD at $T = 300$ K and different supply currents [39].

has been designed on the basis of n^+–p–p^+ InSb S-type MDs (Fig. 5.1d) [9]. This device has been mentioned in §5.1. The reason for the occurrence of a negative resistance in this particular device is the trapping of carriers. The effective lifetime of electrons in the base region is very short ($\sim 10^{-9}$ s). At high injection levels the traps are saturated and the lifetime is effectively increased to 10^{-7} to 10^{-6} s and a filament (domain) appears at a supply current density of 0.2 A cm^{-2}. Due to the conic shape of the filament spread (5.1), an increase in the injection level leads to an increase in the dimensions of the filament, and it can occupy the entire active volume of the structure. This is why the working point (I_0, V_0) in the S-region of the $I(V)$ characteristic plays an important role. This point determines the optimum dimensions of the domain for the triggerable flip-flop. At a total current $I_0 = 10$ mA in a pulsed magnetic field $B = 2$ mT, a current of magnitude 3.8 mA is switched between terminals 1 and 2 within switching times of 2 to $3\mu s$ [9]. A sensor, with a shape analogous to that in Fig. 5.1d, which operates as a dual-base MD manifests linear and polar variations of the differential current $\Delta I_{1,2} = I_1 - I_2 = 12$ mA at a total supply current $I_0 = 15$ mA in a magnetic field $B = 1$ mT. In a circularly symmetric eight-branch MD of the type shown in Fig. 5.1e the filament in the S region starts a circular rotation from contact to contact after a certain threshold value $B > B_0$ of the field is exceeded. The frequency of rotation is $f \approx 2.5$ kHz mT^{-1} in a magnetic field $B = 4$ mT at $T = 77$ K. The frequency f increases linearly with an increase in the magnetic induction [9].

The operation of the orthogonally activated MD sensor from Fig. 5.1f has been described in §5.1. The p^+- and n^+-contacts are formed by means of thermal diffusion in a n-type Si substrate with $\rho \sim 50$ Ω·cm and a thickness 330 μm. Au is deposited by sputtering onto the back surface to make the substrate highly resistive. The magnetic control of the $I_{p1}(V_{p1})$ characteristic at $V_{p2} = 10V$ is polar and to a first approximation is linear, when $B > 0.1$ T. At a current $I = 5$ μA and distance $l = 100$ μm, the sensitivity is 2 V T^{-1}. The resistance variation $\Delta R(B)/R_0$ determined from the $I_{p1}(V_{p1})$ curve is 1.3 at $B = 0.3$ T. If the distance is $l > l_0 = 200$ μm, S-type regions appear in both curves $I_{p1}(V_{p1})$ and $I_{p2}(V_{p2})$, and $\Delta R(B)/R_0 \approx 20$ at $B = 0.1$ T. This type of MD sensor offers good opportunities for IC fabrication with low power dissipation, [10].

Other investigations of orthogonal MD structures have been described in the literature, but their purpose has been the qualitative verification of the influence of the magnetic field B on the $I(V)$ characteristics, provided a magnetodiode effect occurs. Sensor parameters have not been presented, however.

5.5.2. Parallel-field MD devices

Discrete p^+–n MDs with parallel-field activation are fabricated mainly from Ge and Si. No special regions with predetermined surface recombination rates were formed in the earliest devices of this type. The length of the current path can be increased by the sectioning of the injecting p–n junction. This opportunity is illustrated by the device in Fig. 5.1g. This multicontact sensor has, according to [4], a higher transducing efficiency than an MD with an uninterrupted single top p^+–n junction and its sensitivity is $S_V = 10$ V T^{-1}. The conductivity is reduced by the magnetic field irrespective of its polarity. A Ge MD with one injecting junction has been suggested for operation at high injection levels. The current density reaches $i = 10$ A cm^{-2}, and this is why the device is cooled by water which circulates in a radiator capable of 50 to 100 W power dissipation. The magnetosensitivity of this sensor is 20 mA mT^{-1} [4]. The frequency characteristic of the two MD sensors considered above is 15 kHz on average.

The formation of a second injecting junction and regions with different surface recombination rates (Fig. 5.1h) can considerably improve the conditions of operation of a magnetodiode and its sensitivity. The general behavior of the $I(B)$ characteristics, depending on the ratio of the recombination velocities s_1 and s_2, has been analyzed in §5.4 and the experimental results corroborate the semiconductor regime of conductivity. According to the data in [13,15,41], the $V(B)$ dependence of Ge MDs with highly asymmetric surface recombination rates s_1 and s_2 is linear in the neighborhood of point $B \approx 0$ (low fields), and the sensitivity is $S_V = 60$ V T^{-1}. The sensitivity of the Ge MD in Fig. 5.1i with a high recombination rate $s \to \infty$ on the lower surface is 300 V T^{-1} or 30 A T^{-1}, respectively. The noise source in Ge MDs of the types (h) and (i) (Fig. 5.1) is $1/f$ at frequencies $f \le 10$ kHz. Its power spectral density at frequencies near $f_0 = 1$ kHz is 10^{-13} V^2 Hz^{-1} at a current $I = 30$ mA [37]. A summing up of the available data concerning the Ge p^+–n–n^+ MDs whose device structures are shown in Fig. 5.1h and i, leads to the following conclusions: in a magnetic field $B \le 0.05$ to 0.1 T, the $I(B)$ dependence is linear with a polar sensitivity. The response at high frequencies is limited by the minority carrier lifetime $\tau_{\text{eff}} \sim 10$ μs. The corresponding cut-off frequency is in the range 10 to 100 kHz. The lowest detectable field $B_{\min}$ at $S/N = 1$ is in the range 10^{-9} to 10^{-8} T. The use of ferrite conic concentrators of the magnetic field reduces $B_{\min}$ to about 10^{-11} T. The current I in Ge MDs operating at a constant voltage in the field $B \approx 0.1$ T can be abruptly changed, and at $V = 20$ V the transducing efficiency is $S_I \approx 30$ A T^{-1} [13,15]. These results have been achieved by structures of length 4 to 5 mm, width 2 to 3 mm and thickness is 1 to 1.5 mm. The voltage magnetosensitivity is also high: $S_V \approx 300$ V T^{-1}. If a small supply current is required by a particular sensor application, the dimensions of the Ge MD should be reduced, and in this case at a current $I = 5$ mA the parameter S_V reaches 100 V T^{-1} [13,15,37,41].

In the first discrete parallel-field silicon p^+–n–n^+ MDs (Fig. 5.1h), the two active surfaces Σ_1 and Σ_2 are characterized by high recombination rates. At a current $I = 3$ mA and a temperature $T = 300$ K in a magnetic field $B = 0.3$ T, the magnetosensitivity is 40 V T^{-1} [39]. The sensitivity is reduced by a factor of 2 to 3 by an increase in temperature from 20 to 100°C. Neutron irradiation with $K = (3$

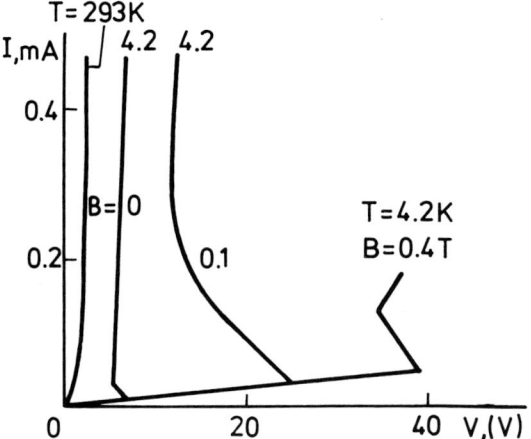

Fig. 5.14. $I(V)$ characteristics of a p–type Si MD at different temperatures [39].

to 5) $\times 10^{12}$ cm^2 per neutron has been reported to reduce S_V by 25 to 40% [39]. These modifications of the $I(V)$ characteristic are caused by the decrease in the length $L_{n,p}$ due to the increase in the defect density in the base region. Annealing at 250°C for one hour can practically restore the $I(V)|_{B=0}$ characteristics of the samples.

A negative resistance region of the $I(V)$ characteristic can be observed in Si MDs (Fig. 5.1h) at low temperatures [42]. The transformation of a standard $I(V)$ curve into an S-type characteristic as the temperature is decreased is shown in Fig. 5.14. The MD is made of p-type Si with $\rho = 30$ kΩ·cm. The length of the base region is $l = 300$ μm. It can be seen from Fig. 5.14 that a typical short-diode $I(V)$ characteristic with a negligible sensitivity is observed at $T = 300$ K. A negative resistance region of the $I(V)$ curve appears as the temperature is decreased to 4.2 K. The resistance increases with increasing magnetic induction B. The most probable reason for the appearance of the S-region is the shunting of the p^+–n junction by a leakage channel. At $T = 300$ K the resistance of this channel is high and the shunting effect of the leakage cannot be observed, but at 4.2 K the resistance of the channel becomes comparable with that of the junction. The voltage V_p increases with increasing magnetic field B $V_p(B) = V_p(0) + S_V B$. The S_V parameter of this MD is 60 V T^{-1}.

A planar version of a MD (Fig. 5.1j) has been fabricated from quasi-intrinsic Si. The p^+- and n^+-contacts were formed by selective diffusion of impurities [16]. The length of the effective base region is $l \approx 100$ μm and the thickness is $t \approx 30$ μm. The lower surface of the device has been subjected to a special abrasive treatment so that $s_2 \to \infty$, and the upper surface is covered by a thermally grown SiO$_2$ ($s_1 \to 0$) to improve the operation of the sensor. At $T = 300$ K the $I(V)$ curve is S-type with polar magnetic-field control. The $I(B)$ dependence is linear in the range $-20 \leq B \leq +20$ mT. A magnetosensitivity $S_V = 10$ V T^{-1} and a frequency response of about 1 MHz have been achieved. An MD sensor (Fig. 5.1j) with highly asymmetric recombination rates on both surfaces and transducing efficiency $S_V \approx 80$ V T^{-1} at a current $I = 3$ mA is reported in [5].

Silicon planar MDs (Fig. 5.1j) with negative resistance and a Au-doped base region are described in [43]. Ion implantation was used for the formation of the p^+–n and n^+–n junctions. The length of the different samples is $50 \leq l \leq 150$ μm. An S-type curve is observed at $T = 300$ K and $l \geq 100$ μm. The polar magnetosensitivity of these sensors is associated with the asymmetry of the surface recombination rates. The magnetosensitivity at a current $I = 0.4$ mA in a magnetic field $B \leq 0.1$ T is 10^5 V A^{-1} T^{-1} and is about 100 times higher than the sensitivity of similar MDs with the device structure shown in Fig. 5.1a, whose base region has been doped with Ni.

The magnetic control of the $I(V)$ curve of the silicon p^+–$i(p)$–n^+ MD from Fig. 5.1l is illustrated by Fig. 5.15. The additional n^+–p junction is used to replace carrier recombination by extraction [18]. The results obtained confirm the favorable effect of carrier extraction on the magnetosensitivity of the device.

The device structure in Fig. 5.1m can be regarded as being obtained by a symmetrical unfolding of the MD in Fig. 5.1c with respect to the longitudinal axis of the device. The differential MD implemented in such a manner is fabricated from n-type silicon with $\rho \sim 7.5$ Ω·cm. This value corresponds to $n_0 \sim 10^{15}$ cm^{-3} and the impurity concentration in the heavily doped p^+ and n^+ regions is 10^{19} cm^{-3}. There is no special treatment of the lower surface of the chip to achieve high recombination velocity. The two outer n^+-contacts are connected to a differential circuit. They are the output terminals of the MD sensor (Fig. 5.16). The trimmer r is used to adjust a zero offset. Polar magnetic control of the $V_{2,3}(B)$ dependencies is obtained experimentally. These characteristics are linear in the interval $-0.6 \leq B \leq +0.6$ T. The magnetosensitivity at $T = 300$ K and total current $I = 10$ mA is 10^2 V A^{-1} T^{-1}. Temperature variation in the range $0 \leq T \leq 80°$C

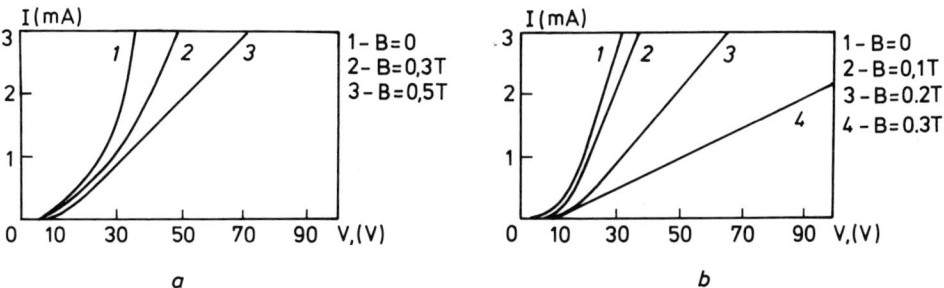

Fig. 5.15. $I(V)$ characteristics of the p^+–$i(p)$–n^+ MD from Fig. 5.1l: (a) the n^+–p collector is floating; (b) the n^+–p collector is operating in the extraction mode [18].

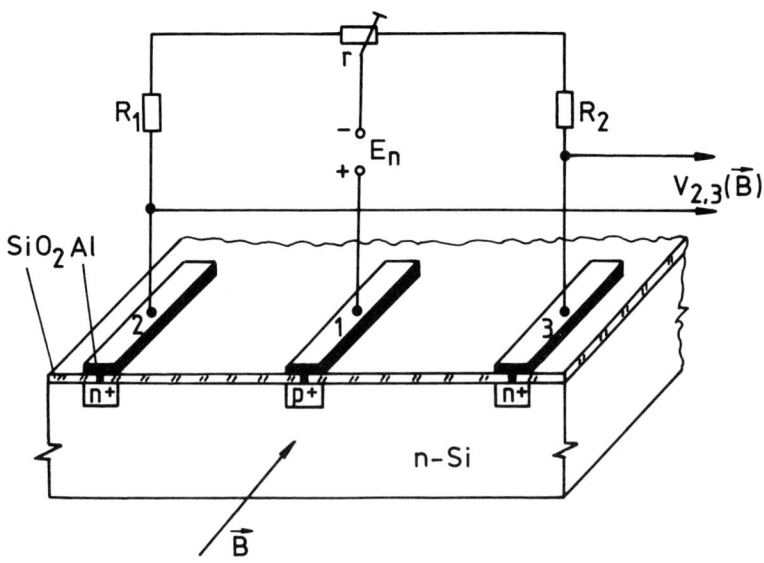

Fig. 5.16. Parallel-field differential Si magnetodiode [19].

does not affect the zero offset already adjusted. The coefficient T.C. is $\sim 0.2\%\ K^{-1}$ [19].

The injection-field MD in Fig. 5.1n is made of a p-type high resistivity silicon with $\rho \sim 10$ kΩ·cm [20]. The p^+- and n^+-contacts are formed by ion implantation and the distance between them can vary in the interval 0.7 to 1.0 mm. A SiO$_2$ layer is used as a gate insulator. Increasing magnetosensitivity with increasing magnitude of the negative voltage, $-V_G$, and of the supply current is manifested by this sensor. Devices with a Al$_2$O$_3$ gate insulator are more sensitive and their S_R can reach 20 V mA^{-1} T^{-1} at a current $I = 3$ mA in a field $B \leq 0.3$ T.

The earliest parallel-field SOS MD (Fig. 5.1o) was suggested in [22]. It is presented schematically in Fig. 5.17a. The n-type base region is characterized by a low impurity concentration and carrier mobilities $\mu_n \sim 400$ cm^2 V^{-1}s^{-1} and $\mu_p \sim 200$ cm^2 V^{-1}s^{-1}. The contribution of the Si–SiO$_2$ and the Si–Al$_2$O$_3$ interfaces to the achievement of highly asymmetric surface recombination rates has been discussed in §5.1 and §5.4. A magnetosensitivity of 10 V T^{-1} has been attained. The noise power spectral density of the same MD has been investigated [37], and its value at $I = 1$ mA is 10^{-12} V^2 Hz^{-1}, hence the minimum detectable field at $S/N = 1$ is $B_{\min} \approx 10^{-7}$ T. The sensitivity of an analogous SOS MD sensor (Fig. 5.1o), is 4.9 V T^{-1} at $n_0 = 7.1 \times 10^{15}$ cm^{-3}, $\tau_{\text{eff}} = 15$ ns and $V = 15$ V [25]. The advance of SOS MDs of the type presented in Fig. 5.1o is described in detail in [15,30,31].

The thickness, t, of the SOS epilayer and the quality of the Si–SiO$_2$ and Si–Al$_2$O$_3$ interfaces cannot be varied freely because of practical difficulties. This is why the

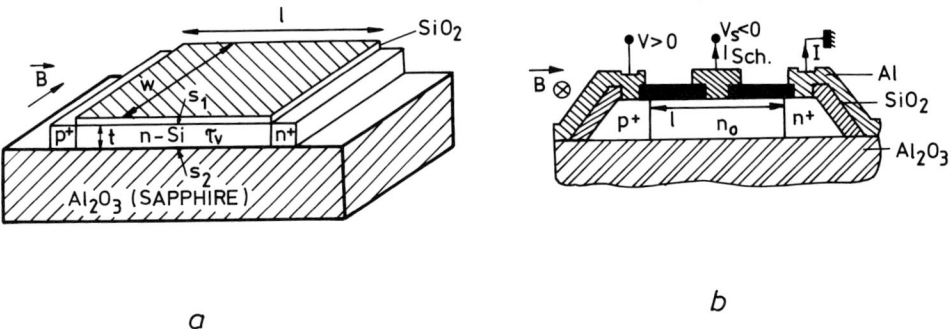

Fig. 5.17. Silicon on sapphire bipolar microsensors: (a) SOS magnetodiode; (b) outline and bias of a SOS Schottky magnetodiode.

recent results in this class of sensors have been achieved by optimizing the initial doping level, n_0, of the base region and the geometrical dimensions l and w. The p^+–n–n^+ structures are fabricated by a conventional MOS process on a (100) SOS substrate obtained by CVD. Ion implantation of phosphorus is employed for the doping of the base region, the epilayer thickness being $t \approx 0.65$ μm. Typical results corresponding to 40 to 60 mW dissipated power at 300 K are presented in Table 5.1. It can be seen from the table that short MDs have better S_I sensitivity whereas long devices exhibit superior S_V.

Gated SOS MDs (Fig. 5.1q) offer the promising opportunity for an additional adjustment of the magnetosensitivity and the mode of operation by the gate voltage V_G [26,27,44]. For instance, at $V = 5$ V, $I = 2$ mA and $V_G = -3$ V relative to the n^+-contact, the magnetosensitivity of a device whose dimensions are $l = 15$ μm, $w = 100$ μm and $t = 0.65$ μm is 10% T^{-1}. This value corresponds to an increase by about 100% of the current variation in the magnetic field $B = 0.35$ T in comparison with the floating-gate case ($V_G = 0$). The noise in this type of MD has been analyzed in detail [44]. It has been found that the source is $1/f$ bulk noise. In the gated MDs in a narrow frequency band $\Delta f \simeq 1$ kHz, the field $B_{\min} = [S(f)\Delta f]^{1/2}/S_I$ equals 5 μT on the average, whereas in a three decade wide frequency range ($f_2/f_1 = 10^3$) this field is $B_{\min} = [S(f)\ln f_2/f_1]^{1/2}/S_I \approx 80$–800 μT.

In p^+–n–n^+ MDs with a lightly doped base region, the bulk nonuniformities and the defects on the Si–Al$_2$O$_3$ interface lead, at high values of the voltage, to negative resistance and to the appearance of a current filament [45]. Carrier heating is the most probable reason for this behavior. The magnetosensitivity of such filamentary MDs is 20 V T^{-1} [31].

The first parallel-field Schottky MD (Fig. 5.1s) was made of 60 Ω·cm Ge and its operation is based on the MC effect. The metal–semiconductor contact was formed upon the surface Σ_1 with the low recombination rate $s_1 \sim 1$ m s^{-1} by evaporation of Au, whereas the recombination rate on the second surface Σ_2 is high: $s_2 \sim 100$ m s^{-1}. The dimensions of the germanium plate are $20 \times 3 \times 1$ mm^3. At a current $I = 3$ mA, the magnetosensitivity of this sensor is polar and is determined by the variation of the Schottky-contact reverse current I_s. At $V_{Sch} = -20$ V, the value of the parameter S_V is about 250 V T^{-1}. The minimum detectable field B_{min} is in the 10^{-7} T range at a frequency $f_0 = 10$ kHz in a $\Delta f = 500$ Hz frequency band, when $S/N = 1$ [28].

The operation of second-generation Schottky MDs (Fig. 5.1t) is based on the magnetodiode effect in SOS p^+–n–n^+ structures, Fig. 5.17b [15,29–31,41]. The two junctions are forward biased to establish the semiconductor regime of conductivity, while the Schottky contact is reverse biased. Its current is the signal that contains the information about the magnetic field. The polar magnetic control of the $I(B)$ characteristic, the output signal linearity in the neighborhood of $B \approx 0$ and a higher magnetosensitivity than that of the respective p^+–n–n^+ SOS MDs are important features of the Schottky SOS MDs. A Schottky SOS MD with dimensions $l = 20$ μm, $w = 50$ μm, $t = 0.65$ μm, for instance, exhibits a sensitivity $S_V = 20$ V T^{-1} at $V = 12$ V, whereas the corresponding p^+–n–n^+ SOS MD has the sensitivity $S_V = 1.5$ V T^{-1}. According to the data in [30,41] the noise power spectral density at $f < 10^4$ Hz in Schottky MDs at the same injection level is lower than that in the respective longitudinal p^+–n–n^+ SOS devices, and the corresponding value of B_{min} is about 10^{-9} T. The optimum device parameters of the MDs shown in Fig. 5.1o and t, are the same.

The integrated MD in Fig. 5.1v is fabricated by a standard p-well CMOS process [32]. A specific feature of this sensor is that a reverse biased p–n junction (collector) plays the role of a region with a high recombination rate, i.e. the recombination of minority carriers injected by the emitter E has been replaced by collection. The device parameters are: the distance $l_{EB} = 126$ μm, the emitter area 18×28 μm^2, the width $w = 48$ μm and the p-well depth $t = 10$ μm. By analogy with the magnetodiode effect, the characteristic $I_E(V_{BE})$ is sensitive to the magnetic field B at high values of the current I_E. At 10 mA supply current, the magnetosensitivity measured by the voltage V_{BE} is $S_V = 25$ V T^{-1}.

Ge parallel-field sensors (Fig. 5.1w, x) are actually resistors whose resistance is magnetoconcentration controlled [33]. A region of high recombination rate s_1 has been formed in the device in Fig. 5.1w. At a current $I = 3$ mA, the $V(B)$ dependence is linear in a magnetic field $B \leq 20$ mT, and the sensitivity is 50 mV mT^{-1} (i.e. 50 V T^{-1}). The initial resistance of these structures is about 24 kΩ and the coefficient T.C. is -0.5% K^{-1} in the interval $+10 \leq T \leq 35$°C. The surface recombination rate can be electrically controlled by means of the field effect, and reproducible metrological characteristics can be achieved by the addition of two gate electrodes G_1 and G_2 (Fig. 5.1x). The magnetosensitivity is 15 V T^{-1} at $V_{G1} = +12$ V, $V_{G2} = -12$ V and current $I = 1.1$ mA, the temperature coefficient being T.C. $\approx -0.5\%$ K^{-1}. The fluctuation processes in MC-based Ge sensors (Fig.

5.1w) have been investigated [46]. It has been established that the $1/f$ source of bulk origin contributes considerably to the noise at low frequencies $f \leq 1$ kHz, and in a magnetic field this contribution is significantly increased.

5.6. Concluding remarks

An analysis of the experimental data in Table 5.1 leads to the conclusion that the optimum performance of p^+-n-n^+ devices is achieved in the semiconductor regime of conductivity. According to the theory in §2.5, in this double injection mode the current is proportional to the minority-carrier lifetime τ_{eff}. This parameter is Lorentz-force controlled by means of the surface recombination, and hence the diode $I(V)$ characteristics are changed drastically and high transducing efficiency results. The principal advantages of integrated MDs lie in their natural processing compatibility with modern IC technology, SOS technology in particular, and in the micron dimensions, improved frequency response, the absence of special treatment of the active surface, high magnetosensitivity, etc. The SOS system has relatively high immunity to different types of radiation and therefore is suitable for certain applications. In the author's opinion, parallel-field MDs are the future trend in the development of sensors because of the advantages of their fabrication technology and the opportunity for the construction of MD vector magnetometers. The formation of two mutually perpendicular MDs of the types shown in Fig. 5.1p, r and u, for instance, with a common central emitter, makes possible the simultaneous and independent registration of the two in-plane components of the field ***B***. Furthermore, the influence of temperature on the output characteristics of these solid-state devices can be weakened considerably by the structural symmetry of the sensors shown in Fig. 5.1m, p, r and u. The on-chip integration of a SOS microprocessor or other circuit and the creation of smart MD systems is not a difficult problem.

References

[1] V.I. Stafeev, Sov. Phys.-Solid State, 1 (1959) 763–768.
[2] E.I. Karakushan and V.I. Stafeev, Sov. Phys.-Solid State, 3 (1962) 1476–1482.
[3] E.I. Karakushan and V.I. Stafeev, Sov. Phys.-Semicond., 9 (1976) 953–955.
[4] V.I.Stafeev and E.I. Karakushan, Magnitodiody, Nauka, Moskwa, 1975 (in Russian).
[5] G.A. Egiazaryan, G.A. Mnatsakanyan, V.I. Murigin and V.I. Stafeev, Sov. Phys.-Semicond. 9 (1975) 829–833.
[6] J. Chretien, Les dispositifs à injection de porteurs en tant que capteurs magnetosensibles, Ph.D. dissertation, I.N.P., Grenoble, France, 1977.
[7] T. Yamada, Proceedings of 9th Int. Conf. Phys. of Semicond., Moscow (1968) 672.
[8] I.M. Vikulin and V.I. Stafeev, Fizika poluprovodnikovich priborov, Sovetskoe Radio, Moskwa, 1980 (in Russian).
[9] I. Melngailis and R.H. Rediker, Proc. IRE, 50 (1962) 2428–2435.

[10] M. Kimura and S. Takahashi, Electron. Lett., 22 (16) (1986) 830–832.
[11] E.I. Karakushan and V.I. Stafeev, Sov. Phys.-Solid state, 3 (1981) 493–498.
[12] Z.S. Gribnikov, G. I. Lomova and V.A. Romanov, Phys. Status Solidi, A 28 (1968) 815.
[13] S. Cristoloveanu, l'Onde Electrique, 59 (5) (1979) 68–74.
[14] S. Cristoloveanu, Phys. Status Solidi, A 64 (1981) 683–695; A65 (1981) 281–292.
[15] A. Chovet and S. Cristoloveanu, Rev. Phys. Appl., 19 (1984) 69–76.
[16] M. Arai and T. Yamada, Proceedings of the Conf. Solid State Devices, Tokyo, Japan, 1970; Journ. of the Japan Soc. Appl. Phys., Suppl., 40 (1971) 93.
[17] S.A. Altunyan, V.V. Airapetyan, M.S. Barkhudaryan, G.A. Egiazaryan and V.I. Stafeev, Sov. Phys.-Semicond. 15 (1981) 12–14.
[18] N.S. Kharshiladze, T.D. Kamushadze, N.I. Maisuradze, T.G. Tabagari et al., Bull of the Acad. Sci. of the Georgian Rep., 123 (2) (1986) 273–276 (in Russian).
[19] Ch.S Roumenin, University Annual (Techn. Phys.) 22(2) (1985) 173–180 (in Bulgarian).
[20] E.I. Karakushan, A.R. Mankulov, E.S. Samsonov and V.I. Stafeev, Sov. Phys.-Semicond., 12 (1978) 39–41.
[21] V.S. Lysenko, R.N. Litovskii, M.M. Lokshin and M.F. Sherbakova, Phys. Status Solidi, A77 (1983) 443–448.
[22] G. Kamarinos, P. Viktorovitch, S. Cristoloveanu, J. Borel and R. Staderini, Proceedings of the IEDM, Washington, D.C. (USA) (1977) 114a–114c.
[23] A. Chovet, Transport, recombination et bruit dans les semiconducteurs ambipolaires en presence de champ magnétique, Ph.D. dissertation, I.N.P., Grenoble, France, 1978.
[24] P. Lilienkamp and H. Pfleiderer, Phys. Status Solidi, A 43 (1977) 479–486.
[25] O.S. Lutes, P.S. Nissbaum and O.S. Aadland, IEEE Trans. Electr. Devices, ED-27 (1980) 2156–2157.
[26] G. Dimopoulos, F. Balestra, A. Chovet, M. Benachir and J. Brini, Solid-State device, Proc. 17th ESSDERC, Elsevier, Amsterdam (1988) 611–614.
[27] G. Dimopoulos and A. Chovet, Physica B, 159 (1989) 219–222.
[28] J. Chretien, G. Kamarinos and P. Viktorovitch, Rev. Phys. Appl., 12 (1977) 1699–1703.
[29] S. Cristoloveanu, A. Mohaghegh and J. de Pontcharra, J. Physique Lett., 41 (1980) L235–L237.
[30] A. Mohaghegh, Fonctionnement et optimisation des micro-magnetodiodes integrees sur silicium sur isolant, Ph.D. dissertation, I.N.P., Grenoble, France, 1981.
[31] A. Mohaghegh, S. Cristoloveanu and J. de Pontcharra, IEEE Trans. Electron Devices, ED-28 (3) (1981) 237–242.
[32] R.S. Popović, H.P. Baltes and F. Rudolf, IEEE Trans. Electron Devices, ED-31 (3) (1984) 286–291.
[33] I. Levitas, J. Pozela and K. Stalioraitis, in: Poluprovodnikovie preobrazovateli, J. Pozela (ed.), Mokslas, Vilnius, 1980, 73–139 (in Russian).
[34] S. Kataoka, Recent development of magnetoresistive devices and applications, Circular of the Electrotechnical Laboratory, N 182 (Tokyo, Japan), 1974.
[35] S. Cristoloveanu, A. Chovet and G. Kamarinos, Solid-State Electron., 21 (1978) 1563–1569.
[36] T. Dilmi, A. Chovet and P. Victorovich, J. Appl. Phys., 50 (1979) 5348–5351.
[37] A. Chovet, S. Cristoloveanu, A. Mohaghegh and A. Dandache, Sensors and Actuators, 4 (1983) 147–153.
[38] F. N. Hooge, Phys. Lett., 29A (1969) 139.
[39] I.M. Vikulin, L.F. Vikulina and V.I. Stafeev, Galvanomagnitniye pribori, Radio i Sviaz, Moskwa, 1983 (in Russian).
[40] V.I. Stafeev and I.M. Vikulin, in: Poluprovodnikoviye pribori i ih primeneniye, I.A. Fedotov (ed.), Sovetskoe Radio, Moskwa, 1974, 23–56 (in Russian).
[41] S. Cristoloveanu, Transport magnétoélectrique dans les sémiconducteurs en présence d'inhomogénéités naturelles òu induites par effects de récombinaison et d'injection, Application aux capteurs magnétiques, Ph.D. dissertation, I.N.P., Grenoble, France, 1981.
[42] E.I. Karakushan, I.G. Ponomarev and V.I. Stafeev, Sov. Phys. -Semicond., 13 (1979) 171–174.
[43] E.I. Karakushan, V.I. Murygin and V.V. Paramonov, Fiz. Techn. Poluprovodn., 15 (1981) 977–979 (in Russian) (Sov. Phys.-Semicond., 15, 1981).

[44] A. Chovet, Ch.S. Roumenin, G. Dimopoulos and N. Mathieu, Sensors and Actuators, A, 21–23 (1990) 790–794.
[45] A.M. Barnet, in: Semiconductors and Semimetals, vol. 6, Acad. Press, New York, 1970, Ch.3.
[46] N. Lukyanchikova, N. Garbar, A. Sasciuk and V. Zebriunaite, Litovsyi Fizicheski Zbornik (1) (1981) 89–97 (in Russian).

Chapter 6

Bipolar magnetotransistors and related sensors

The result of a comparative analysis of the parameters of some common semiconductor magnetosensors is ambiguous concerning purposeful research carried out on bipolar magnetotransistor sensors (BMTs). The first publication associated with such devices came out in 1950 [1]. For instance, the classical orthogonal Hall elements, regardless of their fairly good linearity, have a low transduction efficiency along with a temperature-dependent offset, and are predominantly sensitive to one of the three orthogonal components of the magnetic field B; the output of the MDs is generally nonlinear and inadequately reproducible in different samples, being asymmetrical with respect to the sign of the field B; the output signal of magnetoresistors is a quadratic function of the induction B, hardly sensitive at low values of B, and the polarity of the vector B cannot be detected. At the same time, high magnetosensitivity, the possibility for an integrated realization, linearity of the output, micron dimensions, high spatial resolution, high frequency response, high signal-to-noise ratio, the possibility of building multisensors for magnetic field, temperature and light, and three-dimensional vector sensing are only a few of the unique capacities of BMTs. Furthermore, the BMT approach to sensors stimulates the development of related transducers like unijunction magnetotransistors (UJMTs), carrier-domain magnetometers (CDMs) and other interesting devices.

6.1. General approach to BMT design

Concerning their electrical principle of operation, BMTs are active bipolar devices with a current output. In spite of their great diversity, any BMT has as its essential component a current source in the form of a $p-n$ junction, injected minority carriers into the base region and one or more reverse-biased $p-n$ junctions as collectors that pick up the useful signal.

As a BMT is constructed on a nonmagnetic substrate (Si, Ge, GaAs or some other semiconductor material), the external magnetic field B influences only the kinetic processes in different regions of the transistor structure. As a result, the changes in the output collector current, $I_C(B)$, are a measure of the direction and strength of the field $\boldsymbol{B}$. Because of the effect of transistor amplification, the change in the output voltage $V_C(\boldsymbol{B}) = R_C I_C(\boldsymbol{B})$ may substantially exceed, for instance, that of Hall elements and MDs, where R_C is the collector load resistor. Typically a BMT is an input transducer of the modulating type (§1.2) [2]. Therefore a supply of external power is a necessary condition for its operation. The magnetic field $\boldsymbol{B}$ influences the flow of energy in the structure and modulates that flow; as a result, the output electrical signal changes according to the intensity of the influence. The optimization of the mode of operation and the specific design have always aimed at a high efficiency of magnetic control of the energy distribution in these devices.

Several characteristic features can be distinguished in the familiar BMT versions according to how they have been classified. Depending on the type of carriers injected by the emitter into the base region, the sensor is of the p–n–p or n–p–n type. According to the geometry of the external magnetic field and the design of the device, there are two methods of activation: parallel-field, when the field $\boldsymbol{B}$ is parallel to the plane of the semiconductor substrate, and orthogonal, when $\boldsymbol{B}$ is perpendicular to the surface of the chip. If the carriers move parallel to the surface of the chip, the BMT is lateral, and if the direction of the current is perpendicular to the surface, the device is vertical. When the substrate itself is an active sensor region, the BMT is a bulk device, and in an epitaxial sensor, is surface-merged. Depending on the number of output terminals, the structures are single-collector, differential (dual-collector) or multi-collector, whereas the output signal may be a current, voltage or frequency. A classification according to the BMT action is also possible, but taking into account the ambiguity of its manifestations depending on the device design, this method encounters difficulties. In the literature BMTs are most often divided into lateral and vertical types [3–5]. In general, however, the trajectory of movement of the carriers in BMTs is not a straight line, and it has both a lateral and a vertical component. Therefore, the way that BMTs are activated is more universal as a classification parameter.

BMTs are most often made by means of conventional technologies for the production of integrated circuits (CMOS or bipolar), with some specific technological steps introduced in order to improve the parameters and prospects of achieving intelligent sensors [3–5]. From the point of view of optimization, reproducibility and stability of parameters, greater flexibility and better results for BMTs are achieved rather by the proper selection of the photomask layout than by variations in the two-dimensional distribution of doping impurities in silicon by means of intentional changes in the parameters of the processing steps. As a rule, dual-collector BMTs with relatively high temperature immunity and high transduction efficiency are the most widespread. This is not accidental, because the main requirement for the metrological function of these sensors is the maximum rejection of all parasitic signals which are not associated with the action of the magnetic field. It is in the dual-collector devices that the undesirable in-phase sensitivities are mutually neutralized, while the useful signal is doubled, i.e. a structure compensation is achieved.

The mobility of carriers is of great importance, since it determines the magnetosensitivity of all BMTs. From this point of view silicon is not a typical magnetosensor material when compared with some other semiconductors like InSb, InAs, GaAs and Ge, in which the mobility at room temperature substantially exceeds that in silicon. However, the small bandgaps of these materials (except GaAs) compared to silicon makes them practically unusable at $T > 80°C$ owing to intrinsic behavior. The working temperature of Si devices is $T \leq 150°C$ and for GaAs $T \leq 250°C$. Therefore, mobility is not the only important parameter in BMT design. A comprehensive analysis is needed of the physical and technological properties of the fabrication material. At the present, integrated BMTs are made on bases of silicon (and, in future, on bases of GaAs), whose planar processing is the best studied and developed, as a result of the breakthroughs in IC technology in recent years.

6.2. General principles of BMT operation

6.2.1. Conditions for the occurrence of magnetosensitivity

The fundamental cause of the magnetosensitivity of all semiconductor galvanomagnetic sensors, including BMTs, is the deflecting action of the Lorentz force $F_L = q.v_{dr} \times B + qE$. This law, tempting in its simplicity, meets with surprising and sometimes paradoxical results in BMTs. For instance, the experimentally observed sensitivity covers an extremely wide range from $1\%\ T^{-1}$ to $10^6\%\ T^{-1}$, which can hardly be explained by the lack of optimization alone. In the same way, when there is a unidirectional change in the supply current through the dual-collector drift-aided MT, a change in the sign of the sensitivity is observed, so that there are working points for which the transduction function is eliminated. This fact does not corroborate the idea that in structures with injecting contacts the polarity at the output changes only with the change in the direction of the magnetic field. The problem of the existence of an emitter-injection modulation is not less interesting, and it has been suggested that this activity is similar to the one related to the bulk or surface nature of the $1/f$ noise [6–8].

At present the detailed explanation of various sensor mechanisms in BMTs through the use of an analytical model alone and/or numerical computer simulation of the basic kinetic equations under ideal boundary conditions is extremely difficult. The goal of sensor modeling is to understand the operating principles and, especially, how the design, fabrication and parameters determine, enhance or limit a sensor's sensitivity to the magnetic field. Galvanomagentic phenomena in semiconductor structures take place in a special medium created by the technology: spatially confined, nonuniform and three dimensional in its nature. The chemical and metallurgical aspects of the formation of a device, the spatial distribution of doping impurities and defects in the crystal lattice and their interaction play a substantial role, as well. On the other hand, the galvanomagnetic properties depend on the conditions on the surface, thus making it necessary to introduce empirical parameters in the respective models, which are practically unavailable from the technology. The overall changes in the movement of the carriers, in their concentration and distribution and in the resulting electrostatic potential in the presence of a magnetic field enormously complicate the entirely mathematical approach to the great variety of transducer mechanisms. This is why the interpretation of the operation of this class of bipolar sensors requires clear physical insight. The BMTs should not be considered apart from other semiconductor transducers, e.g., Hall sensors and magnetodiodes, since, in one way or another, the Hall effect and magnetoconcentration exist under definite conditions in the different regions of the same MT structures.

The general physical approach to the analysis of BMT magnetosensitivity assumes the inevitable existence of the electric fields E_{em} in the emitter region, E_B in the base region and E_C in the depletion layer of the collector–base region, [9,10]. This assumption is used to identify the sensor mechanisms in BMTs.

6.2.1.1. Emitter region magnetosensitivity

The emitter regions in bipolar transistors are usually formed by ion implantation. The typical junction depth is 0.5 μm, while the surface concentration of doping impurities in emitter regions is in the 10^{20} cm^{-3} range and decreases by 5 orders of magnitude to reach 10^{15} to 10^{16} cm^{-3} as the depth is increased [11,12]. This means that there is an extremely high concentration gradient in the emitter region, which generates a built-in electric field with magnitude $E_{em} \geq 10^4$ V cm^{-1}. This field has such a direction that under forward biasing conditions it repels minority carriers injected by the base into the emitter region. The charge transport within the emitter region is dominated by majority carriers whose velocity at $B = 0$ is directed perpendicularly to the surface of the device. The combination of a built-in electric field, which results from the doping profile, and an external magnetic field can play a substantial role in magnetotransistor action [9].

The Hall effect under nonuniform conditions, i.e. the gradient $\partial N_D(x)/\partial x$ in the direction of the current I_x, has been analyzed in detail in §2.3.5. It was established that as a result of the circulation current j^* this nonuniformity causes current distortion, (2.49):

$$j_x(y) = \frac{I_x}{wt} \frac{\gamma/2}{\sinh(\gamma/2)} \exp\left[-\gamma\left(\frac{y}{w}\right)\right],$$

where $\gamma \equiv k_x \mu_H B_z w$, $k_x = n^{-1} \partial n/\partial x =$ const, [μm^{-1}].

The meaning of the asymptotic solution (2.49) is that with increasing Hall angle $\Theta_H \sim \mu B_z$ the current $j_x(y)$ is increasingly concentrated towards one of the surfaces of the sample (Fig. 2.12). The nonhomogeneity parameter γ, which is related to the junction width w_E (Fig. 6.1), is the main factor in (2.49). According to

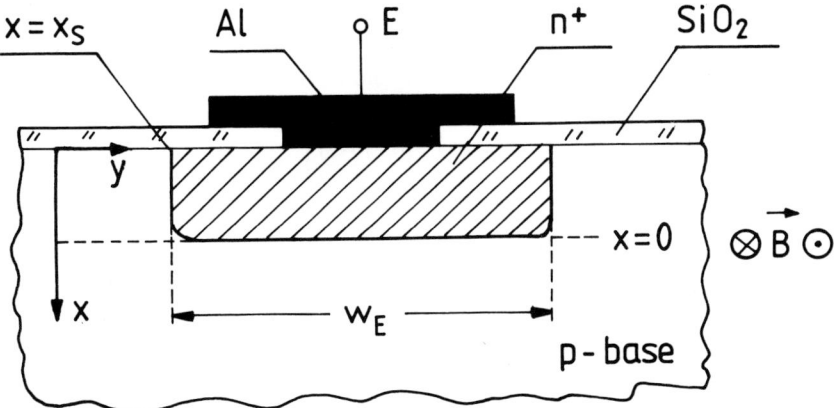

Fig. 6.1. The emitter region of a bipolar transistor.

[12], the excess donor concentration in the n^+–p emitter junction can be assumed to be an exponential function. The appropriate expression for N_D at points between $x = x_s$ and $x = 0$ is:

$$N_D(x) - N_A = N_A[\exp(-k_x x) - 1], \qquad (6.1)$$

where N_D is the net donor concentration in the emitter n^+-region, N_A is the background acceptor concentration in the p-type base region, $x = 0$ is the boundary of the n^+–p interface and $x = x_s$ is the top boundary of the emitter with a donor concentration $N_D \approx 10^{20}$ cm^{-3} on that surface (Fig. 6.1).

Equation (6.1) and numerical data, for instance $N_D \sim 10^{20}$ cm^{-3} and $x_s = 0.5$ μm, can be used to calculate the parameter k_x. In the interval of typical acceptor concentrations $6 \times 10^{14} \leq N_A \leq 1 \times 10^{16}$ cm^{-3} which corresponds to the resistivity $\rho \in (12$ to $0.5)$ Ω·cm [13], the parameter k_x varies from 25 to 20 μm^{-1} [9]. The crystal damage due to ion implantation in the emitter region leads to a relatively low carrier mobility which must be taken into consideration in the analysis of the influence of the external magnetic field B. According to the data in [14], the typical carrier mobilities are $\mu_n = 77$ cm^2 V^{-1}s^{-1} for electrons and $\mu_p = 51$ cm^2 V^{-1}s^{-1} for holes. If a magnetic induction $B = 1$ T, a Hall factor $r_H = 1.6$, $T = 300$ K and $w_E = 10$ μm are assumed, the value of the nonhomogeneity factor of the n^+–p junction will be $-3.1 \leq \gamma \leq -2.5$ [9]. The negative value of γ is associated with the electron current j_x in the n^+–p emitter. The longitudinal component $j_x(y)$ of the carrier density as a function of the y/w_E ratio, with parallel-field orientation of the vector B, is presented in Fig. 2.12. It can be seen that the current density j_E in the emitter junction is a strong exponential function of the position. The exponential distribution of the current density along the y-axis inside the emitter E leads to a modulation of the initial voltage $V_{EB}(B = 0)$ along the y-axis when a magnetic field is applied. For instance, in that part of the emitter area where the current density $j_E(y, B)$ is high the forward bias V_{EB} effectively increases by $\Delta V_{EB}^*(y, B)$. In this part of the emitter–base region the injection of minority carriers increases exponentially in agreement with the familiar diode relationship:

$$j_E(y, B) \sim j_s(0) \exp \left\{ \frac{q[V_{EB}(0) + \Delta V_{EB}^*(y, B)]}{k_B T} \right\}. \tag{6.2}$$

The collector current I_c increases in compliance with this exponential dependency. The transducing mechanism described, which is associated with the emitter region, can in principle generate exponential dependence on the applied magnetic field, i.e. $I_C(B) \sim \exp(B)$ or $\Delta I_{C1,2}(B) \sim \exp(B)$.

6.2.1.2. Base-region magnetosensitivity

Conventional bipolar transistors with a short base region dominated by gradient-diffusive carrier transport manifest a magnetosensitivity which can be neglected to a first approximation [10]. This is direct experimental evidence of the dominating contribution of the electric field E_B in the base region to the operation of BMTs. In this case, the impact of the force $\boldsymbol{F}_L$ on the diffusion current and the distribution of injected carriers can be neglected in weak magnetic fields, $\mu_{\min} B \ll 1$ [15]. Moreover, when $L^* \gg w_{EC}, w_{EB}$, the recombination base current of majority carriers I_B and the related field E_B are negligible, $I_B \ll I_E \sim I_C$, where the gain coefficient $\alpha = I_C/I_E \sim 1$, L^* is the effective diffusion length of the injected carriers, while w_{EC} and w_{EB} are the emitter–collector and emitter–base distances. What is needed for the appearance of the electric field E_B^{maj} in the base region is that the dimensions of the base region should be comparable to or bigger than the diffusion length L^*, $w_{EB} \geq L^*$. In this case the coefficient $\alpha < 1$ and the recombination drift current I_B, which neutralizes the electric charge of the injected carriers not extracted by the collector C, increases. A typical feature of these long transistors, $w_{EB} > L^*$, is that, to a first approximation, the concentration of minority carriers in the base region and, as a result of recombination, their current decrease exponentially with the distance $x > 0$ from the emitter E in the direction of the base contact B. According to §2.6, equation (2.126) and Fig. 2.26, the transport mechanism in the field E_B becomes a drift-diffusion mechanism, which is the necessary condition for the appearance of the Lorentz deflection and the Hall field E_H [16,17]. Another way to generate an electric field in the base region of a BMT is to force an ohmic current $I_{B1,2}$ through the additional base contact B_2. The direction of this field $E_{B1,2}$ is usually chosen so that the injected carriers are forced towards the collector C. As a result of the variety of device structures, specific boundary conditions and modes of operation of transistors with an electric field in the base region, the mechanisms of magnetic control of the output collector current $I_C(\boldsymbol{B})$ can be attributed to one of the following reasons [2–4,10]:

(a) Lorentz deflection of minority carriers in the base region;

(b) magnetotransistor Hall effect in long samples related to any type of influence of the majority-carrier-generated Hall electric field on carriers injected in the base region;

(c) the magnetoconcentration or MD effect, associated with the possibility of magnetic modulation of the base resistance by the conditions on the surface at high levels of emitter injection.

The pure Lorentz deflection of injected carriers (§2.3.2) is a possible sensor mechanism in a BMT with a long base region unconfined in the direction of the Lorentz force. The absence of a boundary eliminates the possibility of the occurrence of a Hall field. The magnetic control of the current $\Delta I_C(B)$ is due to the deflection of the emitter current I_E at the Hall angle $\Theta_{min}^H \sim \mu_{min}B$ with respect to the unchanged direction of the electric field E, generated by the current I_B and/or the interbase current $I_{B1,2}$. The change $\Delta I_C(B)$ can use a deflection of the current I_C if: (a) the distribution of the minority carriers is nonuniform at $B = 0$; and/or (b) the collector region C has such dimensions that a certain number of the minority carriers which, at $B = 0$, reach C are now deflected from it, or vice versa, i.e., the magnetic field B should be capable of modulating the "effective collector area". This sensor mechanism generates a linear output signal $I_C(B) \sim BI_E$ or $\Delta I_{C1,2}(B) \sim BI_E$.

As an illustration, a typical example will be described here of the influence of the Hall field (electric potential gradient) associated with majority carriers upon the magnetosensitivity of a single-collector BMT, while effects related to the structure of the device and the mode of operation will be presented in detail in §6.3. In the case of a long base, $w_{EB} > L^*$, confined in the direction of the force F_L, the drift recombination current I_B in a magnetic field B, generates a Hall voltage with increasing magnitude (§2.6) [16]. Owing to the decreasing profile of the nonequilibrium carrier concentration in the direction from the emitter E to contact B, the base conductivity at $x > 0$ decreases and the electric field $E_B(x > 0)$ increases. Since a uniform magnetic field and carrier mobility independent of the carrier concentration are assumed, the Hall angle remains constant, i.e. Θ_{maj}^H = const. Therefore, the Hall field $E_{maj}^H(x)$ increases in the direction of contact B. On the other hand, the current deflection I_E with respect to the resulting electric field $E = E_x + E_{maj}^H$ in the base is at a Hall angle $\Theta_{min}^H \sim \mu_{min}B$. As a result, the overall deflection of the minority carriers with respect to the x-axis constitutes an angle $\varphi = \Theta_{maj}^H + \Theta_{min}^H \approx (\mu_{maj} + \mu_{min})B$ (Suhl–Schockley effect) (§2.6). The contribution of the Hall field to the pure Lorentz deflection of the injected carriers is often referred to as "indirect" deflection [3]. If one and the same recombination lifetime τ_{min}^* is assumed throughout the base region, the relative change in the concentration of minority carriers on the surface with the collector C located on it will grow as a result of the increase in the Hall field $E_{maj}^H(x > 0)$. At low values of the field B this mechanism allows a linear approximation be made, i.e., $\Delta I_C(B) \sim BI_E$.

In the general case the galvanomagnetic picture in the base region is complicated by the nonhomogeneity-generated circulation current j^* that envelops the entire active region of the sensor (cf. §2.6).

Specific manifestations of base-region sensor mechanisms will be considered below.

(a) Electrical control of the polarity of magnetosensitivity. The recently discovered effect of the electrical control of the polarity of the output signal $\Delta I_{C1,2}(B)$ arises in the base region of differential BMTs with two base contacts B_1 and B_2 [18]. The origin of this transducing mechanism is associated with the existence of a current filament which leads to an S-type negative resistance of the $I_E(V_{EB1})$ characteristic at a current $I_{B1,2} \neq 0$ and a nonlinear carrier trajectories between the planar contacts. The essence of this phenomenon is illustrated by the device in Fig. 6.2.

The emitter E and second base B_2 have similar polarities and are supplied by constant currents I_E and $I_{B1,2}$. The silicon substrate serves as an active base region of the device. The external magnetic field B is applied parallel to the plane of the structure. Figure 6.3a and b represents the dependences of the signal $\Delta I_{C1,2}$ on the emitter, I_E, and base, $I_{B1,2}$, currents. The offset $\Delta I_{\text{off}}(B_x = 0)$ for each of the working points has been compensated in advance. It is clear that, for an increasing current ($I_E > 0$ at $I_{B1,2} = $ const or $I_{B1,2} > 0$ at $I_E = $ const), the output signal changes its polarity. In addition, it is a linear function of the induction for $B < 1$ T. The value of the bias V_{CB} and the temperature T do not influence this effect.

The results of Fig. 6.3 are unexpected, because the polarity of the output signal

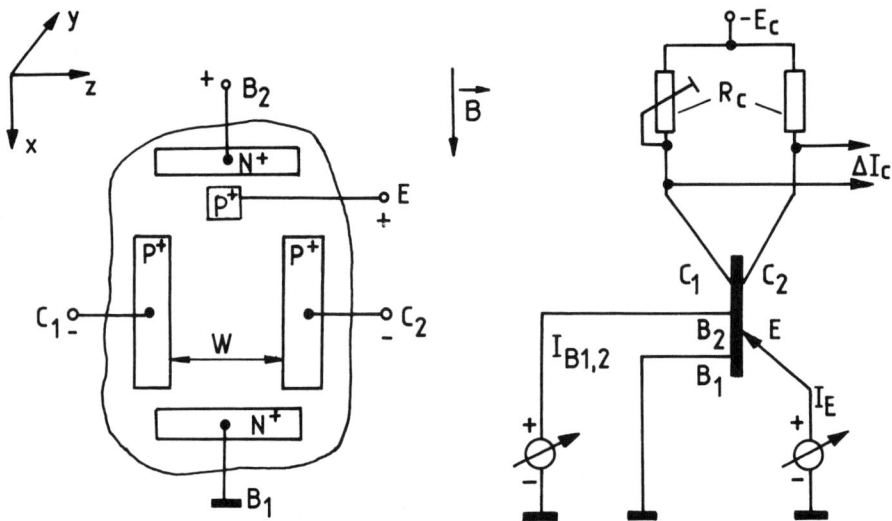

Fig. 6.2. Planar drift-aided dual-collector magnetotransistor (a) and its bias circuitry (b).

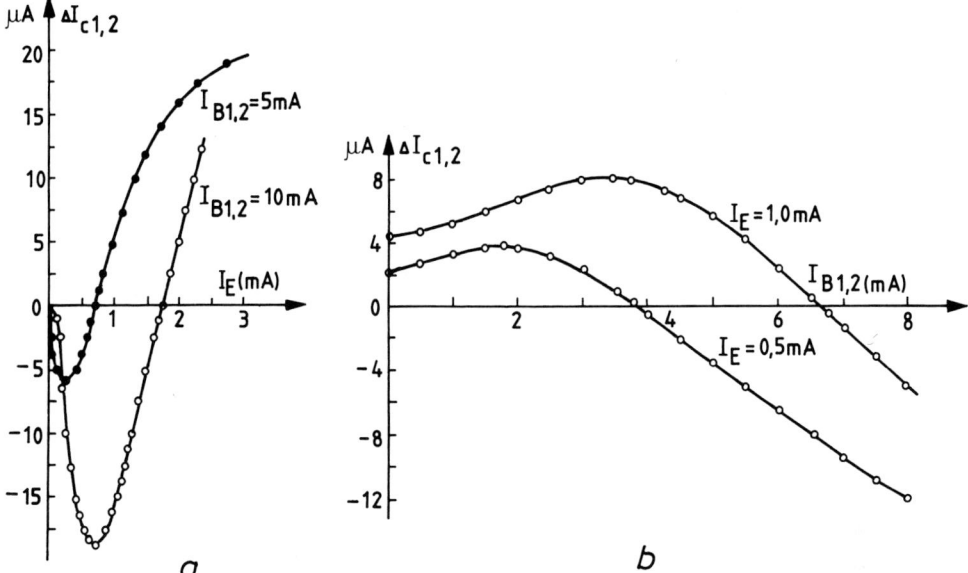

Fig. 6.3. Electrical control of the output signal in the device of Fig. 6.2, with $B_x = 0.7$ T, $V_{CB} = -10$ V and $T = 300$ K: (a) with variation of the current I_E, the current $I_{B1,2}$ as parameter; (b) with variation of the current $I_{B1,2}$, the current I_E as parameter.

of bipolar magnetosensitive devices with injecting contacts has been supposed to be controlled by the direction of the external magnetic field $\boldsymbol{B}$ only. A change in the direction of the supplied current, which modifies the polarity of Hall effect-based sensors, is impossible in this case.

The characteristics of the device in Fig. 6.2a can be explained by the nonlinear trajectories of the injected carriers in a structure with planar contacts [18]. Holes flowing from the emitter E to the base contact B_1 have a velocity whose component v_y is initially directed downwards from the surface, $+v_y$. There is a point A on the trajectory at which the velocity $v_y = 0$ and $v = v_x$. In the section between point A and the contact B_1, the vertical component v_y is directed towards the top surface, $-v_y$. For a parallel field $\boldsymbol{B}_x$ (Fig. 6.4), the Lorentz force $\boldsymbol{F}_L$ influences only the vertical component, v_y, of the velocity. The minority carriers are deflected in the yz plane in opposite directions, according to the section E–A or A–B_1 of the trajectory. This deflection change is the cause of the electrical control of the polarity of the output signal $\Delta I_{C1,2}$. The resulting current magnetosensitivity, S_I^*, of the device is determined by the correlation of the two sensitivities S_I' and S_I'' (associated with the deflections of the injected carriers in the two sections E–A and A–B_1), which are opposite in sign:

$$S_I^* = S_I' - S_I''. \tag{6.3}$$

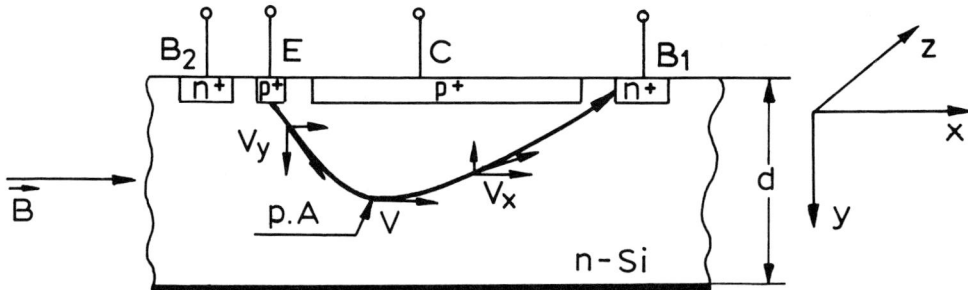

Fig. 6.4. Effective trajectory of the holes at currents $I_{B1,2} > 0$ and $I_E > 0$ in the device of Fig. 6.2a, according to the model.

At B_x, each of the two collectors C_1 and C_2 simultaneously registers the currents of opposite signs:

$$\Delta I'_{C1}(\boldsymbol{B}_x) \text{ and } -\Delta I''_{C1}(\boldsymbol{B}_x);$$
$$-\Delta I'_{C2}(\boldsymbol{B}_x) \text{ and } \Delta I''_{C2}(\boldsymbol{B}_x). \tag{6.4}$$

The control of the effective carriers trajectory, and hence of the two sensitivities S'_I and S''_I, is achieved by the working point, determined by the input currents $I_{B1,2} > 0$ and $I_E > 0$. On the other hand, for a current $I_{B1,2} > 0$, the characteristic $I_E(V_{EB1})$ of the sensor from Fig. 6.2a includes an S-type negative resistance, as in some MDs and unijunction transistors (cf. §5.1). This is why the carriers are nonuniformly distributed over the cross-section of the structure. The current filament spreads along a nonlinear trajectory to contact B_1, with a central conic angle ψ depending on the currents $I_{B1,2}$ and I_E. In sufficiently strong electric fields $E_{B1,2}$ in the region between the base contacts (obtained at high values of the current $I_{B1,2}$ and low levels of emitter injection I_E), the angle ψ is small. In this case the number of minority carriers reaching the collectors C_1 and C_2 is greatest close to contact B_1. Under these operating conditions, the magnetosensitivity S''_I is connected with the mechanism of deflection in the section $A-B_1$ of the trajectory and has a dominant role. With increasing current $I_E > 0$, the electric field $E_{B1,2}$ decreases because of the modulation of the conductivity in the base region. The angle ψ increases and the current filament expands. Therefore, the role of the sensitivity S'_I associated with the deflection of the holes in the section $E-A$ of their trajectory increases. According to equation (6.3), it follows that there are working points determined by the currents $I_{B1,2}$ and I_E in which $S'_I = -S''_I$ and the transduction efficiency of the sensor is eliminated by *electrical* means in a field $B_x \neq 0$ [18]. The electrical control of the sign of the BMT magnetosensitivity can be enhanced by the generation of a Hall field on the top surface with the contacts due to the current $I_{B1,2}$ by analogy with the parallel-field Hall sensors in Fig. 4.26f (cf. §4.3.2). In this case the carrier deflection φ in the two parts of their trajectory will be enhanced by $+\Theta^H_{\text{maj}}$ and $-\Theta^H_{\text{maj}}$, respectively.

(b) Filament magnetosensitivity effect. The current filament in dual-base BMTs is responsible for the so-called filament magnetosensitivity effect (FME) in the base region. The sensor in Fig. 6.2 can be considered rather as a unijunction transistor B_1–E–B_2 (§6.5). It is well known that in such devices the $I_E(V_{EB1})$ curve for current $I_{B1,2} \neq 0$ is of the S type and includes a section of negative resistance (§5.1). The current filament is the basis of a magnetosensitive mechanism in which the dependence of the collector currents $I_{C1}(B)$ and $I_{C2}(B)$ on the magnetic field is exponential [30]. The idea of the FME is based on the fact that the filament with the minority carriers spreads from the emitter E to the main base contact B_1 within a central conic angle ψ, which is determined by the input currents I_E and $I_{B1,2}$ (Fig. 6.2). According to equation (5.1), outside the filament the density of minority carriers decreases exponentially. The orthogonal component B_y controls both the position of the filament and the profile of the hole concentration outside in the region between the collectors. Therefore, at sufficiently small values of the angle ψ, the dependence $I_C(B)$ can be approximated by an exponential function, $I_C(B) \sim \exp(B)$. This effect will be better pronounced in those parts of the collectors which are located most closely to the emitter E. Near the contact B_1 the filament expands, the exponential profile of holes decays and the dependence $I_C(B)$ tends towards linearity. At high levels of the current I_E, the angle ψ is large, the filament is strongly diffused and the curves $I_{C1}(B)$ and $I_{C2}(B)$ are linear. The experimental verification of the FMT together with the respective arrangement are considered in §6.3.1.

(c) Emitter-injection modulation. Under certain conditions, an emitter-injection modulation (EIM) mechanism of magnetosensitivity can be activated in the base region of BMTs [3–10]. Thanks to the fruitful discussions concerning this phenomenon [7,8], nowadays the generally accepted view of EIM is that it is the change in the collector current due to an asymmetrical injection caused by the field B at a current $I_E = $ const or by a change of $I_E(B)$. The injection modulation by the magnetic field B can be observed if:

(i) The base region is confined and electrically biased in such a manner that a long sample Hall effect may occur in it (cf. §2.3.1);

(ii) The Hall voltage generated by the majority carriers is adequately distributed around the emitter junction so that a modulation of the effective emitter–base bias $V_{EB}(V_H)$ can take place.

A change in the collector current $I_C(B)$ or $\Delta I_{C1,2}(B)$ is to be expected as a result. The mechanism discussed in §6.2.1.1 is responsible for the magnetic control of the emitter injection and operates inside the emitter region, whereas the EIM is a mechanism of magnetic control, whose action takes place in the base region. Both mechanisms lead to the modulation of the bias $V_{EB}(y, B)$. The current injected into the base region depends exponentially on the value of $V_{EB}(y, B)$. The dependence of the output collector current, I_C, of the bipolar transistor on the bias V_{EB} is given by the familiar expression:

$$I_C = \alpha I_E \simeq I_s \exp\left(\frac{qV_{EB}}{k_B T}\right), \tag{6.5}$$

where α is the gain coefficient, and I_s is the emitter-junction saturation current.

Owing to the EIM effect, the bias V_{EB} is modulated by the Hall voltage V_H:

$$I_C(B) \simeq I_s \exp\left[\frac{q(V_{EB} \pm V_H)}{k_B T}\right]. \tag{6.6}$$

In the case of a differential BMT with a central emitter around which the Hall voltage is generated, the following relations are valid for the two currents $I_{C1}(B)$ and $I_{C2}(B)$:

$$I_{C1}(B) \simeq I_s \exp\left[\frac{q(V_{EB} + 0.5V_H)}{k_B T}\right] \tag{6.7}$$

and

$$I_{C2}(B) \simeq I_s \exp\left[\frac{q(V_{EB} - 0.5V_H)}{k_B T}\right]. \tag{6.8}$$

The resultant dependence of the output signal $\Delta I_{1,2}(B)$ in the case of the EIM will be:

$$\Delta I_{C1,2} = I_E \alpha \gamma \sinh\left(\frac{qV_H}{k_B T}\right), \tag{6.9}$$

where γ is the emitter injection efficiency, and the other notations are as generally accepted [6,19]. At low values of the argument V_H, the signal $\Delta I_{C1,2}(B)$ makes it possible to make a linear approximation. At high values of the voltage $V_H \sim (I_B + I_{B1,2})B$, the dependence of the injection and that of the current $\Delta I_{C1,2}(B)$ are highly nonlinear owing to the exponential diode dependence $I_E(V_{EB1})$. Since the conditions in the base region are nonequilibrium, the coefficient R_H of the Hall effect is a complicated function of the bulk and surface recombination, the mobility and the concentration of carriers, and the conventional formulae are not suitable for its exact calculation. The substantial role of the Hall voltage generated in the BMT base region, including its effect on the emitter injection, is illustrated by Fig. 6.5 [10].

Emitter injection is only possible at distances $l_{E,B2} > l_{B3,B2}$, otherwise the emitter E remains currentless. The current I_{B2} splits into two components: $I_{B2,B1}$ and $I_{B2,B3-E}$, where $I_{B2,B1} > I_{B2,B3-E}$. Notwithstanding the bipolar nature of the structure, the planar contacts $B2$, $B3$ and $B1$ may be considered as parallel-field Hall elements, like those in Fig. 4.26d and e. The external magnetic field B, perpendicular to the cross-section of the magnetotransistor, generates at least two Hall voltages V_{B2}^H and V_{B1}^H on the upper surface. The voltage V_{B2}^H around contact $B2$ alters the potential distribution between contacts $B3$ and emitter E, as a result of which the injection is modulated. At the same time, the Hall field E_{B1}^H around contact $B1$ concentrates the minority carriers to the collector.

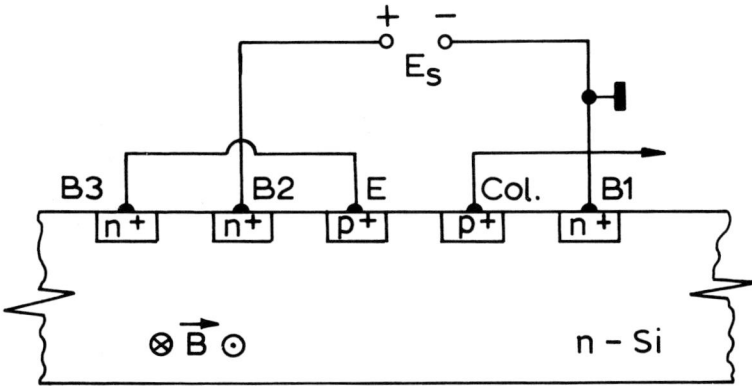

Fig. 6.5. Cross-section of a lateral drift-aided bipolar transistor with functionally integrated parallel-field Hall microsensor.

A novel version of an orthogonal n-p-n BMT with EIM mechanism is proposed in [20]. By means of two emitter contacts, a lateral electric field is applied to the rectangular emitter region. In a magnetic field a Hall voltage is induced across the right and the left emitter edges. This potential difference modulates the right and left emitter-base junction bias and results in the difference of the collector currents. The above idea can be successfully used in parallel-field BMTs. For that purpose parallel-field Hall elements can be fabricated in the emitter regions.

The magnetoconcentration is manifested in some types of differential BMT at very high levels of emitter injection and different rates of surface recombination on the sides to which the force F_L deflects the electron–hole plasma. The magnetic field causes an asymmetrical distribution of carrier concentration within a certain volume of the base region of the BMT, and hence local modulation of the base conductivity occurs. This modulation is further enhanced because of the inherent positive feedback. This entails a considerable change in the current, $\Delta I_{C1,2}(\boldsymbol{B})$.

6.2.1.3. Magnetosensitivity associated with the collector–base depletion region

The electric field $\boldsymbol{E}_C$ in the collector–base depletion region could contribute to the magnetosensitivity in some types of BMT configurations. The cross-section of one of them is presented schematically in Fig. 6.6 [21]. A great share of the collector region has a low doping level and is depleted upon reverse biasing of the junctions C_1 and C_2. The structure in Fig. 6.6 resembles a split-drain MOSFET Hall device, the collectors C_1 and C_2 corresponding to the split drain and the depletion layers of the C_1-base and the C_2-base junctions similar to the MOSFET channel. The only task of the emitter E and the base B in this sensor is the injection of carriers. They are majority carriers in the collector depletion region. The field $\boldsymbol{E}_C$ assists the carrier transport towards C_1 and C_2. The majority-carrier deflection in the depletion layer has the same origin as the Hall current in short samples (cf.

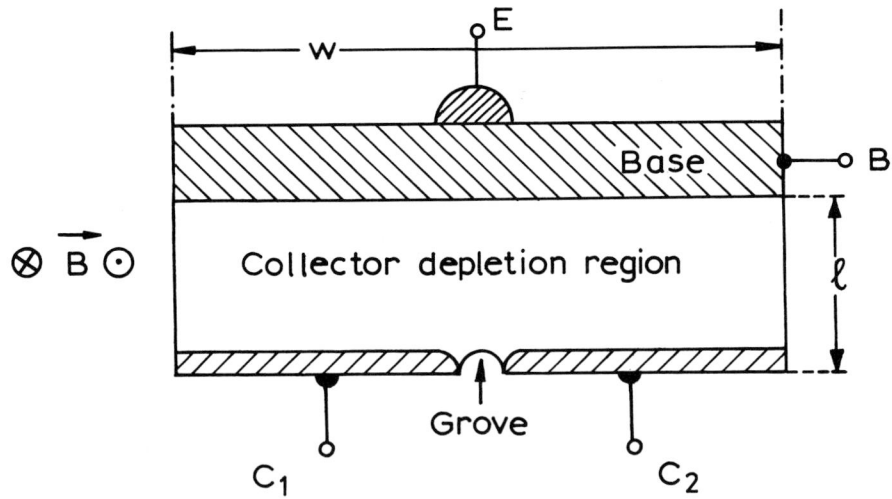

Fig. 6.6. A depletion-layer bipolar magnetotransistor.

§2.3.2). The following relationships are valid for the Hall current and for the total collector current: $\Delta I_H = (l/w)\mu_{\text{maj}} B I_C$; $I_{C1}(0) + I_{C2}(0) = I_C$. The meaning of l and w can be seen from Fig. 6.6. The ionized impurities in the depletion region do not assist the building up of carriers upon its boundaries; therefore no Hall voltage is generated, i.e. $V_H \approx 0$. This is why the differential signal $\Delta I_{C1,2}(B)$ is the Hall current $\Delta I_H(B)$. In such a way this transducing mechanism predetermines the linear output of the respective BMT devices.

In summing up the origin of magnetosensitivity in the different regions considered in §6.2.1, it can be concluded that BMT structures are a unique "store" of all galvanomagnetic phenomena from the pure Hall effect to magnetoconcentration. It is extremely difficult to model the physics of BMT operation, since all mechanisms act synergetically and their identification and distinguishing experimentally are not trivial problems.

6.2.2. Functional connection of BMT devices with Hall sensors

In spite of the extremely wide spectrum of galvanomagnetic phenomena in BMTs, they are associated with the Hall effect in one way or another. Until recently the sensor mechanisms in Hall devices, BMTs, carrier-domain magnetometers, etc., used to be discussed independently without taking into consideration their genealogical connections. The changes in the current and potential pictures in bipolar structures due to the magnetic field as well as the two manifestations of the Hall effect (voltage and deflection) result from the transition from long structure to short sample. In fact, in the case of monopolar conductivity, the transformation

Hall voltage → Hall current is accomplished by a reduction of the distance l. By analogy with this transformation, the long structure is effectively "shortened" if it is filled by injection with electron–hole plasma. Thus the Lorentz deflection becomes dominant when a magnetic field is applied. Therefore, BMTs are, in their essence, functionally integrated transistors with a built-in electric field in the base region and Hall sensors with orthogonal or parallel-field activation. Furthermore, in both types of transducer the output signal is proportional to the induction $|B|$ and exhibits a polar dependence on the direction of the vector B.

If we neglect the nonequilibrium minority carriers in the base region, the same boundary conditions are valid both for magnetotransistors and Hall devices. The general conclusion can be drawn that, from a phenomenological standpoint, a BMT is a superposition of Hall action and nonequilibrium bipolar conductivity, the latter being detected by collector terminals introduced instead of Hall contacts. The effect of the injected carriers is a restricted effective distance l between the supply contacts, which determines the geometrical correction factor, G, of the Hall effect. Therefore, a complicated combination of the Hall voltage V_H and the Hall current ΔI_H is to be expected in the base region, and, assuming very high levels of emitter injection and suitable boundary conditions, magnetoconcentration effects can be anticipated there as well. Moreover, there is a direct experimental evidence of the usual occurrence of the Hall effect in diode and transistor devices (§2.6).

Technical solutions have also been suggested for accomplishing a purposeful integration on one chip of an n-type Hall plate with orthogonal activation and a pair of bipolar $p-n-p$ transistors (DAMS) [22,23]. The emitter p^+-diffusion regions are now placed in the position of the previous Hall contacts and have a direct electrical connection between them. The Hall voltage, $V_H \sim I_{B1,2}B$, generated in the magnetic field B is directly transformed into the difference between the two emitter injections, and a differential collector current $\Delta I_{C1,2}(B)$ results. In fact, the differential transistor serves as an amplifier of the Hall voltage V_H. The evolution of the classical Hall element with orthogonal activation in a BMT is reduced to the formation of a rectangular semiconductor plate, with emitter E, and the replacement of the Hall terminals with collectors (Fig. 6.7a).

The analogous evolution of the parallel-field Hall sensor into a parallel-field BMT is presented in Fig. 6.7b [10]. The idea of the functional connection between a parallel-field BMT and a conformally transformed Hall element can be further developed in the direction of establishing a correspondence between the orthogonal and parallel-field activated BMTs themselves. By conformal transformation along the axis of symmetry of the structures, an orthogonal BMT can be converted into a parallel-field one. This property is shown in Fig. 6.7c for two common types of MT. This approach enriches the family of BMTs by new device structures and can expand their measuring capacities as integrated vector transducers.

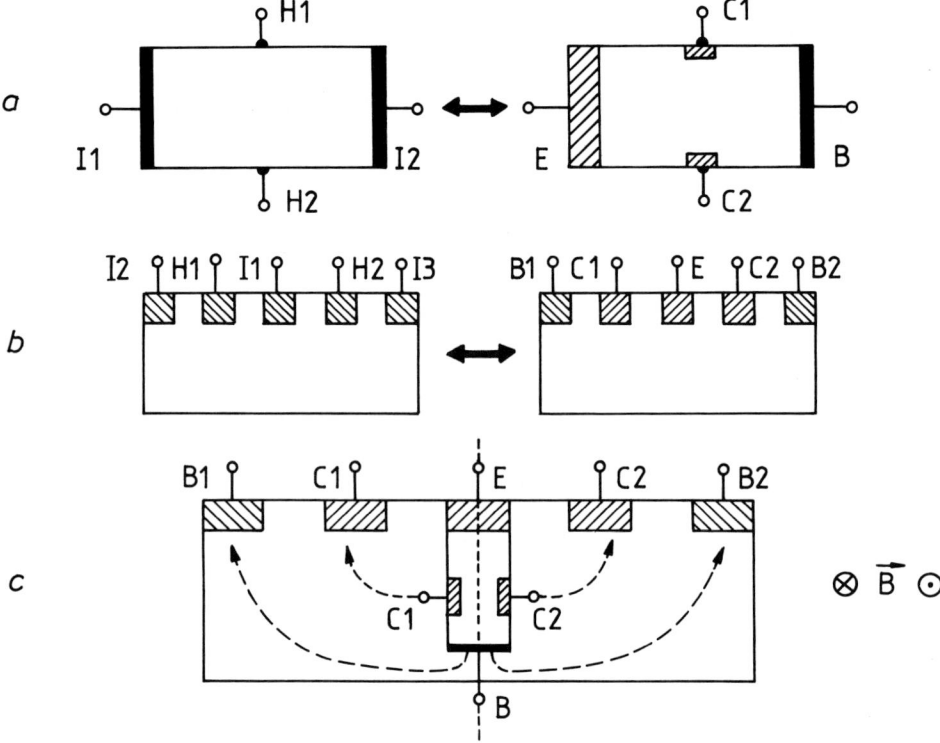

Fig. 6.7. The evolution of: (a) the classical Hall element in an orthogonal BMT; (b) the parallel-field Hall device in a parallel-field BMT; (c) the orthogonal BMT in a parallel-field BMT.

6.3. Device structures and operation of BMT sensors

The well-known orthogonal and parallel-field sensors are analyzed here. Owing to the metrological advantages of differential devices, dual-collector structures have mainly been considered. In view of the more detailed explanation of some of the phenomena of MT operation, whenever necessary, single-collector versions have also been included.

6.3.1. Orthogonal BMT sensors

Figure 6.8, I, shows the evolution of the classical rectangular Hall plate into an orthogonal sensor. The current contacts I_1 and I_2 have been transformed into the emitter E and main base contact B, while the reverse biased collectors C_1 and C_2 have replaced the Hall probes H_1 and H_2, thus providing high output impedance (Fig. 6.8a). In the absence of a magnetic field ($B = 0$), as a result of structural and electrical symmetry, part of the emitter current I_E is divided into

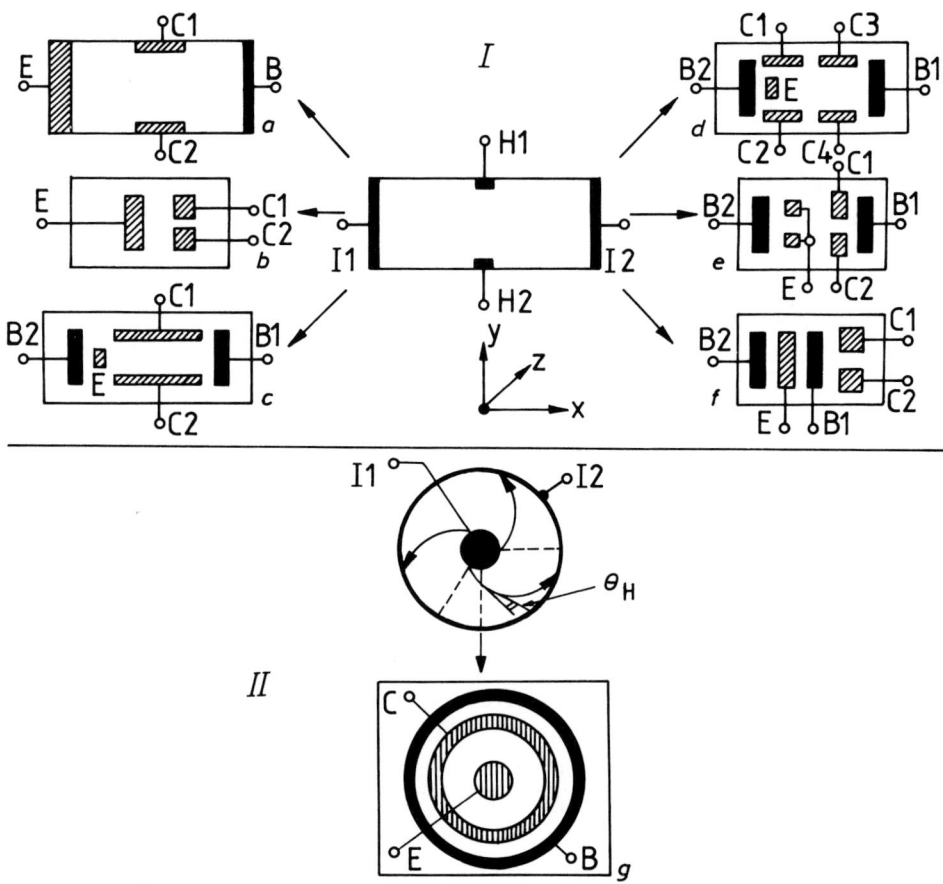

Fig. 6.8. I. (a)–(f) Classification of orthogonal BMT sensors. The respective Hall element is shown. II. (g) Evolution from the Corbino disk into a circular BMT sensor.

two equal components, $I_{C1}(0) = I_{C2}(0)$. In the case of a confined base region, the recombination drift current $I_B = I_E - [I_{C1}(0) + I_{C2}(0)]$ in the presence of a magnetic field generates a Hall e.m.f. and, as described in §6.2.1, the result of the deflection $\psi = \Theta^H_{maj} + \Theta^H_{min}$ is that one of the collector currents increases and the other one decreases. The current $\Delta I_{C1,2}(B)$ is a measure of the magnitude and the direction of B. Such a discrete p–n–p BMT sensor based on Ge was described for the first time in [24]. Its magnetosensitivity S increases with increasing of current I_E. This can be explained by the higher value of the electric field $E_B(I_B)$. The corresponding Si sensors have a lower transduction efficiency, mainly because of the lower carrier mobility. Figure 6.8b shows a dual-collector version with a floating base. The region between the emitter E and the two collectors C_1 and C_2 is covered by a MOS gate, and the gate geometry and the distribution profile of doping impurities have been chosen in such a way so that, with a proper collector–emitter bias V_{CE}, the device may be switched into the avalanche operating mode [25]. As a result of the avalanche breakdown of the collector junction, the electroneutrality of the base is disturbed and the emitter p–n junction is forward biased. The additional minority carriers which have been injected by the emitter and have covered the distance w_{EC} without recombination make an additional contribution to the avalanche multiplication at the collector junction. Thus a positive feedback occurs. Therefore, the overall effect of the transistor action is similar to the mode of operation when there is a base contact and a base current. The breakdown current I_{CE} is limited by the ion-implanted resistor, which has not been shown in Fig. 6.8b. The specific details of the fabrication of this avalanche bipolar MT (ABMT) have been described elsewhere, [25]. The sensor mechanism is similar to the principle of operation of the device shown in Fig. 6.8a. The future development of ABMTs will have to solve the problems associated with the high values of the operating voltage V_{CE} and the critical dependence of the avalanche processes on doping, geometry and temperature, in particular. The extensive experience in 2-D process and device simulation resulting from the rapid progress in VLSI technology can give considerable support to the proper design and fabrication of ABMTs.

In order to obtain a higher transducing efficiency for the device shown on Fig. 6.8a, the field E_B should be increased. However, no such effect can be produced by a simple variation of the current I_E above a given value, because the concentration of nonequilibrium carriers grows rapidly and the field E_B tends towards saturation. This drawback has been eliminated in the sensor shown in Fig. 6.8c and d by the addition of a second base contact B_2 [26–29]. The direction of the applied electric field $E_{B1,2}$ is such as to make the injected carriers drift towards the collectors C_1 and C_2, i.e., $\boldsymbol{E} = \boldsymbol{E}_B + \boldsymbol{E}_{B1,2}$. This is why these BMTs are referred to as drift-aided. If the magnetic field $\boldsymbol{B}$ is applied in a direction perpendicular to the plane of the chip (Fig. 6.8c) in a confined base region, the field $\boldsymbol{E}$ is turned at a Hall angle $\Theta^H_{maj} \sim \mu_{maj} B$ with respect to the x-axis. This is why the magnetosensitivity is determined by the total deflection $\varphi = \Theta^H_{maj} + \Theta^H_{min}$, and the differential collector current is a linear function of the field B:

$$\Delta I_{C1,2}(B) = K(\mu_{maj} + \mu_{min}) B I_E, \tag{6.10}$$

where K is a coefficient depending on the mode of operation, the geometry and electrical properties of the sensor. If the base region is unconfined in the direction of the Lorentz force (bulk sensor), there is no Hall field E^H_{maj} and the output current is reduced to

$$\Delta I_{C1,2}(B) = K \mu_{min} B I_E. \tag{6.11}$$

The Hall field $E^H_{maj}(I_{B1,2})$ in the MT functionally integrated with the Hall element (Fig. 6.8e and f) causes emitter injection modulation (EIM) (§6.2.1.2). The former case is typical of differential devices, and the latter of some single-collector versions.

In sensors of the type shown in Fig. 6.8e and f with a confined base region and at a field $B \neq 0$, the generated Hall field $E_{\text{maj}}^H(I_{B1,2})$ is oriented along the y-axis across the structure. As a result, the potential changes in the substrate along the emitters (Fig. 6.8e) or the emitter (Fig. 6.8f) and EIM appears [22]. Together with the EIM, the mechanism of deflection φ operates in the base region as well. Depending on the device geometry and the mode of operation, one of the two mechanisms would be dominant. The analysis of EIM for BMTs with a confined base region as shown in Fig. 6.8f is analogous. As a result of the asymmetry of the device, the sign of the magnetosensitivity depends not only on the polarity of the field B but also on the direction of the current $I_{B1,2}$ [6].

The well-known Corbino disc is shown in Fig. 6.8, II; its circular shape in an orthogonal field is optimal for the appearance of the geometrical magnetoresistance effect. By extending the functional approach adopted in §6.2.2, proceeding from this device structure it is possible to build a BMT which is nontraditional in its properties. This analog version is shown in Fig. 6.8g and represents a semiconductor substrate with planar emitter E, collector C and contact B which have circular symmetry [31]. As a result of the velocity of the injected carriers, $v_{\text{dr}}^{\text{min}}$, in the field $E_B(I_B)$, the deflection at $B \neq 0$ is at an angle $\Theta_{\text{min}}^H \sim \mu_{\text{min}} B$ with respect to the direction of the current at $B = 0$. As with the magnetoresistance effect, the trajectory of minority carriers becomes longer. Provided the lifetime τ_{min}^* is unchanged, the probability of recombination before their extraction by the collector C grows. As a result, the current $I_C(B)$ decreases, regardless of the polarity of the field B.

The operation of BMTs under cryogenic temperatures is definitely a problem of interest. Unfortunately, no data concerning the performance of orthogonal BMTs at such temperatures, at 77 K for instance, can be found in the publications. The need for such investigations can be justified by the recent discovery of high-T_c superconductivity in metal–oxide ceramics of the Y–Ba–Cu–O type which occurs exactly at these temperatures at $T \approx 77$ K. In the author's opinion the registration of the well-known Meissner effect (§2.7.1) can be accomplished by means of galvanomagnetic sensors, BMTs for instance. In this case the magnetic-field transducer is located very close to the sample under investigation, in order to detect the abrupt change in the magnetic field at the respective "point" when the critical temperature T_C of the transition to superconductivity is reached. It should be mentioned that carrier mobility is higher by an order of magnitude at 77 K than at room temperature $T = 300$ K. Because of this, it may be supposed that the transducing efficiency can reach unexpectedly high values at cryogenic temperatures. In experiments with the orthogonal MT in Fig. 6.8c at $T = 300$ K and $T = 77$ K, it has been established by the author that the device preserves its working capacity at $T = 77$ K. The magnetosensitivity is increased by a factor of 20 in comparison with that at 300 K but the interval of linearity is reduced from $-1 \leq B \leq 1$ T at 300 K to $-0.2 \leq B \leq 0.2$ T at 77 K. The $I_{C1}(B)$ dependence is presented in Fig. 6.8c'.

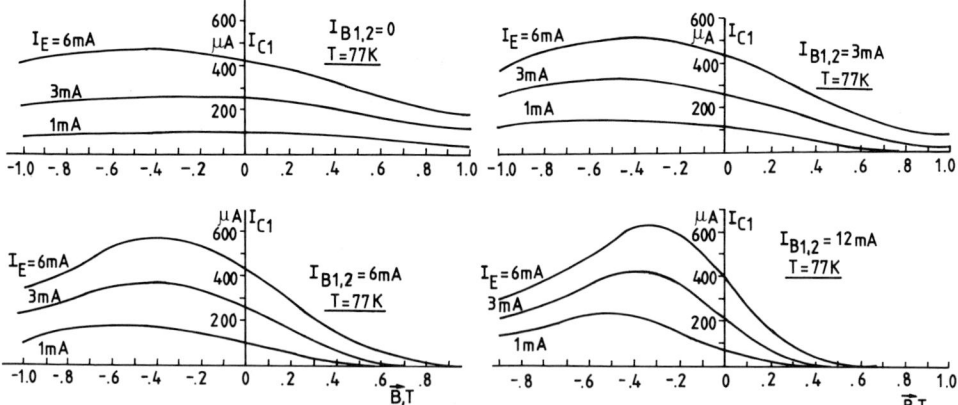

Fig. 6.8 (continued). (c') $I_{C1}(B)$ characteristics of the p^+-n-p^+ BMT from Fig. 6.8c at $T = 77$ K. The emitter current I_E has been taken as a parameter at fixed values of the interbase current $I_{B1,2}$.

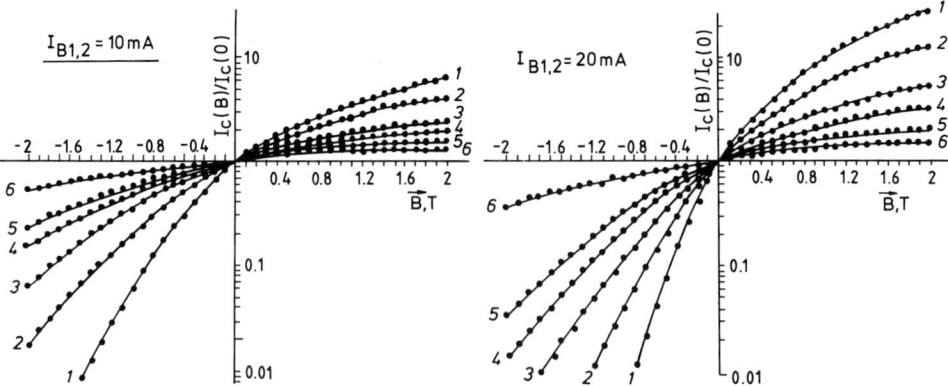

Fig. 6.8 (continued). (d') Dependences of the collector current on the magnetic field obtained for the collector C_1 from Fig. 6.8d at $T = 300$ K; emitter current I_E (mA). *1*: 0.5, *2*: 1, *3*: 2, *4*: 3, *5*: 4, *6*: 2 for collector C_3.

The FME (§6.2.1.2b), can be observed in the BMT in Fig. 6.8c. Depending on the currents I_E and $I_{B1,2}$ and the field **B**, a transition from exponential to linear magnetosensitivity has been observed in an MT prepared for this purpose, Fig. 6.8d [30]. The deflection of the filament and the minority-carrier profile outside enhance the variation in the collector signal $I_C(B)$ (Fig. 6.8d').

6.3.2. Parallel-field BMT sensors

Group I in Fig. 6.9 includes sensors whose magnetosensitivity is determined by the deflection of the majority carriers Θ_{maj}^H in the low-doped collector region. The output current $\Delta I_{C1,2}(B)$ has a Hall origin $\Delta I_{C1,2}(B) \equiv \Delta I_H \sim G(l, w_E)\Theta_{maj}^H I_C(0)$, where G is the geometrical factor and $I_C(0) = I_{C1}(0) + I_{C2}(0)$. Their transduction efficiency, $S = \Delta I_{C1,2}(B)/I_C B$, is $S \sim G(l, w_E)\mu_{maj}$. The device in Fig. 6.9a is the first dual-collector BMT [32]. A p-type substrate, n^+-collectors, a p-type base on top of them and an n^+-emitter on the very top of the structure have been successively fabricated by diffusion. High intercollector resistance is achieved because of the reverse biases V_{CS} and V_{CB} and the separation of terminals C_1 and C_2. However, this sensor is incompatible with standard bipolar technology, which uses an epitaxial layer. The enhancement of sensitivity is achieved in the analogous BMT in Fig. 6.9b, in which the contacts C_1 and C_2 are placed on the lower side of the substrate and have been separated by etching a groove [21]. The collector region must be depleted because otherwise the output voltage $\Delta V_{C1,2}(B)$ would be reduced by the shunting action of the small ohmic resistance of the intercollector groove. The drawback of the solution shown in Fig. 6.9b is that the groove is on the lower side of the chip, which makes the use of conventional microelectronic technology more difficult and produces an uncontrollable offset, because of the difficulties connected with the alignment of the groove with the symmetry axis of the structure.

The two device structures described, as a result of a considerable sophistication of technology, have been modified into the sensors in Fig. 6.9c and d [33,34–37]. These up-to-date transducers are completely compatible with conventional bipolar processing. The BMT in Fig. 6.9c is of the n–p–n type and has two buried n^+-collectors separated by a small distance to eliminate shorting effects [34]. The device in Fig. 6.9d [35–37] is also n–p–n and is realized in an n-type epitaxial layer; an n^+-layer has been deposited under the collectors to enhance sensitivity. The characteristic feature of both sensors (Fig. 6.9c and d) is that their collectors C_1 and C_2 are used as an active load of the differential preamplifier of an op-amp. The current imbalance $\Delta I_{C1,2}(B)$ in the preamplifier is transformed into an amplifier output voltage. The bias circuitry used requires the collectors C_1 and C_2 to be kept at one and the same reverse bias, i.e. $V_{C1} = V_{C2}$, so that the absence of electrical insulation between them might be overcome. If $V_{C1} \neq V_{C2}$, the inevitable offset dominates over the useful signal. Miniature Hall probes H_1 and H_2 [33,35,36], have been made in the base region by analogy with [38] for an experimental test as to whether EIM contributes to the magnetosensitivity of the sensor in Fig. 6.9d. High-precision measurements unambiguously indicate the appearance of a negligible Hall voltage, thus excluding EIM as a dominating mechanism in the action of the sensor.

Figure 6.9, group II, includes discrete BMTs in which the emitter E and the collectors C_1 and C_2 are planar with stripe geometry, and the contact B is on the back of the chip. In the sensor in Fig. 6.9e the sensitivity is determined by the deflection φ, whereby the field $E_{maj}^H(I_B)$ is developed on the upper surface. Magnetic control of the current $I_C(B)$ was first observed in a similar germanium single-collector version with point emitter and collector [1]. Despite its simplicity,

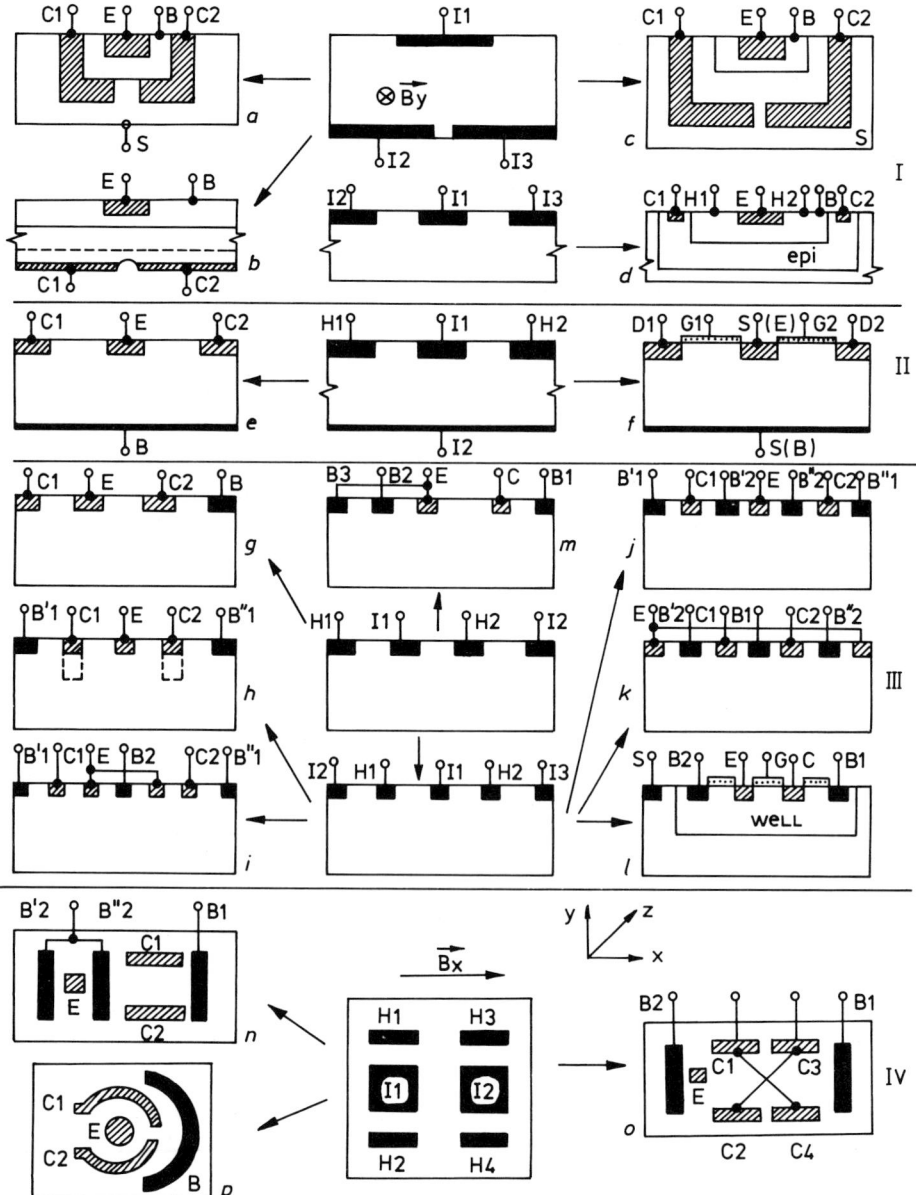

Fig. 6.9. Classification of parallel-field BMT sensors. The respective conformally transformed Hall devices are shown.

the silicon sensor in Fig. 6.9e has satisfactory linearity and sensitivity for a wide range of applications. The dual-drain MOST in Fig. 6.9f operates in the injecting-source (emitter) mode, whereby $l > L^*$ [39]. The indirect deflection φ modulates the channel currents and the coefficients α^L and α^R in the regions to the left and to the right of the source. As a result of different recombination rates in the space-charge region and on the bottom of the chip, the output signal $\Delta I_{D1,2}(B)$ is nonlinear. Magnetosensitive triggers can be built on the basis of this MOST.

The sensors of group III (Fig. 6.9) have planar contacts, and no doubt their advantages lie in their integrated realization. Apart from their functional connection to conformally transformed Hall elements [40], some of these devices have complete orthogonal activation analogies, for instance Fig. 6.8a → Fig. 6.9h and Fig. 6.8c, d → Fig. 6.9i. In order to study in detail the relationship between the MT action and the device structure, a symmetrical sensor has been suggested, similar to the one in Fig. 6.9i, but with different contact locations (Fig. 6.9j and k), as well as versions without structural symmetry (Fig. 6.9g, l and m). The dominant mechanism in the sensor from Fig. 6.9g is the deflection φ, whereby the field $E_{maj}^H(I_B)$ is located on the upper surface. The CMOS bipolar structure in Fig. 6.9l [41], is a drift-aided device with a gate in order to control the conditions on the Si–SiO$_2$ interface. According to [41], the current $I_{B1,2}$ creates a Hall field E_{maj}^H in the "well", which is 12 μm deep, and the deflection φ leads to the deviation of the current $I_C(B)$ and the reverse "parasitic" current $I_S(B)$ through the substrate in two opposite directions. In an analogous CMOS structure with a relatively shallow "well" about 6 μm deep, with no gate and with a floating substrate, no E_{maj}^H field is observed experimentally and an exponential magnetosensitivity appears [42]. The reason for the absence of $E_{maj}^H(I_{B1,2})$ is the profile of the doping impurities in the well. Their concentration has a maximum value near the Si–SiO$_2$ interface. This distribution of the doping impurities in a specific device [42], is responsible for the presence of a built-in electric field and a potential well for majority carriers close to the Si–SiO$_2$ interface and perpendicular to it. This field and the lower resistivity of the crystal in the surface region determine the distribution of the current $I_{B1,2}$ in such a way that most of the majority carriers flow parallel to the Si–SiO$_2$ surface inside the potential well. As a result, the built-in electrostatic field prevents the spatial separation of the majority carriers by the force $\mathbf{F}_L$, and the field $E_{maj}^H \sim 0$. The dependence $I_C(B) \sim \exp(B)$ is connected with the negative resistance of the curve $I_E(V_{EB1})$ and the FME (see §6.2.1.1b).

The BMT with stripe geometry E, C_1, C_2, B_1' and B_1'' (Fig. 6.9h) proposed first in [43] is a typical example of the influence of the device structure and mode of operation on the dominant sensor mechanism. When the silicon substrate plays the role of an active region and the n^+- and p^+-regions are shallow and have equal depths, the magnetosensitivity is linear and is determined by the deflection φ, similar to the device in Fig. 6.8a. The field E_{maj}^H is developed on the upper surface of the chip, as in the corresponding parallel-field Hall element. The optimal position of the collectors is close to the contacts B_1' and B_1''. This trade-off solution is associated, on the one hand, with an increase in the field $E_{maj}^H(x>0)$ and, on the other, with its reduction by the base contacts. The modulation of the coefficients

α^L and α^R through the Hall field $E^H_{\text{maj}}(I_B)$ is responsible for the magnetotransistor effect in the base region of the device from Fig. 6.9h and described in [44]. The conditions for its appearance are moderate levels of injection and deep collector regions, shown with a dotted line in Fig. 6.9h. Along the x-axis, the minority-carrier transport is diffusive. In a magnetic field $\boldsymbol{B} \parallel y$, the Hall field $E^H_{\text{maj}}(I_B)$ is located between the vertical sides of the two deep collector regions. The field $E^H_{\text{maj}}(I_B)$ most probably does not contribute to the deflection of the injected carriers, but controls electrostatically the drift-diffusion distribution of the minority carriers between C_1 and C_2, i.e., the transit time τ^{tr}_{min}. Depending on its direction, the Hall field slows down the diffusion of minority carriers towards one of the collectors and pushes them towards the other; as a result, the coefficients α^L and α^R change and hence so does the current imbalance $\Delta I_{C1,2}(\boldsymbol{B})$.

Magnetosensitivity in the emitter region (cf. §6.2.1.1) is a possible sensor mechanism in the BMT shown in Fig. 6.9h [9]. In this case, similar to EIM, the output signal $\Delta I_{C1,2}(\boldsymbol{B})$ is expected to be an exponential function. The problem is in the experimental identification of each particular mechanism. Another manifestation of the sensor action of Fig. 6.9h is magnetoconcentration [29,43]. It is observed in long-base devices prepared in an epitaxial layer when operating at high injection levels. A reversed-bias base-substrate junction, in analogy with a CMOS MD [45], plays the role of a region of infinite recombination rate. The nonlinearity of the output is a restriction in analog applications. The role of the EIM in the sensor in Fig. 6.9h has been actively discussed in the literature [6–18,19]. Such a mechanism is possible in this structure at $T = 300$ K if the device and the distribution of doping impurities in the emitter are such as to enable the Hall effect to distribute the density of carriers predominantly to the left or to the right. The operation of the sensor (Fig. 6.9i) [46] for a directly coupled emitter and low injection efficiency, is determined both by the deflection φ, as in the device from Fig. 6.8 (e), and by EIM. The field $E^H_{\text{maj}}(I_{B1,2})$ is generated on the upper surface of the substrate. As a result of the diode behavior $I_E(V_{EB})$, the relationship $I_{C1,2} \sim \exp(V_{EB} \pm V_H)$ holds good for the output collector currents. The behavior of a single-collector bulk version B_2–E–C–B_1 (cf. Fig. 6.9l) is analogous at low values of the current $I_C(0)$ and a common-emitter configuration. In either case the change in the output signal in a magnetic field is usually high [46]. The EIM efficiency is determined by the position of the emitters with respect to contact B_2.

The enhanced magnetosensitivity of the sensor in Fig. 6.9j and k [16,47] has a common nature regardless of the different contact geometry. If the base contacts B'_2 and B''_2, whose polarity is the same as that of the emitter, are connected, they remain at the same potential in the magnetic field $\boldsymbol{B}$ as well. Therefore no Hall voltage V_H can appear between them, as in the case of the three-contact planar Hall element [48]. This is why there is no EIM, and the magnetic control of the currents $I_{C1}(\boldsymbol{B})$ and $I_{C2}(\boldsymbol{B})$ is due to the deflection φ. If the mode of operation is chosen in the negative-resistance region of the $I_E(V_{EB})$ curve and if the current $I_{B1,2} \neq 0$, FME appears, and when $I_E = \text{const}$ the dependence $I_C(B)$ is exponential [16]. In [49] an $n-p-n$ sensor similar to the device in Fig. 6.9j [16,47], but manufactured by CMOS technology is described. The contacts B_2' and B_2'' play a threefold role. Firstly, they suppress lateral injection; secondly, their revers bias with respect to E confines the injected carriers to the center of the bottom of the EB junction, and thirdly, an accelerating field is generated in the base region. The unsurpassed linear magnetosensitivity of this MT with a suppressed side-wall injection (SSIMT) is apparently due to the so-called double deflection [49], which includes the following phenomena: the current I_E is divided into three components, two of which determine the currents $I_{C1}(0)$ and $I_{C2}(0)$, and the third I_S is directed downwards and is collected by the well-substrate junction. The force $\boldsymbol{F}_L$ influences the three components and changes the currents I_{C1} and I_{C2}. The deflection $\Theta_{\min}^H \sim \mu_{\min} B$ of the current I_S additionally enhances the variations in the collector currents. The effective base width w_{EC}^* is modulated by varying the potential of contacts B_2' and B_2'', thus the magnetosensitivity is electrically controlled.

When contacts B_2' and B_2'' of the sensors in Fig. 6.9j and k are individually supplied by constant currents and B_1' and B_2'' are connected, by analogy with the three-contact planar Hall sensor [48], a Hall voltage $V_H = V_{B'2, B''2}$ is generated upon B_2' and B_2'' in a magnetic field. Its development close to the emitter or emitters causes an asymmetrical injection and EIM, which together with the deflection φ contributes to the magnetosensitivity of these BMTs. EIM is observed in the asymmetrical device in Fig. 6.9m [50], as well (cf. Fig. 6.5). Under the combined action of EIM and the indirect deflection φ, and in the presence of a current filament, when the value of the current $I_C(0)$ is very low, the magnetosensitivity may reach $10^6 \% T^{-1}$ [50]. It has been established from experiments with the parallel-field BMT in Fig. 6.9m that at 77 K the magnetic control of the current $I_C(B)/I_C(0)$ is increased by a factor of 24 compared with that at $T = 300$ K (Fig. 6.9m′) [50].

Group IV in Fig. 6.9 presents parallel-field BMTs activated by the so-called tangential field $\boldsymbol{B}_x$. Their action is based on the nonlinearity of the trajectories when all contacts are planar (§6.2.1.2a). This feature of carrier transport is the reason for the "parasitic" sensitivity along both the x- and z-axes (Fig. 6.8c) [51], and explains the disagreement between the data from refs. [51] and [52]. The field B_x affects only the component of velocity v_z (Fig. 6.9). The device structure of the sensor on Fig. 6.9n [53] is optimized in such a way that initially the entire current I_E is directed downwards into the substrate. This is achieved by surrounding the emitter E by two connected contacts B_2' and B_2'', with the same polarity as E. As a result, at first the current I_E is directed only along the z-axis, passing deep into the bulk beneath contact B_2''; then it is directed to the upper surface with B_1 and the collectors. The effective distance w_{EC}^*, and hence the sensitivity of this sensor, is electrically controlled by the potential of contacts B_2, in analogy with the sensor described in [49].

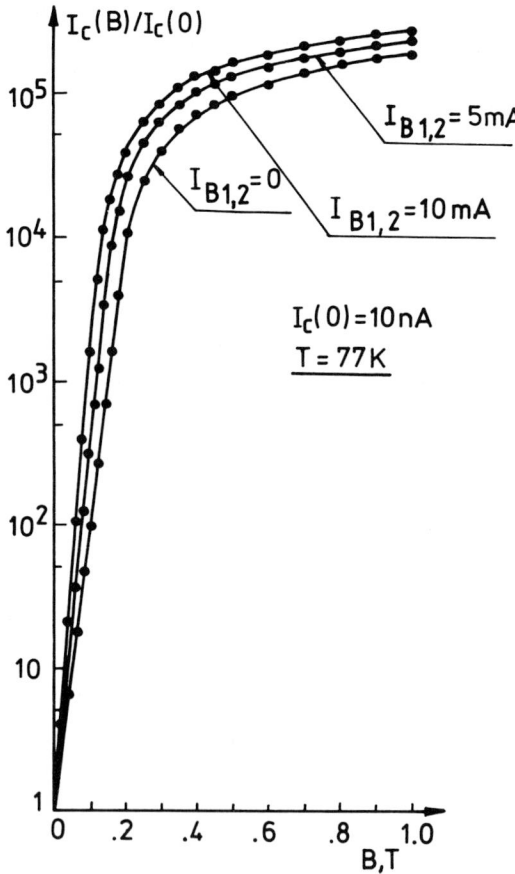

Fig. 6.9 (continued). (m′) Magnetic control of the current $I_C(B)$ of the device from Fig. 6.9m at $T = 77$ K and $I_C(0) = 10$ nA.

The nonlinear trajectory leads to the appearance of the effect of the electrical control of the polarity of the magnetosensitivity in the sensor shown in Fig. 6.8c in a tangential field $\boldsymbol{B}_x$ [18] (§6.2.1.2a). In the structure shown in Fig. 6.9o, the reduced sensitivity of the field $\boldsymbol{B}_x$ is compensated by the cross-coupled pairs of collectors C_1–C_4 and C_2–C_3, thus obtaining the sums of the increments with the same sign, i.e., $\Delta I_{C1} + \Delta I_{C4}$ and $-\Delta I_{C2} - \Delta I_{C3}$ [53]. This is why the transduction efficiency in the field $\boldsymbol{B}_x$ is many times higher than in the device in Fig. 6.8c. The trajectories of the majority carriers are also nonlinear and, in the same way as in the parallel-field Hall element in Fig. 6.9, group IV, on the upper surface two Hall fields of opposite signs are generated [10]. They enhance the respective deflections of minority carriers in either section of the trajectory. The ability to control electrically the sign of the magnetosensitivity underlies the new approach to vector measurements of the field $\boldsymbol{B}$, that is discussed in Chapter 8. The MT action of the bulk semicircular sensor in Fig. 6.9p [54], operating in a mode of onset saturation of the dependence $I_C(V_{CB})$, is the indirect deflection φ, determined by the vertical velocity v_z of the carriers with nonlinear trajectories.

In general, the galvanomagnetic phenomena in the orthogonal and parallel-field BMTs considered are three dimensional, as a result of which there is cross-sensitivity. This worsens the metrological properties of these sensors as one-component transducers. From the scanty information on this problem, it has been established that the orthogonal sensitivity of the sensor in Fig. 6.8b exceeds about 50 times the sensitivity along the other two axes [25]. A similar result has been obtained for the parallel-field device in Fig. 6.9k, operating in a linear mode [16]. If the transduction efficiency for a given axis is more than one order of magnitude higher than for the other two axes, that sensor can be practically discussed as one component.

6.4. Figures of merit of BMT sensors

6.4.1. Magnetosensitivity

The most important figure of merit for BMT sensors is the magnetosensitivity S. There are two methods of defining it. The first considers the variation in the output signal divided by the variation in the magnetic induction ΔB. The useful signal in a single-collector version is the current $I_C(B)$, and in dual-collector devices the current $\Delta I_{C1,2}(B) = I_{C1}(B) - I_{C2}(B)$. The absolute S_A^I and the relative S_A^R current magnetosensitivities at a field $B \to 0$ are defined by their pertinent derivatives [3]: for a single-collector sensor,

$$S_A^I \equiv \left| \frac{\partial I_C(B)}{\partial B} \right| \quad [\text{A T}^{-1}] \text{ and}$$

$$S_R^I \equiv \left| \frac{1}{I_C(0)} \frac{\partial I_C(B)}{\partial B} \right| \quad [\text{T}^{-1}];$$

for dual-collector devices,

$$S_A^I \equiv \left| \frac{\partial I_{C1,2}(B)}{\partial B} \right| \quad [\text{A T}^{-1}] \text{ and}$$

$$S_R^I \equiv \left| \frac{1}{I_{C1}(0) + I_{C2}(0)} \frac{\partial I_{C1,2}(B)}{\partial B} \right| \quad [\text{T}^{-1}].$$

The absolute voltage magnetosensitivity S_A^V characterizes the onset circuitry and depends on the respective collector resistors R_C or $R_{C1} = R_{C2}$, then $S_A^I = R_C S_A^V$, [V T^{-1}]. For sufficiently large values of the collector output resistance $r_C \gg R_C$, the relative voltage sensitivity S_R^V equals the current sensitivity S_R^I. The second method is based on the modulating nature of BMTs as transducers, i.e. the

Table 6.1

Parameters of selected silicon BMT's, whose device structures are shown in Figs. 6.8 and 6.9

No.	Type	Device struct.	Output	Sensitivity S (% T^{-1})	Equiv. noise (T), Δf	Equiv. offset (mT)	Linear range (T)	NL (%)	T.C. of S (% K^{-1})	Freq. resp. (MHz)	Sensor mechanism	Refs.
Orthogonal BMT sensors (Fig. 6.8)												
1	n–p–n	(b)	differ.	56	5×10^{-5}, 1 MHz	7	±0.2			5	deflection	[25]
2	p–n–p	(c)	differ.	7	2×10^{-5}, 5 Hz	36	±0.3	5	−0.7	10	deflection	[26]
3	p–n–p	(c)	differ.	2.6							deflection	[52]
4	p–n–p	(c)	differ.	1.77		45			−0.5		deflection	[27]
5	p–n–p	(c)	differ.	0.6			±1	5			deflection	[29]
6	p–n–p	(d)	differ.	500							filam. eff. sensit.	[30]
7	n–p–n	(g)	single	60 (77 K)							deflection	[31]
8	n–p–n	–	differ.	400							em. inj. mod.	[20]
Parallel-field BMT sensors (Fig. 6.9)												
1	p–n–p	6.8 (c)	differ.	3							deflection	[53]
2	n–p–n	(b)	differ.	8							deflection	[21]
3	n–p–n	(c)	differ.	5	1×10^{-5}, 2 Hz	50	±1	0.5	−0.62	1	deflection	[34]
4	n–p–n	(d)	differ.	6	2×10^{-5}, 5 MHz						deflection	[36]
5	n–p–n	(d)	differ.	1.8							deflection	[35]
6	n–p–n	(d)	differ.	0.8			±0.2	1			deflection	[33]
7	n–p–n	(h)	differ.	7.6	1×10^{-5}, 1 MHz					50	em. inj. mod.	[6]
8	n–p–n	(h)	differ.	7.3	3×10^{-7}, 1 kHz	180	±1	1	−0.34	0.2	deflection	[55]
9	n–p–n	(h)	differ.	1.8			±1	5	−0.5	0.21	magnetoconc.	[29,43]
10	p–n–p	(i)	differ.	400							deflection	[46]
11	p–n–p	(i,l)	single	3000							em. inj. mod.	[46]
12	p–n–p	(j)	differ.	3050					−0.4		em. inj. mod.	[49]
13	p–n–p	(j,k)	differ.	40			±0.03		−0.4		double defl.	[16]
14	p–n–p	(k)	single	30000							deflection	[16]
15	n–p–n	(l)	single	150	4×10^{-5}, 10 Hz		±0.1		−0.5		filam. eff. sensit.	[41]
16	p–n–p	(m)	single	10^6							filam. eff. sensit.	[50]
17	p–n–p	(n)	differ.	3.6			±1		−0.4		em. inj. mod.	[53]
18	p–n–p	(o)	differ.	12		100	±1	1.5	−0.5		deflection	[53]
19	n–p–n	(p)	differ.	8	3.3×10^{-5}, 1 MHz		±0.058			1	deflection	[54]

magnetosensitivity S_R is obtained from a comparison of the output signal with the supply power applied, determined by the currents I_E and $I_{B1,2}$, and ΔB. In this case $S_R^{tot} \equiv \left(1/(I_E + I_{B1,2})\right) \partial I_C/\partial B$, $[T^{-1}]$.

Both methods are used in different works on BMTs, and this makes a precise comparative analysis more difficult. The parameter S_R^{tot} is more suitable for characterizing MT action. Table 6.1 contains data about S. The extremely wide range of variation of S results, above all from the variety of galvanomagnetic effects involved in MT operation.

6.4.2. Noise

Noise is the fundamental parameter which determines the lowest value, B_{min}, of the magnetic field that can be detected by a BMT (§3.3.1). The sensor noise will be defined (and measured) by its power spectral density, i.e. by the frequency spectrum of the mean square value of the fluctuating (noisy) quantity [55]. The magnetic field B_{eq} equivalent to the noise in a frequency range Δf for a signal-to-noise ratio = 1 is defined as $B_{eq} = N/S$, where N is the noise (current or voltage) and S is the corresponding magnetosensitivity. The study of the intrinsic noise in BMTs is far from complete and the available data are sporadic, although it is an important parameter. Relatively satisfactory investigations of noise have been performed for the devices from Fig. 6.9h [55], and Fig. 6.9l [41]. For the sensor in Fig. 6.9h it has been found that between 1 Hz and 1 kHz the noise is $1/f$ and varies in the same way as I_E^2 (Fig. 6.10).

The $1/f$ noise may be interpreted as resulting from fluctuations in the bulk conductivity (most probably due to fluctuations in the mobility) in the emitter–collector space. For the $1/f$ noise the value of the Hooge α-parameter is 5×10^{-5} to 10^{-3}. At a frequency $f \geq 1$ kHz, the shot noise appears to be proportional to the current I_E (and I_C). It affects the carriers injected by the emitter p–n junction, and its amplitude exceeds that of the $1/f$ noise. On the other hand, the voltage V_{CB}, the offset, the temperature $0 \leq T \leq 70°C$ and the field $B < 1$ T in the bulk device in Fig. 6.9h do not actually influence the spectrum of the noise. In the CMOS sensor in Fig. 6.9l, the shot noise has been studied in detail as a function of the device structure and mode of operation [41], and the results are similar to those in ref. [55]. The intrinsic noise can be positively correlated in the CMOS SSIMT of Fig. 6.9j by the careful choice of the device geometry and the current $I_{tot} \leq 1$ mA [36,37]. Owing to the differential coupling, the noise is suppressed by about four orders of magnitude which determines a resolution $\approx \mu T$ for a frequency band $\Delta f \sim 5$ MHz. When the MT operates at high base currents, however, this property is lost and a lower S/N ratio is measured [37]. A comparison of the data on noise in BMTs from Table 6.1 with those of other semiconductor magnetosensors, such as SOS MDs, MOS and MAGFET Hall devices, leads to an important conclusion: regardless of the great differences in sensitivity, the wide variety of structures and

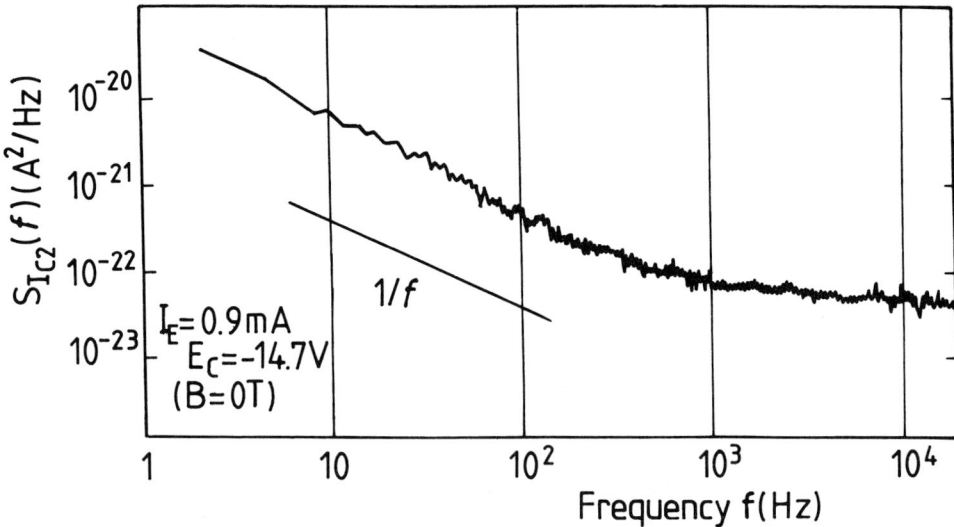

Fig. 6.10. Power spectral density of collector-current fluctuations $S_{IC}(f)$ for the BMT from Fig. 6.9h at $I_E = 0.9$ mA, $I_{C1} = I_{C2} = 0.23$ mA.

transducer mechanisms, the noise parameters are of the same order of magnitude [55]. This fact needs an explanation.

6.4.3. Offset

By definition, an offset is a static or very weakly varying output voltage, current or frequency in a differential BMT under field $B = 0$. In the absence of information about the nature of the sensor or the field B, the offset $\Delta I_{C1,2}(B = 0)$ interferes with the useful signal $\Delta I_{C1,2}(B)$ and cannot be distinguished from it [3,4]:

$$\Delta I_{C1,2}(B) = BS_1 + \Delta I_{C1,2}(0). \tag{6.12}$$

The offset can be presented as an equivalent magnetic field $B_{0,eq} = \Delta I_{C1,2}(0)/S_A^I$. The reason for the appearance of an offset are associated with the imperfection of the technology, the inevitable geometrical asymmetry due to misalignment of the photomask and the piezo-resistance effect. The misalignment problem in most lateral BMTs with carrier deflection is not severe since the same lithography is used for the formation of both the emitter and the collectors. Therefore the equivalent offset is relatively low (about 50 mT). The misalignment of the emitter with respect to the base contacts contributes considerably to the offset. The asymmetry of the positions of these junctions results in an asymmetrical distribution of the emitter base bias along the emitter, thus generating an offset. The BMT base region is, in essence, a Hall plate and all considerations in §4.2.4.2b concerning the reasons for the offset remain valid in this case. An important cause of the offset in the BMTs from Fig. 6.9c and d is the presence of impedance imbalance in the collector leads, which may be introduced by imperfections in the electronic circuitry connected to these leads. An impedance imbalance in the collector leads to a higher current of the collector with the smaller impedance. The fact that the collectors of the magnetotransistor do not behave as perfect current sources is the underlying cause of this phenomenon. (This is not valid for devices in which the collectors are electrically isolated from each other.) There is a finite intercollector resistance between the collectors which depends on the gap between the collectors, the thickness of the epitaxial layer and its resistivity [3]. On the other hand, this parasitic signal depends on the supply current, temperature, the mechanical and thermal stress introduced during packaging, aging, and the nonuniform distribution of the thermal flow in the chip and the positive feedback associated with it. In the general case, the temperature drift of the offset is nonlinear and irreproducible during a heating–cooling cycling, and its value depends on the rate of temperature variation. This drift is also a function of the magnitude of the field B, because at $B \neq 0$ the symmetry of both the currents in the structure and of the thermal flow is disrupted. The traditional approaches aimed at the elimination of the offset, like calibration, compensation, improvement of the process parameters, the chopper-device method, etc., are not sufficiently effective in BMTs. Detailed information on the negative impact of the offset due to mask misalignment is available [56]. A new sensitivity-variation offset-reduction method has been proposed [57] for the device in Fig. 6.9c (§9.5.1). A comparison of the offset in BMT sensors (Table 6.1) and in Hall devices shows that in the second case, the values of $B_{o,eq}$ are lower by about one or two orders of magnitude. This is due to the great number of processing steps necessary to fabricate MT devices.

6.4.4. Linearity

The linearity of the output signal in BMTs depends mainly on the respective transduction mechanism. Devices based on carrier deflection tend to have a linear output [3,4]. By definition, the nonlinearity error is given by NL $\equiv (V_{\text{out}} - V_{\text{out}}^{(0)})/V_{\text{out}}^{(0)} 100\%$, where V_{out} is the measured value of the output voltage and $V_{\text{out}}^{(0)}$ is the measured slope of the straight line best fitting the output characteristic obtained experimentally (§3.2.2). Another equation is often used as well, i.e. equation (3.6).

6.4.5. Temperature coefficient of magnetosensitivity

The temperature coefficient of magnetosensitivity, T.C. $\equiv (1/V_{\text{out}}) (\partial V_{\text{out}}/\partial T) 100\%$, in BMTs characterizes the temperature variation of the transduction efficiency itself (§3.3.3). This phenomenon is based on the temperature dependence of the carrier mobility, the latter decreasing together with the sensitivity as a result of the temperature increase. To a first approximation, the mode of operation and the MT action do not influence the T.C. [3]. According to the data from Table 6.1, essentially all devices have similar values of the T.C., which correlate with the respective values for Hall sensors. The reduction of the T.C. is one of the serious problems in sensor electronics. Three methods are known for BMTs. The first is based on the sensitivity-variation offset-reduction method, in which the T.C. is annulled [57]. The second one uses an epilayer resistor for I/V conversion, placed on the same chip as the sensor [34]. The third approach is based on a functional principle and allows the achievement of T.C. compensation in a very wide range of ΔT. It will be discussed in Chapter 8.

6.4.6. Frequency response

The frequency response (§3.2.5) of BMTs is limited by the capacitance charging times and base transit times, similar to those of conventional bipolar transistors. Since MTs usually have a long base, they have a worse frequency response than do classical transistors [3]. The dynamic behavior of a BMT with regard to an a.c. magnetic field is of interest. It has been established that a frequency-independent output signal in the few devices investigated is obtained in the range of $f \leq$ 100 kHz. With increasing frequency f, the effect of the electric vector in the generated electromagnetic field should be taken into account. In [54] it is reported that the amplitude of the output signal of the BMT in Fig. 6.9p at 1 MHz is half the value at $f = 100$ kHz. In the sensor shown in Fig. 6.9h, the frequency of the a.c. magnetic field at which the output signal drops by a factor $1/\sqrt{2}$, i.e. by $3db$ is 210 kHz [29], and it is claimed that an a.c. magnetic field with $f \sim 5$ MHz can be detected with the sensor from Fig. 6.8b [25]. Drift-aided BMTs have a better frequency response.

Unfortunately the other figures of merit discussed in Chapter 3, have not been investigated at all, or the relevant data are contradictory. Such an important parameter as the stability, for instance, which to a large extent determines the applicability of BMTs in smart solutions, has practically not been studied. Provided the technology is perfect and the consumed power is low, the BMTs can be expected to be at least as stable as Hall sensors.

6.5. Unijunction magnetotransistor sensors

Like the bipolar magnetosensors considered so far, the unijunction magnetotransistor (UJMT) is also a typical example of the superposition of the Hall effect and a nonequilibrium electron–hole plasma. The functional integration of UJTs with other transistor versions, for instance the high sensitivity devices presented in Fig. 6.8c, d, and Fig. 6.9i–k, m–o, etc., is of considerable importance for progress in the field of BMTs. The deliberate electrical generation of negative differential resistance in UJTs offers new opportunities for the construction of advanced solid-state magnetosensors.

6.5.1. UJMT device and operation

The UJT is a three-terminal semiconductor element with lateral injection of minority carriers into the base region by the emitter p–n junction. The base region is lightly doped and has two ohmic contacts B_1 and B_2 [13]. A bulk and two planar device structures are presented schematically in Fig. 6.11a–c, respectively. They can be used also as magnetosensors. Typical current–voltage curves $I_{EB1}(V_{EB1})$ as well as the equivalent circuit of the UJT device are presented in Fig. 6.11d and e [58–62]. The resistance between point A and the main contact B_1 and the resistance between point A and the contact B_2 are denoted by R_{B1} and R_{B2}, respectively. R_N in the circuit in Fig. 6.11e is the equivalent negative resistance. The following relationships between the resistances are valid: $R_N + R_S = R_{B1}$ and $R_{B1,2} = R_{B1} + R_{B2}$. When an external voltage $V_{B1,2}$ is applied, a longitudinal electric field $E_{B1,2}$ appears and a current $I_{B1,2}$ flows via the base region. Such a potential distribution is established as a result of the current flow that a voltage whose value is approximately $\eta V_{B1,2}$ is added to the potential of the n-side of the emitter–base p–n junction. η is given by the formula $\eta \approx R_{B1}/(R_{B1} + R_{B2})$. The polarity of the additional voltage $\eta V_{B1,2}$ reversely biases the emitter junction E. If the forward bias V_{EB1} across the emitter p–n junction is smaller than $\eta V_{B1,2}$, no injection takes place. The space-charge region of the emitter junction at a current $I_{B1,2} \neq 0$ and voltage $V_{EB1} < \eta V_{B1,2}$ is presented in Fig. 6.12. Point M has a higher potential than point N. The junction surface can be virtually partitioned into elementary diodes, each one assocoiated with a quasi-equipotential n-type base region, for instance the neighborhood of points M and N. If $V_{EB1} \geq \eta V_{B1,2} = V_p$, minority carriers are injected into the base. The injection begins with the elementary diode N whose potential is V_p. Owing to the field $E_{B1,2}$ the minority carriers drift towards the main contact B_1. The conductivity between the emitter E and the contact B_1 is progressively increased. The voltage V_{EB1} drops with the further increase of the current I_{EB1}, the conductivity increases again, and, as in the case of S-diodes (cf. §5.1), an S-type region of negative resistance appears in the $I_{EB1}(V_{EB1})$ characteristic as a result of this positive feedback (curves 2, 3 and 4 in Fig. 6.11d).

The transition from "normal" to "negative" resistance, the so-called critical point C, represented by the "vertical" line (curve 2 in Fig. 6.11d), i.e. the zero differential resistance [63], is of practical importance for the operation of some types of UJMTs. Three parameters determine the critical point C: the voltage $V_{B1,2C}$ at which the dynamic resistance of the $I_{EB1}(V_{EB1})$ curve is zero, i.e. $\partial V_{EB1}/\partial I_{EB1} = 0$; the voltage V_{EB1C} which corresponds to the vertical region; and the current I_{EB1C} corresponding to the critical point C. This point divides the $I_{EB1}(V_{EB1})$ characteristic at a current

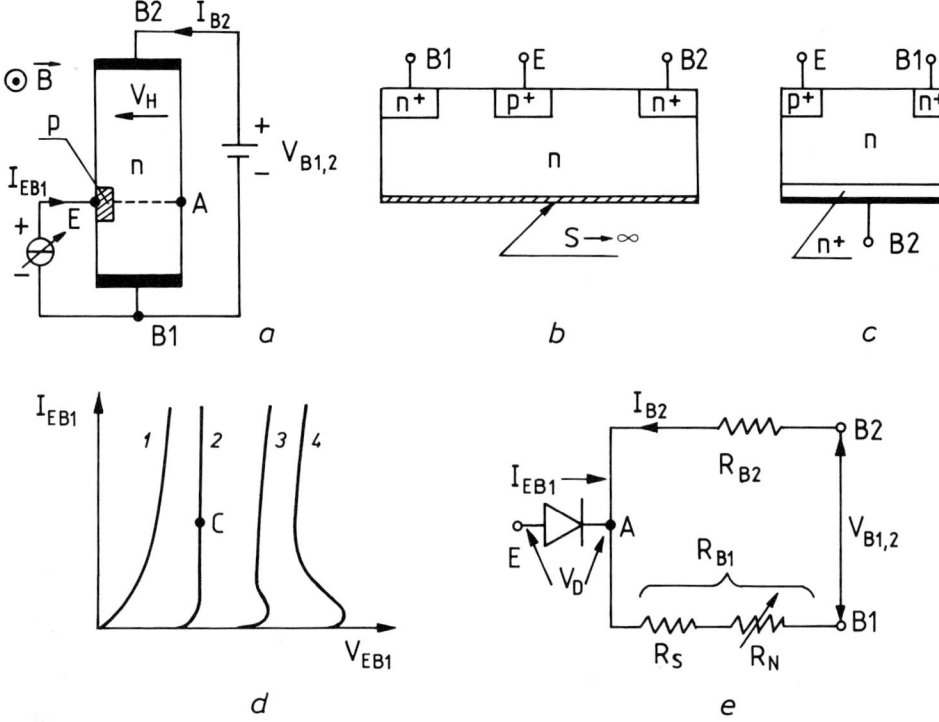

Fig. 6.11. Unijunction transistor structures used as magnetic sensors: (a) bulk version; (b), (c) planar versions; (d) typical UJT current–voltage characteristics corresponding to an increasing current $I_{B1,2}$; curve *1*: $I_{B1,2} = 0$, curve *2*: zero differential resistance–critical point C; (e) UJT equivalent circuit.

$I_{B1,2}$ = const into a stable region and an unstable region with negative resistance. Current and voltage oscillations may arise in the second case. There is an analogy between this electrical transition in UJTs and the second-order phase transitions (e.g. critical point of a fluid, Curie point of a magnet) [63]. Another important feature of this instability is the formation of a current filament which starts from point N and widens in the direction of the base contact B_1. As result of the charge neutrality in the base region, the electron and hole concentrations in the filament are approximately equal, and thus a plasma domain is formed. At a moderate injection level determined by a longitudinal sectioning of the UJT base, two regions are obtained:

(a) a region with a very high conductivity confined within the current filament;

(b) a region of almost equilibrium conductivity determined by the concentrations n_0 and p_0.

If the emitter is operating in the constant-voltage mode V_{EB1} = const, the valley point V_V on the $I_{EB1}(V_{EB1})$ curve corresponds to a state when the lower part of the base region is filled with the domain.

The influence of the external magnetic field B on particular regions of the $I_{EB1}(V_{EB1})$ characteristic of the UJMT has been investigated and is discussed in

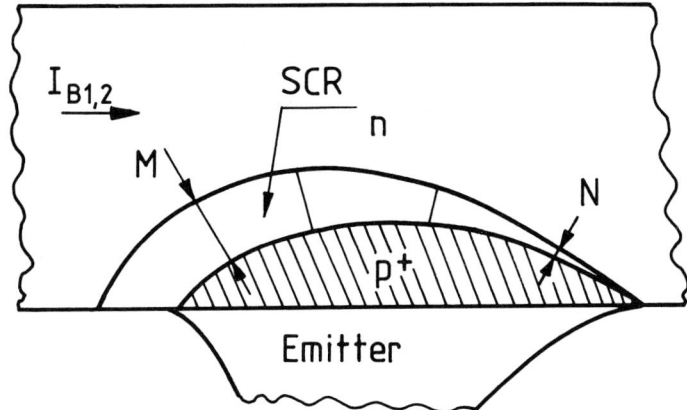

Fig. 6.12. Space-charge region of the UJT emitter junction just before injection sets in.

[58–62]. In accordance with Fig. 6.11e the peak-point voltage V_p in a magnetic field $B = 0$ is given by the expression:

$$V_p = \left[\frac{V_{B1,2} R_{B1}}{R_{B1} + R_{B2}}\right] + V_D, \qquad (6.13)$$

where V_D is the cut-in voltage of the p–n junction.

If a magnetic field B is applied perpendicularly to the plane of Fig. 6.11a at a voltage $V_{EB1} < V_p$ and current $I_{B1,2} \neq 0$, the influence of B upon the p–n junction is negligible [60]. In this case, because of the reverse biasing of the emitter E, a magnetoresistance effect occurs in the base, which modulates the resistance $R_{B1,2} = R_{B1} + R_{B2}$, and a Hall field arises in a direction perpendicular to the vector B and to the current $I_{B1,2}$, Fig. 6.11a. Because of these two galvanomagnetic effects, the new peak-point voltage will be:

$$V_p' = \left[\frac{V_{B1,2} R'_{B1}}{R'_{B1} + R'_{B2}}\right] + V_D + V_H^*, \qquad (6.14)$$

where R'_{B1} and R'_{B2} are the respective values of R_{B1} and R_{B2} modulated by the magnetic field; the voltage V_H^* is developed between the base region and the emitter E of the UJMT. Actually the full Hall voltage V_H^* is not going to contribute to V_p'.

Since the percentage variations in R_{B1} and $R_{B1} + R_{B2}$ due to the magnetoresistance effect are approximately equal, the field B, to a first approximation, does not change the ratio $R_{B1}/(R_{B1} + R_{B2})$ in equation (6.14). Therefore the main contribution to the voltage V_p' in a magnetic field belongs to the Hall voltage V_H^*, cf. (6.14).

In the emitter injection $V_{EB1} > V_p$, the E–B_1 region can be regarded as a long magnetodiode with one injecting contact (Fig. 6.11a), whose operation has been analyzed in §2.5.3.1. In this case the conditions for the occurrence of the MD effect are: the effective distance l_{EB1}^* exceeds the diffusion length of the injected carriers, i.e. $l_{EB1}^* > L_p$; a high injection level is provided. The magnetic control of the UJMT current–voltage characteristic $I_{EB1}(V_{EB1})$ is determined by the relationship (2.107). Therefore, at a fixed current $I_{B1,2} = \text{const}$ and an adequately chosen injection level $I_{EB1} = \text{const}$, the voltage $V_{EB1}(B)$ is the magnetosensitive output of the UJMT sensor. The transducing efficiency of the device can be improved by the formation of a region with a high surface recombination rate, $s \to \infty$, on the opposite side of the emitter p–n junction.

The $I_{B1,2}(V_{B1,2})$ current–voltage characteristic at $V_{EB1} = const$ also includes a negative resistance, but, unlike the $I_{EB1}(U_{EB1})$ curve, this is an N-type characteristic. This dependence is again controlled by the magnetic field due to the MD effect in the base and can be used for magnetic field registration [61].

If n^+–n junctions are used as contacts B_1 and B_2, a double injection with a favorable influence upon the MD effect and the sensitivity occurs in the UJMT. In this case the performance of the UJMTs is analogous to that of MDs with two injecting contacts (cf. §2.5.3.2).

Typical UJMT device structures are shown in Fig. 6.11a–c. High resistivity Ge and Si are the most widely used materials. It is worth mentioning that conventional UJTs manifest very high magnetosensitivity, although they are designed for another purpose [62–65].

Unfortunately, there have been almost no investigations of the noise characteristics nor identification of the source of noise in UJMTs. An analysis of the noise in UJTs can be found in [66]. It has been established that at $B = 0$ there is a large increase of the fluctuation amplitude at the critical point C together with a decreased regression rate of the fluctuation characteristics. Satisfactory information concerning the frequency response and the T.C. of the magnetosensitivity is also unavailable. The T.C. of the Russian KT-117 commercial UJT has been investigated at $B = 0$ [67]. The T.C. of the resistance $R_{B1,2}$ was found to be 0.5 to 0.9% °C^{-1}. A strong temperature dependence of the UJMT parameters can be expected at the critical point C. Therefore a very efficient compensation technique is required.

6.5.2. Review of UJMT sensors

Ge was used for the fabrication of the first UJMT devices with an n-type base region with resistivity $\rho \sim 40$–50 $\Omega \cdot \text{cm}$. The device structure is shown in Fig. 6.11a. The magnetic control of the $I_{EB1}(V_{EB1})$ characteristic of such a discrete Ge magnetosensor at $I_{B1,2} = 0.8$ mA is illustrated in Fig. 6.13a [61]. The voltage sensitivity, defined as $\Delta V_p / \Delta B$ when $B \leq 0.2$ T, is 20 mV mT^{-1}. This transducer does not include a special region with a high surface recombination rate. The magnetic control of the $I_{EB1}(V_{EB1})$ characteristic of a KT-117 UJT fabricated from n-type Si with $\rho \approx 200$ $\Omega \cdot \text{cm}$ is illustrated in Fig. 6.13b. The device structure is shown in Fig. 6.11c. The relative sensitivity of the negative resistance region at $I_{EB1} = 4$ mA is $S_R = 5 \times 10^3$ V A^{-1}T^{-1} and almost equals that of Si MDs [62]. Planar UJMTs, Fig. 6.11b, prepared from 20 k$\Omega \cdot$cm p-type Si exhibit an

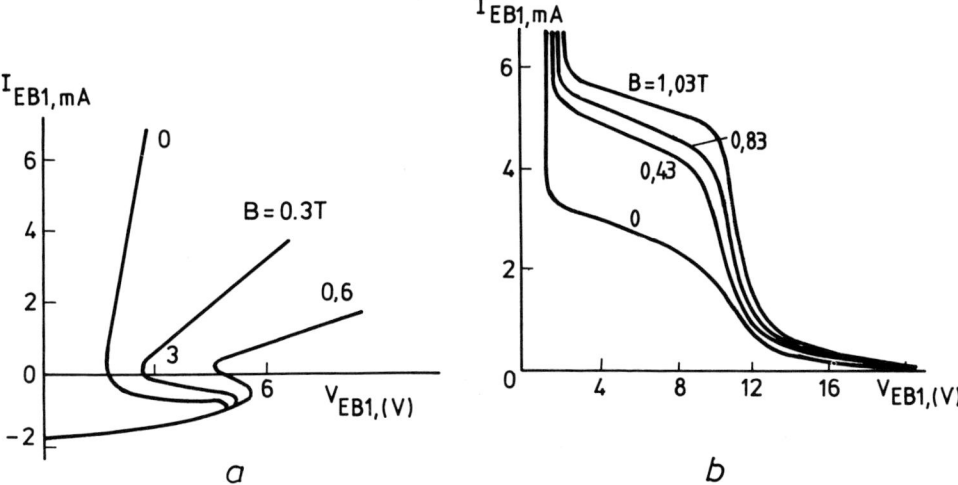

Fig. 6.13. Magnetic control of UJMT $I_{EB1}(V_{EB1})$ characteristics: (a) Ge version from Fig. 6.11a; (b) Commercial Si device from Fig. 6.11c [61,62].

analogous current–voltage curve [68]. At currents $I_{B1,2} = 0.6$ mA, $I_{EB1} = 2$ mA and in a magnetic field $B = 0.3$ T, their magnetosensitivity is $S_V = 60$ V T^{-1}. The lower surface of the sample has been subjected to a special abrasive treatment to achieve $s \to \infty$. The $I_{B1,2}(V_{B1,2})$ characteristic has been used for magnetic-field registration [68]. At $I_{EB1} = 1.5$ mA, $V_{B1,2} = 16$ V and $B = 0.3$ T, the sensitivity is 2.4 mA T^{-1}. The planar UJMT in Fig. 6.11b is fabricated from 20 kΩ·cm p-type Si by means of a planar process [69]. A polar magnetosensitivity similar to that of MDs has been achieved in this device by different treatments of the two surfaces towards which the plasma is deflected by the magnetic field $\boldsymbol{B}$.

A commercially available 2N1671 UJT (Fig. 6.11a) has been successfully used as a high-sensitivity magnetic-field transducer for the registration of a low a.c. magnetic field [70]. The critical point, C, was chosen as a working point of the device. The sensor was connected to a conventional RC relaxation oscillator and was excited by a low-frequency magnetic field $0 \leq B \leq 1200$ nT. This frequency does not differ much from the free-oscillation frequency of the circuit. The amplitude is a linear function of the magnetic induction B. The resultant magnetosensitivity of the circuit is 3 μV nT^{-1}, the field $B_{\min}$ being ≈ 100 nT at $S/N = 1$. This value can be compared with the minimum detectable field in SOS MDs, which is about 10 nT. An important feature of the circuit presented in [70] is the high selectivity.

A possible solution of the problem of inductance implementation in a solid-state form as well as its deliberate modification is suggested in [71]. For that purpose the inductance behavior of UJTs in a magnetic field was studied. It was found that, depending on the injection level, the UJT inductance is increased by a factor of 3 as the magnetic field is varied in the interval $0 \leq B \leq 1.5$ T. Similar to the MD effect, the above result is associated with the magnetic-field-dependent carrier lifetime. Owing to the negative resistance in UJMTs, it is possible to design simple generators of relaxation oscillations whose amplitude and frequency can be controlled by a d.c. magnetic field. Such generators with a sensitivity of 20 Hz mT^{-1} and T.C. $\approx$ 1 kHz °C^{-1} are described in [61,62]. The switching behavior of a relaxation oscillator circuit that employs a 2N2646 UJT in a d.c. magnetic field has been studied and is reported in [65]. Because of the magnetoconcentration effect, the turn-off and the turn-on times vary, and at $B = \pm 1$ T their variation is about 1.5 to 2 times the value at $B = 0$.

Although research into UJMT physics and its application is in its initial stage, UJMTs are a promising type of sensor, as they are capable of presenting information concerning the magnetic field in a digital form and their compatibility with modern microprocessor systems can be easily achieved.

6.6. Carrier-domain magnetometers

The carrier domain (CD), also termed a filament, is, actually, a highly nonequilibrium quasi-neutral electron–hole plasma, generated under certain doping, temperature and biasing conditions in solid-state devices with an adequate structure. The S-type or N-type regions of negative differential resistance in their current–voltage curves is an indication of the existence of the plasma filament. Some electric and galvanomagnetic properties of the three-layered filament structures concerned with S-MDs, BMTs and UJMTs have been discussed in §5.1, §5.5, §6.3 and §6.5. A specific feature of the domain is its strict localization within the bulk of the device. In the ideal case, when there are no structural defects and external fields, the domain is in neutral equilibrium. The absence of an abrupt filament boundary readily allows its migration. Most often displacement of the domain results from an electric or magnetic field. The Lorentz deflection, for instance, is in one and the same direction both for holes and electrons, and hence the filament can move from one of the surfaces of the structure to the other one. If suitable probes are properly positioned to register the migration of the filament, information can be obtained concerning the original cause of the migration, i.e. the field B. The principle of magnetic-field registration by appropriate devices lies at the basis of carrier-domain magnetometers (CDMs). Since the Lorentz force is the reason for their operation, the genetic connection of these sensors with the Hall effect in short samples is obvious.

6.6.1. Three-layer carrier-domain magnetic-field sensors

An orthogonal CD device based on a p-type InSb MD operating at $T = 77$ K has been described in §5.5.1. Two other orthogonal magnetotransistor sensors with a circular geometry and frequency output will be presented here.

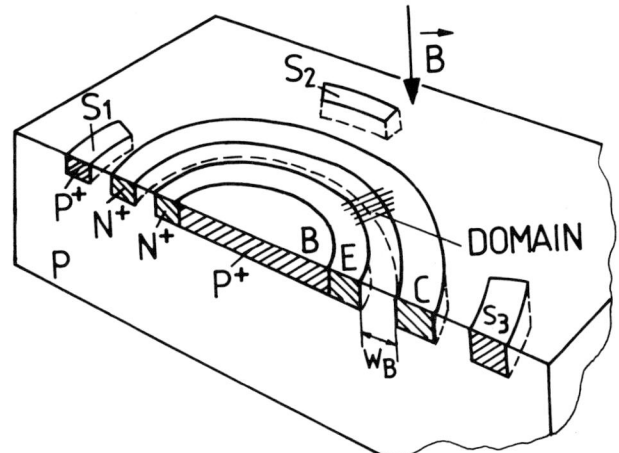

Fig. 6.14. Circular three-layer carrier-domain silicon magnetometer [72].

The device structure of a CDM is presented in Fig. 6.14. This device is a circular lateral bipolar transistor with four voltage probing p^+-contacts S_1, S_2, S_3 and S_4 [72]. The sensor operates in a collector–emitter breakdown regime with short-circuited emitter and base contacts (the internal circle in Fig. 6.14). The typical dimensions are a collector diameter of 500 μm and a width of the base region $w_B = 8$ μm; the doping concentration is $\sim 10^{16}$ cm^{-3} in the base region and $\sim 10^{20}$ cm^{-3} in the emitter and collector regions. The sensor is biased by an external, constant current source and the working point is chosen to be at the sustaining voltage, where the resistance is very low or even negative, i.e. $dV_{CES}/dI_C \sim 0$, where V_{CES} is the collector–emitter voltage, the emitter and base contacts being short circuited. The operation of this device is based on internal feedback. It is involved in the transistor-breakdown mechanism and confines the current domain within the structure to the narrow sector of the base region. The shrinking of the filament in the particular structure considered takes place at 30 mA.

According to [72], mainly intrinsic thermal effects are responsible for the confinement of the domain as well as for its spontaneous rotation. The frequency of the carrier-domain rotation is modulated by an orthogonal magnetic field **B**. The frequency is increased or decreased by the Lorentz force, depending on the relative directions of **B** and the rotation. At $B = 0$, the initial spontaneous frequency is $f_+ = f_- = 280$ kHz. The output signal is registered by the probe contacts S_1, ..., S_4. The magnetosensitivity of this CDM around zero magnetic induction is approximately 250 kHz T^{-1}.

The second CDM can be regarded as a functionally integrated UJT and a BJT (Fig. 6.15) [73]. In general, this CDM is a circular lateral bipolar p^+–n–p^+ transistor with two base contacts, B_1 and B_2. The four collectors Col 1, ..., Col

4 serve to register the filament rotation in time, and the pulses generated are the output signals for the sensor. The typical dimensions of the realized prototype are the emitter diameter, d_E, of 30 μm, $d_{B1} = 60$ μm and $d_{B2} = 360$ μm. The resistivity of the n-type silicon substrate is 7.5 Ω·cm ($n \sim 10^{15}$ cm^{-3}). The circuitry of the circular transistor, i.e., of the CDM, is a common-base B_2 configuration (Fig. 6.15c). The (emitter E–base B_2) and (base B_1–base B_2) circuits operate at constant currents $I_{EB2} = $ const, and $I_{B1,2} = $ const. The polarities of the emitter E and the contact B_1 coincide. The reverse voltages $V_{C1,B2}, \ldots, V_{C4,B2}$ are the output signals of the sensor.

The operation of the CDM in Fig. 6.15 is based on the S-type $I_{EB2}(V_{EB2})$ characteristic at a current $I_{B1,2} > 0$. The formation of a negative-resistance section on the dependence $I_{EB2}(V_{EB2})$ is associated with the variation of the voltage on the contacts. In this case the current in the device's cross-section is nonuniform. Since the emitter E and the contact B_1 have the same polarity, the filament is initially directed downwards; then its direction below the contact B_1 becomes parallel to the chip surface, and finally the direction turns towards the planar contact B_2. By decreasing the injection of the emitter E at a current $I_{B1,2} = $ const, the filament narrows; by increasing this injection, the filament expands. In a similar way, the effective cross-section of the filament can be controlled by the current $I_{B1,2}$ at $I_{EB2} = $ const. For fixed geometrical sensor dimensions and by selecting the respective working point $(I_{EB2}; I_{B1,2})$, the filament cross-section in the region of the collectors Col 1 to Col 4 can be kept to a minimum close to the dimensions of the collectors. The location of the collectors along the CDM radius ensures a better resolution capability for the individual pulses caused by the rotation of the filament. The most probable reason for the spontaneous rotation of the current filament at $B = 0$ is similar to that described in ref. [72]. Despite the fact that the sensor does not employ a breakdown regime, the high current density in the filament

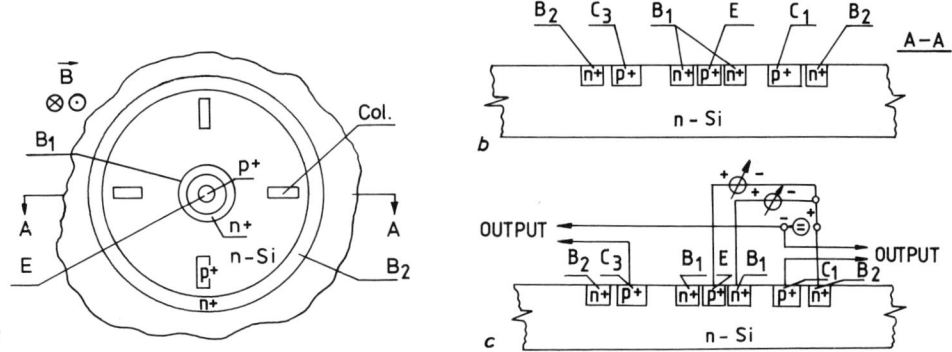

Fig. 6.15. Device structure of a digital silicon CDM: (a) horizontal plane; (b) cross-section; (c) biasing circuitry [73].

leads to a local increase of the temperature which is nonuniformly distributed along the trajectory $E \to B_2$. At the same time, the trajectory itself is asymmetrical with respect to the radial contact B_1. Initially, the velocity of the carriers below the emitter E contains only a vertical component, below contact B_1 only a lateral component, and, thereafter, both lateral and vertical components. The appearance of an incidental fluctuation in the temperature distribution or in the radial electric field, violating the quasi steady-state condition of the filament, is enough for the latter to rotate spontaneously in a clockwise or counterclockwise direction. The switching off and on of one of the collectors might also serve as a turning-on effect [73]. Depending on the relative directions of the field B, the Lorentz force will modulate (enhance or lower) the initial rotation of the filament.

Figure 6.16 shows the overall behavior of the dependence of the rotation frequency f on the external magnetic field B. In some samples, in the absence of a field, as soon as the respective supply currents I_{EB2} and $I_{B1,2}$ are switched on, rotation starts. In other samples the rotation is excited after switching one of the collectors on or off. The direction of the rotation is either clockwise or counterclockwise, determines the sequence of the collector pulses. At the concrete geometrical dimensions of the sensor, the CDM is able to function provided that the working point is in the negative-resistance section of the $I_{EB2}(V_{EB2})$ curve at a total current $I_{tot} = I_{EB2} + I_{B1,2} \geq 20$ mA. The frequency of the spontaneous rotation of the filament in a field $B = 0$ does not depend very much on the operating conditions of the CDM, and in the individual samples is 120 to 150 kHz.

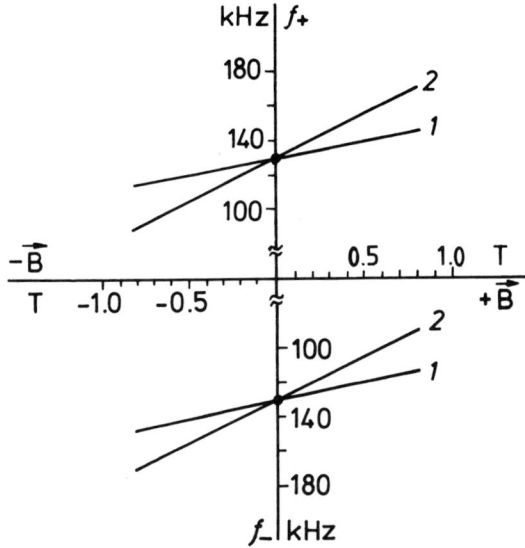

Fig. 6.16. Rotation frequency $f_+(B)$ and $f_-(B)$ of the CDM from Fig. 6.15 as a function of the magnetic field. *1*: $I_{tot} = 25$ mA; *2*: $I_{tot} = 32$ mA; $V_{CB2} = -5$ V, $f(B = 0) = 130$ kHz, $T = 300$ K.

The $f_+(B)$ and $f_-(B)$ dependences are proportional to the induction B. According to the experimental results (Fig. 6.16), by increasing the current I_{tot} the sensitivity tends to increase; at $I_{\text{tot}} = 32$ mA, it is ~ 50 kHz T^{-1}.

In conclusion, we should say that despite the complex character of the processes occurring in the two digital CDMs, the experimental results obtained are very promising.

6.6.2. Four-layer carrier-domain magnetic-field sensors

6.6.2.1. Carrier-domain formation in four-layer structures

The multiterminal four-layer $p-n-i-p-n$ device whose strip geometry is used for the formation of a current filament was first suggested in [74,75]. It is presented schematically in Fig. 6.17. The p^+- and n^+-emitters are biased by a current $I = $ const. The n^-- and the p^--base regions have a sufficiently high distributed resistance. They are separated by a lightly doped i-region. The lateral contacts to the base regions are kept at one and the same voltage V. Because of this voltage source V, the $n-i-p$ diode is reverse biased and the i-region is depleted. The total current $2I$ across these areas consists of two components (injected electrons and holes), and each component equals I. The electron current reaching the n-base is divided into two parts that flow towards the external base contacts, thus generating a nonuniform potential distribution along the x-axis in the n-base. If the p^+-emitter is regarded as an equipotential plane, the hole injection current sharply peaks at the center owing to the exponential injection law by analogy with the filament formation in UJTs (§6.5.1). There is only a moderate lateral widening of the current density distribution in the i-region, and this current, almost unchanged, reaches the p-base. The same scenario is valid for the electron-injection profile. A stable state is established when the two filaments coincide with the central axis of the structure. When the filament is strictly confined to the center, the base resistance acts very much like the resistor R in series with the source V, and a negative resistance from terminal 1 to terminal 2 will similarly result. The lateral majority-carrier currents are split in accordance with the direction of the lateral electric field. The splitting takes place at a point x where $E_x = 0$. At this point the n-base potential is a minimum and the p-base potential a maximum.

When the magnetic field B is applied perpendicularly to the plane of the device, the tightly confined current profiles deflect to the right of the center owing to the Lorentz force. The carriers traverse the i-region used to increase the deflection distance. It depends on the biasing method whether the filament will be entirely shifted to the right or not. If fixed base potentials are maintained at both sides, the position of the filament is stabilized by a "restoring" force [74]. This force is proportional to the displacement of the domain relative to the center, whereas the Lorentz force remains constant. The analysis in [74] leads to the conclusion that the position x_p of the peak-current density of holes arriving at the p-base no

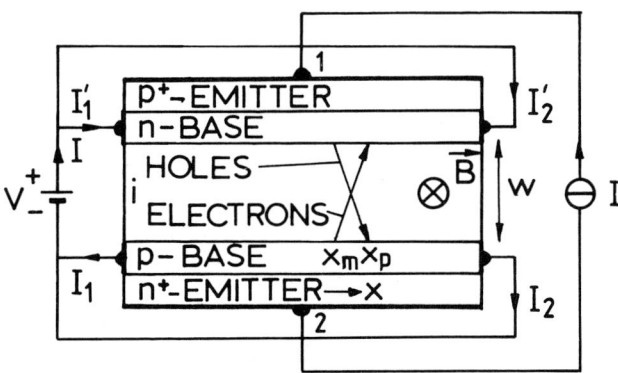

Fig. 6.17. Current-filament forming p–n–i–p–n structure.

longer coincides with the point x_m of the p-base voltage maximum, at which point the current is split. Therefore, the electron-injection center is shifted to the left with respect to x_p (Fig. 6.17). Exactly the same happens in the n-base region, and, therefore, unequal currents $I_1 < I_2$ and $I'_1 < I'_2$ flow across the side contacts of the two bases when an external magnetic field is applied. According to the above model, displacement of the domain in a magnetic field should be linear, i.e. the output signals are linear functions of the induction B. If a sharp current–density profile is assumed, the formula which describes the differential base current and the sensitivity can be determined as follows:

$$S_I = \frac{I_2 - I_1}{IB} \approx \mu_H \left(\frac{w}{l}\right)\left(\frac{2V_{B0}}{3V_T}\right), \tag{6.15}$$

where w and l are the width and length of the i-region, respectively, and V_{B0} is the p-base voltage drop from the center to the edge in the absence of a magnetic field.

Equal Hall mobilities $\mu_{Hn} \approx \mu_{Hp}$ are assumed, as well. The factor $2V_{B0}/3V_T$ is proportional to the current I and is the sensitivity enhancement with respect to a Hall plate with $l \geq 4w$. The value of this factor is more than 10 in the real structures in Fig. 6.17 [74,75].

6.6.2.2. Review of four-layer CD devices

A practical implementation of the lateral analog device in Fig. 6.17 is described in [74,75]. Si was used for the fabrication of the sensor. The resistivity of the base sheet is 10 kΩ/□, the width of the base is 12.5 μm, the i-region is about 50 μm wide and the length is 200 μm. According to equation (6.15) the magnetosensitivity is 22% T^{-1} at a current $I = 10$ μA and voltage $V = 11$ V. The lateral geometry is limited by the small current-device capacity in a magnetic field.

The silicon parallel-field four-layer analog CDM whose cross-section is shown in Fig. 6.18 was suggested in [76,77].

The n–p–n–p structure consists of merged n–p–n and p–n–p transistors that

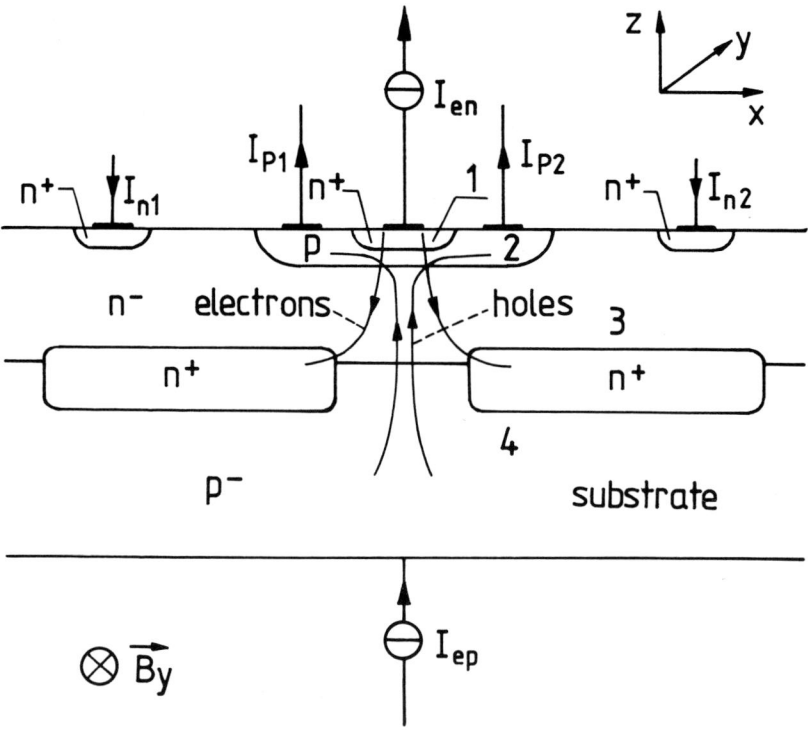

Fig. 6.18. A parallel-field carrier-domain magnetometer. The filament is formed by inverse current crowding in the bases of the merged n–p–n and p–n–p transistors.

share a common base–collector junction. This junction is always reverse biased when the sensor is operating. The two transistors are biased by an emitter current source and operate in the forward active regime. Electrons are injected into the base of the n–p–n transistor (layer 2) and are collected by the p–n–p transistor base (layer 3), which serves at the same time as a collector of the n–p–n transistor. In the same way, holes are injected by the emitter of the p–n–p transistor (layer 4) and collected by the base of the n–p–n transistor (layer 2). Because of the I_x current along the x-axis which generates an $I_x R$ voltage drop in the base regions, the current injection is localized over a very small portion of the base–emitter junction. Two current filaments are formed, the first one consisting of electrons that flow from layer 1 to layer 3, and the second filament consists of holes that flow from layer 4 to layer 2. By analogy with the case considered in §6.6.2.1, the carrier-domains are located along the x-axis. The holes injected by the emitter of the p–n–p transistor provide for a base current which is collected by the base of the n–p–n transistor. In the same way, the electron flow in the n–p–n transistor is the base current of the p–n–p transistor. These base currents flow out of the end terminals of the two base regions. When an external magnetic field $\boldsymbol{B}_y$ is applied, because of the Lorentz force, the two filaments deflect in one and the same direction along the x-axis and, like the sensor described in [74,75], generate a differential output current. If the domain is deflected, for instance, to the right, the two right-hand base currents I_{p2} and I_{n2} increase. This current modulation indicates the filament displacement and serves as a measure for the magnetic induction B [76,77]:

$$\Delta I_T = B(\mu_{Hn}L_n + \mu_{Hp}L_p)I_T\gamma, \tag{6.16}$$

where ΔI_T is the total current difference, I_T is the total sum of the base currents, $\mu_H B$ is the respective Hall angle, and L_n and L_p are the distances over which deflection takes place; the factor γ depends on the width of the domain and its value is in the interval $10 \leq \gamma \leq 100$.

At a device current of 10 mA and equivalent noise level as low as 5×10^{-6} T, the magnetosensitivity of the CDM in Fig. 6.18 is 30% T^{-1}, when $0 \leq f \leq 1$ kHz. The linearity is preserved up to $B = 0.3$ T [77]. The high temperature sensitivity of the parameters of this CDM can be reduced by the integration of a flat metal coil into the sensor. The a.c. current in this coil generates an a.c. magnetic field which is superimposed on the external field B. In this way the output signal of the sensor includes a component which is proportional to the local magnetic field. The additional field is subsequently used to control the gain of the amplifier and maintain a constant transduction efficiency [78].

A circular four-layer silicon CDM is shown in Fig. 6.19. The device has a digital output. The limited filament displacement has been eliminated in this sensor [79,80]. The radial domain formation is analogous to the familiar mechanism of domain formation in the device in Fig. 6.18. If an orthogonal magnetic field B is applied, the filament travels round the circumference of the structure. The frequency of rotation is proportional to the induction B. The actual position of the domain is detected by monitoring the current pulses in the 12 outer segmented collectors, S.C., any time they are traversed by the domain. The frequency response is given by the expression:

$$f_r = \frac{d\mu_p B}{2\pi t_p r}, \tag{6.17}$$

where d is the radial spacing between the emitter region of the n–p–n transistor and the base region of the p–n–p transistor, $\mu_p B$ is the Hall angle, t_p is the charging time of the p–n–p transistor base region, and r is the radius of the outer edge of the n–p–n transistor emitter [79,80].

A linear dependence of the frequency of rotation of the domain, f_r, on the magnetic induction B has been established experimentally. A sensitivity of about 100 kHz T^{-1} has been achieved at a supply voltage of 5 V and a total current of 10 mA. Serious limitations are imposed on the performance of this CDM by the existence of a threshold value of the magnetic field $B_t \sim 0.1$ to 0.4 T below which no domain rotation occurs. This disadvantage is explained by the space

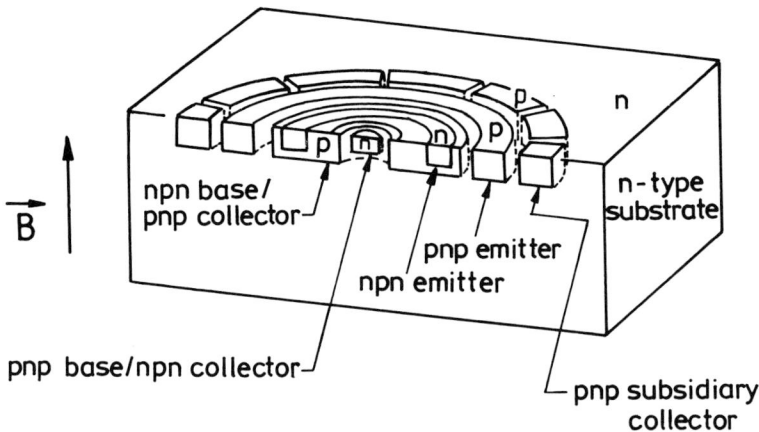

Fig. 6.19. Rotating carrier-domain magnetometer [79,80].

variation of the common-base current gain of the lateral p–n–p transistor [81]. $B_t \approx 100$ mT has been reported [82], and a correlation has been observed [83] between B_t and the misalignment of the emitter diffusion region of the n–p–n transistor with respect to the base diffusion region of the latter. An efficient circuit for the compensation of the strong temperature dependence of the CDM sensitivity is suggested in [84]. With a magnetosensitivity of 57 kHz T^{-1}, the error due to its temperature dependence is about 2% in the interval $-18°C \leq T \leq 102°C$.

Two other thyristor-like devices must be mentioned to complete the review. The first device is a circular n^+–p–n–p structure with a digital output [85]. The sensor is fabricated from 1 Ω·cm n-type Si by means of a planar-diffusion process. The n^+-emitter is in the center of the device and the p-type collector is sectioned into four separate radial symmetric regions at a distance d from the emitter, which serve as output terminals. It has been found that at $T = 300$ K with a working point in the negative resistance region of the $I(V)$ characteristic, an orthogonal field $B \geq B_t$ leads to a filament rotation whose frequency increases linearly with B. The linear velocity of the rotating filament at $B = 1.8$ T is 3500 cm s^{-1}, the value of B_t being 0.9 T.

The second thyristor-like magnetosensitive transducer is shown in Fig. 6.20a. It consists of two functionally integrated magnetothyristors with a common anode and a common n-base [62]. This device is similar to the dual-collector BMT in Fig. 6.8c. The difference between the two device structures is that the thyristor includes two additional n^+-emitters inside the p-type collector regions. There is an internal connection between the two thyristors. If, owing to the orthogonal magnetic field, the injected carriers are deflected from C_2 towards C_1, for instance, at the same time the turn-on voltage of the first thyristor decreases and the turn-on voltage of the second thyristor increases (Fig. 6.20b). The sensor in Fig. 6.20 can be used

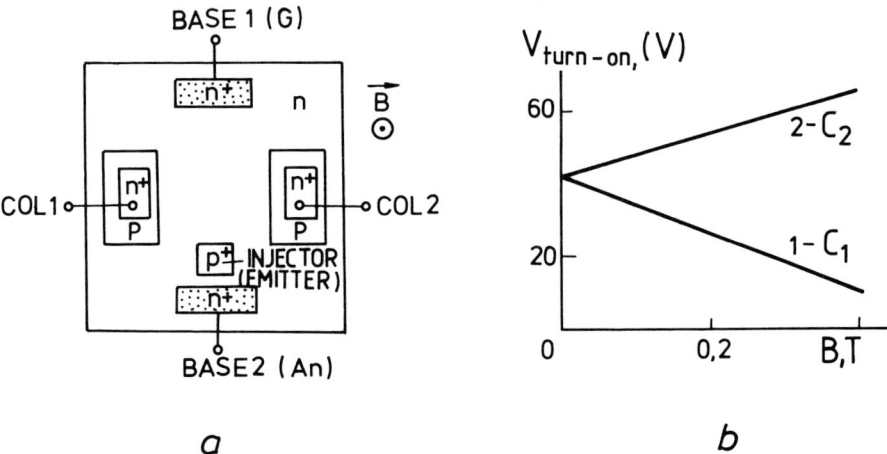

Fig. 6.20. (a) Thyristor-like magnetic-field transducer. (b) Dependence of the collector (cathode) turn-on voltages on the magnetic induction; curve *1*: turn-on voltage of collector C_1, curve *2*: turn-on voltage of collector C_2.

both as a magnetic switch and as an analog differential magnetosensor with load resistors connected to the cathodes (collector regions). In the second case the two transistors operate as amplifiers of the hole current injected by the common anode, thus providing for an additional enhancement of magnetosensitivity.

6.7. Concluding remarks

In summing up the properties of BMT sensors, the question naturally arises: which one of them is the most promising? At present it is impossible to give an unambiguous answer for the following reasons. We do not yet have complete knowledge of the secrets of MT action, or of the proper approaches to its simulation. It is sufficient to point out such "unplanned" properties in the general picture of BMT sensor operation as the filament sensitivity effect and the electrical control of the sign of the magnetosensitivity. All this seriously impedes the obtainment of maximum metrological parameters. On the other hand, the possible applications of BMTs are so wide that they could hardly be satisfied by one single type of MT. For instance, for contactless switches, collectorless d.c. motors, keyboards, tachometers, position sensors, read-out of information stored on magnetic tapes, discs, credit cards, etc., the high value of the output signal, typical of most BMTs, is important; in magnetometry, in electronic compasses, vector measurements, detection of the earth's magnetic field, potential-free current detection, watt meters, cryo-electronics and other applications, the basic requirements are linearity and temperature independence, which are characteristic of deflection-dominated BMT sensors. If amplification of the sensor signal is necessary, the signal-to-noise ratio and the drift of the parameters are critical. Low power consumption is important in smart BMT sensors, thus making CMOS technology indispensable. CDMs are the most suitable option when a direct conversion of magnetic information into digital form is required.

The future trends in the development of BMT and CDM sensors are electrical control and the general improvement of the performance, especially studies of noise and of frequency response. Flexible numerical models should be developed, covering not just one type of sensors, but the entire variety of device structures and boundary conditions. In this respect the results in [86,87] can be regarded as an encouraging start. The finite-element technique has been applied in the analysis of magnetotransistor operation. It has been found that carrier magnetoconcentration at the insulating boundary where the emitter–base junction is terminated may be the mechanism responsible for magnetosensitivity [86]. The first attempt at a 2-D numerical modelling of the complete device structure of a Si dual-collector magnetotransistor in its full extension was presented in [87–89]. The results of this simulation indicate emitter-efficiency modulation apart from Lorentz deflection.

The functional connection between orthogonal or parallel-field Hall sensors and BMTs is not only a convenient phenomenological approach, it is a manifestation of the common physical genesis of the two groups of magnetosensitive devices. Owing to the fundamental significance of the Lorentz deflection, the sensor action in both cases results from the Hall effect under monopolar or bipolar conductivity conditions. It is based on a wide variety of transducing mechanisms. Besides Si, the introduction of GaAs is to be anticipated as an alternative material for integrated sensors, in combination with further sophistication of the technology.

References

[1] C. Bradner Brown, Electronics, 23(7) (1950) 81–83.
[2] S. Middelhoek and D.J.W. Noorlag, Sensors and Actuators, 2 (1982) 211–228.
[3] H.P. Baltes and R.S. Popović, Proc. IEEE, 74 (1986) 1107–1132.
[4] S. Kordić, Sensors and Actuators, 10 (1986) 347–378; Offset reduction and Three Dimensional Field Sensing with Magnetotransistors, Ph.D. dissertation, Delft University of Technology, The Netherlands, 1987.
[5] S. Cristoloveanu, J. Korean Inst. Elect. Eng., 2 (1986) 86–95
[6] A.W. Vinal and N.A. Masnari, IEEE Electron Device Lett., EDL-3 (1982) 203–205.
[7] V. Zieren, S. Kordić and S. Middelhoek, IEEE Electron Device Lett., EDL-3 (1982) 394–395.
[8 A.W. Vinal and N.A. Masnari, IEEE Electron Device Lett., EDL-3 (1982) 396–397.
[9] A.G. Andreou, The Hall Effect and Related Phenomena in Microelectronic Devices, Ph.D. dissertation, Johns Hopkins University, Baltimore, Maryland, USA, 1986.
[10] Ch.S. Roumenin, Sensors and Actuators, A, 24 (1990) 83–105; A, 30 (1992) 77–87.
[11] D. Redfield, IEEE Trans. Electron Devices, ED-27 (1980) 766–771.
[12] A. Fortini, A. Hairie and M. Gomina, IEEE Trans. Electron Devices, ED-29 (1982) 1604–1610.
[13] S.M. Sze, Physics of Semiconductor Devices, John Wiley and Sons, New York, 1981.
[14] W.S. Johnson, Solid-State Electron., 13 (1970) 951–956.
[15] W. Shockley, Electrons and Holes in Semiconductors, Van Nostrand, Princeton, NJ, 1950.

[16] Ch.S. Roumenin and P.T. Kostov, Sensors and Actuators, 6 (1984) 19–33.
[17] Ch.S. Roumenin, C.R. Acad. Bulg. Sci., 38 (1985) 1501–1504.
[18] Ch.S. Roumenin, Sensors and Actuators, 14 (1988) 177–190.
[19] A.W. Vinal and N.A. Masnari, IEEE Trans. Electron Devices, ED-31 (1984) 1486–1494.
[20] Seung-Ki Lee, Kwang-Hoon Oh et al., IEEE Trans. Magn., MAG-28 (5) (1992) 2193–2195.
[21] J. B. Flynn, J. Appl. Phys., 41 (1970) 2750–2751.
[22] S. Takamiya and K. Fujikawa, IEEE Trans. Electr. Devices, ED-19 (1972) 1085–1090.
[23] R.M. Huang, F.S. Yeh and R.S. Huang, IEEE Trans. Electron Devices, ED-31 (1984) 1001–1004.
[24] I.M. Vikulin, M.A. Glauberman, L.F. Vikulina and Yu. Zaporozhchenko, Sov. Phys.-Semicond., 8 (1974) 369–370.
[25] A.W. Vinal, IBM J. Res. Develop., 25 (3) (1981) 196–201; Magnetically sensitive transistor utilizing Lorentz field potential modulation of carrier injection, U.S. Patent 4 654 684 (Mar. 31, 1987).
[26] L.W. Davies and M.S. Wells, Proc. IREE Aust., (June) (1971) 235–238.
[27] L. Halbo and J. Haraldsen, Congr. Exposition Soc. Automative Eng., Detroit, MI, USA, 1980
[28] I.M. Vikulin, N.A. Kanisheva and M.A. Glauberman, Sov. Phys.-Semicond., 11 (1977) 340.
[29] G.I. Rekalova, D.M. Kozlov and T.V. Persiyanov, IEEE Trans. Magn. MAG-17 (1981) 3373–3375.
[30] Ch.S. Roumenin, Sov. Phys.-Semicond., 22 (1988) 1040–1042.
[31] Ch.S. Roumenin and N.D. Smirnov, Magnetosensitive element, Bulg. Patent 35 162 (June 19, 1981).
[32] E.C. Hudson, Semiconductive magnetic transducer. US Patent 3 389 230 (June 18, 1968).
[33] A. Nathan, K. Maenaka, W. Allegretto, H.P. Baltes and T. Nakamura, IEEE Trans. Elecron Devices, ED-36 (1989) 108–117.
[34] V. Zieran and B.P.M. Duyndam, IEEE Trans. Electr. Devices, ED-29 (1982) 83–90.
[35] K. Maenaka, T. Ohsakama, M. Ishida and T. Nakamura, Proc. 6th Sensor Symp., Japan, 1986, 47–50; Sensors and Actuators, 16 (1989) 101–108.
[36] A. Nathan and H.P. Baltes, Sensors and Actuators, A, 21–23 (1990) 758–761; Sensors and Actuators, A, 21–23 (1990) 780–785.
[37] R. Castagnetti, H. Baltes and A. Nathan, Sensors and Actuators, A, 25–27 (1991) 363–367.
[38] Ch.S. Roumenin, C.R. Acad Bulg. Sci., 38 (1985) 579–581.
[39] N.D. Smirnov, Ch.S. Roumenin and I.G. Stoev, Sensors and Actuators, 2 (1981/82) 187–193; Ch.S. Roumenin, N.D. Smirnov, V.S. Lysenko and R.N. Litovskii, Rev. Phys. Appl., 18 (1983) 87–92.
[40] Ch.S. Roumenin, C.R. Acad Bulg. Sci., 41(11) (1988) 49–52.
[41] R.S. Popović and R. Widmer, IEEE Trans. Electron Devices., ED-33 (1986) 1334–1340.
[42] Ch.S. Roumenin and N.P. Georgiev, C.R. Acad. Bulg. Sci., 42(6) (1989) 59–62.
[43] I.M. Mitnikova, T.V. Persiyanov, G.I. Rekalova and G. Shtyubner, Sov. Phys.-Semicond., 12 (1978) 26–28.
[44] A.G. Andreou and C.R. Westgate, Electron. Lett., 20 (1984) 669–701.
[45] R. S. Popović, H.P. Baltes and F. Rudolf, IEEE Trans. Electron Devices, ED-31 (1984) 286–291.
[46] R.S. Popović and H.P. Baltes, Sensors and Actuators, 4 (1983) 155–163; IEEE Electron Device Lett., EDL-4 (3) (1983) 51–53.
[47] Ch.S. Roumenin and N.D. Smirnov, Magnetosensitive sensor, Bulg. Patent 34 017 (May 5, 1982).
[48] Ch.S. Roumenin and P.T. Kostov, Planar Hall effect devices, Bulg. Patent 37 208 (Dec. 26, 1983); C.R. Acad. Bulg. Sci., 38 (1985) 1145–1148.
[49] L. Ristić, T. Smy and H.P. Baltes, IEEE Trans. Electr. Devices, ED-36 (1989) 1076–1086; Sensors and Materials, 2 (1988) 83–92.
[50] Ch.S. Roumenin and M.M. Atanasov, C.R. Acad. Bulg. Sci., 39 (8) (1986) 45–48; Ch. Roumenin, University Annual (Techn. Phys.) 25(1) (1988) 235–246 (in Bulgarian).
[51] D.M. Kozlov, G.I. Rakalova, I.M. Mitnikova, D.A. Tairova and A.A. Shakhov, Sov. Phys.-Semicond., 13 (1979) 1263–1265.
[52] I.M. Vikulin, M.A. Glauberman, G.A. Egiazaryan, N.A. Kanisheva, Yu. S. Manvelyan and I.P. Shnaider, Sov. Phys.-Semicond., 15 (1981) 274–275.
[53] Ch.S. Roumenin, Sov. Phys.-Semicond., 20 (1986) 887–888.
[54] J.L. Lopez and J.C. Licini, J. Appl. Phys., 53 (1982) 8389–8391.
[55] A. Chovet, Ch. S. Roumenin, G. Dimopoulos and N. Mathieu, Sensors and Actuators, A, 21–23

(1990) 790–794
[56] V. Zieran, Sensors and Actuators, 5 (1984) 199–206.
[57] S. Kordić, V. Zieren and S. Middelhoek, Sensors and Actuators, 4 (1983) 55–61.
[58] I. Karakushan, V.I. Stafeev, and A.P. Schtager, Radio Eng. Electron. Phys., 6 (1964) 846.
[59] I.M. Vikulin, A. Zaporozhchenko, M.A. Glauberman and L.F. Vikulina, Radio Eng. Electron. Phys., 17 (1972) 1347.
[60] S.L. Agrawal and R. Swami, J. Phys. D., Appl. Phys., 14 (1981) 283–291.
[61] V.I. Stafeev and E.I. Karakushan, Magnitodiody, Nauka, Moskwa, 1975 (in Russian).
[62] I.M. Vikulin, L.F. Vikulina and V.I Stafeev, Galvanomagnitniye pribory, Radio i Sviaz, Moskwa, 1983 (in Russian).
[63] J. Brini, C.R. Acad. Sci., Paris, B290 (1980) 5–8.
[64] S.L. Agrawal and R. Swami, Sensors and Actuators, 11 (1987) 157–172.
[65] S.L. Agrawal and R. Swami, Paramana, J. Phys., 27 (1986) 459–468.
[66] D. Berlan, Etude des fluctuations au voisinage d'une transition électrique (bruit du transistor unijunction), Ph.D. dissertation, I.N.P., Grenoble, France, 1977.
[67] I.G. Nedolujko and E.F. Sergienko, Odnoprehodniye transistory, Energiya, Moskwa, 1974 (in Russian).
[68] E.I. Karakushan, A.R. Mankulov, E.S. Samsonov and V.I. Stafeev, Sov. Phys.-Semicond., 12 (1978) 35–38.
[69] L.S. Gasanov, Sov. Phys.-Semicond., 12 (1978) 587–588
[70] J. Brini and G. Kamarinos, Sensors and Actuators, 2 (1981/82) 149–154.
[71] S.L. Agrawal and R. Swami, Phys. Stat. Sol., (a) 94 (1986) k159–k161.
[72] R.S. Popović and H.P. Baltes, Sensors and Actuators, 4 (1983) 229–236.
[73] Ch.S. Roumenin and V. Dorovska, Sensors and Actuators, A, 32 (1992) 661–664.
[74] G. Perski and D.J. Bartelink, Bell Syst. Techn. J., 53(3) (1974) 467–502.
[75] D. J. Bartelink and G. Perski, Appl. Phys. Lett., 25 (1974) 590–592.
[76] J.I. Goicolea, R.S. Muller and J.E. Smith, IEEE Solid-State Device Res. Conf., Santa Barbara, CA, USA, June 1981; Abstr. in IEEE Trans. Electron Dev., ED-28 (1981) 1252.
[77] J. I. Goicolea, R.S. Muller and J.E. Smith, Sensors and Actuators, 5 (1984) 147–167.
[78] J.I. Goicolea, R.S. Muller and J.E. Smith, 3rd Int. Conf. on Solid-State Sensors and Actuators, Philadelphia, PA, USA, 1985, Dig. Techn. Papers, 300–303.
[79] B. Gilbert, Electron. Lett., 12 (1976) 608–610.
[80] M.H. Manley and G.G. Bloodworth, Electron. Lett., 12 (1976) 610–611.
[81] M.H. Manley and G.G. Bloodworth., Solid-State and Electron Dev., 2 (1978) 176–184.
[82] M.H. Manley and G.G. Bloodworth., Radio Electron. Eng., 53 (1983) 125–132.
[83] D.R.S. Lucas and A. Brunnschweiller, Sensors and Actuators, 4 (1983) 33–43.
[84] S. Kirby, Sensors and Actuators, 3 (1983) 337–384; 4 (1983) 25–32.
[85] M.E. Alekseev, K.F. Komarovskih, V.I. Ribalchenko, V.P. Sondaevsky and G.I. Fursin, Microelectronika, 3(4) (1974) 360–362.
[86] M.G. Guvenc, IEEE Trans. Electron Devices, ED-35 (1988) 1851–1860.
[87] C. Riccobene, G. Wachutka, J. Burgler and H. Baltes, Sensors and Actuators, A, 31 (1992) 210–214.
[88] A. Nathan, H. Baltes and W. Allegretto, IEEE Trans. Comp.-Aided Design, 9 (1990) 1198–1208.
[89] W. Allegretto, A. Nathan and H. Baltes, IEEE Trans. Comp.-Aided Design, 10 (1991) 501–511.

Chapter 7

SQUID sensors

Superconductivity Quantum Interference Devices (SQUIDs) are the most sensitive magnetic-field transducers of all known magnetosensors, including those considered in Chapters 4–6. Actually a SQUID is a converter of the external magnetic flux Φ_{ext}; a voltage V is generated at the output of the SQUID which is a periodic function of Φ_{ext} with a period proportional to the quantum flux $\Phi_0 = h/2q \approx 2.07 \times 10^{-15}$ Wb (§2.7). These sensors have wide functional capabilities and allow the registration of practically all physical quantities that can be unambiguously transformed into a magnetic field; magnetic induction and its gradient, electric current and voltage, magnetic permeability, and displacement are examples. Therefore SQUIDs find an extremely wide variety of applications that include the registration of the extremely low magnetic fields generated by the human brain and heart, as well as in unique instruments for the detection of gravitational waves or spin noise. This chapter is dedicated to the analysis of the SQUID system and its components, the different SQUID structures, their characteristics and other related problems.

7.1. The SQUID system

The complete SQUID system includes the sensors (DC and RF versions), the signal-input coupling, read-out circuits and the sensor periphery.

7.1.1. Josephson junctions and their characteristics

The quality of the Josephson junctions is of great significance for the performance of SQUIDs. Detailed descriptions of different types of Josephson contacts, either those used nowadays or the promising candidates for future applications, can be found, for instance, in [1–4]. The tunnel junction, Fig. 7.1a, is the earliest Josephson contact known. It is a superconductor–insulator–superconductor (*S–I–S*) sandwich in which the electrical coupling of the superconducting electrodes (*S*) is accomplished by electron-pair tunneling across a very thin (2–5 nm) insulator film (*I*). Thin (0.1 to 1.0 μm) superconducting films are usually used as electrodes and the dielectric is an oxide layer grown upon the surface of the substrate. It must be pointed out that the relationship $J_s(\varphi)$ (2.138), for tunnel junctions, is practically in all cases a sine-wave function. Besides, the junction area is several μm^2 and this allows modern lithography to be used for its fabrication. The current–voltage characteristics of the tunnel contacts are shown in Fig. 2.31. The normal current J_N of electrons which are not coupled in Cooper pairs is small at voltages lower than the so-called threshold voltage $V_g \sim 2\Delta(T)/q$, where V_g is 2 to 3 mV. The relatively high value of the junction capacitance C and the relatively low stability of tunnel oxide layers under repeated temperature cycles of the transition from normal to the superconducting state and back to normal conductivity, ($N \Longleftrightarrow S$), are the main disadvantages of tunnel junctions. In spite of these shortcomings they are among the most widely used Josephson contacts.

The capacitance, C, of the contact can be reduced by the use of a normal metal. The structures of this type are usually denoted as S–N–S and are illustrated by Fig. 7.1b. Unlike tunnel junctions, the S–N–S sandwiches can retain a relatively high critical current when the N-thickness is $\approx$100 nm. The superconductivity current I_s decreases exponentially as the thickness is increased. This is due to the penetration of Cooper pairs into the normal metal to a depth of the same order of magnitude as that of the coherence length ξ_{0N}. In pure metals ξ_{0N} is several times the size, ξ_0, of Cooper pairs together (2.130), whereas in metals containing impurities, where the mean free path of electrons is $l_0 \ll \xi_0$, the parameter ξ_0 has another value: $\xi_{0N} \sim (\xi_0 l)^{1/2}$. The S–N–S sandwiches manifest a high $N \Longleftrightarrow S$ stability. Their main disadvantage is their low resistance in the normal state ($R_N \leq 1\ \Omega$), even if layers of high resistivity materials, degenerate semiconductors with contact area of several μm^2 for instance, are used. The S–N–S system sets severe requirements for the fabrication technology.

The low-resistance Josephson-junction sandwich structures can be avoided by bridge-type configurations of variable thickness, Fig. 7.1c. In this case the superconducting thin-film electrodes are connected by a narrow (about 1-μm wide) link (bridge) of a normal metal or a superconductor, with thickness 10 to 100 nm. Owing to the very small cross-section its resistance R_N can reach values up to 10 Ω. The principal disadvantage of these structures is that a very small separation (30 to 100 nm) between the superconducting electrodes, i.e. the microbridge length, must be guaranteed by the fabrication process. The high reproducibility needed by S–N–S sandwiches requires a more sophisticated technology. This is why microbridges are usually applied in RF SQUIDs. In Josephson structures with a small intrinsic capacitance C (S–N–S sandwiches and microbridges) in a resistive state ($V \neq 0$) the time variation of the phase φ may be highly nonlinear. This is why the superconducting current averaged over time does not equal zero and makes a significant contribution to the current–voltage characteristic (Fig. 2.30.)

The Josephson contact in Fig. 7.1d is similar to that in Fig. 7.1a but the parameters of superconductivity are altered by the addition of a thin ($\approx 10^3$ nm)

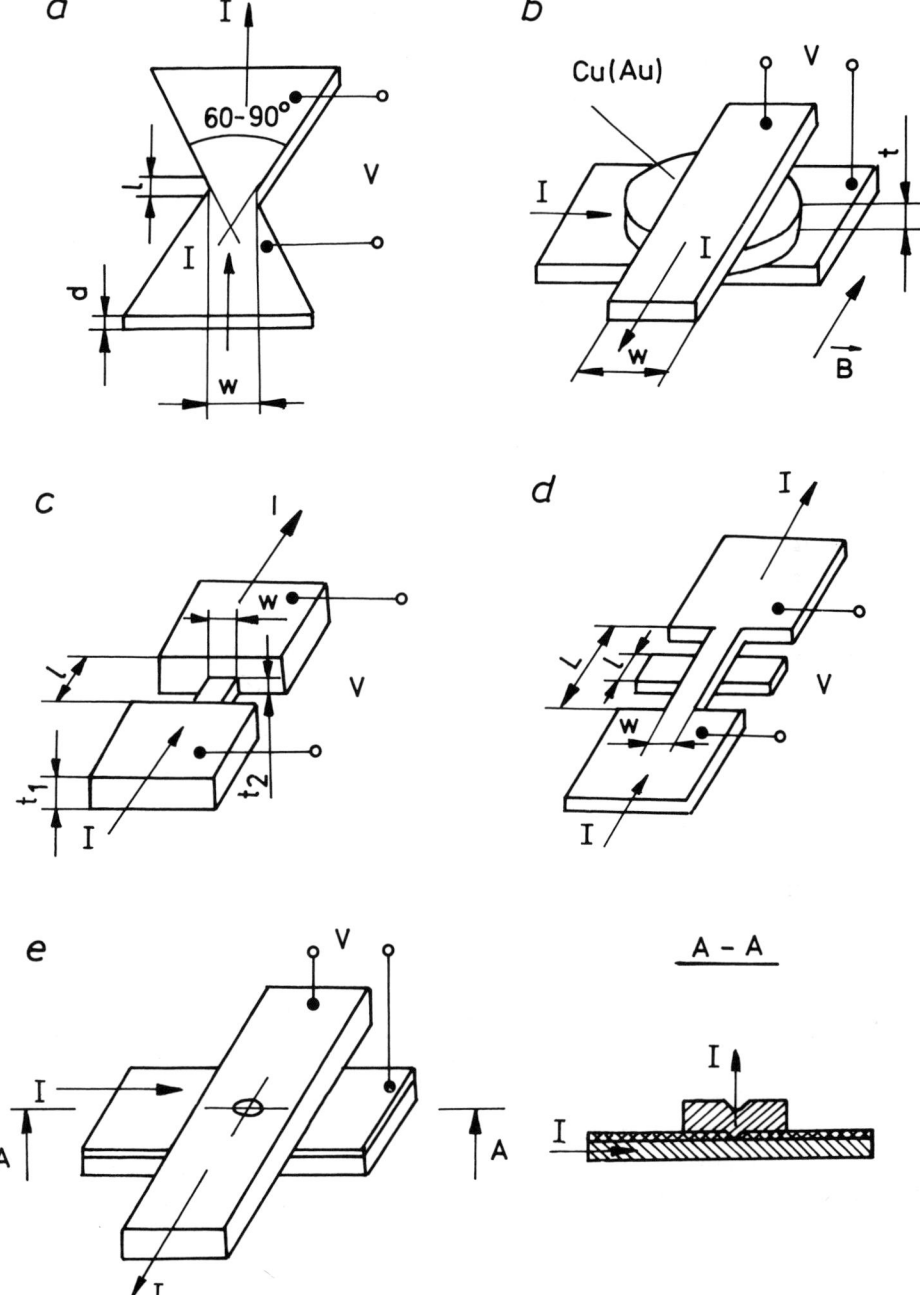

Fig. 7.1. Various superconducting bridge contacts applied in SQUIDs.

film of normal metal deposited by sputtering. Superconductivity is attenuated in the narrow link (bridge) and its becomes possible to use a relatively large microbridge with a length and width more than 10^3 nm. The resistance R_N in this system is comparable with that of the device in Fig. 7.1a.

Microbridges formed by the burn-in technique, filling the opening in the insulator layer with a metal, are illustrated in Fig. 7.1e. The main advantage of such junctions is the simple technology involved in their fabrication. The resistance achieved is $R_N \leq 0.1\ \Omega$.

Mechanical point contacts are another version of Josephson junctions. A superconducting needle is pressed to the contact area upon the plane of another superconductor. One and the same metal is often used both for the plane and for the needle. It can be concluded from the temperature variation of the contact resistance $R(T)$ and from the $V(B)$ dependence that a weak link is formed at several points inside the contact area, i.e. the connection of several tunnel junctions in parallel is possible. The advantage of this system is the small value (several μA) of the critical current I_0 which can be obtained by adjusting the mechanical pressing force. This is commonly achieved by means of a screw. The principal difficulties, however, arise from the high accuracy required when adjusting the pressure exerted upon the plane by the needle. Other shortcomings of such contacts, when used in cryostats, arise from their problematic stability with respect to vibrations, to periodic heating and cooling as well as to long-term storage of the samples at room temperature [1].

The techniques employed in semiconductor-device technology, such as lithography, selective etching and thin-film deposition, can be used for the fabrication of modern Josephson tunnel junctions. The methods for thin-film deposition in the case of superconducting materials include electron-beam evaporation, plasma sputtering, ion-beam sputtering, molecular-beam epitaxy, etc. [5,6]. Silicon is often used as a substrate.

Josephson junctions and SQUID sensors operating at $T \geq 77$ K [2,8,9] have been implemented on the basis of high-T_c superconductivity ceramics of the $YBa_2Cu_3O_{7-\delta}$(YBaCuO) type, for instance. A layer of Au is deposited upon the YBaCuO substrate during fabrication of the contacts. The regions which have not been covered by Au are subjected to ion implantation. They behave as insulating regions at low temperatures. Microbridges formed on the boundaries of the randomly oriented YBaCuO crystal grains play the role of Josephson junctions. The polycrystalline high-T_c material can be regarded as consisting of many superconducting grains connected by weak links across the grain boundaries. The technology of high-T_c Josephson-junction fabrication is in a state of rapid development. Detailed information on the lithographic and thin-film deposition techniques employed by modern Josephson tunnel junction technology can be found, for instance, in [2,9].

The main geometrical dimensions and the extremely high magnetosensitivity of Josephson junctions impose special requirements on the experimental techniques and operating conditions. The weak links can be destroyed if a high-density current happens to be forced through the device. The mounting of SQUIDs into Dewar designs similar to MOS IC packaging must be done by gloved hands or by employing special appliances for the removal of static charge. Josephson junctions must necessarily be placed on a grounded metal plate. The electrical characteristics of the weak links are measured only after the sensor has been cooled down to the liquid helium temperature or, for high-T_c Josephson junctions, the liquid nitrogen temperature. At such temperatures the stability of the junction to current loading increases owing to the superconductivity and to the high thermal conductivity of liquid and gaseous helium. Special care is taken to avoid the use of materials with magnetic impurities, and the presence of magnetic dust or particles in the neighborhood of the sensor. Before the cooling of Josephson junctions, of their planar versions in particular, the atmospheric air should be removed from the Dewar; it is then filled with gaseous helium and finally the liquid helium is poured in. If this sequence of operations is not observed, residual gases and water vapour may freeze upon the junction itself. Removal of the sensor from the Dewar can only be carried out after the liquid helium has been completely evaporated. Then the sample temperature is gradually increased in gaseous helium until room temperature is reached. Direct contact of the liquid with the weak links and sharp drops in their temperature should be avoided when the helium is poured into the Dewar [1].

7.1.2. Signal–input coupling

SQUID sensors are commonly encapsulated in a small cylindrical package, containing the SQUID itself mounted together with an input coupling coil and either a RF tank circuit (for an RF SQUID) or a matching transformer (for an DC SQUID). The external magnetic field is detected by means of a pickup coil connected to the input coil via two superconducting screw contacts on the package of the sensor device. The flux transformer formed in this way consists of two superconducting coils. One of them is used for picking up the signal and the other is used for coupling the signal to the SQUID loop via the mutual inductance M between the input coil inductance and the SQUID loop inductance (Fig. 7.2). As the whole flux-transformer configuration is superconducting (including the screw connections), the principle of flux quantization also applies to it (§2.7.2). A magnetic flux caught by the large pickup loop drives a circulating current which, in turn, produces a flux in the input coil coupling it to the SQUID loop. Quantum signal transfer is achieved if the inductances of the pickup coil and the input coil are matched (i.e., equal) [2]. SQUID inductances are usually small in order to obtain an optimum signal-to-noise ratio, but then it is difficult to couple the magnetic flux into the SQUID ring efficiently without a flux transformer. Another advantage of using a flux transformer is that with the help of damping filters it is possible to reduce the band width of the SQUID input according to experimental

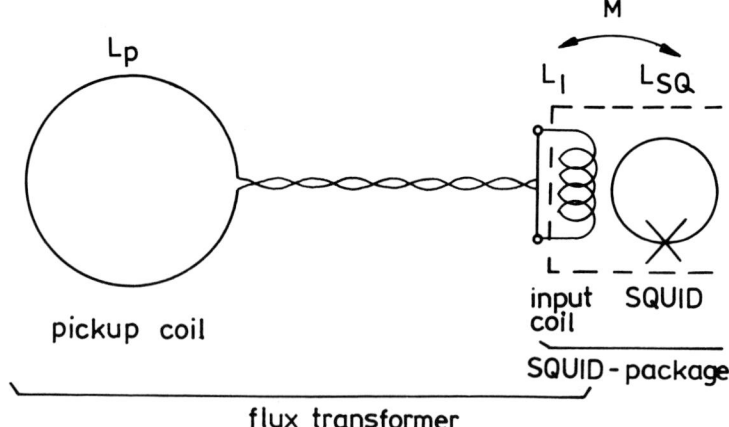

Fig.7.2. Flux transformer coupled to a SQUID loop.

requirements. High-frequency noise may easy deteriorate the signal-to-noise ratio owing to down-mixing in the nonlinear characteristics of the Josephson junctions.

Furthermore, the configuration of the pickup loop may be individually tailored to extract the optimum signal level from the individual type of signal source investigated (size, shape of samples, etc.). Very frequently, pickup loops are arranged to detect the gradient of the magnetic field. These are coil configurations with two or more connected coils, having opposite windings, arranged at certain distances ("baseline") from each other [2]. A pickup loop with a special shape is used for the investigation of the gradient of the magnetic field (§7.7).

7.1.3. Read-out schemes

It is possible to electronically read-out both the DC SQUID and the RF SQUID by taking advantage of the periodic $V(\Phi)$ responses. The dynamic range would be small if a direct read-out were employed. Counting the flux periods would not be accurate enough for most applications. Better resolution and a large dynamic range can be obtained with a linearization scheme termed the "flux-locked loop" [2,10]. In this scheme, an a.c. flux (usually at 100 kHz) is applied to the SQUID, in addition to the input signal to be detected. Modulation techniques such as this one are often used in dedicated measurements to detect small changes in noisy low-frequency signals (lock-in detection).

Figure 7.3 shows the resulting output-voltage waveform for three different choices of DC flux bias. If the modulated flux is DC biased at the minimum (or maximum) of the "triangle" pattern, a frequency-doubled voltage develops (Fig. 7.3a). If the DC component of the flux deviates from this working point, a voltage component with the fundamental frequency develops which is in-phase or out-of-

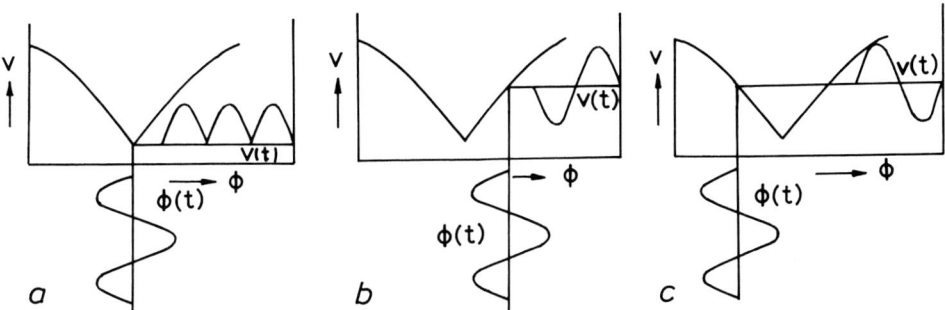

Fig. 7.3. The three basic output-voltage waveforms obtained for the respective DC bias values of the modulating flux Φ [2].

phase with the modulating flux, depending on the sign of the flux change (Fig. 7.3b and c) [2].

A circuit exploiting this property may be used for both types of SQUID, as indicated in the block diagram in Fig. 7.4 (Fig. 2.37a). The circuit employs an oscillator with two purposes: (a) via a modulation coil, it modulates the flux threading the SQUID loop (in the case of an RF SQUID the tank circuit inductance may be used as a modulation coil); and (b) it provides the reference for the phase-sensitive detector.

The latter detects and amplifies the signal output from the modulation amplifier. A high-gain integrating amplifier connected to the phase-sensitive detector provides the DC feedback to the SQUID circuit and thus a linearized voltage drops across the feedback resistor R. This voltage is a direct measure of the magnetic flux applied to the SQUID [2].

Instead of feeding back the modulation signal to the SQUID loop itself, it may be advantageous to couple the signal to a section of the flux-mode transformer circuit in order to null the current. This mode of operation reduces cross-talk problems in multi-channel SQUID applications. In addition to this most popular read-out scheme, several others have been proposed: e.g., (a) a relaxation oscillator read-out suitable for hysteresis SQUIDs., (b) a digital read-out consisting of a comparator, an up-down counter and a digital-to-analog converter in a feedback loop. This concept appears to be very promising for future "on-chip" integration to provide a "smart sensor". An alternative design of a digital read-out requires only two leads per SQUID — a considerable advantage in multisensor systems with their need for minimal heat loss per channel [2,11,12].

7.1.4. SQUID sensor periphery

The successful operation of SQUIDs depends, to a great extent, on their peripheral apparatus. The sensor is necessarily cooled down to T_c. Low temperature

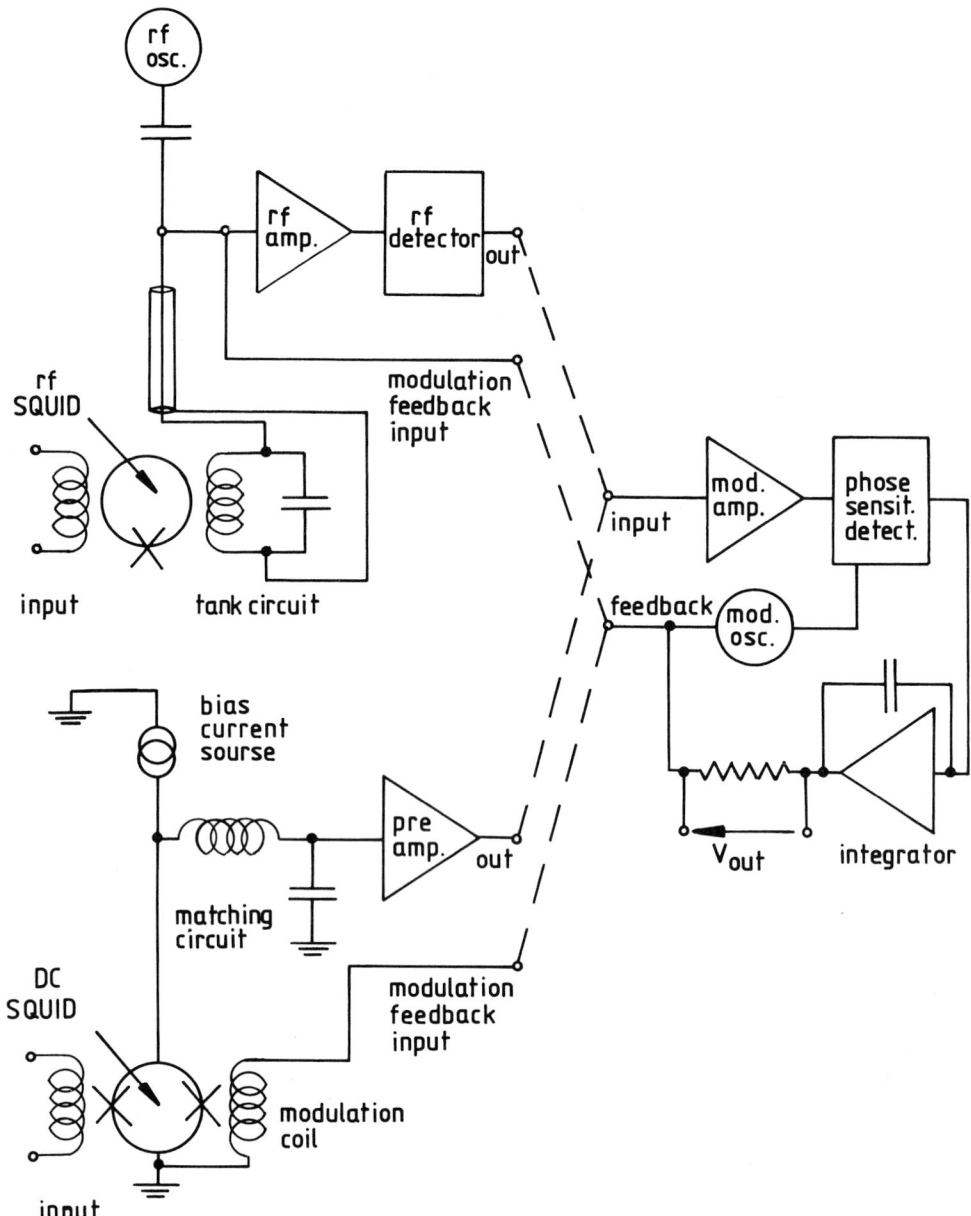

Fig. 7.4. Flux-locked loop read-out scheme [2].

SQUIDs are immersed in liquid helium ($T = 4.2$ K). Liquid nitrogen is used for high-T_c SQUIDs, $T = 77$ K. Special Dewar vessels are designed for the purpose. The mechanical stability of the fixing of the SQUID in the cryostat acquires particular significance because of the extremely high magnetosensitivity. For instance, in the geomagnetic field, whose magnitude is $\approx 0.5 \times 10^{-4}$ T, the SQUID oscillations change the magnetic flux by $\delta \Phi = A_0 \, B \sin \alpha \approx A_0 B \alpha$, where A_0 is the loop area. In the simplest case of a homogeneous field $\delta B = \delta \Phi / A_0 \approx B \alpha$, a variation of 1 angular second corresponds to $\delta B = 0.25 \times 10^{-9}$ T. From the fact that the SQUID sensitivity is in the range 10^{-14} to 10^{-12} T it follows that there must be practically no vibrations. With such an unique level of transducing efficiency, the sensor's orientation and stability in space acquire considerable importance. Depending on the particular SQUID application, the device's active axis (the normal to the plane of the superconducting ring) must be adequately oriented relative to the vertical of the plane or to the total vector of the geomagnetic field with an accuracy of 1 angular minute or even 1 second [1].

The low temperature needed for the stable operation of the SQUID is provided for by a coolant (liquid helium or nitrogen) in a cryostat. It is kept there as long as possible (until it evaporates completely). The specific character of the highly sensitive registration of the components of the magnetic vector demands that the cryostat itself does not perturb the magnetic field and does not include magnetic components. In the Dewar there should be no electric currents due to temperature gradients or eddy currents resulting from motion of the Dewar in a magnetic field. Because of these limitations, nonmetal designs of glass and fiber plastics should be used. In some particular operational case which do not involve moving the Dewar, metal cryostats made of materials with a low magnetic permeability (Cu, Ti) or combinations of such metals and insulating composites can be used. The cryostat must have such a shape that the pickup coils of the SQUID sensor may be located as close as possible to the source of the magnetic field being measured. The stability of the orientation of the SQUID inside the cryostat depends on the reliability of its mechanical fixture. The thin walls (screens) of the Dewar sections should not swing when the coolant is boiling and there should be no vibrations of the seating (or of the ground, in the case of measurements of the geomagnetic field). Portable magnetometers need light and small cryostats. The flux of external heat towards the liquid helium or nitrogen should be a minimum. The small mass and dimensions of the cryostat are usually achieved by reducing the volume of the coolant which, as a result, reduces the period of uninterrupted operation till the next filling of the Dewar vessel. For some special SQUID applications, in submarines, for instance, the precise spatial orientation of the cryostat is accomplished by appliances made of nonmagnetic materials.

The electronic circuitry can be located either outside the Dewar (at the temperature of the environment) or partially in the coolant. Usually, the tank circuit or the matching transformer whose primary coil is connected to the SQUID is cooled. In this case, the coil can be made of a superconducting material. The electronic components which are heated during operation should not be a source of electromagnetic disturbance. This is achieved by the use of integrated circuits without packages and by the proper choice of the distance between the sensor and the signal processing circuitry. The extremely low intrinsic noise of SQUIDs imposes severe demands concerning the noise of the electronic circuitry, of the input amplifiers in particular. The use of amplifiers whose input noise level is in the range 10^{-9} to 10^{-8} V Hz$^{-1/2}$ is imperative. Since the electrical connections to the SQUID are usually on the top side of the cryostat, it is recommended that the electronic preamplifiers are placed there. The electronic circuitry should be mechanically fixed with respect to the sensor. Quite often the $X - Y$ recorders and oscilloscopes used for the registration of signals are separated from the SQUID by dozens of metres [1]. In addition, the measurements should be performed in shielded rooms.

7.2. Noise characteristics and sensitivity of SQUIDs

The noise level is a fundamental parameter of SQUID sensors; it determines the minimum detectable value of the magnetic flux. This problem is analyzed in detail in [2,9,13–16]. The sensitivity of a SQUID to a magnetic flux can be characterized by an equivalent noise flux $\Phi_N(t)$ whose spectral density is determined by the expression:

$$S_\Phi(f) = \frac{S_V(f)}{V_\Phi^2}. \tag{7.1}$$

where f is the frequency and $S_V(f)$ is the spectral power density of the SQUID noise voltage at a constant current.

The noise flux energy which corresponds to a given spectral density, however, is a more appropriate characteristic:

$$E(f) = \frac{S_\Phi(f)}{2L}, \tag{7.2}$$

where L is the loop inductance.

There are two limitations of the SQUID parameters determined by the existence of thermal noise. The first limitation is that the energy $I_0\Phi_0/2\pi$ of each quantum transition should considerably exceed k_BT. From the analysis in [13] it follows that:

$$\frac{I_0\Phi_0}{2\pi} \geq 5k_BT. \tag{7.3}$$

From (7.3) it follows that at $T = 4.2$ K, the critical current I_0 is $I_0 \geq 0.9$ μA. In addition, the root-mean-square value, $\langle\Phi_n^2\rangle^{1/2} = (k_BTL)^{1/2}$, of the loop noise flux should be substantially less than Φ_0. The following expression is valid, according to [13]:

$$L \leq \frac{\Phi_0^2}{5k_BT}. \tag{7.4}$$

This is why $L \leq 15$ nH at $T = 4.2$ K.

As shown in §2.7.3, each SQUID is characterized by its critical current I_0, intrinsic capacitance C and shunting resistance R. These parameters are selected in such a way that the coefficient α which characterizes the hysteresis satisfies the condition $\alpha \equiv 2\pi I_0 R^2 C/\Phi_0 \leq 1$, cf. §2.7.4.1. In agreement with the Nyquist formula, uncorrelated noise currents are generated in each resistor. In the DC SQUID, for instance, the shunting resistors are R_1 and R_2 and their corresponding noise currents are $I_{N1}(t)$ and $I_{N2}(t)$. At $T =$ const they have identical spectral densities $S_I(f) = 4k_B T/R$. According to [14–16], the phase differences $\delta_1(t)$ and $\delta_2(t)$ satisfy the equations:

$$V = \frac{\hbar}{4q}\left(\frac{\partial \delta_1}{\partial t} + \frac{\partial \delta_2}{\partial t}\right); \tag{7.5}$$

$$J = \frac{\Phi_0}{2\pi L}\left(\delta_1 - \delta_2 - \frac{2\pi \Phi}{\Phi_0}\right); \tag{7.6}$$

$$\frac{\hbar}{2q}\frac{\partial^2 \delta_1}{\partial t^2} + \frac{\hbar}{2qR}\frac{\partial \delta_1}{\partial t} = \frac{I}{2} - J - I_0 \sin \delta_1 + I_{N1}; \tag{7.7}$$

$$\frac{\hbar C}{2q}\frac{\partial^2 \delta_2}{\partial t^2} + \frac{\hbar}{2qR}\frac{\partial \delta_1}{\partial t} = \frac{I}{2} + J - I_0 \sin \delta_2 + I_{N2}. \tag{7.8}$$

Equation (7.5) gives the connection between the voltage V and the mean variation of the phase. Equation (7.6) describes the flux quantization, and equations (7.7) and (7.8) are of the Langevin type. They are mutually connected by the term J. Relations (7.5)–(7.8) can be solved numerically within limited intervals of values of the noise parameter $g = 2\pi k_B T/I_0 \Phi_0$, the inductance parameter $\beta = 2L I_0/\Phi_0$ and the hysteresis parameter α. For a typical DC SQUID, $g = 0.05$ is obtained at $T = 4.2$ K. At $\Phi =$ const, the voltage V_Φ slowly increases with increasing supply current. If the noise voltage as a function of the current I is changed at a fixed value $\Phi =$ const, it can be established that at frequencies much lower than the Josephson generation frequency, the spectral density corresponds to the white noise spectrum. The noise voltage increases with increasing I to reach the maximum value V_Φ. According to [13–16], the noise energy reaches its maximum at $\beta \approx 1$. At $\beta = 1$ the values of g and Φ are $g = 0.05$ and $\Phi = (2n+1)\Phi_0/4$. If the current I corresponds to the maximum voltage V_Φ, the following relationships are valid:

$$V_\Phi \approx \frac{R}{L}; \tag{7.9}$$

$$S_V(f) \approx 16 k_B T R; \tag{7.10}$$

$$E(f) \approx \frac{9 k_B T L}{R}. \tag{7.11}$$

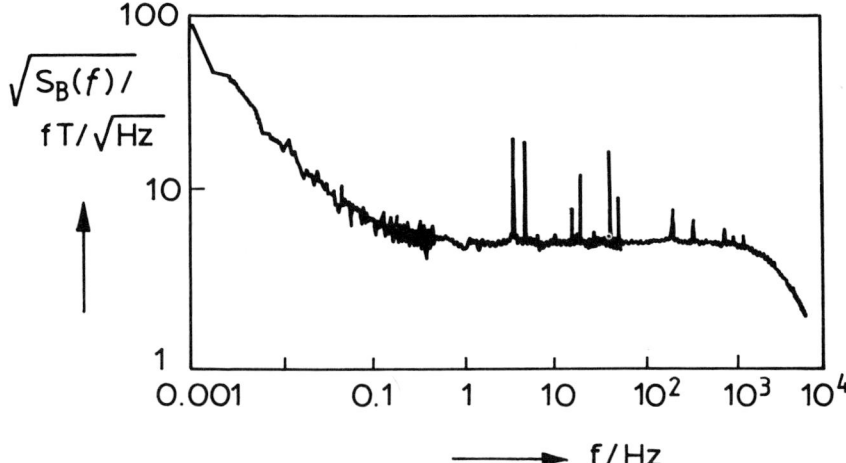

Fig. 7.5. Typical frequency spectrum of the magnetic-field noise of a DC SQUID in a flux-locked loop circuit.

Since $R = (\alpha \Phi_0/2\pi I_0 C)^{1/2}$, equation (7.11) can be written in the form:

$$E(f) \approx 16 k_B T \left(\frac{LC}{\alpha}\right)^{1/2}; \quad \alpha \leq 1. \tag{7.12}$$

Therefore, the parameters C, L and T should be minimized in order to increase the SQUID sensitivity.

The SQUID-loop output signal is transferred to a spectral analyzer to determine the noise power spectral density $S_B(f)$ within a narrow frequency interval $\Delta f = 1$ Hz about a fixed value f_0 (§3.3.1). Hence the total noise power in the entire frequency range is given by:

$$\langle B^2 \rangle = \int_0^\infty S_B(f) df. \tag{7.13}$$

A typical SQUID noise spectrum is presented in Fig. 7.5. The well-known $1/f$ noise is observed at low frequencies and the "white" regime appears as the frequency is increased. The cut-off frequency, at which the spectral density begins to drop, is determined by the read-out circuitry instead of the sensor itself.

The field noise $\sqrt{S_B(f)}$, measured in T Hz$^{-1/2}$, is a convenient parameter when the performance of SQUID sensors is to be compared. Typical values of the flux noise expressed by the field noise $\sqrt{S_B(f)} = \sqrt{\langle B^2 \rangle}$ are in the range 10^{-14} to 10^{-13} T Hz$^{-1/2}$.

The well-known relationships between the energy, the current and the flux ($E = 1/2 L I^2$ and $I = \Phi/L$) can be used to introduce the equivalent flux noise per unit bandwidth and its corresponding energy E_n. It is measured in J Hz^{-1} and characterizes the detection sensitivity of the SQUID. The minimum detectable flux energy E_{min} per 1 Hz bandwidth is thus defined as $E_n = \langle\Phi_n^2\rangle/2L \sim B_n^2/2L\Delta f$, where $\langle\Phi_n^2\rangle$ is the value of the flux which generates at the output a signal equal to the RMS noise, in a frequency bandwidth $f = 1$ Hz. Typical values of the energy sensitivity of RF SQUIDs are $E_n < 10^{-29}$ J Hz^{-1} and of DC SQUIDs are $E_n < 10^{-30}$ J Hz^{-1} [2].

The SQUID sensitivities are the highest values manifested so far by any known magnetosensors irrespective of their type. The low noise level of SQUIDs allows the achievement of a magnetosensitivity which corresponds to a signal smaller by several orders of magnitude than the magnetic-flux quantum Φ_0. It is noteworthy that the energy transducing efficiency of SQUIDs can approach the value 10^{-34} J Hz$^{-1} \sim \hbar = h/2\pi$. This, according to the uncertainty principle of Heisenberg, is the energy limit of the sensitivity of any quantum system. The output voltage per unit magnetic flux Φ_0 has also been accepted in the literature as a definition of SQUID magnetosensitivity: $S_{\Phi_0} \equiv \Delta V_{\text{out}}/\Phi_0$, [$\mu$V per 2.07×10^{-15} Wb].

7.3. DC SQUID sensors

7.3.1. DC SQUID structures

Modern DC SQUIDs are multilayer structures and their thin films are patterned by photolithography. Usually these interferometers include two Josephson contacts of the $S-N-S$ type, microbridge structures of variable thickness or tunnel junctions with low resistance shunts R_1 and R_2 used to obtain a $I(V)$ characteristic without hysteresis (§2.7.3). Some of the most widely used device designs of DC SQUIDs can be seen in Fig. 7.6. They can be classified as: (a) bulk devices; (b) nonplanar thin-film devices; (c) hybrid sensors.

Bulk DC SQUIDs (a) are fabricated from a bulk material (Nb is most widely used) and include point Josephson junctions [17–19]. Their principal advantage is that sophisticated processing steps are not necessary. The adjustment required to achieve identical parameters of the two Josephson junctions as well as problems associated with the reproducibility of the characteristics of the individual samples limit the application of such devices.

Nonplanar DC SQUIDs (b) are commonly fabricated by applying a superconducting film onto a cylindrical quartz rod. Shunting resistors (Pb is most often preferred) are used to smooth the $I(V)$ characteristic of the Josephson junction [1,20]. Hybrid DC SQUIDs (c) are manufactured by planar thin-film techniques for the preparation both of the Josephson junctions and the SQUID loop. The junctions and the loop, together with the input and output wires, are mounted upon a superconducting body of special shape [1].

The advance of DC SQUID designs is the result of the wide application of planar thin-film technology. Because of these techniques, planar input and output

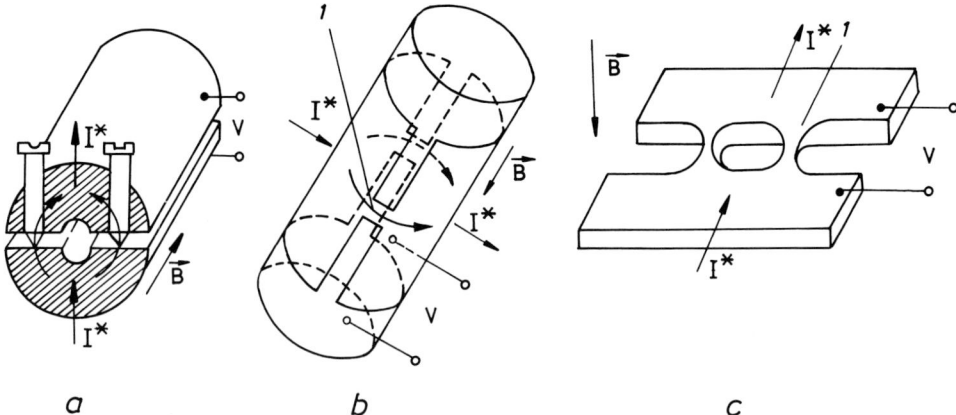

Fig. 7.6. Schematic presentation of SQUID structures; *1* = microbridge contact.

modulation coils or flux transformers, as well as read-out circuitry, can be integrated, thus considerably improving the performance of the SQUID. In this way the magnetosensitivity is enhanced and a better coupling is achieved, since the inductance of the thin-film strip is substantially reduced when it is positioned upon a ground plane [21]. Nonhysteresis operation ($\alpha \sim 1$) and a better signal-to-noise ratio can be more easily achieved by these integrated SQUID versions. Various DC SQUID modifications fabricated by planar thin-film technology are known at present. According to [2], the following types of SQUID can be distinguished: small loop-area devices with microbridge junctions [22,23]; microbridge contacts formed by the burn-in technique; S–N–S junctions [24]; multiloop DC SQUIDs [25,26]; DC SQUIDs with double loops and a resistive shunt [27]; sensors with a relaxation generator; Nb_3Ge-based DC SQUIDs with S–N–S Josephson junctions or microbridge configurations [28]; DC SQUID magnetometers with integrated input coupling coils [6,29]; DC SQUIDs with a double transformer [30,31]; and integrated DC SQUID magnetometers [2,32]. Detailed information about these modern devices can be found, for instance, in [2].

DC SQUIDs are usually made of Pb, Nb, Sn and their alloys. High-T_c SQUID versions are fabricated mainly from YBaCuO metaloceramics [2,9]. The integrated sensor in Fig. 7.7. serves as an illustration of the advance of DC SQUID technology. Its main components are: a thin-film, 3, of a superconducting material, Nb for instance, deposited by sputtering upon a dielectric substrate, 1, with contact pads, 2, formed in advance. The central region of the film is covered by a normal metal, 4. These films are used to pattern the lower electrode by lithography. Then the structure is covered by a dielectric layer, 5, e.g. Al_2O_3. Two windows whose diameters are several micrometers are opened in layer 5 with a distance of tens of micrometers between them. After that a second superconducting layer, 6, is deposited by sputtering to form the second electrode. This leads to the formation

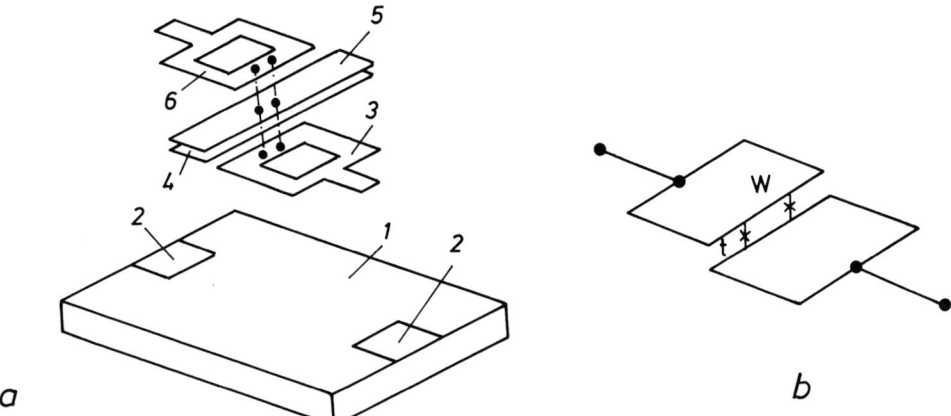

Fig. 7.7. (a) Schematic presentation of an integrated DC SQUID; (b) Electric circuit of the same integrated DC SQUID; w is the separation between the Josephson contacts.

of two S–N–S Josephson junctions in the two windows of the dielectric between the two superconductors. Thus a DC SQUID is obtained with two external loops that form the input coil (Fig. 7.7). Special care is taken to provide for highly efficient inductance coupling of the SQUID and the input coil when the device structure is designed. A satisfactory solution of this problem is given in [5,6]. Its essence is the deposition of a square-shaped spiral coil upon the SQUID itself. The coil is separated from the sensor by a dielectric layer. These planar devices are fabricated upon silicon wafers covered with SiO_2. A 30-nm thick Au–Cu layer is deposited first, in which resistive shunts are patterned (§2.7.3). Then a Nb layer with a thickness of 100 nm is deposited by sputtering, and the SQUID loop and a strip used as a connection to the internal end of the spiral coil are formed lithographically by selective etching. The third layer is 200-nm thick SiO_2. Windows of large dimensions are opened in it for the contacts to the Au–Cu shunt and there are windows to the ends of the Nb strip, which are used as connections to the spiral coil. The next step is the formation of a Nb spiral coil patterned in the 300-nm thick deposited layer. The coil may have 4, 20 or 50 windings. Then the wafer is scribed into chips, each one with a single SQUID upon it. Subsequent processing is different for every chip. There are two methods to form the oxide barrier. The first method includes Nb oxidation with subsequent deposition of the so-called counterelectrode (300-nm thick Pb with In) as an input. Then follows the formation of the contacts to the shunt. The second processing step includes the deposition of approximately 6 nm of Al and oxygen treatment to obtain Al_2O_3 [7]. The typical value of the shunting resistance obtained by this technique is $R \sim 8\,\Omega$, the capacitance is $C \sim 0.5$ pF and the inductance is $L \sim 0.4$ nH.

7.3.2. Characteristics of the $1/f$ noise in DC SQUIDs

High sensitivity at frequencies $0.1 \leq f \leq 3$ Hz is required for many applications, for instance, in investigations into variations in the geomagnetic field, biomagnetism, etc. Because of this fact, the low-frequency $1/f$ noise is a pressing reality in such measurements, and therefore at present it is the subject of intensive research. A typical curve of $1/f$ noise is presented in Fig. 7.5. According to [33,34] there are at least two independent $1/f$ noise sources in DC SQUIDs. The first source is associated with the noise generated by the critical current I_0 in the Josephson junction. In the process of tunneling across a barrier, an electron can be trapped and subsequently released. When a trap is occupied, a local modification of the potential barrier takes place. Hence the current density is changed in that area of the junction. Therefore, if there are isolated traps in the junction region, the critical current is spontaneously switched on between two fixed values, thus generating the random signal. The spectral density of this process is determined by the mean time $\langle \tau \rangle$ between two successive pulses. It is given by $S(f) \sim \tau/[1 + (2\pi f \tau)^2]$. It can be deduced that at frequencies $f > 1/2\pi\tau$, the process exhibits the $1/f^2$ dependence of the white-noise spectrum.

In many cases electron trapping is a thermally activated process described by $\tau = \tau_0 \exp(E/k_B T)$, where E is the height of the potential barrier. Several types of trap are observed as a rule in the Josephson junction region. Each type is described by its characteristic time τ. The acts of electron trapping and releasing, instead of being correlated, are believed to be statistically independent [35]. At a fixed temperature, the dominant contribution to the $1/f$ noise belongs to traps whose energy levels are within $k_B T$ of the value of E for the junction barrier [9,35]. The level of the $1/f$ noise in the critical current is strongly dependent on the quality of the Josephson contact and can be characterized by the leakage current at a voltage smaller than $(\Delta_1 + \Delta_2)/q$, where Δ_1 and Δ_2 are the respective bandgap widths of the two superconductors. Owing to the existence of traps in the junction region, electron tunneling takes place in this voltage region, as a result of which a leakage current and $1/f$ noise arise. That is why the technology of the fabrication of Josephson junctions should guarantee a low leakage current in order to achieve a low $1/f$ noise level.

The second source of the $1/f$ noise in DC SQUIDs is most probably associated with a motion of the magnetic flux confined to the bulk of the sensor [33]. Essentially, this noise acts as an external noise source applied to the SQUID. In such a way the noise spectral density of the flux changes linearly with $V_\Phi = 0$ and becomes zero at points $\Phi = (n \pm 1/2)\Phi_0$, when $V_\Phi = 0$. The level of this $1/f$ noise depends strongly on the microstructure of the superconducting film. For instance, the intrinsic noise of a DC SQUID whose loop has been prepared by Nb sputtering is about $10^{-10}\Phi_0^2$ Hz^{-1} at $f = 1$ Hz [33], if special requirements have been observed during preparation of the coil. At the same time a DC SQUID with the same geometry but with a Pb loop manifests a $1/f$, noise level of about $2 \times 10^{-10} \Phi_0^2$ Hz^{-1} at $f = 1$ Hz. In this device the noise arises because of critical-current fluctuations. A noise level about $3 \times 10^{-13} \Phi_0^2$ Hz^{-1} in a Nb DC SQUID is reported in [36]. These results indicate a pronounced dependence of the $1/f$ noise level on the quality of the superconducting material used.

From a practical point of view, there is a fundamental difference between the two sources of the $1/f$ noise. The noise generated by critical-current fluctuations can be minimized by an adequate modulation circuit, while the flux noise cannot be reduced by such methods [9,23].

7.4. RF SQUID sensors

7.4.1. RF SQUID structures

The operation of RF SQUIDs is based on the quantization of the magnetic flux in a ring with a single weak link (§2.7.4.1). Any type of Josephson contact described in §7.1.1 can be used in the device. Depending on their structure, RF SQUIDs can be divided into three basic groups: bulk, nonplanar thin-film, and hybrid SQUIDs [2]. Bulk versions can be: single-hole point Josephson contact [37]; two-hole point contact [38]; multi-hole point contact [39]; toroidal point Josephson contact [4], etc. The thin-film nonplanar RF SQUID is usually a cylindrical quartz structure with a microbridge Josephson contact [40,41]. A whole class of commercial versions of hybrid RF SQUIDs has been developed [42]. A bulk RF SQUID with two-holes and point Josephson contact is shown in Fig. 7.8a to complete the presentation of RF SQUIDs [38].

The magnetic field B is applied orthogonally to the flux transformer. Nb is used for the preparation of the device, and the point contact is symmetrical with respect to the two holes. The main advantage of this structure is the satisfactory mechanical stability of its components and its relatively low inductance. In addition, this RF SQUID is protected from external parasitic magnetic fields. The measured magnetic field B is introduced by means of a wire-wound superconducting signal coil placed in one of the symmetrical openings of the SQUID. The generating coil used for SQUID excitation by a frequency of about 30 MHz is placed in the other opening. The required resistance of the point contact is achieved by an adjustment screw outside the cryostat at $T = 300$ K. A special calibration instrument is used for this purpose.

In the RF SQUID described in [4], the bridge contact is connected to a thin-film input coil fixed upon a quartz cylinder. Minimization of the inductance, which leads to a minimum noise level, high sensitivity and stability of the critical current, is achieved if $I_0 L_0 < \Phi_0$. This condition corresponds to a quantization area of several mm^2. The toroidal version of an RF SQUID with a single-point contact made of Nb

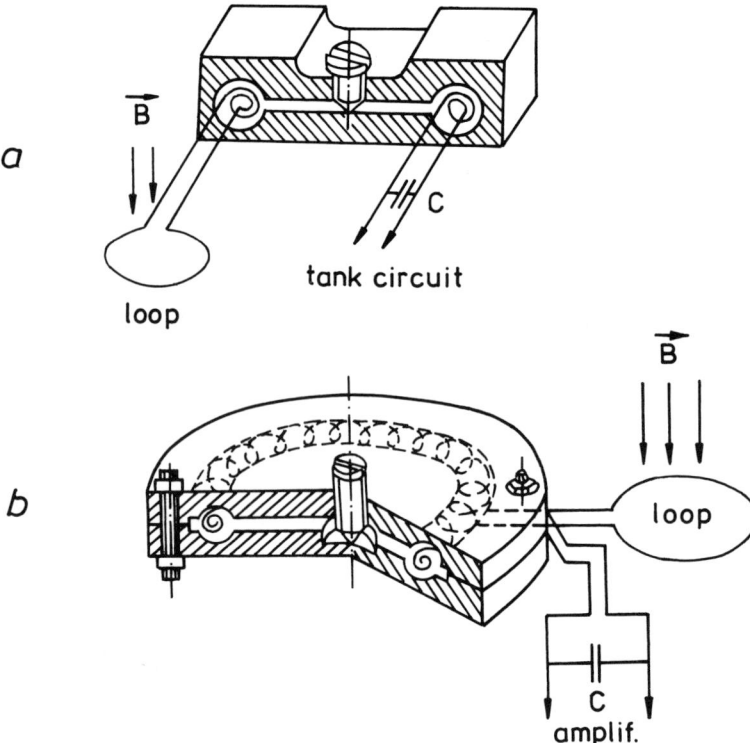

Fig. 7.8. (a) Two-hole and point-Josephson-contact bulk RF SQUID. (b) Toroidal RF version with single-point contact.

(Fig. 7.8b) meets the above requirements. The modulation coil and flux transformer are toroidal. The diameter of the input coil is 10^{-2} m. Quite often this type of RF SQUID is commercially available.

7.4.2. Noise properties of RF SQUIDs

A detailed analysis of RF SQUID noise can be found, for instance, in [2,9,43–48]. In this case, unlike DC sensors, the noise of the tank circuit and the preamplifier should be taken into account. Two effects that influence the intrinsic noise of the SQUID occur because of thermal fluctuations. The first effect is associated with noise in the RF resonant coil. The noise is described by the spectral power density of the intrinsic flux equivalent noise $S_{\Phi,\text{int}} \approx [(LI_0)^2/w_{rf}](2\pi k_B T/I_0\Phi_0)^{4/3}$ [47,48]. Secondly, the existence of such a noise leads to the appearance of a finite slope of the "steps" of the current–voltage characteristic in Fig. 2.37b. This slope, η, is connected with $S_{\Phi,\text{int}}$ by $\eta^2 \approx (S_{\Phi,\text{int}} w_{rf})/\pi \Phi_0$ [44].

The noise temperature T_α of typical RF amplifiers at $T = 300$ K is considerably higher than that of amplifiers with an operational frequency $f \leq 500$ kHz. This is why the noise temperature should not be neglected when the operation of RF SQUIDs at $T = 4.2$ K is considered. In addition, the coaxial connection of the tank circuit with the preamplifier is kept at $T = 300$ K. Since the intrinsic capacitances of this line and the preamplifier determine the main component of the capacitance of the tank circuit, a part of the resistance which shunts the tank circuit itself has a temperature which substantially exceeds that of the Dewar. This is why an additional noise is generated which, together with the preamplifier noise, determines the effective noise temperature $T_{\alpha,\text{eff}}$ of the system. According to [44], the noise energy generated by the sources discussed above is $2\pi \eta k_B T_{\alpha,\text{eff}}/w_{rf}$. The summation of these components together with the internal noise sources yields $E \approx (1/w_{rf})(\pi \eta^2 \Phi_0^2/2L + 2\pi \eta k_B T_{\alpha,\text{eff}})$ It can be seen that the energy E is proportional to $1/w_{rf}$, and the influence of T_α, which increases as w_{rf} is increased, should be taken into account as well. The main conclusion from the analysis of RF SQUID noise is that a considerable increase in the frequency of operation w_{rf} is required in order to improve the performance of the sensor. This frequency should exceed the commonly used values of w_{rf} in the 20- to 30-MHz interval. Decreasing T_α by cooling the preamplifier as well as decreasing the tank-circuit temperature to approach 4.2 K is another way of noise minimization.

7.5. Review of SQUID sensors

The unique magnetosensitivity of SQUIDs has motivated a wide variety of ingenious technical solutions, including the use of high-T_c superconducting materials. Besides commercially available DC and RF SQUIDs there are a number of other versions whose performance guarantees the reliability of cardiomagnetograms and the impeccability of submarine, missile and satellite navigation systems. Basic types of SQUID sensors which can be used for scientific investigations and multiple practical applications will be described below.

7.5.1. DC SQUID sensors

The DC SQUIDs known so far can be classified into two groups depending or whether the SQUID itself serves as a magnetometer or whether it is included in a combined system together with a superconducting flux transformer. One of the earliest magnetometers was fabricated from bulk Nb and includes two point contacts (Fig. 7.6a) [18,50]. The diameter and length of the SQUID are 2×10^{-2} m. The sensor is immersed in liquid helium ($T = 4.2$ K). A copper screen inside the Dewar design is used to eliminate the influence of external high-frequency electromagnetic fields. The critical current, I_0, of the contacts in the 0.1- to 1.0 μA range is achieved by adjusting from outside the cryostat the point-contact pressure.

The resistance R_N is tens of Ω, and the quantization area is $A_0 = 2 \times 10^{-5}$ m^2, which corresponds to a period $\Delta B \sim 10^{-10}$ T of the oscillations of the $V(B)$ characteristic.

DC SQUIDs similar to those in Fig. 7.6 are described in [1]. Mainly thin-film Josephson contacts have been used. They are prepared by the burn-in technique which forms microscopic regions of breakdown in the dielectric layer deposited upon a Nb or Ta superconducting wafer. Another superconductor is deposited in the openings by sputtering. The design approach in this case is aimed at: decreasing coil inductance by means of two planar superconductors laid one upon the other and applying the magnetic field being measured in the space between them; the absence of complex elements with large dimensions used for the adjustment of the contact pressure and of the current I_0 across the contacts; total DC SQUID miniaturization. According to [1], the characteristic areas of quantization determined by the distance between the two superconducting films is in the range 10^{-9} to 10^{-8} m^2, the critical currents I_{01} and I_{02} are between 10 and 10^3 μA, and the resistance value is $R_N \approx 0.01$ to 10 Ω. Such a low inductance SQUID is presented schematically in Fig. 7.9a.

A typical Nb–In–Pb-based DC SQUID with stable tunnel junctions and a quantization area $A_0 = 10^{-6}$ m^2 is described in [20] (Fig. 7.9b). The input coil was deposited by sputtering upon a cylindrical quartz rod whose diameter is 3 mm and is connected with the two contacts. A thin-film Au shunt is connected in parallel to

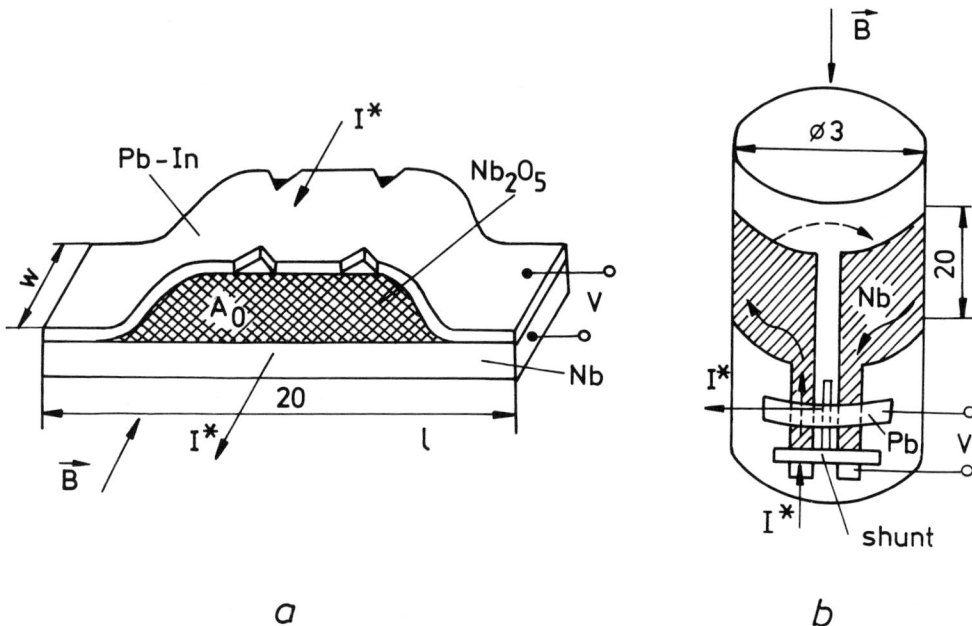

Fig. 7.9. (a) Low-inductance version with two bridges. (b) A cylindrical interferometer with two Nb–Nb$_2$O$_5$–Pb tunnel junctions [1].

the Josephson junctions to obtain a stable $I(V)$ characteristic. The shunt resistance is selected in such a way that the relationship $R \leq (\Phi_0/4\pi I_0 C)^{1/2}$ is valid, where C is the Josephson contact capacitance. A magnetic flux of $1/10^4 \Phi_0$ or $(1/10^4)\Delta B$ has been registered, which corresponds to a sensitivity limit of 10^{-13} T within a frequency interval $\Delta f = 1$ Hz. It has been established that the sensor noise is considerably increased by temperature variations in the helium Dewar. The noise level can be reduced several times by the stabilization of the helium vapour pressure, thus achieving temperature stability with an accuracy of approximately 5×10^{-4} K. This relatively simple DC SQUID is remarkable in its parameters. They are not inferior to the parameters of the best RF SQUIDs. For instance, with $I_0 L_0 = 100\Phi_0$, $R_0 = 100$ Ω, $A_0 = 3 \times 10^{-4}$ m^2, $L_0 = 2 \times 10^{-8}$ H and a symmetry coefficient $a = 1$ (§2.7.4.2), the sensitivity limit is $\sim 10^{-16}$ T Hz$^{-1/2}$ in the $0 \leq f \leq 1$ Hz interval. Such a sensitivity limit at $T = 4.2$ K cannot be achieved under real conditions because of the amplifier noise, temperature instability in the Dewar, the mechanical instability of the cryostat and the experimental setup, as well as the stray magnetic field. Satisfactory screening from external electromagnetic fields is achieved by the use of superconducting shields. The noise level of the input circuitry can be reduced to that of the intrinsic electrical noise in the SQUID sensor [1,51].

It is very difficult to minimize the temperature instability over long time intervals (hours and days). In most DC SQUIDs, the temperature sensitivity of the current I_0 and of the geometrical area of quantization is a serious problem. This temperature sensitivity is due to thermal expansion and to the variation with temperature of the depth, λ, of penetration of the magnetic flux into the superconductor. If the low-frequency random fluctuations of temperature due to atmospheric pressure variations are taken into consideration, their values δT being in the range 10^{-5} to 10^{-4} K, the variation with temperature of the DC SQUID output voltage will be $\delta V = \delta T \alpha^* R_N = 5 \times 10^{-9}$ V when $dI_0/dT \equiv \alpha^* = 5$ mA K^{-1} and $R_N = 0.1$ Ω. This value of the voltage variation is three orders of magnitude higher than the intrinsic noise of the DC SQUID itself. With $\alpha^* = 10^{-7}$ A K^{-1}, the voltage variation is $\delta V = 10^{-11}$ V at $R_N = 10$ Ω, this value is approaching the intrinsic electrical noise of the SQUID [1]. Therefore the real value of δT of the DC SQUID and the change, δP, in the atmospheric pressure are of primary importance for the magnetosensitivity limit. Furthermore, the temperature dependence of the coefficient of thermal expansion of the superconducting materials of the SQUID, the thermal dependence $\lambda(T)$, the parameters of the stray magnetic field and the characteristics of the sensor itself should be known in order to determine the level of magnetic noise in this class of magnetometers. These are the main factors that determine the magnetosensitivity limit. Its value in conventional magnetometers with optimized geometrical and electrical parameters is 10^{-13} to 10^{-12} T [20].

Another typical configuration is that of a DC SQUID with a flux transformer. It is presented schematically in Fig. 7.10 [19]. The external magnetic field B, which is applied perpendicularly to the plane of a superconducting pickup loop, generates a

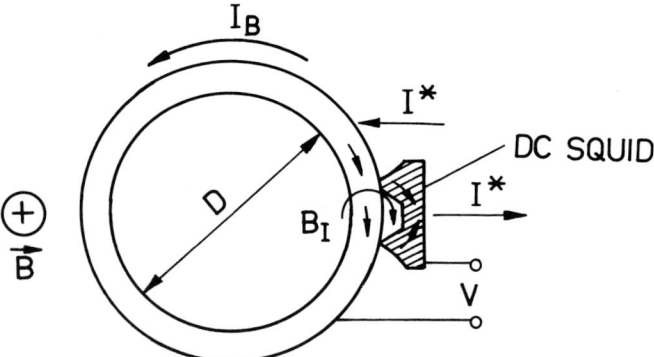

Fig. 7.10. A combined DC SQUID: the superconducting magnetometer has been coupled with a flux transformer to form a galvanometer.

current I_B which does not disappear, even at $B = $ const. The current excites a field B_1 on the loop surface. It is measured by a DC SQUID which is mounted directly upon the loop. This magnetometer version is known also as a superconducting galvanometer. The loop and the galvanometer taken together form the magnetic flux transformer. The flux transformer converts the unknown field B_x into the magnetic field B_1, which has already been enhanced by a calibrated factor n. It has been established that $n \approx D/5r$, where D is the diameter of the loop and r is the radius of the wire. The value of n for a thin-film pickup coil combined with a thin-film DC SQUID is $n \approx D/w$, where w is the width of the film in the part of the sensor which serves as a galvanometer [1]. The technical solution described allows an improvement of the DC SQUID sensitivity limit. The region where the SQUID is located must be screened by a superconducting shield.

A loop with a big diameter, $D = 50$ mm, has also been prepared with bulk Nb. The loop was connected to a thin-film DC SQUID whose Josephson contacts are microbridges. A sensitivity of 10^{-12} T Hz$^{-1/2}$ has been obtained, the total loop inductance being 10^{-8} H. The galvanometer can register currents as low as 10^{-7} A. This example proves that the transducing efficiency can be increased by increasing the diameter, D, of the pick-up loop. This is why at one and the same B_x, the current I_B increases together with the sensitivity.

Another important factor that limits the transducing efficiency of a DC SQUID is the magnetic noise of the pick-up loop which is added to the sensor's intrinsic noise. In the entire interval of frequencies of the current which can circulate in the coil the relationship $L\langle i^2_{\text{noise}}\rangle/2 = k_B T$ at $T = 4.2$ K is valid. Hence $\langle i^2_{\text{noise}}\rangle = k_B T/L$, i.e. the current noise decreases with increasing coil diameter. In the general case the diameter D is limited by the Dewar dimensions. Quite often a screen of a normal metal, Cu for instance, is used for the protection of DC SQUIDs from high-frequency electromagnetic fields. The magnetometer described in [18] is placed in a 1 m high glass Dewar filled with 3 l liquid helium. The device can operate for 24 hours without interruption. The registration circuitry is at a distance of 30 m from the Dewar vessel. This instrument was first used for the measurement of the vertical component of the geomagnetic field. A sensitivity limit of 10^{-13} T has been achieved in the frequency interval $0 \leq f \leq 10^{-3}$ Hz.

The integration of a DC SQUID with a pickup loop and spiral input coil on one and the same substrate is the most promising approach to the design of modern supersensitive magnetometers. Its evolution towards the integration of part of the read-out circuitry and other electronics upon the same substrate is the basis of smart solutions. The sensitivity of such an integrated SQUID magnetometer made by a Nb thin-film technique with two Nb–SiN–Nb tunnel junctions and operating in a magnetically shielded room is about 5×10^{-15} T Hz$^{-1/2}$ (i.e. 5 fT), when operated without a matching transformer with a conventional flux-locked loop circuit [2].

7.5.2. RF SQUID sensors

A complete RF SQUID system must meet the same requirements as DC SQUID magnetometers. The device in Fig. 7.8a) exhibits a sensitivity of approximately 10^{-13} T in the frequency interval $0 \leq f \leq 1$ Hz [38]. The version with a thin-film Josephson contact and a pick-up coil upon a quartz cylinder, whose diameter is 2 mm, has a magnetosensitivity of 5.6×10^{-14} T [4]. The modulating frequency of this sensor is 30 MHz, the inductance being $L_0 \approx 10^{-9}$ H. A RF SQUID similar to that in [4], with a quartz body of length 20 mm and diameter 2 mm and transducing efficiency of 10^{-14} T, is described in [1]. The development of RF SQUIDs is aimed at inductance minimization and stabilization of the current I_0 under the condition $I_0 L_0 < \Phi_0$. The device in Fig. 7.8b with an inductance $L_0 \approx 10^{-10}$ H at a current $I_0 \approx 1$ to 5 μA and $T = 4.2$ K manifests a sensitivity of 10^{-14} T. Its flux transformer has $L \approx 10^{-6}$ H; the dimensions of this sensor are a diameter of 10^{-2} m and a height of 0.6×10^{-2} m. A cylindrically shaped Nb thin-film RF SQUID with a sensitivity 10^{-12} T at $\Delta f = 30$ MHz has been reported [1]. A RF SQUID system with the following parameters is described in [2]: SQUID inductance of 1 nH; pickup loop inductance and input coil inductance of 10 nH; operating frequency of the tank circuit of 20 MHz, while the transducing efficiency is 5×10^{-15} T Hz$^{-1/2}$. This magnetometer was designed for biological experiments. In typical RF SQUIDs the white-noise energy is 5×10^{-29} J Hz^{-1}, while the $1/f$ noise energy is about 10^{-28} J Hz^{-1} at a frequency of 0.1 Hz [9].

Data collected from different periodicals and catalogues concerning planar RF SQUID magnetometers can be found in ref. [52], thus rounding off the general picture of these instruments. The magnetosensitivity of the devices reported in [52] is between 10^{-16} and 10^{-13} T Hz$^{-1/2}$ and the tank-circuit operating frequency is between 1 MHz and 350 MHz.

In general, the sensitivity of DC SQUIDs is higher than that of RF SQUIDs.

7.6. High-T_c SQUID sensors

The discovery of high-T_c superconductivity has motivated the creation of SQUID magnetometers operating at the temperature of liquid nitrogen ($T = 77$ K) [53]. Generally speaking, the new high-T_c materials both of bulk samples and polycrystalline films, are oxidized copper with a granulated structure. Their electrical conductivity is due to the overlapping of Cu and O atomic orbitals and is highly anisotropic because of the layered structure of the material. The charge transfer is determined by the hole pairs along the Cu–O layers, but the possibility of electrical conductivity along other layers cannot be eliminated, either. In the superconducting state, the carrier concentration is about 10^{21} cm^{-3}. Depending on the doping concentration, these new compounds behave like insulators, conventional conductors or superconductors. Such a wide range of variation of the electronic properties resulting from trivial chemical substitution of atoms in a given superconductor is a unique phenomenon and can serve as a basis for the creation of new devices. Polycrystalline high-T_c ceramics can be regarded as composed of a great number of superconducting grains connected by weak links across the grain boundaries. This three-dimensional set of weak links is a property which can be used for SQUID operation.

By analogy with low-temperature SQUIDs, thermal noise determines the sensitivity limit in the white-noise region of high-T_c devices. The thermal-noise level is increased by a factor of $\sqrt{77/4.2}$ as the temperature is increased from 4.2 K to 77 K [2]. When SQUIDs operate at $T = 77$ K, the limitations of the critical current I_0 and the inductance L associated with the thermal noise can be expected. According to §7.2 and equations (7.3) and (7.4), at $T = 77$ K, values of $I_0 > 16$ μA and $L < 0.8$ nH are obtained. The inductance value $L(77K) = 0.2$ nH and the critical current $I_0(77K) = 20$ μA are assumed in order to compare the factors that determine the SQUID performance at $T = 77$ K and at $T = 4.2$ K. The sensitivity limit is achieved at $2LI_0/\Phi_0 = 4$ for DC SQUIDs and at $LI_0/\Phi_0 = 2$ for RF SQUIDs [9]. As the temperature is increased from 4.2 K to 77 K, the $1/f$ noise sources are thermally activated. All the high-T_c superconductive SQUIDs known so far exhibit a $1/f$ noise level 4 to 5 orders of magnitude higher than the white-noise energy, [54,55]. These first results are encouraging since the metallurgy of the thin-films used in these sensors has not yet been optimized by far. The practical implementation of these SQUIDs is based on Josephson junctions naturally formed at the grain boundaries of the superconducting material. The development of a technology for the reproducible formation of Josephson junctions will apparently improve the SQUID performance at $T = 77$ K [56].

The DC SQUID noise energy is determined according to (7.11). Since the experimentally determined resistance R is, typically, about 5 Ω [54], the value $E = 4 \times 10^{-31}$ J Hz^{-1} is obtained for the noise energy at $L = 0.2$ nH and $T = 77$ K. This value is a factor of 10 higher than the noise energy of thin-film Nb DC

SQUIDs at $T = 4.2$ K. The intrinsic noise energy of a RF SQUID, according to §7.4.2, is approximately 6×10^{-29} J Hz^{-1} at a current $I_0 = 20$ µA, $LI_0 = 2\Phi_0$, $w_{rf}/2\pi = 20$ MHz and $T = 77$ K. This value is in good agreement with the experimental results of a device operating at $T = 4.2$ K, in which the effective noise temperature, $T_{\alpha,\text{eff}}$, of the preamplifier and the tank circuit is well above $T = 4.2$ K (the preamplifier operates at $T = 300$ K) (§7.4.2). This is why, when a RF SQUID is operating at $T = 77$ K, no considerable increase of $T_{\alpha,\text{eff}}$ is to be expected and the level of the noise energy should be near that at $T = 4.2$ K.

A number of DC SQUIDs have been developed on the basis of YBa$_2$Cu$_3$O$_{7-\delta}$ [2,9,54]. Such a planar version with an inductance of 80 pH is presented in Fig. 7.11. Two microbridges formed at the boundaries of the randomly oriented YBaCuO crystalline grains serve as Josephson junctions. The device structure in Fig. 7.11 can be manufactured by photolithography and ion etching. The magnetic flux applied perpendicularly to the plane of the structure modulates the junction $I(V)$ characteristics. Quite often, however, the $V(\Phi)$ relationship exhibits a hysteresis and nonperiodic behavior, most probably as a result of the magnetic flux freeze-out inside the YBaCuO loop. The minimum level of the noise energy at $f = 1$ Hz is 4×10^{-27} J Hz^{-1} at $T = 41$ K and 2×10^{-26} J Hz^{-1} at $T = 77$ K [9,54].

A TlBaCaCuO-based DC SQUID for operation at 77 K is described in [57]. It has been observed that in separate samples the noise level is considerably higher than in YBaCuO devices. At low frequencies, the noise spectral density manifests a $1/f$ dependence. The energy sensitivity of the devices with inductance of 80 pH at $f = 10$ Hz is $E = 5 \times 10^{-29}$ J Hz^{-1}, while those with inductance of 5 pH exhibit the energy sensitivity $E = 2 \times 10^{-29}$ J Hz^{-1}. There is an important fact: the low-frequency noise is not generated by the flux or by the current I_0. The most probable origin of this noise is the redistribution of the flux captured by the microbridges. The lower noise level results from the increased size (10 to 40 µm) of the crystal grains in the TlBaCaCuO films in comparison with the grains (≈ 1

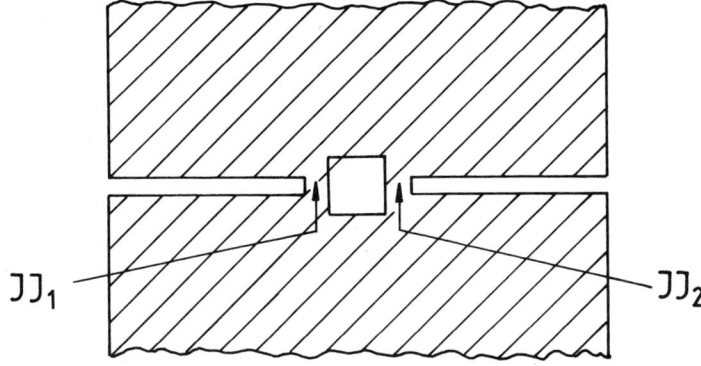

Fig. 7.11. A planar high-T_c DC SQUID with a thin YBaCuO film.

μm) in YBaCuO films. Many TlBaCaCuO SQUIDs have the structure shown in Fig. 7.11, and their $V(\Phi)$ characteristic is also nonperiodic and exhibits hysteresis. It has been established that devices with very small pickup loops have a periodic $V(\Phi)$ characteristic and negligible hysteresis [58].

At present only a few high-quality thin-film high-T_c SQUIDs are selected from amongst the numerous sensors fabricated by microtechnology. The active area of the pickup loop of the SQUID in Fig. 7.11, which operates at $T = 77$ K, is an order of magnitude smaller than that of integrated SQUIDs operating at $T = 4.2$ K. This is explained by the higher fluctuation intensity at $T = 77$ K and the need for reducing the inductance L. YBaCuO films have a higher level of intrinsic noise. This is probably due to the motion of flux quantums captured at the film grain boundaries. It can be reliably assumed that this mechanism is the source of the $1/f$ noise which, at $T = 4.2$ K, is observed in Nb SQUIDs as well. The favorable trend of a decrease in the level of this noise as the film microstructure is improved is an indication that the noise levels achieved so far are not a limitation to a further reducing of the noise. Therefore, thin-film technology should necessarily be perfected in order to improve the performance of DC SQUIDs and their flux transformers on the basis of high-T_c superconducting materials. The sensor described in [59] has optimum characteristics compared with the other available high-T_c RF SQUIDs. Its sensitivity is $4.5 \times 10^{-4} \Phi_0$ Hz$^{-1/2}$ at $f = 50$ Hz. This value corresponds to the noise energy $E = 1.6 \times 10^{-27}$ J Hz^{-1} with an inductance of 0.25 nH. Two-hole toroidal [60] and point-contact [61] RF SQUIDs have been suggested and investigated.

Another type of YBaCuO superconducting magnetosensor that indirectly uses Josephson weak links at $T = 77$ K has been developed. This device, termed a super magnetoresistor, is described in [62]. As for the physical mechanism of these characteristics, we consider the following. Inherently there will be very thin insulators between grains or point contacts of grains in a ceramic superconductor and they may operate as weak couplings. When the magnetic field is weak, the superconducting current flows through these weak couplings with zero resistance by the Josephson effect. However, when a certain value of magnetic field is applied, it becomes difficult for superconducting carriers to flow through them owing to the decrease of the Josephson critical current, and a finite resistance suddenly appears as a result. The characteristics of this novel solid-state magnetic sensor are completely different from those of conventional sensors. A magneto-resistive element made of semiconductor or magnetic materials shows a slight increase in resistance, ΔR, at a low magnetic field, while it has a relatively large R_0 at a zero magnetic field. Thus the figure of merit $\Delta R/R_0$ amounts to only about 1% at best at 5 mT. However, $\Delta R/R_0$ of the novel magnetic sensor described above is infinitely high because $R_0 = 0$, leading to an extremely high transduction efficiency when the magnetic field is below 5 mT. The limit sensitivity reaches ~ 10 nT.

The solution of two problems emerges as a future trend in the evolution of high-T_c SQUIDs. The first problem is related to the creation of reliable high-grade Josephson junctions with reproducible characteristics and the development of an adequate technique for their fabrication. The second problem is associated with reducing the adverse effect of hysteresis and noise in thin-films of high-T_c materials.

7.7. SQUID gradiometer sensors

7.7.1. Basic configurations of SQUID gradiometers

Superconducting SQUID gradiometers measure the variation of the gradient of each of the components, B_x, B_y and B_z, of the external magnetic field $\boldsymbol{B}$ as a function of the position of the sensor relative to a given inertial system. In this case the basic matrix of magnetic field gradients is [1]:

$$M_{Gr}(x, y, z) = \begin{bmatrix} \dfrac{\partial B_x}{\partial x} & \dfrac{\partial B_x}{\partial y} & \dfrac{\partial B_x}{\partial z} \\ \dfrac{\partial B_y}{\partial x} & \dfrac{\partial B_y}{\partial y} & \dfrac{\partial B_y}{\partial z} \\ \dfrac{\partial B_z}{\partial x} & \dfrac{\partial B_z}{\partial y} & \dfrac{\partial B_z}{\partial z} \end{bmatrix}. \tag{7.14}$$

A special design of the SQUID pickup coil is used for the registration of the gradient of the field $\boldsymbol{B}$ along a given axis (x, y or z) with a fixed (standard) length b. Figure 7.12 illustrates schematically the pickup coil configurations used for the registration of selected components of the matrix (7.14), i.e. $\partial B_x/\partial b$, $\partial B_y/\partial b$ and $\partial B_z/\partial b$. The length b is oriented along one of the coordinate axes ($b \| x$, $b \| y$ or $b \| z$). The two sections of the pickup coil have equal areas but opposite winding directions; therefore, the currents in them excite magnetic fields with opposite orientations. Initially, they balance each other in such a way as to generate a total magnetic flux which equals zero in the transformer, when the applied external magnetic field is homogeneous. When there is a gradient of the magnetic field, the total flux differs from zero and determines the SQUID output signal. Since these coil configurations register the first derivative of the field $\boldsymbol{B}$, they are termed first-order gradiometers. When a gradient $\partial B/\partial b$ of the field component is measured, which is orthogonal to the plane of the two coil-sections, a current I arises, which is not attenuated. The relationship $I \sim bA/L$ is valid, where L and A are, respectively, the inductance and the area of the coil-section into which the field $\boldsymbol{B}$ is penetrating. If the field between the two sections, separated by a distance b, is linearly increasing in the direction parallel to b, the following relationship is valid: $I \approx A(B_1 - B_2)/L$, where B_1 and B_2 are the values of the magnetic induction inside the first and the second sections of the coil respectively [1].

The current I is measured by a SQUID galvanometer. In order to eliminate the influence of the external field B in the gradient-field registration, the sensor is placed inside a screen which is usually made of a superconducting material. The distance b is determined by the dimensions of the Dewar and by the size limitations on the SQUID set by the fabrication technology. In most of the SQUID gradiometers used the distance b is between 3 and 20 cm. The approach to the design of superconducting gradiometers illustrated in Fig. 7.12 exhibits satisfactory efficiency despite its simplicity. Insensitivity to an increase in the uniform magnetic field is the main advantage of these device configurations [1,2,9,63]. The second-order gradient, i.e. $\partial^2 B_x/\partial x^2$; $\partial^2 B_z/\partial z^2$, etc. can be measured by means of a pickup coil modification. Its practical implementation is presented in Fig. 7.13.

The coil consists of three sections, the ratio of the winding numbers being $1:-2:1$ (the windings with a clockwise current flow are assumed to be positive and

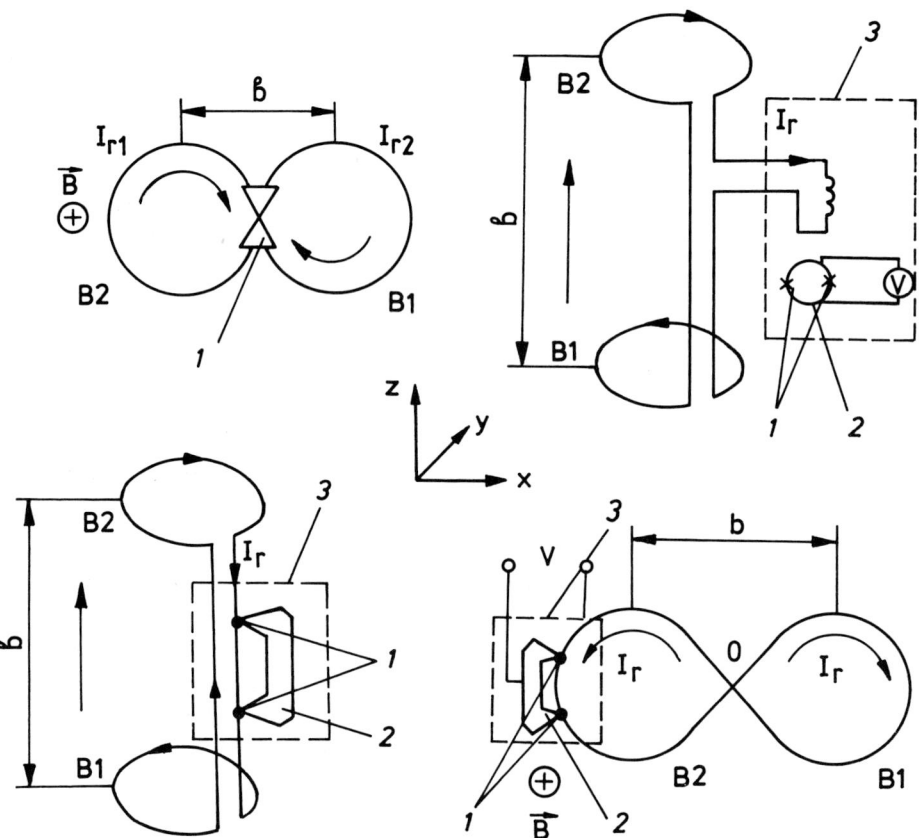

Fig. 7.12. Various pickup coils for SQUID gradiometers; *1*: Josephson junctions, *2*: DC SQUID, and *3*: magnetic shield.

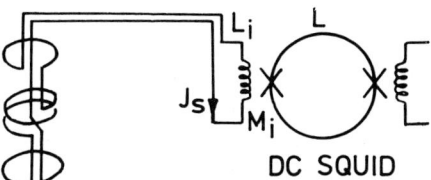

Fig. 7.13. Second-order SQUID gradiometer [9].

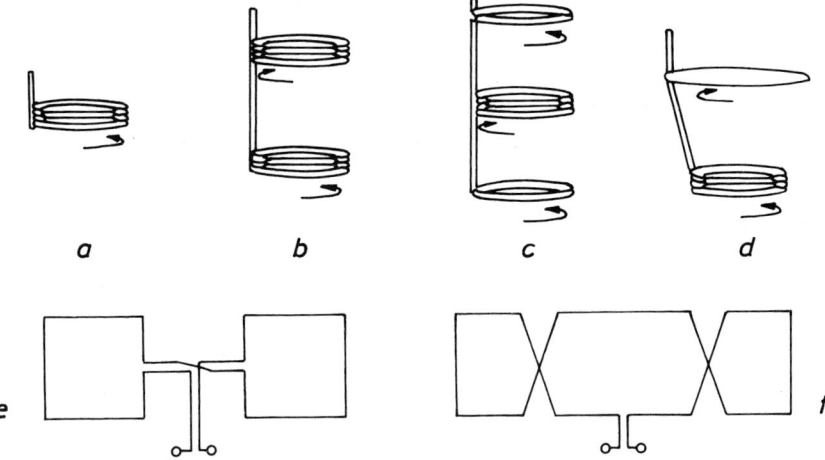

Fig. 7.14. Various pickup coil configurations: (a) magnetometer coil; (b)–(f) gradiometers. (b) First order, (c) second order, (d) asymmetric first order, (e) planar first order, (f) planar second order [2,63].

those with an counterclockwise current flow, negative). The third-order gradient $(\partial^3 B_x/\partial x^3, \ldots,$ etc.) can be measured if the pickup coil has 4 sections in the proportion $1 : -3 : +3 : -1$. In addition to bulk pickup coil modifications, planar implementations are possible as well. First- and second-order gradiometric configurations are shown in Fig. 7.14.

7.7.2. Requirements of SQUID gradiometer configurations and their characteristics

The basic requirements of SQUID gradiometer configurations are a minimum pickup coil inductance in order to achieve a high magnetosensitivity and a high coefficient of structural symmetry of the coil configuration. The coefficient F of structural symmetry is a dimensionless factor. It can be defined by the equality of the magnetic flux which corresponds to the sensitivity limit, S, to the flux that pierces the SQUID coil in a homogeneous field B, i.e.: $\delta\Phi = SbA = B\delta A$, where δA is the difference between the areas of the gradiometer coil arising from coil imperfections. In this case the parameter F will be:

$$F = \frac{A}{\delta A} = \frac{B}{Sb}. \tag{7.15}$$

A high value of the parameter F can be achieved by precise manufacturing of the components of the gradiometer and by means of special adjustment elements. These elements are tiny superconducting bodies whose positions in the neighborhood of the SQUID can be changed. Because of the Meissner effect, they can influence the magnetic flux in the regions around the pickup coils. The flux inequality in the separate sections is thereby removed by a perturbation in the field, i.e. by a local increase of the field B inside one section of the pickup coil and a decrease inside another one. Equality of the respective magnetic fluxes results [1].

The superconducting gradiometer includes all basic elements of the SQUID system (§7.1); RF and DC SQUIDs are used as sensors. The sections of the pickup coils are manufactured from a Nb wire whose diameter is 0.1 to 0.2 mm. A sensitivity limit of 10^{-13} T m^{-1} in the frequency interval $0 \leq f \leq 10$ Hz with $b = 0.2$ m has been achieved by SQUID gradiometers measuring the vertical gradient $\partial B_z/\partial z$. Josephson-point contacts made of Nb are commonly used as SQUID sensors [63].

The sensitivity limit of planar gradiometers used for the registration of the gradients $\partial B_z/\partial x$, $\partial B_x/\partial z$, etc. is 10^{-12} T m^{-1} in the interval $0 \leq f \leq 1$ Hz with $b = 0.1$ m. The dimensions of thin-film coils are 50×50 mm^2 and the value of the parameter F is $10^{-5} \div 10^{-4}$ [64]. A DC SQUID with Nb–In–Pb Josephson junctions shunted by a Au film has been used as a galvanometer. The sensor is directly immersed in liquid helium. The best result $F \approx 10^{-6}$ has been achieved by a multicomponent gradiometer whose sensitivity limit is 10^{-13} T m^{-1} in the frequency interval $0 \leq f \leq 1$ Hz, with $b = 0.3$ m [1]. A 19-channel DC SQUID magnetometer system for neuromagnetic investigations is described in [65]. The first-order gradiometers for sensing the signal are placed in a hexagonal configuration. DC SQUIDs based on niobium/aluminium technology have been developed leading to a field sensitivity of about 5 fT Hz$^{-1/2}$. SQUID read-out is realized with a resonant transformer circuit at 100 kHz. The multichannel control and detection electronics is compactly built.

Magnetocardiography and magnetoencephalography are two of the most important fields of application of SQUID gradiometers [63]. The characteristic magnitude of the magnetic field of the human heart is 10^{-11} T at a distance of 2 to 3 cm from the chest and the field of the brain is 10^{-13} to 10^{-12} T. Therefore if $b = 10$ cm, a sensitivity of 10^{-12} to 10^{-11} T will be necessary in the first case and 10^{-13} to 10^{-12} T in the second. Magnetocardiograms can be used for the early diagnosis of cardiac diseases. Infarction, for instance, can be predicted 15 to 20 days in advance, when conventional electrocardiography does not detect any symptoms. Another application of SQUID gradiometers is the early diagnosis of cancer by obtaining the precise topology of the magnetic field of certain organs, including the circulation of the blood. In this case complex multiparameter information is obtained by means of SQUID gradiometers. With this data, the disadvantages of single-parameter tests are avoided. SQUID systems with 50 to 100 gradiometric channels are needed for medical applications. Therefore, there are no alternatives to integrated thin-film sensors in such cases.

7.8. Concluding remarks

When SQUIDs are compared with other types of solid-state magnetosensor the following characteristics are of interest: the unsurpassed sensitivity, which is achieved both in extremely low d.c. magnetic fields and in constant fields that exceed by 6 to 9 orders of magnitude the minimum detectable variation ΔB. Therefore, the registration of extremely low magnetic fields, biomagnetic fields in particular, is an important application of SQUIDs that contributes to our knowledge about different phenomena in nature. The sensors described in Chapters 4 to 6 cannot be practically used in such investigations. The second advantage of a SQUID is the high stability of the zero output signal, i.e. $V_{\text{off}} = 0$. After several hours of uninterrupted operation, the drift of the zero does not exceed 10^{-13} T [1]. So far, there have been no reports of Hall sensors or magnetotransistors achieving comparable results. The third remarkable feature of SQUIDs is the existence of a magnetosensitivity axis, whose direction can be precisely determined. Therefore the problem of space orientation can be easily and efficiently solved. The fourth favorable property of SQUIDs is the possibility of combining a magnetometer and a gradiometer into a miniature system and simultaneously obtaining all parameters of the magnetic field at a given "point".

Owing to their unsurpassed transducing efficiency, SQUIDs are a unique instruments with numerous applications, such as medical experiments and examinations, space research, geophysical investigations, geological surveys, etc. Several years ago, the superconducting compound Nb_3Ge was a promising candidate for a transition from helium to hydrogen temperatures. However, the sudden discovery of high-T_c materials has upset our concept of the nature of superconductivity. Therefore, SQUID sensors based on these new composites are the forthcoming new wave in magnetometry.

References

[1] S.I. Bondarenko and V.I. Sheremet, Primeneniye sverhprovodimosty v magnitnich izmereniya, Energoatomizdat, St. Petersburg, 1982 (in Russian).
[2] H. Koch, in: Sensors, vol. 5, R. Bolk and K.J. Overshott (eds.), VCH, Weinheim, 1989, 381–445.
[3] R.C. Jaklevic, J. Mercereau and A.H. Silver, Phys. Rev., A140 (1965) 1628–1637.
[4] N. Nissenoff, Rev. Phys. Appl., 5 (1970) 21–24.
[5] J.M. Jaykox and M.B. Ketchen, IEEE Trans. Magn., MAG-17 (1981) 400–403.

[6] M.B. Ketchen and J.M. Jaykox, Appl. Phys. Lett. 40 (1982) 736–738.
[7] M. Gurvitch, M.A. Washington and H.A. Huggins, Appl. Phys. Lett., 42 (1983) 472–474.
[8] R.H. Koch, C.P. Umbach, G.I. Clark, P. Chaudhari and R.B. Laibowitz, Appl. Phys. Lett., 51 (1987) 200–202.
[9] J. Clarke, Proc. IEEE, 77 (1989) 1208–1223.
[10] J. Clarke, W.M. Goubau and M.B. Ketchen, J. Low Temp. Phys., 25 (1976) 99–144.
[11] D. Drung, Cryogenics, 26 (1986) 623–627
[12] N. Fujumaki, H. Tamura, T. Imamura and S. Hasuo, IEEE Trans. Electron Devices, ED-35 (1988) 2412–2418.
[13] J. Clarke and R.H. Koch, Science, 242 (1988) 217–223.
[14] C.D. Tesche and J. Clarke, J. Low Temp. Phys., 27 (1977) 301–331.
[15] J.J.P. Bruines, V.J. de Waal and J.E. Mooij, J. Low Temp. Phys., 46 (1982) 383–386.
[16] V.J. de Waal, P. Schrijner and R. Liurba, J. Low Temp. Phys., 54 (1984) 215–232.
[17] A.H. Silver and J.E. Zimmerman, Phys. Rev., 157 (1967) 317–341.
[18] R.L. Forgacs and A. Warnick, Rev. Sci. Instrum., 38 (1967) 214–220.
[19] J. Clarke, Philos. Mag., 13 (1966) 115.
[20] J. Clarke, J. Low Temp. Phys., 25 (1976) 99–144.
[21] W.H. Chang, J. Appl. Phys., 50 (1979) 8129–8134.
[22] W. Richter and G. Albrecht, Cryogenics, 15 (1975) 148–149.
[23] H. Ohta, IEEE Trans. Magn., MAG-23 (1987) 1072–1075
[24] E. Houwman et al., IEEE Trans. Magn., MAG-25 (1989) 1147–1150.
[25] M.W. Cromar and P. Carelli, Appl. Phys. Lett., 38 (1981) 723–725.
[26] M.F. Sweeny, IEEE Trans. Magn., MAG-21 (1985) 656–657.
[27] H. Koch, in: SQUID'85 — Superconducting Quantum Inerf. Devices and Their Applications, H.D. Hahlbohm and H. Lübig (eds.), W. de Gruyter, Berlin, 1985, 773–777.
[28] M.S. Dilorio and M.R. Beasley, IEEE Trans. Magn., MAG-21 (1985) 532–535.
[29] J. Clarke, Physica, 126 B (1984) 441–448.
[30] B. Muehlfelder et al., Appl. Phys. Lett., 49 (1986) 1118–1120.
[31] J. Knuutila et al., J. Low Temp. Phys., 68 (1987) 269–284.
[32] F. Wellstood et al., Rev. Sci. Instrum., 55 (1984) 952–957.
[33] R.H. Koch et al. J. Low Temp. Phys., 51 (1983) 207–224.
[34] C.T. Rogers and R.A. Buhrman, Phys. Rev. Lett., 53 (1984) 1272–1275.
[35] P. Dutta and P.M. Horn, Rev. Mod. Phys., 53 (1981) 497–516.
[36] C.D. Tesche, Proc. 17th Int. Conf. on Low Temp. Phys., LT-17 (1984) 263–264.
[37] A.H. Silver and J.E. Zimmerman, Phys. Rev., 157 (1967) 317–341.
[38] J.E. Zimmerman, P. Thiene and J.T. Harding, J. Appl. Phys., 41 (1970) 1572–1580.
[39] J.E. Zimmerman, J. Appl. Phys., 42 (1971) 4483–4487.
[40] J.E. Mercerau, Rev. Phys. Appl., 5 (1970) 13–20.
[41] C.M, Falko and W.H. Parker, J. Appl. Phys., 46 (1975) 3238–3243.
[42] R.L. Fagaly, Sci. Prog. (Oxford), 71 (1987) 181–201.
[43] J. Kurkijarvi, Phys. Rev., B6 (1972) 832–835.
[44] L.D. Jackel and R.A. Buhrman, J. Low Temp. Phys. 19 (1975) 201–246.
[45] H.N. Hollenhorst and R.P. Giffard, J. Appl. Phys., 51 (1980) 1719–1725.
[46] R.P. Giffard and J.N. Hollenhorst, Appl. Phys. Lett., 32 (1978) 767–769.
[47] J. Kurkijarvi and W.W. Weeb, in: Proceedings of Applied Superconductivity Conf., 1972, 581–587.
[48] J. Kurkijarvi, J. Appl. Phys., 44 (1973) 3729–3733.
[49] H. Ahola, G.H. Ehnholm, B. Rantala and P. Ostman, J. Physique, 39 (1978) 1184–1185.
[50] A.H. Silver and J.E. Zimmerman, Phys. Rev., 157 (1967) 317–341.
[51] V. Danylov, K.K. Liharyov and O.V. Snigiryov, Radiotechnika i Elektronika, 22 (1977) 18–26 (in Russian).
[52] V.N. Alfeev, Poluprovodniky, sverhprovodniky i paraelektriky v crioelektronike, Sovetskoe Radio, Moskwa, 1979, Ch.3 (in Russian).
[53] J.G. Bednorz and K. Muller, Z. Phys., B64 (1986) 189–193.

[54] R.H. Koch, C.P. Umbach, G.J. Clark, P. Chaudhari and R.B. Laibowitz, Appl. Phys. Lett., 51 (1987) 200–202.
[55] R.L. Sandstrom, W.J. Gallagher, T.R. Dinger, R.H. Koch et al., Appl. Phys. Lett., 53 (1988) 444–446.
[56] M. Decroux, Sensors and Actuators, A, 21–23 (1990) 9–14.
[57] R.H. Koch, W.J. Gallagher, B. Bumble and W.Y. Lee, Appl. Phys. Lett., 54 (1989) 951–953.
[58] R.H. Koch, Bull Am. Soc., 34 (1989) 573.
[59] J.E. Zimmerman, J.A. Beall, M.W. Cromar and R.H. Ono, Appl. Phys. Lett., 51 (1987) 617–618.
[60] Y. Zhang et al., IEEE Trans. Magn., MAG-25 (1989) 869–871.
[61] T. Ryhanen and H. Seppa, IEEE Trans. Magn., MAG-25 (1989) 881–884.
[62] S. Kataoka, S. Tsuchimoto, H. Nojima, R. Kita et al., Sensor Mater., 1 (1987) 7–12.
[63] Y. A. Holodov, A.N. Kozlov and A.M. Gorbach, Magnitniye polya biologicheschih obektov, Nauka, Moskwa, 1987 (in Rusian).
[64] M.B. Ketchen, W.M. Goubau, J. Clarke and G.B. Donaldson, J.Appl. Phys., 44 (1978) 4111–4116.
[65] J. Flokstra, H.J.M. der Brake, E.P. Houwman, D. Veldhuis et al., Sensors and Actuators, A, 27 (1991) 781–788.

Chapter 8

Functional magnetic-field sensors

Device integration by means of proper processing techniques is the basic principle in the construction of both conventional ICs and integrated magnetosensors. It imposes the need for the miniaturization of individual active and passive elements. Regions with transistors, diodes, resistors, Hall devices or other types of transducers, capacitors, inductance elements, etc. can be identified in such an integrated circuit. The number of devices and connections between them grows as the complexity of the functions performed is increased. If a multisensor IC for the registration of more than one nonelectrical quantity is to be manufactured, individual sensor regions for the measurement of magnetic fields, temperature, chemical composition, etc. must be formed in the substrate. The integration of several hundred thousand elements on a single chip is a very difficult design and manufacturing problem, since the layout and the physical limitations of the chip's density and complexity become factors of primary importance.

The functional approach to the construction of ICs and microsensors allows the achievement of the desired electronic properties and new characteristics by making use of the physical effects in solid-state materials. The conventional discrete elements are identified by local volumes in the chip having the necessary properties to perform the required functions. In the case of multisensing, for instance, one and the same active transducing region is used for the measurement of more than one nonelectrical quantity like a magnetic field, including its components B_x, B_y, B_z, the magnetic-field gradient, temperature, light intensity, etc. In the author's opinion some unexpected results in the field of magnetosensors have led to the question as to what else, in addition to the conversion of magnetic energy into an electric signal, is "hidden" in the active regions of solid-state transducers: what parallel processes take place that might yield new information by signal processing or by the formation of additional output terminals? This functional approach expressed in concrete forms has led to the invention of multisensors, 2-D and 3-D vector magnetometers, high-resolution galvanomagnetic gradiometers and the other types of devices considered in this chapter.

8.1. Functional multisensors

8.1.1. Multisensors for magnetic fields and temperature

(a) Device structure and performance principle. Figure 8.1 illustrates schematically some of the most frequently considered versions of differential BMTs (§6.3). A possible onset circuitry of the device is shown in Fig. 8.1a. The analysis of the bipolar microdevices in Fig. 8.1 has shown that they include a *p–n* thermometer, the output signal being the emitter–base bias $V_{EB}(T)$ [1]. This important and useful property of the BMTs was discovered recently, but the linear dependence of the forward bias of a *p–n* junction, V_{p-n}, on the temperature, T, when supplied with a constant current is well known [2,3]. For example, the structures (Fig. 8.1) contain three *p–n* junctions, but only the emitter–base junction is forward biased and may perform a thermometric function at a constant current supply. In this case the forward bias in the space-charge region of the emitter *p–n* junction, V_{p-n}, decreases linearly with increasing temperature:

$$V_{p-n} = \frac{\Delta E_g}{q} - \left[\left(\frac{k_B T}{q}\right)\ln\left(\frac{I_{s\infty}}{I_{EB}}\right)\right] = \varphi_0 - \left[\frac{T}{11600}\ln\left(\frac{I_{s\infty}}{I_{EB}}\right)\right], \tag{8.1}$$

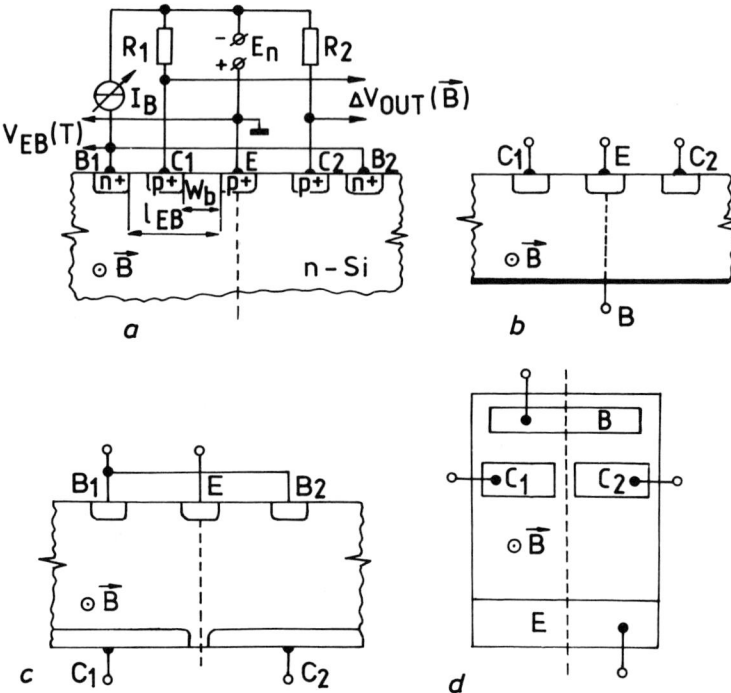

Fig. 8.1. Differential bipolar magnetotransistors for multisensing [1].

where $\varphi_0 = \Delta E_g/q$ is the band gap in Si, $I_{s\infty}$ is the saturation current at $T \to \infty$, $I_{EB} \ll I_{s\infty}$, T is the temperature of the crystal and the other notations are as usual.

The coefficient $K_{p-n} = \partial V_{p-n}/\partial T$ is negative; $K_{p-n} = -(k_B/q)\ln(I_{s\infty}/I_{EB})$. The voltage drop V_B on the resistance R_B of the base magnetotransistor region should be considered for the typical case $V_B = I_{EB}R_B$, where $R_B \sim \rho_B \sim 1/qn_0\mu_n$. In fact, the resistance R_B increases linearly with temperature T because of decreasing carrier mobility, $\mu_n(T)$. The temperature coefficient K_B is positive, and therefore the resultant coefficient $K_{EB} = \partial V_{EB}/\partial T$, which depends on the distance l_{EB} and the current I_{EB}, is smaller than the standard value $K_{p-n} \approx -2$ mV °C^{-1}.

The above conclusions refer to structures where $I_E = I_B = I_{EB} = $ const. Only the base recombination current I_B is constant for magnetotransistors, for example in the common-emitter configuration (Fig. 8.1a):

$$I_B = I_E - 2I_{C1,2} = \text{const.} \tag{8.2}$$

This is why, depending on the current gain factor, $\alpha_E = I_C/I_E$, of the magnetotransistor, the emitter and base currents will differ in value, i.e. $I_E \neq I_B$. However, the resistances of the reverse-biased p–n collector junctions C_1 and C_2 are sufficiently high, and we may also assume that, for $\boldsymbol{B} = 0$, the current $I_{C1} \simeq I_{C2} = $ const. Therefore, according to equation (8.2), the emitter current of the device is constant, i.e. $I_E = $ const. This general property of the BMT in Fig. 8.1a provides grounds for considering the bias $V_{EB}(T)$ as a linear thermometric signal. In this case, the variation with temperature of the collector saturation current I_S is neglected. This is justified if $I_S(T_{\max}) \ll I_{C1,2}$, where $T_{\max}$ is the maximum working temperature of the sensors. Regardless of the transduction mechanism of the magnetic field for the devices in Fig. 8.1, in the presence of a linear output signal $\Delta V_{out}(\boldsymbol{B})$ and because of their differential character, the equation $\Delta I_{C1}(\boldsymbol{B}) = -\Delta I_{C2}(\boldsymbol{B})$ is valid for the changes in the collector currents $I_{C1}(\boldsymbol{B})$ and $I_{C2}(\boldsymbol{B})$. This is why the external magnetic field $\boldsymbol{B}$ does not change the emitter current I_E either.

(b) Conditions for temperature and magnetic-field multisensing. In order to reveal the metrological properties of the selected thermometric signal $V_{EB}(T)$, the disturbing effect of the external magnetic field $\boldsymbol{B}$ on it should be analyzed. In general, the main galvanomagnetic phenomena associated with the devices in Fig. 8.1 are the magnetoresistance and magnetodiode effects, basically related to the pure Lorentz deflection, and the generation of Hall voltage by the majority carriers. We will examine separately the effect of each of these on the thermometric signal $V_{EB}(T)$.

The well-known magnetoresistance effect (§2.4) does not depend on the polarity of the magnetic field $\boldsymbol{B}$, and simply results in an increase of the resistance of the emitter–base region. This is why the component $\Delta V_{MR} \sim \mu_{p,n}^2 B^2$ is further added to the bias V_{EB}, i.e. $V_{EB} + \Delta V_{MR}$. However, owing to nonequilibrium bipolar conductivity and the diffuse character of the processes in the structures at $l_{EB} \sim L_{p,n}$, the effective velocities of the electrons and holes, respectively, are very low [4]. The values of μ_n and μ_p in Si are not high either, and it has been found experimentally that, for magnetic fields $0 < B < 1$ T in short Si diodes, i.e. $l_{EB} \sim L_{p,n}$, the variation of the resistance is very small and does not exceed $\leq 0.4\%$ [5,6].

At very high levels of emitter injection, when $n \gg n_0$ and $p \gg p_0$, the Lorentz deflection can generate an MD effect (§2.5). A crucial requirement for magnetodiode operation is that the difference between the recombination rates is large, i.e. $s_1 \gg s_2$, and the thickness d should be of the order of the ambipolar diffusion length $L_{p,n}$, $d \geq L_{p,n}$. Therefore the magnetic field causes a polar change in the bias V_{EB}, i.e. $V_{EB} \pm V_{MD}(B)$. In the general case $V_{MD}(+B) \neq V_{MD}(-B)$. This is why, even for structures similar to that in Fig. 8.1a, c with short circuit contacts B_1 and B_2, the difference $(\Delta V_{MD}^L - \Delta V_{MD}^R)$ will be added to the bias V_{EB}. This would disturb the metrological properties of the thermometric signal. The construction and mode of operation of BMTs can always be optimized so that there is no possibility of a magnetodiode effect being exerted.

The sensor mechanism in magnetotransistors is also defined by the emitter-injection modulation (§6.2.1.2c). The magnetosensitivity for the specific type of differential BMT in Fig. 8.1a will be defined by the redistribution of the emitter current I_E to the left and right of the emitter E, due to the Hall voltage V_H generated by the base drift currents I_{B1} and I_{B2}. At low values of the argument $V_H \sim I_{EB}B$ in equation (6.9), we can adopt the linearization expression $I_E^L/I_E^R = 1 + qV_H/k_BT + \ldots$. When this condition is satisfied the connection of contacts B_1 and B_2 provide the redistribution of the emitter current I_E to B_1 and B_2 in such a way that the sum of the respective components remains the same, i.e., $I_E = I_E^L + I_E^R = $ const. Actually, the values of the effective Hall voltage, V_H^*, generated by the base drift currents I_{B1} and I_{B2} are small. The influence of V_H^* on the contacts B_1 and B_2 is opposite in sign and, because of its relatively small value with respect to the voltage $V_{EB} \sim 0.8$–1.0 V, i.e., $V_H^* \ll V_{EB}$, it is symmetric. Therefore, with the assumed linear approximation of the disturbing effect of the emitter-injection modulation on the bias V_{EB}, the thermometric signal $V_{EB}(T)$ may be considered to be independent with regard to the examined transduction mechanism for both polarities of the field B. For specifically designed BMT structures providing for a high value of the voltage V_H^*, a nonlinear variation of the thermometric signal $V_{EB}(T)$ may be expected because of the magnetic field (§6.2.1.2c).

The galvanomagnetic effects explained above exert a combined influence on the thermometric signal $V_{EB}(T)$. Proceeding from the analysis of these effects, for example for the element in Fig. 8.1a, we may conclude that there are certain practical conditions under which, within the range $0 \leq B \leq 1$ T, the external magnetic field B does not affect the bias $V_{EB}(T)$. In order to minimize the magnetoresistance and to eliminate the MD effect, the base length of the sensor $l_{EB1} = l_{EB2}$ should be 1 to 2 times $L_{p,n}$ and the thickness $d > L_{p,n}$, the structure should be symmetric about the emitter E, and the two base contacts B_1 and B_2 should be directly connected. For such a version of the BMT, the output signal $\Delta I_{C1,2}(\boldsymbol{B})$ is a linear function of the magnetic field, since the Lorentz deflection of the minority carriers and the small values of the effective Hall voltage in the emitter-injection modulation determine the linear transduction efficiency. In a similar way, the differential BMTs in Fig. 8.1b–d may be optimized in order to make the thermometric bias $V_{EB}(T)$ independent of the magnetic field $\boldsymbol{B}$.

The metrological properties of the linear thermometric bias $V_{EB}(T)$ reveal the interesting possibility of reducing the temperature dependence of the output signal $\Delta V_{C1,2}(\boldsymbol{B}, T)$ in an appropriate way (described in Chapter 9).

In general, differential BMTs of the type shown in Fig. 8.1, under conditions where the magnetoresistance effect is minimized, the MD effect is absent and the emitter–base circuitry has been supplied by a constant current, will permit the simultaneous measurement of magnetic field and temperature. In this case, BMTs may be considered as a class of functional multisensors without any other additionally formed contacts. According to the concept in §1.6, multisensors can be regarded as smart solutions.

(c) Multisensor characteristics. Figure 8.2 illustrates the temperature dependence of the bias $V_{EB}(T)$ for silicon p^+–n–p^+ version of the device in Fig. 8.1a with a base current $I_B = I_{B1} + I_{B2}$ as a parameter in the temperature interval $0 \leq T \leq 80°C$. The lateral base width $W_b = l_{EC1} = l_{EC2}$ is 30 μm, the distance between the emitter and the base contacts is $l_{EB1} = l_{EB2} = 70$ μm and $L_p \sim 40$ μm [1]. The characteristics obtained are linear, the maximum deviation from linearity reaching no more than $\sim$1.2%. The coefficient K_{EB} decreases with increasing current I_B. The variation with temperature of the collector currents $\Delta I_{C1}(T)$ and $\Delta I_{C2}(T)$ is monotonous throughout the entire range $0 \leq T \leq 80°C$ and represents $\sim$2% for $I_{C1,2}(T = 20°C) \geq 250$ μA.

The influence of the polarity and induction of the external magnetic field $\boldsymbol{B}$ on the thermometric signal $V_{EB}(T)$ within the range $0 \leq T \leq 80°C$ has been investigated. It has been found that this influence is nonlinear, while, for magnetic fields $B \leq 0.6$ T, it is not observed. However, for fields $B > |0.6T|$, the variation of the bias $V_{EB}(T)$ does not depend on the polarity of $\boldsymbol{B}$, and for $B = \pm 1$ T, the maximum variation of $V_{EB}(T)$ is 0.5 mV. Therefore, in the case of $B \leq 0.6$ T, $\partial K_{EB}/\partial B = 0$. For magnetic fields $0.6 < B < 1$ T, for example at the current $I_B = 2mA$, when $\Delta V_{EB} = V_{EB}(0°C) - V_{EB}(80°C) \simeq 100$ mV, the disturbing effect of the field $B = \pm 1$ T constitutes $\sim$0.5% of ΔV_{EB}. When summing up the metrological properties of the thermometric signal $V_{EB}(T)$, one could note that, in a magnetic field $B \leq 0.6$ T, the main error is determined by the nonlinearity of the temperature characteristic $V_{EB}(T)$ and constitutes $\sim$1.2%, while for a magnetic field in the range $0.6 < B \leq 1$ T, the error is combined and reaches $\sim$2%.

For more detailed information, Fig. 8.3 shows the magnetic control of the

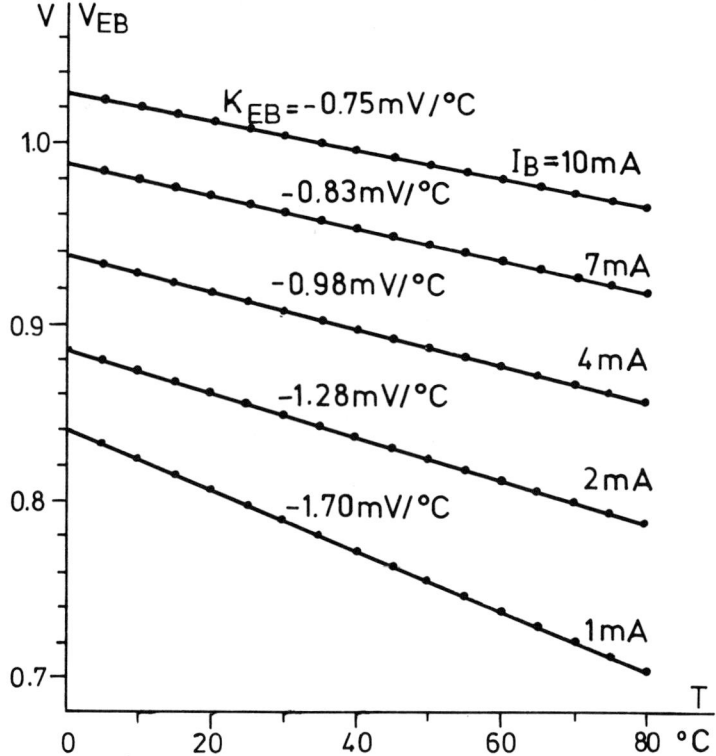

Fig. 8.2. Temperature dependence of the bias V_{EB} for the device in Fig. 8.1a with the current I_B as a parameter and $B = 0$ [1].

differential output signal $\Delta I_{C1,2}(B)$ of the multisensor. One can see that the device is a linear transducer. The coefficient T.C. for the range $0 \leq T \leq 80°C$ is constant: T.C. $= -0.34\%$ °C^{-1} [1].

The functional multisensor in Fig. 8.1a has also been used for experiments at cryogenic temperatures ($T = 77$ K). All samples have been subjected beforehand to training which includes cooling–heating cycles. In such a way it has been established that in spite of the mechanical strain and thermal stress with a high negative gradient, multisensors preserve their working capacity under such "abnormal" conditions. When the temperature is decreased to $T = 77$ K the voltage–current characteristic is shifted in the direction of the growth of the voltage V_{EB}. The magnetic control of the currents $I_{C1}(B)$ and $I_{C2}(B)$ at $T = 77$ K with a fixed polarity of B is illustrated in Fig. 8.4a.

At $T = 77$ K the output characteristics are nonlinear in general, unlike the linearity observed at $T = 300$ K and $B \leq 1$ T. This fact is clearly shown in Fig. 8.4b as well. The $I_{C1}(B)$ with the two polarities $\pm B$ is presented in the figure. At $B \leq 0.4$ T the dependence can be readily approximated by an exponential law, i.e.

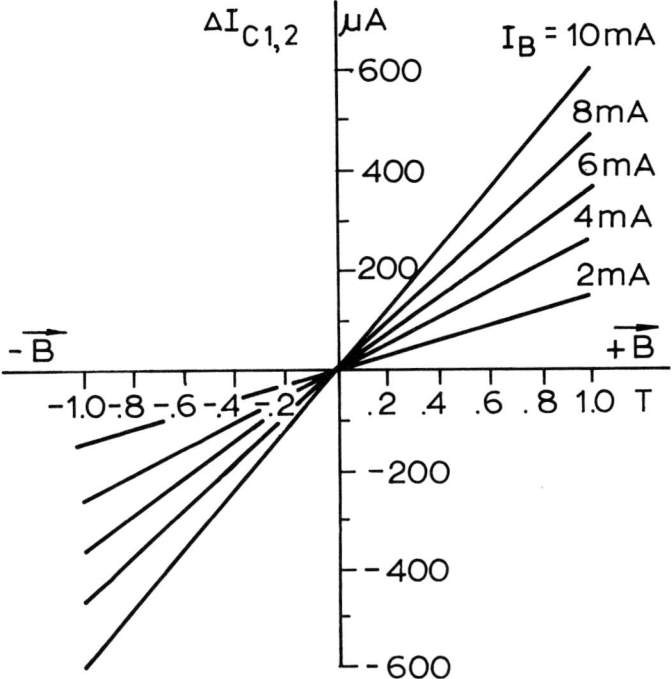

Fig. 8.3. Output collector signal $\Delta I_{C1,2}$ as a function of the external field B, $T = 20°C$ [1].

$I_C(B) \sim \exp(B)$. There is an interval, however, i.e. $-0.25 \leq B \leq +0.25$ T, within which the output $\Delta I_{C1,2}(B) = I_{C1}(B) - I_{C2}(B)$ is a linear function of the field B. With increasing $B > 0.3$ T, the characteristics $I_C(+B)$ tend to saturation (Fig. 8.4a). The multisensor in Fig. 8.1a, at $T = 77$ K in the interval of output-signal linearity with respect to the field B, exhibits a magnetosensitivity an order of magnitude higher than that at $T = 300$ K. It has been established by experiment that at $T = 77$ K an external magnetic field of magnitude $B \leq 0.12$ T does not influence the thermometric signal $V_{EB}(T)$. At $B > 0.12$ T a magnetoresistance component of the voltage V_{EB} is generated. Satisfactory reproducibility of the data taken for one and the same sample has been observed during a great number of cooling–heating cycles. The principal conclusion from the experimental results is that, at $T = 77$ K, the multisensor function of the device is preserved and its magnetosensitivity is considerably increased.

8.1.2. Multisensors for magnetic fields, temperature and light

The next step in the development of the multisensing functional approach is to increase the number of measured variables. It has been established that the transducing region of the device in Fig. 8.1a can be used to measure a radiant

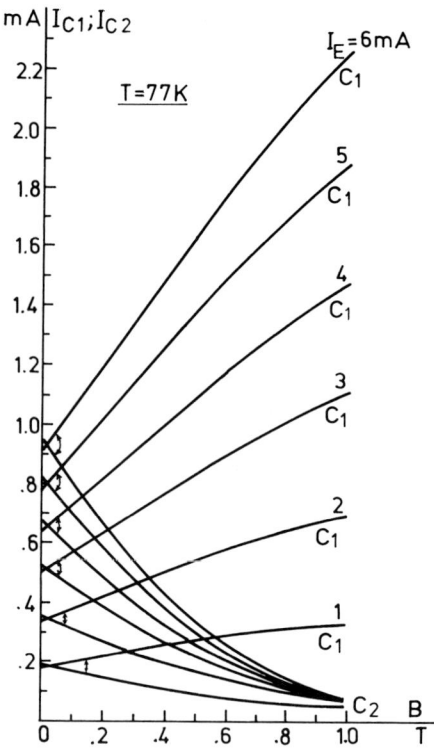

Fig. 8.4. (a) Magnetic control of the collector currents $I_{C1}(B)$ and $I_{C2}(B)$ of the BMT shown in Fig. 8.1a with a fixed polarity of B at $T = 77$ K.

flux in addition to magnetic fields and temperature [7]. Useful information can be derived by means of simple signal processing.

A circuit that includes a BMT structure like that in Fig. 8.1a is presented in Fig. 8.5. The data concerning the homogeneous magnetic field and temperature are obtained in the same way and under the same conditions as discussed in §8.1.1. This is why only the registration of the radiant flux will be considered here together with related details. The principle of light detection is analogous to the mechanism of the action of the well-known photodiode sensors operating in the reverse bias mode [8]. The two reverse biased p^+–n collector junctions C_1 and C_2 in Fig. 8.5 can be regarded as such diodes. The radiant flux arriving at the top surface, provided $h\nu > E_g$, will generate a nonequilibrium electron–hole plasma, where $h\nu = hc/\lambda$ is the energy of the incident photons, ν is the frequency of the electromagnetic radiation, c is the speed of light in vacuum and λ is the wavelength of the radiation; E_g is the bandgap. If the radiation is assumed to be completely absorbed within a layer of thickness w in the surface region of the quasi-neutral n-type base and if w is smaller than the diffusion length L_p,

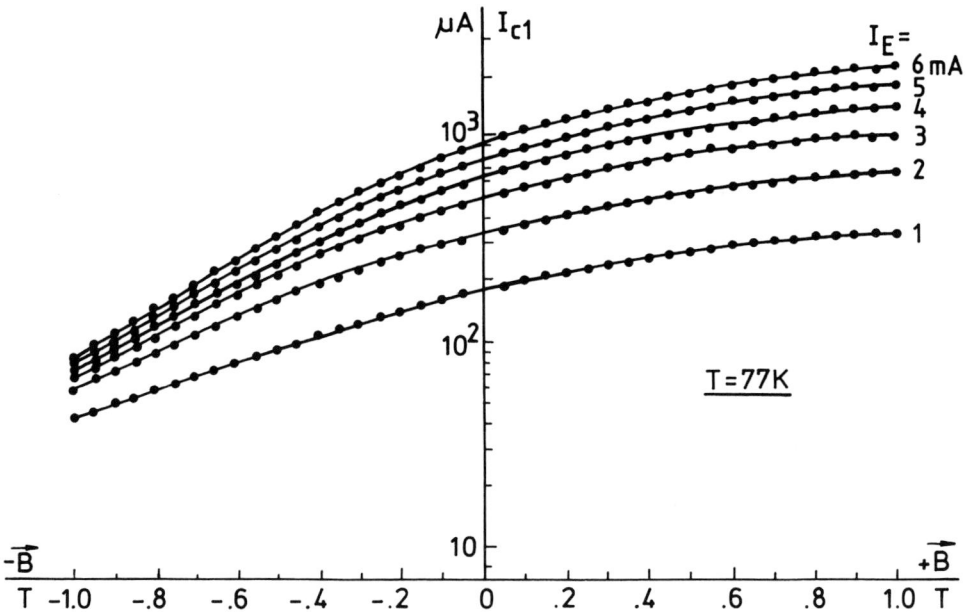

Fig. 8.4 (continued). (b) $I_{C1}(B)$ characteristics of the BMT shown in Fig. 8.1a with the two polarities of B at $T = 77$ K.

i.e. $L_p > w$, the nonequilibrium holes will be able to reach the two p^+-n collector junctions C_1 and C_2. As they enter the depletion layers of the collector junctions, the reverse collector currents are increased to values that exceed the initial (without radiation) currents $I_{C1}(0)$ and $I_{C1}(0)$. At sufficiently high voltages V_{C1B} and V_{C2B}, the current–voltage characteristics of the two collector junctions with a radiant flux $\Phi_l \neq 0$ are given by the expression:

$$I_{C1,2}(\Phi_l) = -[I_{C1,2}(0) + I_{\Phi_l}] = -I_{C1,2}(0) - qK\beta A\Phi_l, \tag{8.3}$$

where $I_{C1,2}(0)$ are the respective dark currents of the collectors C_1 and C_2 at $\Phi_l = 0$, and $I_{C1,2}(0) = \alpha I_E/2$; $I_{\Phi l}$ is the photo-current due to the nonequilibrium holes generated by the flux Φ_l; β is the internal quantum efficiency defined as the number of excess carriers generated per incident photon; K is a dimensionless coefficient that characterizes the portion of radiation absorbed in the n-base region, $K < 1$; A is the effective photosensitive area of the top surface of the structure [9].

According to (8.3) both collector currents $I_{C1}(\Phi_l)$ and $I_{C2}(\Phi_l)$ are proportional to the concentration of excess minority carriers generated by the light flux Φ_l: $I_{C1,2}(\Phi_l) = S_i \Phi_l$, where S_i is the integral photosensitivity, i.e. the two currents are linear functions of the flux Φ_l. The bandgap of Si is $E_g \approx 1.12$ eV, and this value corresponds to the absorbtion of those wavelengths of visible light within the interval $0.4 \mu m \leq \lambda \leq 0.76$ μm [8]. Therefore the silicon version of the device

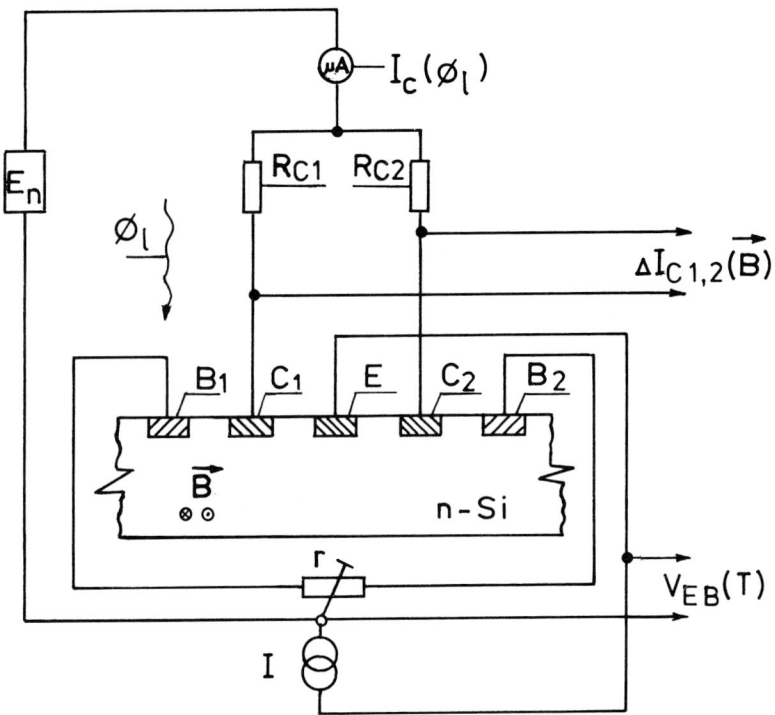

Fig. 8.5. Device structure and circuitry of a multisensor for magnetic fields, temperature and radiant flux.

in Fig. 8.5 can be used to measure the radiant flux Φ_l by means of the collector currents I_{C1} and I_{C2}. This problem needs a solution which allows the summation of the two variations $\Delta I_{C1}(\Phi_l)$ and $\Delta I_{C2}(\Phi_l)$, i.e. $\Delta I_{C1}(\Phi_l) + \Delta I_{C2}(\Phi_l)$. It can be seen from Fig. 8.5 that the sum of the two currents can be measured between the two identical resistors, $R_{C1} = R_{C2}$, in the circuit and the constant voltage source E_n. In the absence of light at $I_E = 0$, the two saturation currents of the collector junctions are very small (10^{-14} to 10^{-11} A) and can be neglected. Such small values of p–n junction leakage currents are guaranteed by modern IC technology. Thus the information concerning the radiant flux is represented by the difference between the total collector current $I_C(\Phi_l) = I_{C1}(\Phi_l) + I_{C2}(\Phi_l)$ and its dark value $I_C(0) = I_{C1}(0) + I_{C2}(0)$ at $\Phi_l = 0$.

There is a basic problem, however, which concerns to what extent the results can be influenced by cross-sensitivity when B, T and Φ_l are simultaneously measured. Experiments performed with the circuit in Fig. 8.5 lead to the conclusion that in the general case, at a radiant flux $\Phi_l = 0$, the total current I_C, is disturbed by the magnetic field B. This fact naturally makes the metrological function of the current $I_C(\Phi_l)$ problematic. It has been unexpectedly established by analyzing the experimental data that the $I_C(B)$ dependence results from a BMT offset. The output BMT signal $\Delta I_{C1,2}(B)$ in a homogeneous field B is given by the general expression $\Delta I_{C1,2}(B) = SB + I_{\text{off}}(0)$, where S is the resultant transducing efficiency. Owing to the structural symmetry of the device, the sensitivities S_L, to the left of the emitter, and S_R, to the right of the emitter, are assumed to be equal, i.e. $|S_L| = |S_R|$; hence the resultant transducing efficiency will be $S = |S_L| + |S_R|$.

There is another aspect of the offset phenomenon which has not been discussed in published works so far. The electrical asymmetry of differential bipolar structures, i.e. $I_{\text{off}} \neq 0$, inevitably leads to galvanomagnetic asymmetry. This means that the results of the action of the Lorentz force action to the left and to the right of the emitter will be different owing to the different spatial distributions of majority and minority carriers in the BMT base region at $B = 0$. The considerations are simplified by the assumption that the Lorentz deflection of holes is the dominating transducing mechanism of the BMT. This is why the values of the magnetosensitivities $|S_L|$ and $|S_R|$ are different in the general case. Therefore, the variations of the two currents $I_{C1}(B)$ and $I_{C2}(B)$ also differ in a homogeneous field B, i.e. $\Delta I_{C1}(B) = S_L B \neq \Delta I_{C2}(B) = S_R B$, and, moreover, S_L and S_R have different signs. When there is an offset, $I_{\text{off}} \neq 0$, in a homogeneous field B, since $|S_L| \gtrless |S_R|$, the inequality $I_{C1}(0) + I_{C2}(0) \gtrless I_{C1}(B) + I_{C2}(B)$ is valid for the sum of the collector currents. Only in the case of zero offset ($I_{\text{off}} = 0$) is the equality $I_{C1}(0) + I_{C2}(0) = I_{C1}(B) + I_{C2}(B)$ satisfied owing to the validity of $|S_L| = |S_R|$ and $|\Delta I_{C1}(B)| = |\Delta I_{C2}(B)|$. The experimental verification of the above conclusions by means of the BMT in Fig. 8.5 completely confirms their validity and the adequacy of the model. The purposeful variation of I_{off} is accomplished by a low-resistance trimmer r connected between the base contacts B_1 and B_2. By changing the interbase currents I_{B1} and I_{B2}, the collector currents $I_{C1}(0)$ and $I_{C2}(0)$ are varied. The control of I_{off} by means of the resistors R_{C1} and R_{C2} is practically impossible since the resistances of the reverse biased p^+–n junctions of collectors C_1 and C_2 are several orders of magnitude higher than R_{C1} and R_{C2}. In addition, the collector junctions themselves play the role of current sources. This is why it is the trimmer r which can be used for the variation of $I_{C1}(0)$ and $I_{C2}(0)$. Typical behaviors of the currents $I_{C1}(B)$, $I_{C2}(B)$ and of the sum $I_C(B) = I_{C1}(B) + I_{C2}(B)$ as functions of the field B are presented in Fig. 8.6 [7].

Only when $I_{\text{off}} = 0$, i.e. $|S_L| = |S_R|$, is the sum $I_C(B)$ invariant under a homogeneous magnetic field. Therefore, a zero offset must necessarily be adjusted by the trimer r at $\Phi_l = 0$, in order to make the total current $I_C(\Phi_l)$ a source of trustworthy information about the radiant flux Φ_l. R_{C1} and R_{C2} in Fig. 8.5 are assumed equal. If at $I_{\text{off}} = 0$, a radiant flux Φ_l is incident at the top surface of the chip, the current I_C is a linear function of Φ_l (Fig. 8.7), and the cross-sensitivity associated with the field B is eliminated. The effect of the radiant flux Φ_l upon the metrological characteristics of the magnetosensitive and thermometric signals has been studied as well. It was found that, up to a certain threshold

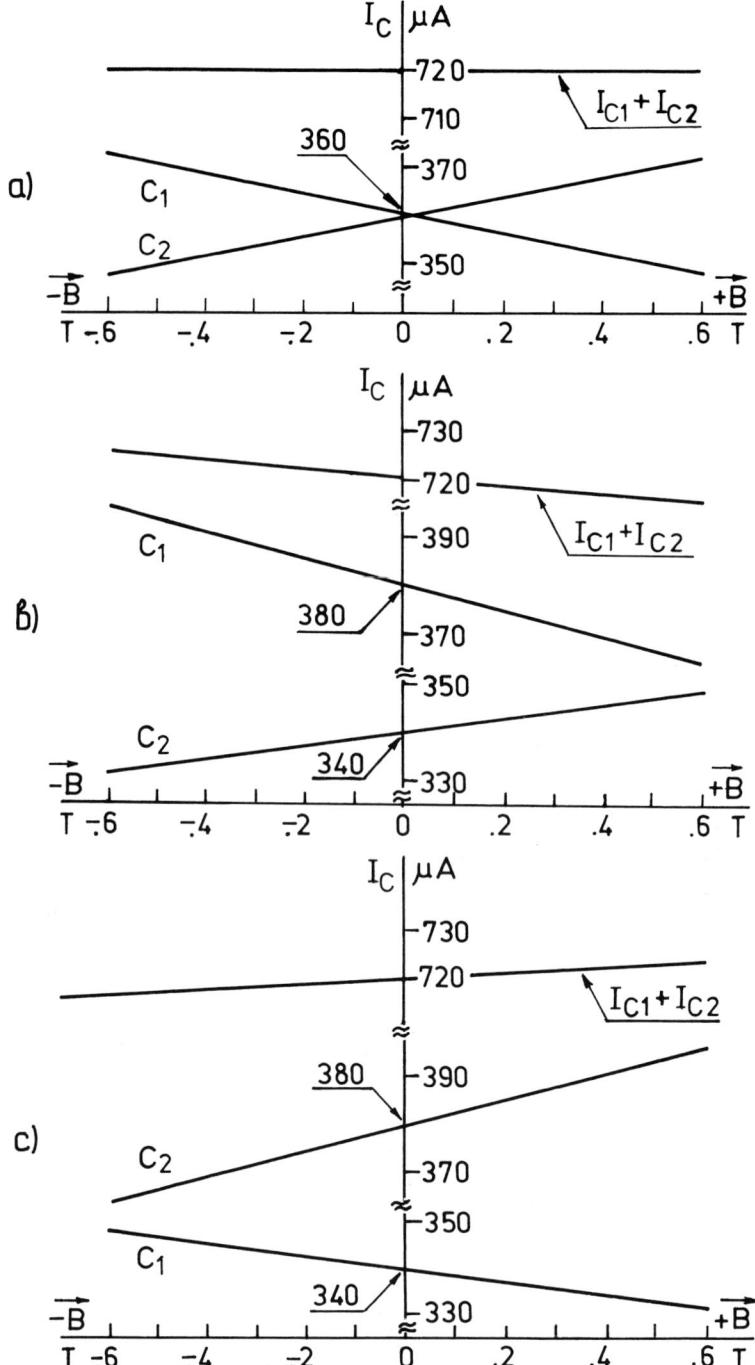

Fig. 8.6. Typical $I_{C1}(B)$, $I_{C2}(B)$ and $I_C(B) = I_{C1}(B) + I_{C2}(B)$ characteristics of the silicon p^+–n–p^+ sensor shown in Fig. 8.5; (a) $I_{\text{off}} = 0$; (b) $I_{\text{off}} = +40$ μA; (c) $I_{\text{off}} = -40$ μA.

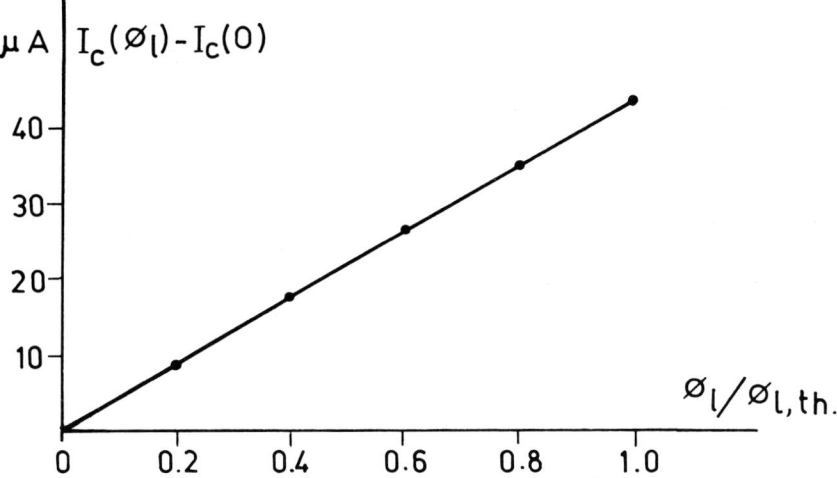

Fig. 8.7. The current $I_C(\Phi_l) - I_C(0)$ as a function of the radiant flux of visible light; $I_{EB} = 5$ mA, $V_{CB} = -10$ V.

value of the luminous flux $\Phi_{l,\text{th}}$, light does not noticeably influence $\Delta I_{C1,2}(\boldsymbol{B})$ and $V_{EB}(T)$. However, V_{EB} and the magnetosensitivity depend on the flux Φ_l when $\Phi_l > \Phi_{l,\text{th}}$. This adverse effect is less pronounced at higher values of the emitter current I_E. This kind of cross-sensitivity is caused by the considerable increase in the nonequilibrium bipolar conductivity in the surface region of the device, which changes the electrophysical characteristics of the multisensor. The behavior of the $I_C(\Phi_l) - I_C(0)$ signal at $\Phi_l \leq \Phi_{l,th}$ is illustrated in Fig. 8.7. The results discussed so far definitely confirm the applicability of the functional approach to multisensing.

8.2. Functional gradiometer sensors

The magnetic induction must be measured when nonhomogeneous magnetic fields are investigated (§7.7). In the general case, the gradient of the vector $\boldsymbol{B}$, i.e. grad $\boldsymbol{B} \equiv \nabla \boldsymbol{B}$, is a tensor with nine components [10]. There are two types of components $\partial B_i/\partial e_i$, i.e. $\partial B_x/\partial x$; $\partial B_y/\partial y$; $\partial B_z/\partial z$ and $\partial B_k/\partial l_k$, i.e. $\partial B_z/\partial x$; $\partial B_z/\partial y$; $\partial B_x/\partial z$, etc. Gradiometer sensors usually include two identical input transducers with equal sensitivities. They are connected in a differential circuit and have parallel orientations. The two sensors are separated by a fixed distance b. They can be Hall elements, BMTs, etc. In these cases the gradient is defined as $\Delta B/b$. The relatively low resolution determined by the dimensions of the sensors themselves is the limitation of this method. The gradient can be measured also by a single device which is moved at a fixed distance b in a given direction. The disadvantage of this solution is that it is not always possible to move the device.

The resolution of the magnetogradiometers can be enhanced by reducing the distance b in the direction of the measured difference ΔB and by increasing the magnetosensitivity of the solid-state sensor. The principle of functional multisensing requires an approach which allows the use of one and the same structure with a fixed spatial position for the determination of the magnetic-field gradient. In this case, the parameter b represents the dimension of the active transducing region, which can be tens of micrometers. Two functional solutions have been suggested so far. The first one is based on the Hall effect, while the second solution employs the recently discovered magnetogradient effect in differential BMTs with a linear output.

8.2.1. Hall-effect gradiometer sensors

Registration of the magnetic gradient by means of a single orthogonal Hall device, by contrast with the trivial differential circuit, is based on the summation of the two output signals V_{H1} and V_{H2} from terminals H_1 and H_2 (Fig. 8.8) [7].

The useful signal proportional to the magnetic-field gradient $\partial B_z / \partial x$ is the difference $(V_{H1} + V_{H2}) - V_{1,2}$, where $V_{1,2}$ is the voltage at the supply contacts C_1 and C_2. The signal processing is accomplished by the circuit presented in Fig. 8.8. The Hall structure must necessarily be long, i.e. $l \gg w \equiv b$, in order to eliminate the geometrical magnetoresistance. Moreover, a zero offset $V_{\text{off}}(0) = 0$ is assumed. The distribution of the magnetic induction is a linear function:

$$B_x(x) = B_{zo} + l(z)x, \tag{8.4}$$

where $l(z)$ is the measured gradient, $\partial B_z / \partial x = l(z)$.

Depending on the component of the magnetic field which is perpendicular to the plane of the device, the Hall field between terminals H_1 and H_2 can be written as:

$$E_H = R_H B_z(x) J, \tag{8.5}$$

where J is the supply current density, the other notations being as usual.

The Hall voltage between a point in the center of the device and terminal H_1 is

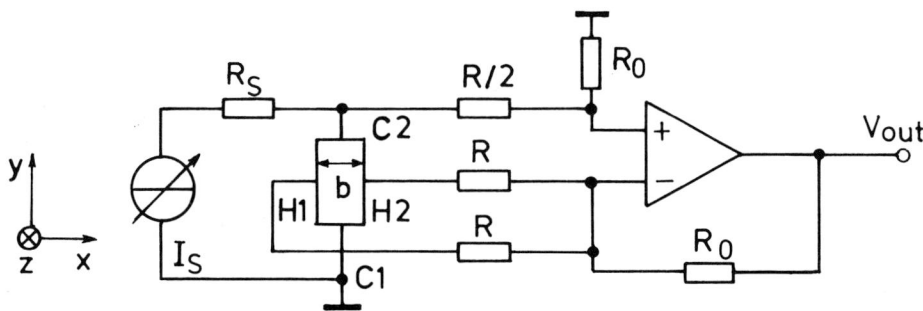

Fig. 8.8. A circuit for the measurement of the magnetic-field gradient by means of a Hall sensor.

determined by the integration of (8.5) over the interval $[0, b/2]$ in accordance with (8.4):

$$V_{H1} = \int_0^{b/2} R_H B_z(x) J \, dx = \frac{R_H}{t} I B_{z0} + \frac{R_H}{t} I l(z) \frac{b}{8}, \tag{8.6}$$

where t is the thickness of the Hall plate.

The Hall voltage between the center and the other Hall terminal (H_2) is obtained by the integration of (8.5) between 0 and $-b/2$:

$$V_{H2} = -\frac{R_H}{t} I B_{z0} + \frac{R_H}{t} I l(z) \frac{b}{8}. \tag{8.7}$$

The subtraction of (8.7) from (8.6) yields the well-known expression for the Hall voltage:

$$V_{H1} - V_{H2} = V_H = \frac{R_H}{t} I B_{z0}. \tag{8.8}$$

The summation of (8.6) and (8.7), however, leads to:

$$V_{H1} + V_{H_2} = \frac{R_H}{t} I \frac{b}{4} l(z). \tag{8.9}$$

Therefore the sum of the voltages $V_{H1} + V_{H2}$ determined with respect to the center of the Hall sensor is proportional to the x-component of the gradient of B_z, i.e. to the B_z gradient along the axis which connects the Hall terminals.

The summation of the two output voltages V_{H1} and V_{H2} determined with respect to one of the current contacts C_1 or C_2 is necessary for the practical implementation of the above property of the Hall sensor:

$$V_{H1} + V_{H2} = Ir + \frac{R_H}{t} I \frac{b}{4} l(z), \tag{8.10}$$

where r is the resistance of the Hall plate. The output voltage V_{out} is obtained by the subtraction of $Ir = V_{C1,C2}$ from (8.10):

$$V_{\text{out}} = V_{H1} + V_{H2} - V_{C1,2} = \frac{R_H}{t} I \frac{b}{4} l(z). \tag{8.11}$$

This signal processing is accomplished by the circuit in Fig. 8.8. The Hall sensor magnetosensitivity S can be used to write equation (8.11) in another form:

$$V_{\text{out}} = S \frac{b}{4} I l(z). \tag{8.12}$$

The high resolution determined by the distance b between terminals H_1 and H_2, which is usually 50 to 100 μm, is an important advantage of the Hall-effect gradiometer described. For instance, if the width b of a discrete Hall device is 0.5 mm and the sensitivity at a supply current of 0.1 A is $S = 1.0$ VA^{-1}T$^{-1} = 10^{-3}$ V A^{-1} mT^{-1} in a magnetic field whose gradient is $l(z) = 10$ mT mm^{-1}, the output voltage will be 250×10^{-6} V [11]. The main function of the Hall transducer for measuring the magnetic induction B_{z0} by means of the Hall voltage V_H [equation (8.8)] is preserved too. Therefore, long Hall structures can be regarded as functional sensors which measure independently and simultaneously the magnetic induction, the direction of the field $\mathbf{B}$ and the magnetic-field gradient.

8.2.2. BMT gradiometer sensors

BMT gradiometer sensors employ the magnetogradient effect (MGE) arising in linear output differential BMTs [12]. One of the possible device structures and the appropriate circuitry used to observe the MGE is the same as that in Fig. 8.5 except for the registration of the radiant flux Φ_l. At $B = 0$ the zero offset is adjusted as described in §8.1.2 to obtain the magnetosensitivities $|S_L| = |S_R|$. The MGE is, in fact, the linear control of the sum of the two collector currents, $I_C = I_{C1} + I_{C2}$, by the gradient $\nabla \mathbf{B}$ in a nonuniform magnetic field $\mathbf{B}$, when the gradient $\partial B_z/\partial x$ is oriented along the length $l_{C1,2}$ of the BMT. If the sign of the gradient changes, the polarity of $\Delta I_C(\mathbf{B})$ changes together with it. In a homogeneous magnetic field $\mathbf{B}_{z0}$, when $I_{\text{off}} = 0$, the current $I_C(\mathbf{B})$ remains constant, i.e. $I_C(B) = I_C(0)$. If a linear distribution of the magnetic induction B_z along the x-axis is assumed, i.e. $B_z(x) = B_{z0} + (\partial B_z/\partial x)x$, the corresponding variations of the two collector currents are $\Delta I_{C1} = S_L B_{z1}$ and $-\Delta I_{C2} = -S_R B_{z2}$. The values B_{z1} and B_{z2} correspond to the induction $B_z(x)$ at the collectors C_1 and C_2 separated by the distance $l_{C1,2}$. The change $\Delta I_C(\mathbf{B}) = I_C(\mathbf{B}) - I_C(0)$ of the current in a magnetic field with a linear distribution is a measure of the gradient, i.e. of the parameter $\partial B_z/\partial x$, since $\Delta I_C = S_L B_{z1} - S_R B_{z2} = S_{L,R}(B_{z1} - B_{z2})$. Therefore the magnetic induction gradient B_z in the region with a length $\Delta x = l_{C1,2}$ is given by the expression:

$$\frac{\Delta B_z}{\Delta x} = \frac{\Delta I_C(\mathbf{B})}{S_{L,R} l_{C1,2}}, \tag{8.13}$$

where $S_{L,R} = S/2$, and S is the BMT magnetosensitivity determined in a homogeneous magnetic field at $I_{\text{off}} = 0$.

This is why the change, ΔI_C, of the current is a linear function of $\nabla \mathbf{B}$ and does not depend on the induction B_{z0}. The high resolution of the BMT gradiometer is determined by the distance between the two collectors which can be tens of micrometers in integrated versions. It has been established from experiments with a BMT gradiometer sensor like the device in Fig. 8.5, for instance, that, with $l_{C1,2} = 0.15$ mm (on the mask) and sensitivity $S_{L,R} = 300$ μA T^{-1} in a magnetic field whose gradient $\partial B_z/\partial x$ is 40 mT mm^{-1}, 57 mT mm^{-1}, 64 mT mm^{-1} and 78 mT mm^{-1}, the respective values of the current variation ΔI_C are 1.8 μA, 2.6 μA, 2.9 μA and 3.5 μA [12].

Another version of a one-component magnetogradiometer based on a differential drift-aided BMT is presented in Fig. 8.9. Its specific feature is the sectioning of the base contact B_1 into two identical parts B'_1 and B''_1. A zero offset is adjusted by the trimmer r at $I_E = $ const, $I_B = $ const and $B = 0$.

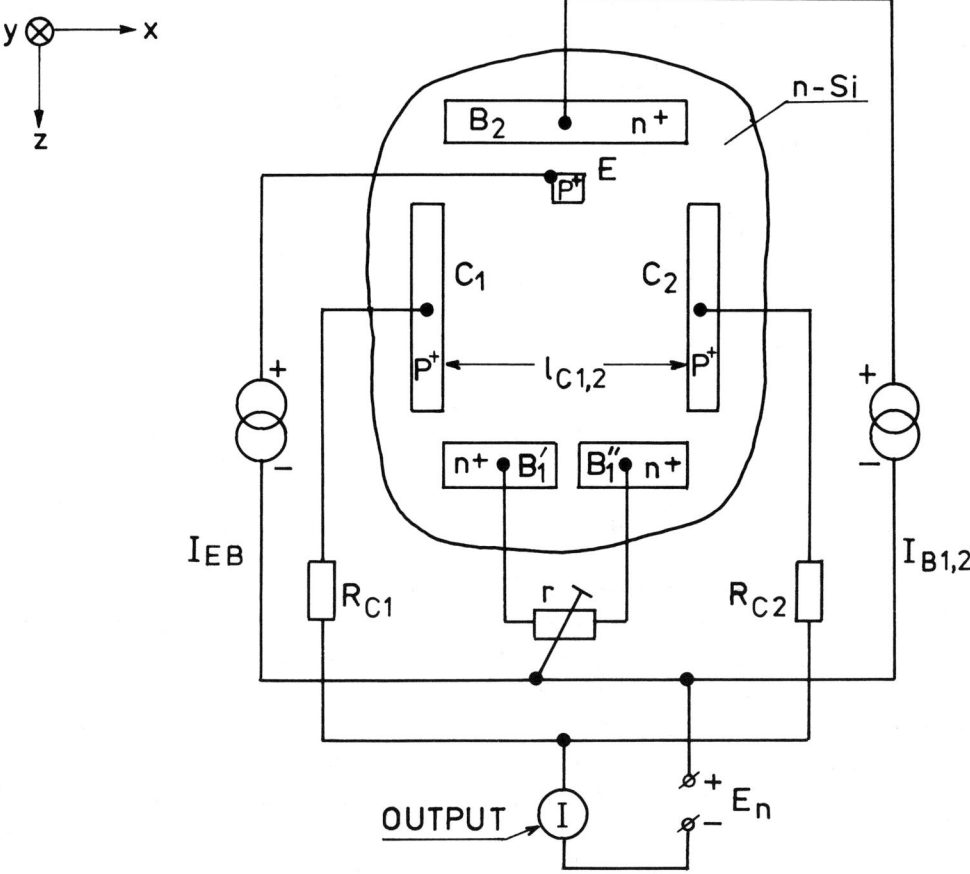

Fig. 8.9. One-component $\partial B_y(x)/\partial x$ gradiometer using a drift-aided BMT.

In addition to the registration of a magnetic-field gradient the sensor described above can be used to perform its ordinary function and determine the magnetic induction B_{z0} by means of the differential signal $\Delta I_{C1,2}(B)$. Thus the multisensor in Fig. 8.5 can simultaneously and independently measure the magnetic induction B, its gradient along a given axis, and the temperature of the environment.

8.3. Magnetic-field vector sensors

The integration of more than one measuring function in the active region of the transducer is a novel trend in detecting the orthogonal components of the magnetic-field vector by means of one and the same microstructure. The principal applications that determine the development of this approach to vector magnetometry are the following: detection of the vector $\boldsymbol{B}$ and the magnetic-field topology; investigation of the properties of magnetic materials and instruments; measuring of the geomagnetic field in geophysical research and geological surveys; sea, air and space navigation; proximity switches; contactless determination of angular displacements, slope detectors, etc. The sophisticated device structures and fabrication technology of modern integrated 2-D and 3-D vector transducers guarantee the following: an extremely high resolution (the possibility of measuring the vector components B_x, B_y and B_z "at a spot"); improved orthogonality when detecting the magnetic vector components, because of the precision of planar technology; that the position of the 3-D sensor with respect to the magnetic source is not as critical as in the case of 1-D magnetic transducers; optimum electrical, thermal and processing compatibility of the B_x, B_y and B_z channels, etc. This section presents a systematic review of the vector magnetometers known so far, in which the functional principle has been applied.

8.3.1. Methods for the registration of the components of the magnetic vector

Two mutually complementary techniques (simultaneous and successive [13]) can be used for the determination of the components B_x, B_y and B_z of the magnetic vector $\boldsymbol{B}$ in terms of a given inertial reference system (§1.3). In its classical form the first technique implies the existence of two or three magnetosensors with perpendicular orientations, placed as close to each other as possible. The information from the B_x, B_y and B_z channels is simultaneously (in real time) transferred for registration with the purpose of attaching the data to the "point" of the transducer's location. The conventional solution of this problem by means of solid-state devices is to place three identical galvanomagnetic sensors with equal sensitivities, like magnetoresistors, Hall elements or magnetotransistors, upon the three neighboring faces of a quartz cube. The main shortcomings of such a device are its unsatisfactory resolution, especially in a highly divergent field $\boldsymbol{B}$, and the fact that IC technology can only be partially used at certain stages of manufacture. Miniaturization and high resolution of the registration of B_x, B_y and B_z can be achieved by the application of IC technology and the functional principle for the design and fabrication of the new generation of vector sensors.

The second technique is based on the time-extension of the metrological task. The standard version employs a one-component magnetosensor for the successive measurement of the orthogonal components of the magnetic vector. Constancy of the field $\boldsymbol{B}$ is necessarily required during the measurements. The resolution of this method is limited by the dimensions of the transducing element itself. Vector microsensors have been developed whose operation is based on the electrical control of the sign of BMT magnetosensitivity (§6.2.1.2). In such devices, a single structure with a fixed spatial position is used for the successive measurement of the three orthogonal components B_x, B_y and B_z.

The integrated magnetometers for the simultaneous and successive registration of the components of the magnetic vector employ the Hall effect and the specific features of the magnetotransistor action. Owing to this fact, the transducing mechanism, the method of registration (successive or simultaneous) and the number of detected components (2-D or 3-D measurements) are the most suitable criteria for the classification of such devices.

8.3.2. 2-D and 3-D Hall-effect vector magnetometers

Parallel-field Hall transducers (§4.3 and Fig. 4.26) are the basis of the functional 2-D and 3-D vector sensors. Owing to the planar positions of the contacts, the trajectories of the majority carriers are curves, and therefore the majority-carrier velocity has both a lateral and a vertical component. A summary of the Hall-effect 2-D and 3-D magnetometers is presented in Fig. 8.10 [14–18].

The detection of a magnetic field $\boldsymbol{B}_y$, normal to the surface of the device, is based on its effect upon the lateral velocity v_x (Fig. 4.26f). This velocity component has opposite signs to the left and to the right of contact C_1 (Fig. 8.10a–b), i.e. $\pm \boldsymbol{F}_L \sim \pm v_x \times \boldsymbol{B}_y$. The cross coupling of the current contacts C_2 with C_5 and C_3 with C_4 (Fig. 8.10a) in a magnetic field B_y leads to the summation of the Hall-current variations with identical signs. Thus a differential output signal $\Delta I(\boldsymbol{B}_y)$ arises. The role of the sectioned contacts C_2 and C_3 as well as C_4 and C_5, is analogous to that of split-drains in FETs. This is why the Hall current is the sensor mechanism of this device. The $\boldsymbol{B}_x$ component of the magnetic vector is measured by means of contacts H_1 and H_2 in the Hall-voltage mode of operation of the device from Fig. 8.10a, i.e. $V_{H1,2}(\boldsymbol{B}_x)$ is the output signal. It is determined by the y-component of the carrier drift velocity. The cross-sensitivity of the 2-D vector sensor from Fig. 8.10a, i.e. the influence of the components $\boldsymbol{B}_x$ and $\boldsymbol{B}_z$ upon the $\Delta I(\boldsymbol{B}_y)$ channel and that of components $\boldsymbol{B}_y$ and $\boldsymbol{B}_z$ upon the $V_{H1,2}(\boldsymbol{B}_x)$ channel is eliminated to a first approximation thanks to the symmetry of the device with respect to contact C_1, and because of C_2–C_5 and C_3–C_4 coupling [15]. The lateral parasitic conductance between the output terminals due to the different potential levels of the offset and the Hall voltage is avoided by means of circuits for the alternative implementation of the nulling-potentiometric techniques, whereby the output terminals are subject to conditions of equal voltage. Suitable circuits that include this type of sensor are described in Chapter 9. If contacts C_2–C_3 and C_4–C_5 are connected according to the circuit in Fig. 8.10b, the vector transducer will measure the B_x and the B_z components simultaneously and independently. Therefore, the same Hall microstructure (Fig. 8.10a–b) can be used to determine the three components of $\boldsymbol{B}$ by two successive measurements. The magnitude of $\boldsymbol{B}$ is given by the familiar formula $|\boldsymbol{B}| = (B_x^2 + B_y^2 + B_z^2)^{1/2}$.

Only two conformally transformed Hall elements are needed for the simultaneous registration of the B_x and B_z components of the vector $\boldsymbol{B}$ which is parallel to the xz plane of the chip. They are functionally integrated, have a common central

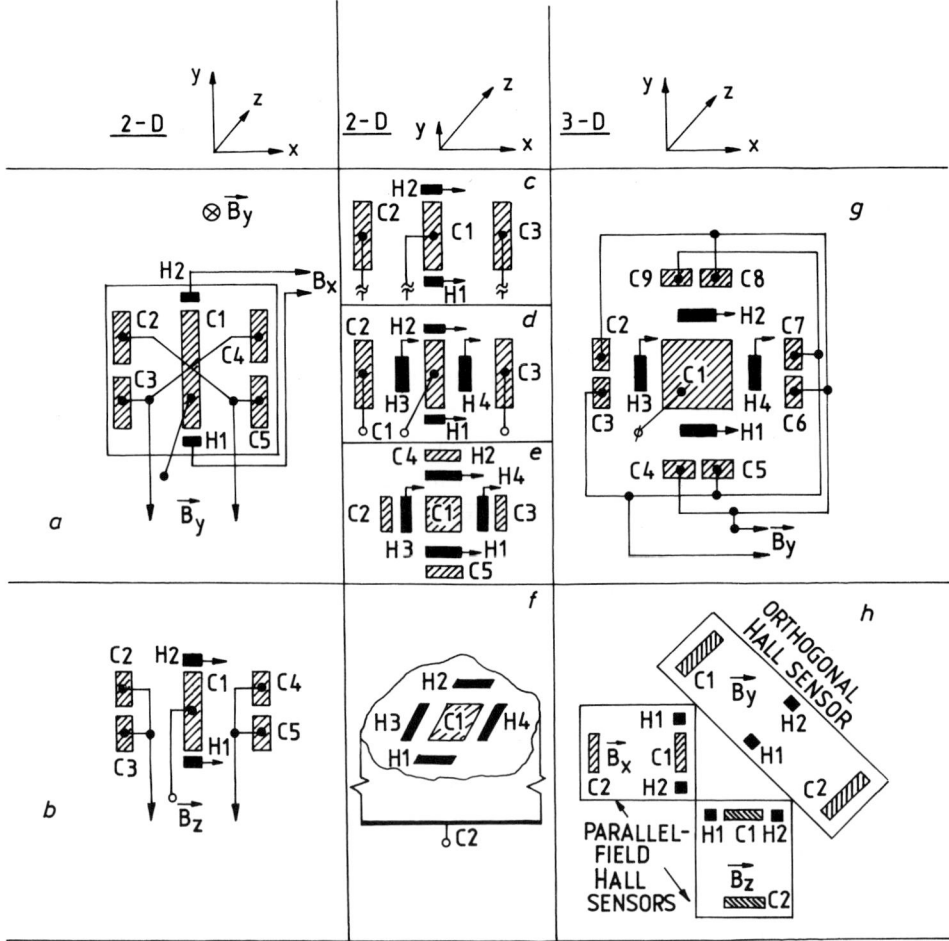

Fig. 8.10. 2-D and 3-D vector microsensor structures using the Hall effect.

current contact C_1 and mutually perpendicular orientations. Such integrated 2-D vector magnetometers are presented in Fig. 8.10c–f [14]. An interesting device is shown in Fig. 8.10g. It detects simultaneously and independently the three components B_x, B_y and B_z [15]. Generally speaking, it includes two parallel-field Hall devices with mutually perpendicular orientations. They are used for the measurement of $\boldsymbol{B}_x$ and $\boldsymbol{B}_z$. The connections of ohmic contact pairs C_2–C_3, C_4–C_5, C_6–C_7 and C_8–C_9 are analogous to those in Fig. 8.10a. The cross-sensitivity is eliminated to a first approximation by means of proper connections, structural symmetry and a circuit which keeps the output terminals under equal voltages. The magnetic control of the output differential currents of the three sensor channels of the n-type Si 3-D magnetometer from Fig. 8.10g is illustrated in Fig. 8.11.

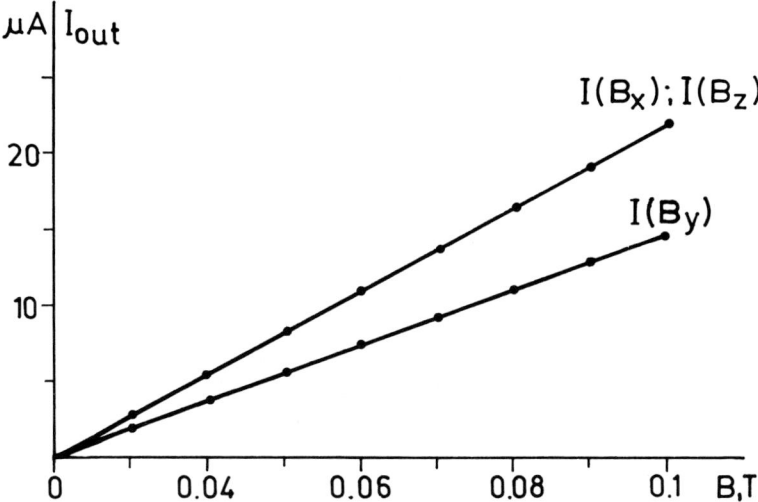

Fig. 8.11. Magnetic control of the currents $I(B_x)$, $I(B_y)$ and $I(B_z)$ in the three channels of the Hall vector magnetometer from Fig. 8.10g.

The silicon substrate itself serves as an active transducing region of the 2-D and 3-D functional Hall-effect vector sensors shown in Fig. 8.10. The spatial confinement of the active volumes of the above devices by means of deep p^+-rings with suitable layout leads to a partial isolation of the sensor from the rest of the substrate. Spreading currents will be limited in this way. The use of n-type GaAs is an alternative for increasing the transducing efficiency of the Hall effect 2-D and 3-D vector magnetometers.

An integral combination of a conventional orthogonal Hall microsensor and two identical parallel-field transducers like that in Fig. 4.26g is presented in Fig. 8.10h. The whole device has been fabricated in an n-epilayer [16–18]. The spatial resolution of this sophisticated solution is basically determined by the dimensions of the structures involved: it is a few hundred microns on the side and the epitaxial layer is 17 μm thick. The processing circuit is on the same chip. The three amplified signals are squared using the Gilbert multiplier circuit for differential current output [19]. Experiments confirm that the integrated sensor achieves an omnidirectional magnetic measurement with a maximum error of $\pm 4\%$.

8.3.3. 2-D and 3-D BMT vector magnetometers

8.3.3.1. BMT vector sensors for simultaneous 2-D and 3-D registration

The synthesis of BMT vector magnetometers for the simultaneous registration of the components of B is based on the functional integration of orthogonal and parallel-field BMTs into a single microstructure. Two parallel-field differential

BMTs with a central emitter, a common active region and mutually perpendicular orientations of the other contacts are sufficient for the in-plane detection of the components of $\boldsymbol{B}$. Presented in Fig. 6.9 (a, c, d, e, h, i) are devices which can be used for the construction of such dual-collector 2-D magnetometers [13].

The earliest integrated $n-p-n$ 2-D vector sensor designed for the simultaneous registration of the in-plane B_x and B_y components consisted of two identical parallel-field differential BMTs (see Fig. 6.9c) with a common emitter, [20–22]. Its operation is analogous to the 2-D vector transducer in Fig. 8.10e and f, but the Hall terminals have been replaced by collectors, and the contact C_1 by emitter E. The Lorentz deflection of the carriers injected in a vertical (z) direction is the dominating transducing mechanism. The adverse effect of the collector–base voltage, V_{CB}, upon the magnetosensitivity of both sensor channels is analyzed in detail in [20–22]. The drop in sensitivity is mainly the result of the reduced Hall angle Θ_H in high electric fields (§2.3.3). The transistor is operated at a constant emitter current and all four collectors are kept at the same reverse bias voltage, V_{CB}, with respect to the base. Each pair of opposite collectors is connected to a signal-processing channel, which supplies an output voltage proportional to the differential collector current of the pair. The approach adopted in [20–22] to the design of a 2-D magnetometer for the in-plane components of the magnetic vector can be successfully applied to the other parallel-field BMTs presented in Fig. 6.9a, d, e, h and i. The 2-D sensor described in [23] includes two mutually perpendicular BMTs of the type shown in Fig. 6.9d. The collectors are used as an active load of the differential amplifier stage of an operational amplifier. The sensitivity for the output voltage can be changed by external resistors. The linear relationship between the applied magnetic field and the output signal is observed for the measured range of ±0.3 T. The input offset drift is a problem for the 2-D BMT magnetometers and improvements in this direction are in progress [23].

The structure which has been analyzed possesses a p-type base diffusion so that the current can be injected into the double contact epilayer by the forward-biased EB-junction. The disadvantageous effects of the CB-junction on the sensitivity would be avoided if the sensor could be made without a base diffusion. In that case, the injector would simply be an ohmic top (or emitter) contact, so that the structure would basically consist of two three-terminal resistors. The magnetic current splitting properties of such a baseless device are outlined in [22]. The magnetosensitivity of this 2-D Hall magnetometer is lower by a factor of 2 than the sensitivity of the BMT version.

3-D magnetic-field sensing is accomplished by the one-chip integration of an in-plane 2-D BMT sensor with an orthogonally activated magnetotransistor. This approach leads, in addition, to improved spatial resolution and miniaturization of the transducing region. The first 3-D vector magnetosensor on the basis of an $n-p-n$ BMT was suggested in [24–26]. It is analogous to the 2-D version in Fig. 6.9c. In fact, the current flow in a 2-D in-plane magnetotransistor is not completely

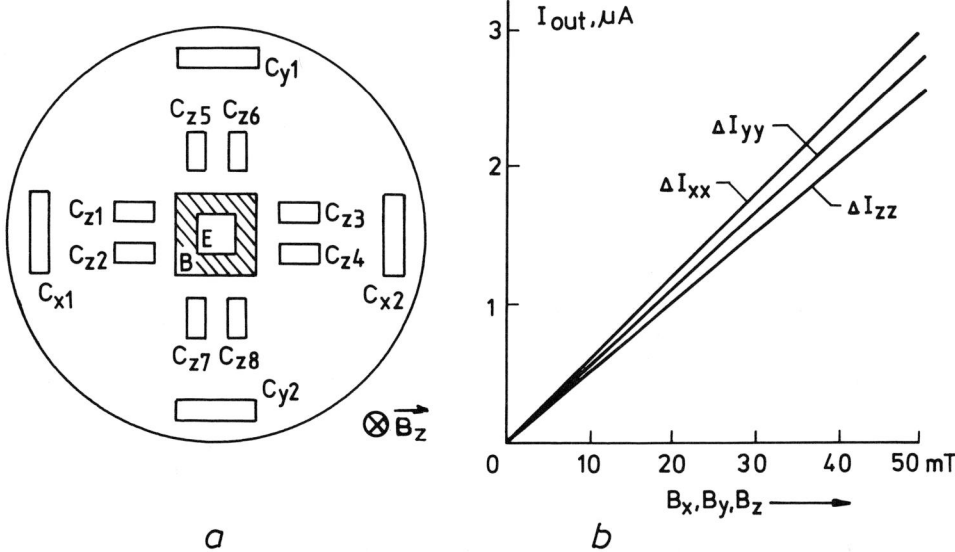

Fig. 8.12. (a) Schematic diagram of a silicon 3-D BMT vector sensor for simultaneous measurement of the x-, y- and z-components of the field $\mathbf{B}$. (b) Response of the collector to the x-, y- and z-components of $\mathbf{B}$ [8,25].

vertical; there are also significant lateral components of the total current in the n-epitaxial collector region. The vertical current component is responsible for the sensing of the B_x and B_y components of the vector $\mathbf{B}$ in exactly the same way as in the 2-D magnetometer. The lateral current component is deflected by the respective Lorentz force and is used for the registration of the B_z component. The collector arrangement of a Si magnetosensor for the simultaneous measurement of the x, y and z components of the field $\mathbf{B}$ is presented schematically in Fig. 8.12a.

The redistribution of these horizontal currents can be detected by means of additional collector contacts. In a well-designed sensor for 3-D field mapping, the current difference measured by the additional collector contacts should only respond to the z-component of the field $\mathbf{B}$ [8]. When only two additional collectors are employed, for example C_{z1} and C_{z2} (Fig. 8.12a), this is usually not the case. The current difference registered by this channel also depends on the B_x and B_y components, thereby showing undesirable cross-sensitivity. The reason for this is the fact that the current drawn by the z-collectors is not purely lateral; it also contains a significant vertical component. This current can be deflected by the x- and y-components of the field vector. Maximum possible suppression of the cross-sensitivity and enhanced magnetosensitivity of the y-channel can be achieved by making use of four collector pairs instead of one (C_{z1}–C_{z2}; C_{z3}–C_{z4}; C_{z5}–C_{z6}; C_{z7}–C_{z8}) and their cross coupling, e.g. C_{z1} to C_{z4} or C_{z6} to C_{z7}, etc. In fact in the magnetic field $\mathbf{B}_z$, the cross-coupled collectors are used for the summation of

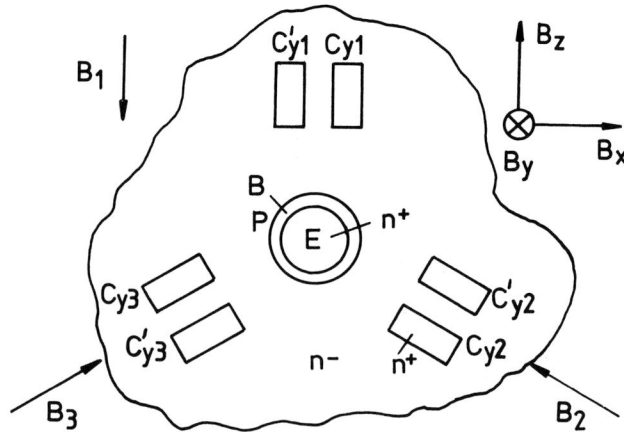

Fig. 8.13. A BMT structure without buried layers and with three collector pairs oriented at an angle of 120° relative to each other is sensitive to all three magnetic components [25].

collector current variations with the same signs. At the same time the z-channel responses to the $\boldsymbol{B}_x$ or $\boldsymbol{B}_y$ field (cross-sensitivity) will be subtracted and the resulting signal will be much smaller than that of a single z-collector pair. In Fig. 8.12b the current difference components are plotted as a function of the field component in the relevant direction. The achieved spatial resolution of the measurement is $8 \times 10 \times 20$ μm^3 [24–27].

Figure 8.13 depicts another structure which is sensitive to all three components of the magnetic-field vector. This device has no buried layers at all, unlike the foregoing 3-D sensors. The functioning of the individual collector pairs is the same as described in Fig. 8.12. Each one of the pairs is sensitive to the vertical component of the magnetic-flux density B_y. They are also sensitive to the in-plane vectors B_1, B_2 and B_3 in the same manner in which the z channels exhibit cross-sensitivity. The in-plane field vectors are in turn functions of B_x, B_z and the angular displacement between the collector pairs. The angular displacement is in this case 120° [25].

An $n-p-n$ Si 3-D magnetometer fabricated by a standard 3 μm CMOS process has been investigated (Fig. 8.14) [28]. The differential BMT device from Fig. 6.9i has been used.

The structure comprises a single emitter, a common base region, and eight symmetrical collectors. The device has four base contacts on each of the emitter sides designed in such a way as to ensure the symmetry of the sensor. These contacts are connected together by the metallization process. There are also four p^+-regions placed on the four corners of the emitter. These regions serve to suppress the lateral emitter injection of electrons through the corner area. The p-well base region is formed by an ion implantation. The n^+-emitter and all eight n^+-collectors were realized using the standard doping process of the source and

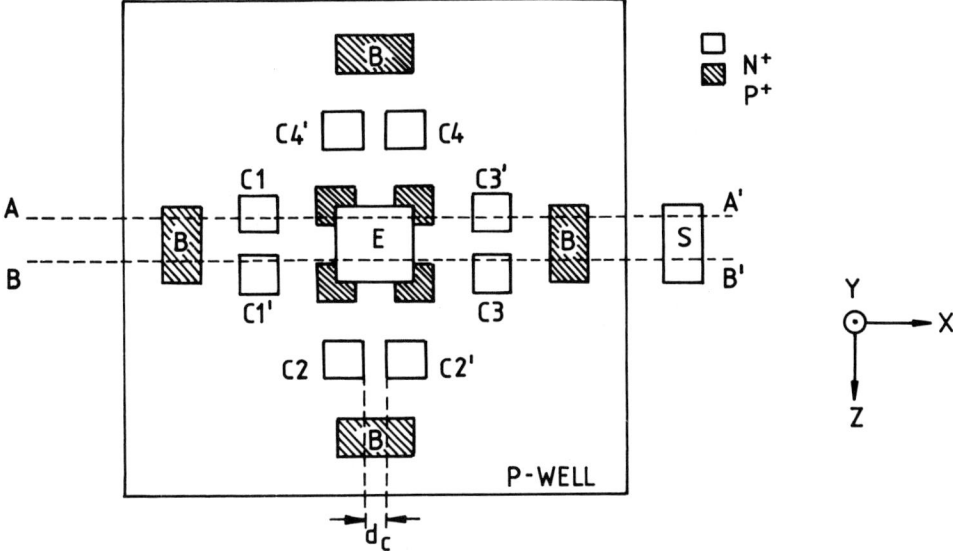

Fig. 8.14. 3-D magnetometer fabricated by a standard CMOS process [28].

drain of n-channel MOSTs. The p^+-regions on the four corners and the base contacts were formed using the standard doping process for p-channel MOSTs. The operation of the 3-D sensor, according to [28], is as follows: the electrons injected into the neutral base region are divided into four lateral components along the x- and z-axes and one vertical component along the y-axis (Fig. 8.14). The vertical flow of carriers injected by the bottom of the emitter is collected by the substrate and does not contribute to the sensitivity of the device. The carriers of these four lateral flows are collected by eight collectors (two collectors for each lateral flow). In the absence of a magnetic field, these collector currents are equal because of the device symmetry. When a magnetic field with all three components B_x, B_y and B_z is applied, the collector currents will change as a result of the Lorentz deflection. For example, the B_y component, which is perpendicular to the chip surface, will cause a deflection of the carriers. As a result, the collector currents $I_{Ci'}$ will increase and I_{Ci} will decrease ($i = 1$ to 4). This means that the collector pairs on one and the same side of the emitter can be used to detect the y-component of the magnetic field. It should be noted that, for this condition, the difference between I_{C3} and I_{C1}, or I_{C4} and I_{C2}, is zero because of the symmetry of the device (the same is true for $I_{C3'}$ and $I_{C1'}$, etc.) In a similar manner, the deflection mechanism can be used to determine the B_x and B_z components. Namely, the B_z component will cause a deflection of carriers flowing in the x direction. As a result I_{C3} and $I_{C3'}$ will increase and I_{C1} and $I_{C1'}$ will decrease. Therefore, the difference between I_{C3} and I_{C1} (not affected by B_y) can be used to detect B_z. It is worth mentioning that in the ideally case B_z should not affect I_{C2}, $I_{C2'}$, I_{C4} and $I_{C4'}$ because the flow of these components is parallel to B_z. But these four collector currents will be affected by B_x, in a similar way that I_{C3}, $I_{C3'}$, I_{C1}, and $I_{C1'}$ were affected by B_z. In this case I_{C4} and $I_{C4'}$ will increase and I_{C2} and $I_{C2'}$ will decrease. The difference either between I_{C4} and I_{C2} or between $I_{C4'}$ and $I_{C2'}$ can be used to detect B_x. From to these explanations, is obvious that the spatial resolution of the structure is determined practically by the space bounded by the eight collectors. The sensitive part of the device is $0.8 \times 36 \times 36$ μm^3, which is much smaller than the overall dimensions of the chip [28].

An integrated 3-D DMOS magnetic-vector sensor has been suggested [29]. The basic structure is obtained by combining a 1-D split-drain orthogonal DMOS, sensitive to the B_y component, with a 2-D parallel-field DMOS, sensitive to the B_x and B_z components. The spatial resolution of the magnetometer is $10 \times 16 \times 32.5$ μm^3 if only one split-drain pair is used. The sensor, according to [29], exhibits a good linearity in magnetic fields lower than 0.5 T, and the measured T.C. is less than -0.2%. A 3-D magnetic-field transducer, whose structure and operation are similar to the 3-D magnetometer in Fig. 8.12a, is described in [30]. The device was manufactured by a three-mask bipolar technology. According to [30] the cross-sensitivity can be eliminated by the establishment of a proper configuration of the collector depletion regions by individual reverse biasing of each collector junction. A high degree of accuracy of the respective vertical or lateral current components subjected to the impact of B_x, B_y or B_z is achieved thanks to this adjustment.

An opportunity which has not been used so far is the obtainment of a thermometric signal, $V_{EB}(T)$, from the emitter–base junction when biased by a current $I_{EB} = $ const. This information can be used to compensate the T.C. of the sensitivity at the next stage of signal processing.

8.3.3.2. BMT vector sensors for successive 3-D registration

The successive registration of the three components of the magnetic-field vector ***B*** by one and the same BMT microstructure with a fixed spatial position is based on the electrical control of the sign of the magnetosensitivity (§6.2.1.2) [31,32]. A p^+–n–p^+ drift-aided BMT which is used by this class of magnetometers is shown in Fig. 8.15a.

Analysis of the electrical control of the polarity of the magnetosensitivity leads to the conclusion that the successive registration of the magnetic-vector components B_x, B_y and B_z by the structure shown in Fig. 8.15a is possible if the following conditions are fulfilled:

(a) the collectors C_1, C_2, C_3 and C_4 must be differentially coupled in pairs;
(b) the working point with $I_E > 0$ and $I_{B1,2} \equiv I_B > 0$ must be properly chosen.

The curved carrier trajectories determined by the planar contacts play a key role in the explanation of the operation of this magnetometer. The B_x component does not cause a Lorentz deflection of carriers that move in a lateral direction at

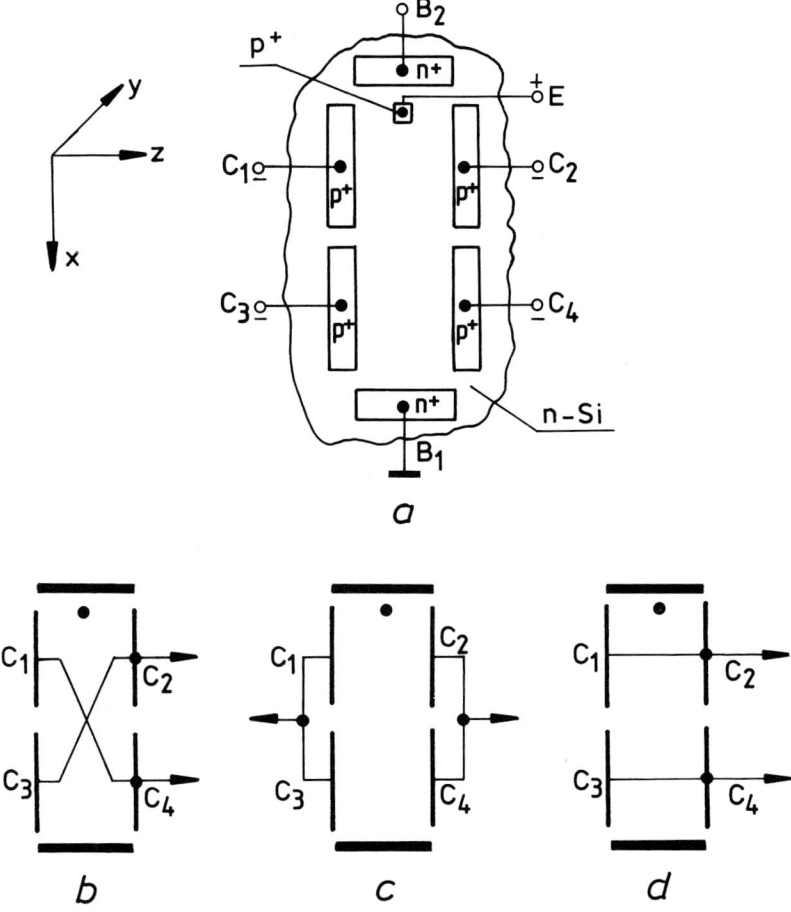

Fig. 8.15. Sensor for 3-D vector measurement of the field B. (a) Layout of the base structure; (b)–(d) the three different methods of coupling the collectors for successively measuring the components B_x, B_y and B_z, respectively.

a velocity v_x, since $F_L = qv_x \times B_x = 0$ owing to the collinearity of B_x and the velocity v_x. The Lorentz force $F_L = qv_y \times B_x$ which is applied to the minority carriers in the emitter E–point A and point A–base contact B_1 regions of their trajectories causes a deflection of the injected holes in two opposite directions (Fig. 6.4). The B_y component together with v_x determines the Lorentz force $F_L = qv_x \times B_y$ and the respective hole deflection in a direction parallel to the xz plane and does not affect the velocity component v_y. Two forces $F'_L = qv_x \times B_z$ and $F''_L = qv_y \times B_z$ result from the field component B_z. They act together to distort, by shrinking or extending the lines of current flow in the xy plane. The effective trajectory is, in the general case, asymmetric with respect to point A, hence different distortions of the lines of the current flow are caused by the B_z component in the E-point A and point A–B_1 sections. That is why a differential output signal $\Delta I_{C1}(\boldsymbol{B}_z) + \Delta I_{C2}(\boldsymbol{B}_z) \neq \Delta I_{C3}(\boldsymbol{B}_z) + \Delta I_{C4}(\boldsymbol{B}_z)$ is generated. It follows from the galvanomagnetic properties described above of the structure in Fig. 8.15a that the B_x component can be measured by connecting collectors C_1 with C_4 and C_2 with C_3, respectively (Fig. 8.15b) (this case is illustrated by Fig. 6.9o); the B_y component can be measured by connecting C_1 to C_3 and C_2 to C_4, respectively (Fig. 8.15c); collectors C_1 and C_2 must be connected together as well as C_3 with C_4 to register the B_z component (Fig. 8.15d). In fact, the summation of current increments with the same sign is accomplished by means of these connections into the collector pairs.

In the general case, magnetosensitivity along the three axes is observed simultaneously. Therefore the cross-sensitivity should necessarily be eliminated in order to measure correctly the three components of the magnetic vector $\boldsymbol{B}$. This problem is solved by a proper choice of the working point determined by the emitter I_E and the interbase current $I_{B1,2}$. The working point must be selected in such a way as to eliminate the sensitivity along one of the coordinate axes. The choice of a working point which guarantees the elimination of the sensitivity along the x-axis in a field $\boldsymbol{B}_x$ with collector connections as shown in Fig. 8.15c has been illustrated in §6.2.1.2, Fig. 6.3. The y-sensitivity in the field $\boldsymbol{B}_y$ can be eliminated in a similar way by the connections shown in Fig. 8.15b with a suitably chosen working point. In this case $\Delta I_{C1}(\boldsymbol{B}_y) = -\Delta I_{C4}(\boldsymbol{B}_y) = -\Delta I_{C2}(\boldsymbol{B}_y) = \Delta I_{C3}(\boldsymbol{B}_y)$, and therefore no output signal is generated by $\boldsymbol{B}_y$. The successive registration of components B_x and B_y with the appropriate collector connections shown in Fig. 8.15b and c is not influenced by the component B_z, which generates signals with equal magnitudes at the differential output of the device. When B_z is measured (Fig. 8.15d), the cross-sensitivity to B_x and B_y is eliminated by the appropriate collector couplings. Figure 8.16 presents the output characteristics of the 3-D magnetometer corresponding to the three mutually perpendicular components of the field $\boldsymbol{B}$. The collectors have been connected as shown in Fig. 8.15b–d and the working points have been adequately chosen to eliminate the cross-sensitivity.

The 3-D magnetometer described above shows a high resolution because one and the same magnetosensitive region is used for the registration of the three components B_x, B_y and B_z. The different transducing efficiency along the three axes x, y, and z (Fig. 8.16) can be made equal at the next stage of the signal processing. The working points which eliminate the cross-sensitivity can be selected also by the variation of one of the currents at a fixed value of the other, for example $I_E > 0$ at $I_B = \text{const}$ or $I_B > 0$ and $I_E = \text{const}$. Because of this fact, the establishment of the necessary operational mode becomes easier by making use of suitable circuitry.

An improvement of the BMT sensor's structural symmetry can reduce the cross-sensitivity without establishing particular working points when the components of $\boldsymbol{B}$ are measured successively. This approach is discussed in [33].

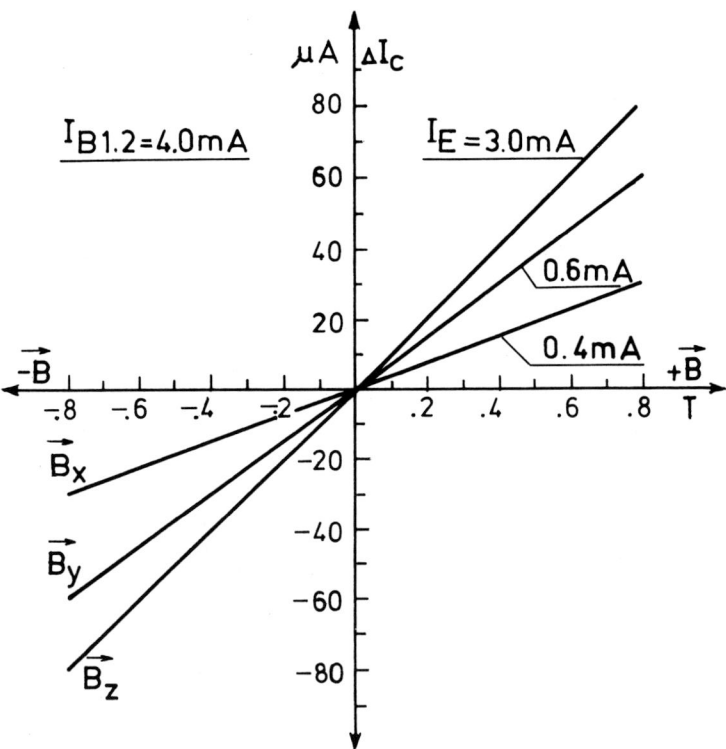

Fig. 8.16. Successively measured components B_x, B_y and B_z, according to Fig. 8.15b–d, at the respective working points where the cross-sensitivity along the x- and y-axes is eliminated, $T = 20°C$.

8.4. Concluding remarks

The implementation of the functional principle in the form of particular magnetosensitive microdevices has proven its usefulness and applicability even to other types of transducers of nonelectrical signals. The need for such investigations is associated with the advance of smart solutions which improve the metrological characteristics. The author believes that functional organization lies at the basis of the biosensors of living organisms.

References

[1] Ch.S. Roumenin and P.T. Kostov, Sensors and Actuators, 8(1985) 307–318; C.R. Acad. Bulg. Sci., 41(5) (1988) 71–74.
[2] D.I. Zaks, Parametry teplovova regima poluprovodnikovich mikroshem, Radio i Sviaz, Moskwa, 1983 (in Russian).
[3] S. Middelhoek and D.J. Noorlag, Sci. Instrum., 14 (1981) 1343–1350
[4] P.S. Kirejew, Physika poluprovodnikov, Vishaya Shkola, Moskwa, 1975 (in Russian).

[5] V.I. Stafeev and E.I. Karakushan, Magnitodiody, Nauka, Moskwa, 1975 (in Russian).
[6] I.M. Vikulin, L.F. Vikulina and V.I. Stafeev, Galvanomagnitniye pribori, Radio i Sviaz, Moskwa, 1983 (in Russian).
[7] Ch.S. Roumenin, Multisensors, Bulg. Patent 95 949 (Febr. 20, 1992); Sensors and Materials, 5 (5) (1994) 285–300.
[8] S. Middelhoek and S.A. Audet, Silicon Sensors, Academic Press, London, 1989, Chapters 2 and 5.
[9] I.M. Vikulin and V.I. Stafeev, Fizika poluprovodnikovich priborov, Sovetskoe Radio, Moskwa, 1980, Chapter 2 (in Russian).
[10] Y.V. Afanasiev, N.V. Studentzov, V.N. Horev, E.N. Chechurina and A.P. Shelkin, Sredstva izmereniya parametrov magnitnova polia, Energia, St. Petersburg, 1979 (in Russian).
[11] A.Y. Shihina, Ispitanie magnitnich materialov i system, Energoatomizdat, Moskwa, 1984 (in Russian).
[12] Ch.S. Roumemin, C.R. Acad. Bulg. Sci., 42(12) (1989) 63–66; Magnetosensitive Device, Bulg. Patent 83 788 (April 18, 1988).
[13] Ch.S. Roumenin, Sensors and Actuators, A, 24 (1990) 83–105.
[14] Ch.S. Roumenin, C.R. Acad. Bulg. Sci., 42(4) (1989) 59–62.
[15] Ch.S. Roumenin, Hall Device, Bulg. Patent 85 596 (Oct. 4, 1988); Magnetic-field Vector Sensor, Bulg. Patent 79 426 (April 21, 1987).
[16] Lj. Ristic and M. Paranjape, Sensors and Materials, 5 (5) (1994) 301–316.
[17] K. Maenaka, T. Ohgusu, M. Ishida and T. Nakamura, 4th Int. Conf. on Solid-state Sensors and Actuators, Tokyo, Japan, 1987, Dig. Techn. Papers, 523–526; 7th Sensor Symp., Japan (1988) 43–46.
[18] K. Maenaka, M. Tsukahara and T. Nakamura, Sensors and Actuators, A, 21–23 (1990) 747–750.
[19] B. Gilbert, Electron. Lett., 11 (1975) 14–16.
[20] V. Zieren and B.P.M. Duyndam, IEEE Trans. Electron Devices, ED-29 (1982) 83–90.
[21] V. Zieren and S. Middelhoek, Sensors and Actuators, 2 (1982) 251–261.
[22] V. Zieren, Integrated silicon multicollector magnetotransistors, Ph. D. dissertation, Delft Univ. Techn., The Netherlands, 1983.
[23] K. Maenaka, H. Fujiwara, T. Ohsakama, M. Ishida et al., 5th Sensor Symp., 1985, Japan, 179–183.
[24] S. Kordić, IEEE Electr. Dev. Lett., EDL-7(3) (1986) 196–198.
[25] S. Kordić, Offset reduction and three-dimensional field sensing with magnetotransistors, Ph.D. dissertation, Delft Univ. Techn., The Netherlands, 1987.
[26] S. Kordić, Proc. 16th European Solid-state Device Res. Conf., ESSDERC'86, Cambridge, U.K., 1986, 97–98; IEDM Techn. Dig., 1986, 188–191.
[27] S. Kordić and P.J.A. Munter, IEEE Trans. Electron Devices, ED-35 (1988) 771–779.
[28] Lj. Ristić, M.T. Doan and M. Paranjape, Sensors and Actuators, A, 21–23 (1990) 770–775.
[29] L. Zongsheng, L.Yi, W. Guangli and J. Hao, Sensors and Actuators, A, 21–23 (1990) 786–789.
[30] S.M.A. Nailah and K.N. Khalis, Solid-state Electron., 33 (1990) 1119–1124.
[31] Ch.S. Roumenin, Magnetotransistor Vector Sensor, Bulg. Patent 79 616 (May 6, 1987).
[32] Ch.S. Roumenin, C.R. Acad. Bulg. Sci., 41(1) (1988) 59–62.
[33] Ch.S. Roumenin, C.R. Acad. Bulg. Sci., 41(7) (1988) 51–54.

Chapter 9

Interface and improvement of solid-state magnetosensor characteristics

In this chapter solid-state magnetosensors are considered as electronic devices in interaction with their direct circuit environment such as the bias circuitry and the first amplifier stages. Some promising devices like flip-flop magnetometers are also considered. The methods for the correction of typical sensor defects, such as offset, temperature dependence and nonlinearity are described in detail. In addition to our considerations there is an enormous amount of literature on signal conditioning, analog and digital circuits and instrumentation systems, data storage, etc. This is why particular attention is paid to the specific character of magnetosensitive transducers as electronic components in order to improve their performance.

9.1. Biasing circuitry

Generally, low output resistance of voltage outputs and high output resistance of current outputs are needed when output efficiency and interfacing convenience are desired. Two basic modes of biasing are possible, depending on the particular magnetosensor: constant voltage and constant current. Hall sensors, for instance, can operate in both supply modes, while changing, to a certain extent, their characteristics and parameters (§4.2.4.2g). The temperature coefficient, for example, is negative at $V = $ const, while at $I = $ const it is positive. The MOS Hall device free from short-circuit effects requires distributed current sources [1,2]. Modern Hall sensors, MDs and magnetoresistors are preferably biased by a constant current.

The biasing of BMTs, CDMs and other related devices needs special comment. The BMT sensors from Fig. 6.8a and g and from Fig. 6.9a–e, g and h can operate in both modes: $V_{EB} = $ const and $I_{EB} = $ const. In the latter case, $V_{EB}(I = $ const) can, under certain conditions (§8.1.1), play the role of a thermometric voltage. If necessary, the emitter circuit can be excited by an additional a.c. signal. Then the d.c. operating point is established by means of a current source, $I_{EB} = $ const, and an a.c. signal is applied through the decoupling capacitor C from a constant-amplitude signal generator.

The biasing of the S-type MDs and the other BMTs in Figs. 6.8 and 6.9 should be in compliance with the fact that the input emitter–base characteristic at interbase current $I_{B1,2} \neq 0$ or voltage $V_{B1,2} \neq 0$ includes a region of negative resistance. The control over the operating points in the S-curve where the most interesting effects of magnetosensitivity occur can be achieved by biasing of the emitter–base circuit by a constant current. Otherwise, when $V_{EB} = \text{const}$, the current I_{EB} increases abruptly with a "vertical" slope and runs out of control until the whole structure is occupied by the filament. CDMs, being thyristor-like devices, also operate in negative-resistance mode. Depending on the device structure, two current sources are necessary to bias the emitter junctions and, like in the case of BMT collector biasing, at least one voltage source is needed for the base contacts. The biasing of UJMTs leads to a similar problem.

There is a group of magnetosensors which perform the function of a magnetic switch. Unijunction sensors, like magnetothyristors, belong to this group. Their biasing circuits need special comment [3]. Such devices have been presented schematically in Fig. 9.1a, b, and Fig. 9.1c shows a typical current–voltage characteristic illustrating the device's performance. The $I-V$ curve has been recorded in a constant-current mode. The constant supply voltage is an important feature of the sensors shown in Fig. 9.1. The supply voltage E_s is determined with respect to the contact B_1 and is distributed between the resistor R and the diode contact $E-B_2$ (Fig. 9.1a) or $E-B_3$ (Fig. 9.1b) according to the expression $E_s = V_D + V_R = V_D + I_s R$.

The current in the circuit and the voltages across the diode and the resistor are determined by the points of intersection of their $I-V$ characteristics. Since $V_R = E_s - V_D$, the load line is determined by points E_s and E_s/R (Fig. 9.1c). If the

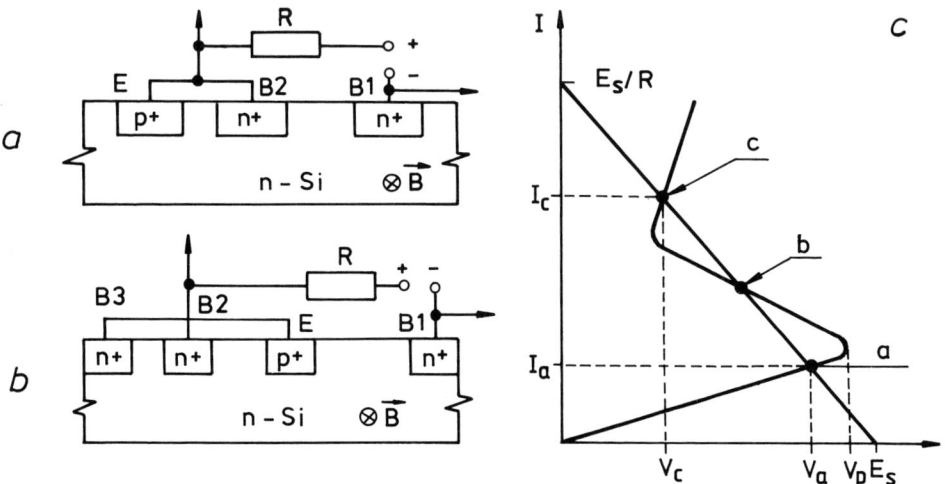

Fig. 9.1. Two S-type device structures used as magnetic switches (a), (b); and their typical $I-V$ characteristic (c).

diode resistance in the negative region is $R_{D,N} > R$, three points of intersection a, b and c of the load line with the diode $I-V$ curve are possible at certain values of the bias E_s. When the voltage source E_s is connected, the working point a is established and a relatively small current I_a flows in the circuit. If an external magnetic field B is applied with such a magnitude and polarity that the generated Hall voltage satisfies the condition $V_H > V_p - V_a$, the voltage V_D exceeds V_p and the diode is turned on. The current I_s is sharply increased and an abrupt transition of the device to the other stable working point c takes place. A high current $I_c > I_a$ is established in the circuit. Point b is a point of instability. The device can be switched from I_c to I_a by a magnetic field with the opposite polarity.

The biasing circuitry of the S-type BMT from Fig. 6.8c is shown in Fig. 9.2. The device can be regarded as a one-chip functional integration of the transistor of Fig. 6.8a and the UJT in Fig. 6.11. The UJT serves as a relaxation generator. The oscillation period is determined by the charging time of the capacitor C. A pulsed current I_{EB} of period t results. The differential output signal $\tilde{V}$ between the collectors C_1 and C_2 has the same shape. In the absence of a magnetic field, this voltage is $\tilde{V} \approx 0$. With a positive polarity of $+B$, the output pulsations are positive, and the reverse polarity $-B$ leads to a negative amplitude of the output signal. If the sensor in Fig. 9.2 is excited by an a.c. magnetic field, the amplitude of the output a.c. signal is modulated [4].

No external biasing is necessary when Hall sensors operate in the ultra-high frequency range. The electric component E of the external field penetrates into the semiconductor plate and generates an a.c. current of majority carriers, while as a result of the Lorentz force the magnetic component B, $(B \perp E)$, causes their deflection.

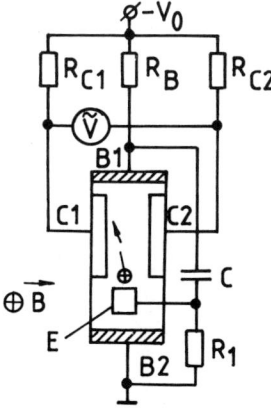

Fig. 9.2. Biasing circuit of an S-type p–n–p BMT with a frequency output.

9.2. Signal-processing electronics for solid-state magnetosensors

The processing of the output signals of long Hall sensors depends, to a certain extent, on the biasing method. At a current $I = $ const, for instance, in the general case the first amplification stage operates under unfavorable conditions. The temperature-dependent voltage drop over the Hall plate or, worse, the application of an a.c. current can lead to considerable variations of the common-mode Hall voltage at the sensor output. This fact requires the amplification stage, to whose input the voltage V_H is transferred, to be more complex [5,6]. The problem can be circumvented by the biasing method illustrated in Fig. 9.3 [7]. In this case, the common-mode voltage is determined only by the input offset voltage of the op-amp OA, and hence a conventional amplifier can be used. The left sensor terminal is virtually grounded and the full Hall signal appears at the right output contact. In addition, the circuit in Fig. 9.3 allows the current to be doubled via the Hall device.

When the differential current of split-drain MAGFETs, vector Hall microsensors and other devices operating in the Hall-current mode (short samples) is measured, the scheme shown in Fig. 9.4 can be conveniently used. The two op-amps $OA1$ and $OA2$ are low-noise and low-offset devices. They are connected as transconductance amplifiers with feedback resistors R_1 and R_2 which determine their gains. The outputs of the two op-amps are fed to an instrumentation amplifier A whose output voltage is proportional to the differential current of the two output terminals of the sensor. The $OA1$ and $OA2$ are connected with their virtual ground input at some potential V_{SD}, which is the bias potential for the device. The two output contacts must be kept at the same potential to eliminate the spreading currents and parasitic conductance in Hall-effect vector magnetometers. Three identical setups to the one shown in Fig. 9.4 are used.

The well-known nulling potentiometric circuit in Fig. 9.5 is an alternative solution to the problem. When a magnetic field $B \neq 0$ is applied, a current flows via the galvanometer G_1. This current is then nullified by means of a voltage source V_p, which is introduced into the circuit by closing the switch $SW1$. As a result of this compensation, the current generated due to the magnetic field is measured

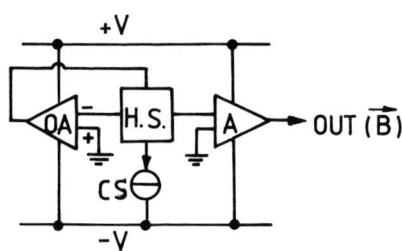

Fig. 9.3. Biasing and signal amplification circuitry for a Hall sensor (H.S.) using an opamp (OA), amplifier (A) and current source bias (CS).

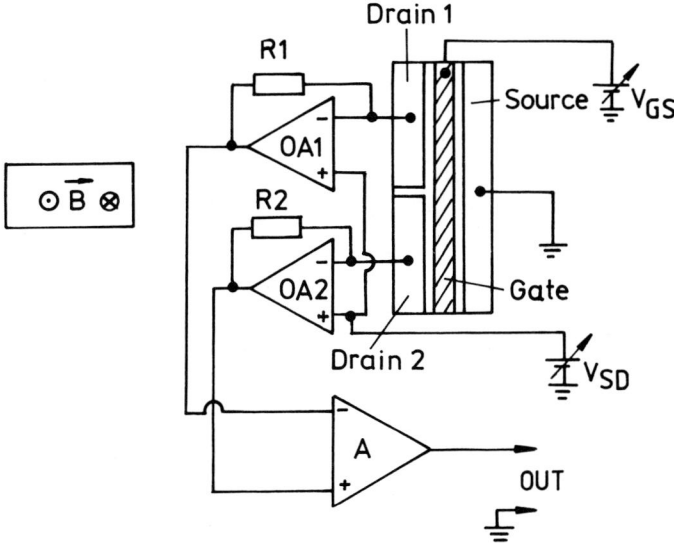

Fig. 9.4. Preferable circuit for measurement of Hall current using op-amps.

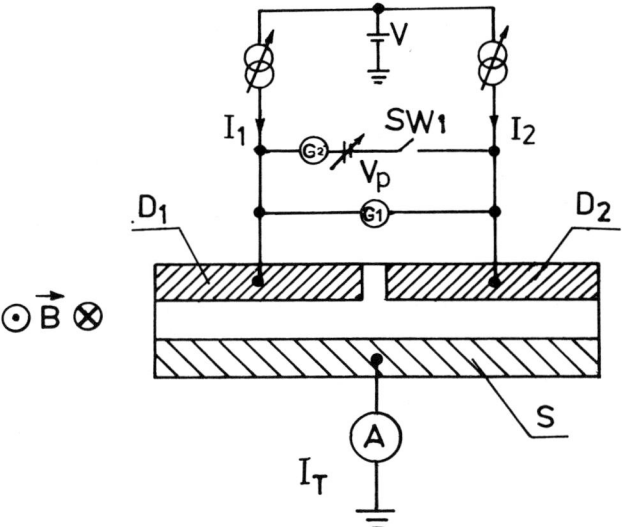

Fig. 9.5. Nulling potentiometric circuit for measurement of Hall current; a split-drain device has been taken as an example.

by keeping the terminals at the same potential. The Hall current equals twice the current displayed by the galvanometer G_2 [8].

The best metrological characteristics of the signal processing from resistive sensors like magnetoresistors and MDs are obtained by means of Wheatstone-

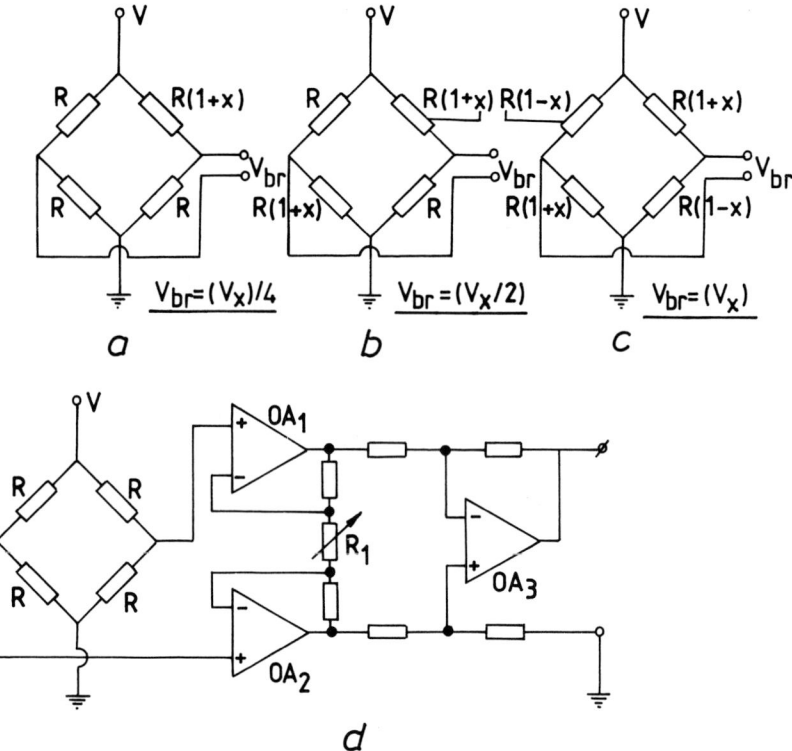

Fig. 9.6. Bridge output signals of: (a) one, (b) two, and (c) four varying resistors. (d) Magnetosensor bridge amplifier with a differential stage.

bridge circuits (Fig. 9.6). When the four resistors R_i ($i = 1, \ldots, 4$) exactly equal each other, the bridge output is zero. If one of the resistors is subjected to the magnetic field, it changes from R to $R(1 + x)$, and the output voltage V_{br} of the bridge is $V_x/4$ (Fig. 9.6a). If two opposite resistors are influenced by the magnetic field, the output signal is doubled and its value is $V_x/2$ (Fig. 9.6b). If the bridge consists of four sensors in such a way that two of them increase their resistance and the resistance of the other two is reduced when a magnetic field is applied, a maximum output voltage V_x is obtained (Fig. 9.6c). Such a magnetodiode arrangement is presented in Fig. 5.1c. A magnetosensor bridge amplifier with a differential stage is shown in Fig. 9.6d.

The simplest magnetotransistor circuits are presented in Fig. 9.7. When $R_{C1} = R_{C2}$, a zero offset $V_{off} = 0$ is adjusted by means of the trimmer r. Figure 9.8 presents signal-processing circuitry for dual-collector BMTs [8,9]. The scheme provides for a stable and adjustable bias for the sensor, measures and compensates the offset signal, and converts the magnetically controlled collector-current difference into an output voltage independent of the bias conditions. The base contact B is grounded

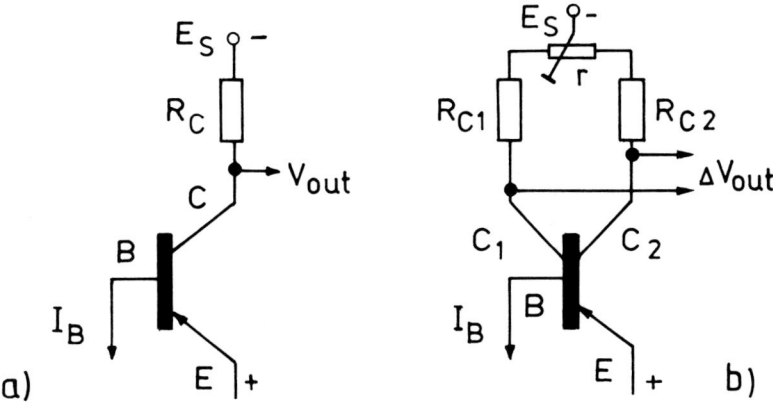

Fig. 9.7. Simple magnetotransistor circuits: (a) for a single collector BMT, and (b) for a differential BMT version.

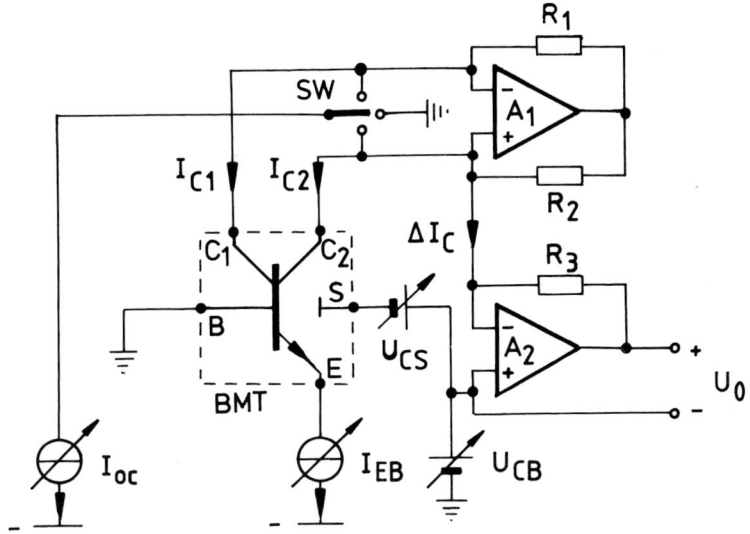

Fig. 9.8. Diagram of bias and measurement circuitry for differential BMTs using op-amps.

and the emitter E is connected to an adjustable current sink I_{EB}. If the collectors C_1 and C_2 of the BMTs in Fig. 6.9c and d are not mutually isolated, spurious output currents will flow between the collector contacts when they are not kept at the same voltage. It becomes necessary, therefore, that the intercollector resistance be virtually short-circuited by the low input impedance of the first electronics stage. A current mirror is chosen for this purpose, which includes the op-amp A_1 and the feedback resistors R_1 and R_2. The two collectors C_1 and C_2 are directly connected to the mirror inputs. Equal voltages arise across the mirror resistors, thus pushing the differential current onto the second stage, $\Delta I_{C1,2} = (R_1/R_2)I_{C1} - I_{C2}$ [10].

This stage around the op-amp A_2 provides for a collector–base voltage V_{CB} and at the same time converts the single-sided current into the output V_0 by the resistor R_3. For the purpose of eliminating the influence of the mirror-resistor values it is preferable to choose $R_1 = R_2 = R$, thus yielding $V_0 = -R_3(I_{C1} - I_{C2})$.

In the absence of magnetic fields ($\boldsymbol{B} = 0$), the output voltage V_0 may differ from zero if the BMT shows an offset current. A second current sink I_{0C} connected to one of the collectors by means of the switch SW is used to compensate the offset. Two or three identical circuits like that in Fig. 9.8 are used together with BMT vector sensors for the processing of the x, y and z signals. The direct transfer of the BMT differential signal $\Delta I_{C1,2}$ or $\Delta V_{C1,2}$ to the input of the op-amp is suitable for numerous applications [11].

Irrespective of the signal-processing, a boosting circuit for the amplification of the relatively small output signal must necessarily be integrated onto the chip with the magnetosensor. Conventional and merged amplifiers can be used for this purpose [6]. Conventional amplifiers are to be preferred when the sensor is self-contained, i.e. when an appropriate bias is sufficient for its operation and a signal suitable for amplification is available. Hall sensors can be used as examples (Fig. 9.3).

In an analog circuit, an op-amp is widely used for many functions, e.g. linear amplification, oscillation, filtering, switching, etc. The reason is that the op-amp has a high-impedance differential input and a sufficiently large gain amplifier within itself, i.e. the output voltage of the op-amp, V_0, is $V_0 = A(V_{in}^+ - V_{in}^-)$, where A is the gain (sufficiently large), and V_{in}^+ and V_{in}^- are the differential input voltages. For example, some applications of the extensive uses of op-amps are shown in Fig. 9.9 [12,13].

Merged arrangements should be designed when the magnetotransducer itself can be a part of the amplifier. Split-drain MAGFETs operated in the saturation mode and BMT versions can serve as examples. The output resistance of these devices is extremely high and a large output voltage at the collectors or drains can be obtained by means of high-load resistors. The differential amplification magnetic sensor (DAMS) is a classical solution [14], and the devices presented in Fig. 9.10 are its modifications. The two interbase currents between contacts B_1–B_2' and B_1–B_2'' generate Hall voltages which are amplified by a differential transistor pair functionally integrated in the structure (Fig. 9.10b). A completely integrated magnetosensitive amplifier is suggested in [15,16]. The circuit is reminiscent of CMOS differential amplifiers with dynamic loads. Each of the usual two complementary MOST pairs is replaced by a single split-drain MAGFET. The MAGFET drains are cross-coupled. Magnetic-field sensing, working point control, differential-to-single-ended conversion and gain are performed simultaneously by this scheme.

The universality of magnetosensors is one of the aspects that entails their wider applicability. The op-amp is now a most useful analog device because it can be employed to realize different functions, such as amplification, a Schmitt trigger, an

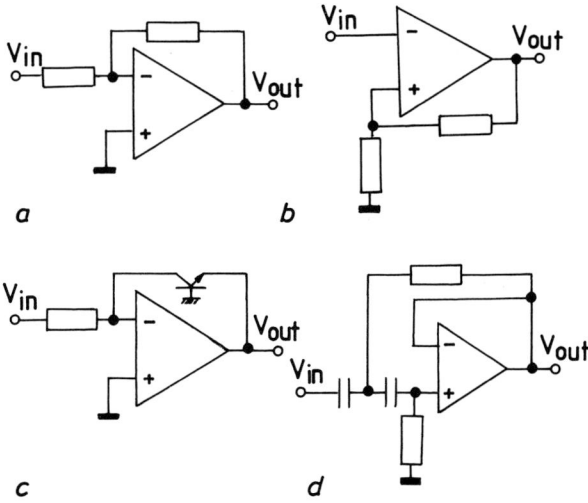

Fig. 9.9. Applications of op-amp: (a) inverting amplifier; (b) comparator with hysteresis; (c) log-amplifier; and (d) high- pass filter [12].

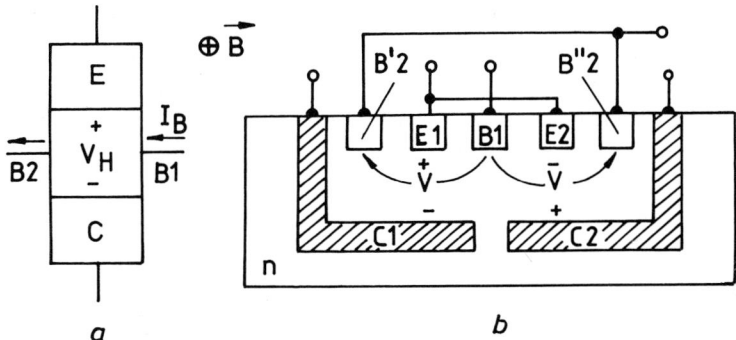

Fig. 9.10. Versions of differential amplification magnetic sensors (DAMS) using BMTs with Hall field: (a) discrete DAMS; (b) integrated sensor.

active filter, an integrator, an oscillator, etc. This gives rise to the idea to make magnetic sensors similarly adaptable. The new approach has been materialized in the magneto-operational amplifier (MOP) [12,17].

The concept of the MOP is just based on an op-amp. Figure 9.11 shows the diagram of the MOP. Like an op-amp, the MOP has a pair of input electrodes and a sufficiently large gain. The difference is that a magnetic-field-induced voltage, V_m, is added to the electrical inputs. Therefore the output voltage of the MOP is $V_{out} = A(V_{in}^+ - V_{in}^- + V_m)$; $V_m = S_b B$, where V_m is the magnetically induced voltage, which is proportional to the applied field B, while S_b is the system's sensitivity. When the feedback circuit, whose transfer function is F, is formed, as

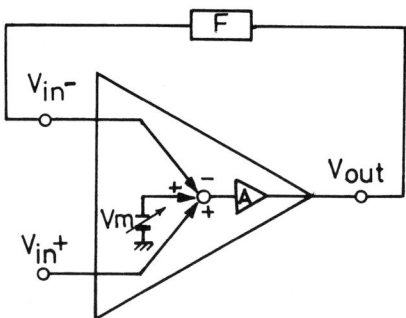

Fig. 9.11. Concept of a magneto-operational amplifier [17].

shown in Fig. 9.11, the output voltage is determined as $V_{out} = (V_{in}^{+} + V_m)/(1/A + F)$, which is similar to the case of the normal op-amp circuit except for the term V_m. Since the term V_m is of the same order as the electrical input, the operation of the magnetic signal is realized like the electrical operation of the op-amp. The MOP has the function of a normal op-amp and a magnetic sensor. Similarly to the normal op-amp, the MOP offers many applications for magnetic signals, e.g. as a linear magnetic sensor with variable sensitivity, a magnetic switch, magnetic filtering, magnetic integration, etc., when the external circuits (which may include few passive and or active elements) being analogous to the usual op-amp circuits. The application of an electric feedback signal to a magnetically sensitive device based on the MOP concept is in [18]. A silicon magnetosensitive IC with an optimized parallel-field BMT (Fig. 6.9h) and a differential CMOS amplifier is described in [19]. The overall sensitivity is 10 V T^{-1}.

The growing importance of microprocessor-aided systems brings the need for magnetosensors with a digital output. The first step in this direction is the creation of transducers with a frequency output. On-chip frequency conversion of the magnetosensor signal can serve several purposes:

(a) A frequency modulated (FM) signal can be regarded as an intermediate step between analog and digital signals since its frequency can easily be "counted" by a comparator and a counter. A semidigital bus interface is possible;

(b) The FM signal is highly immune to noise and interference signals. It can be recovered by filtering, phase-lock-loop techniques and synchronous detection. It can be frequency converted and multiplexed into a frequency band suitable for transportation over long distances;

(c) Frequency modulation does not need a fixed relation between internal and external voltages or current references. On the other hand, a fixed relation between the internal time constant and the external clock frequency is needed [5].

Described in [20] is a GaAs magnetosensitive IC which includes a GaAs Hall sensor, differential input and output amplifier with a nominal gain of 10, a level shifter commonly used in d.c.-coupled GaAs FET ICs, a comparator and an output buffer which has been designed to drive standard CMOS and TTL loads. This magnetosensitive switch is operable between $-30°C$ and $+200°C$ and its magnetic threshold is 450 G or less.

A magnetically controlled oscillator (MCO) has been developed for direct connection with digital systems. N-channel split-drain MAGFETs, an amplifier and a voltage-controlled or current-controlled oscillator have been integrated onto one-chip [21,22]. The sensor outputs drive two voltage-controlled oscillators which produce a frequency-modulated signal [21]. The output characteristic $f_{out}(B)$ is linear. A sensitivity $S = \Delta f_{out}/(f_0 B) \approx 7.8\%$ T^{-1} has been achieved by means of CMOS MCO [22]. In a smart transducer system, the differential frequency output Δf_{out} may be decoded by a microcomputer mounted adjacent to the sensor chip itself.

A parallel-field Hall element (Fig. 4.26e), with frequency output formed in an n-type Si epilayer and two I^2L ring oscillators has been suggested in [23]. The starting point is the magnetic cell transforming the field into a Hall voltage V_H. The V_H is further transformed into a difference of the collector currents by the p–n–p transistors. The p–n–p transistors provide the injection current for the ring oscillators, and, in the end, a difference between the two oscillator frequencies f_1 and f_2 in response to the magnetic field is obtained. The frequency difference yields the magnetosensitivity $S_f = \Delta(f_1 - f_2)/\Delta B \simeq 1$ kHz mT^{-1} [23].

An integrated magnetic-field silicon sensor with a frequency output is described in [24]. The chip consists of buffer elements and two ring-oscillators with a magnetic-field-sensitive part of the injector. The frequencies of the output signals of both oscillators depend on the magnitude of the field $\boldsymbol{B}$. The characteristics of the device are: a power supply of $(0.5$ to $5)$ V $\times$ $(0.3$ to $0.8)$ mA; an absolute sensitivity of 2.5 to 10 kHz T^{-1}; a working range of fields of $0 \leq B \leq 0.4$ T; a maximum nonlinearity of 0.1%.

A magnetic-field multivibrator sensor based on a silicon BMT (Fig. 6.9h) has been proposed and tested [25,26]. A version of this circuit is shown in Fig. 9.12. The output frequency is a linear function of the variation in the magnetic field from 1 mT up to 1 T. The device's sensitivity is 60 Hz mT^{-1} with an error of less than 0.05% °C^{-1} in the $0 \leq T \leq 70°C$ interval.

The addition of a frequency conversion circuit to a solid-state magnetic sensor so that they are combined on one chip potentially results in a number of desirable features that will assist the development of smart solutions [5]: improvement of sensitivity, linearity, accuracy and interference resistivity; standardization of output signal formats, such as an analog voltage at low impedance, an analog current at high impedance, a frequency, a digital signal, or a bus interface; and reduction of the size and cost of the measurement system and of the number of its components.

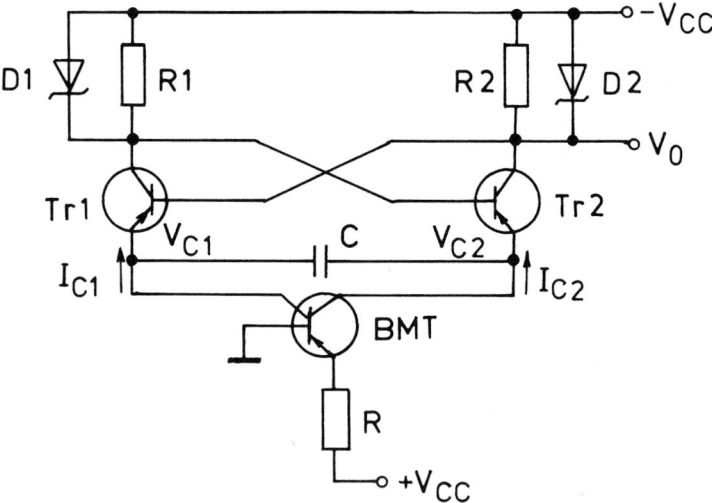

Fig. 9.12. Magnetosensitive multivibrator sensor using a silicon differential p^+–n–p^+ BMT [25,26].

9.3. Magnetic flip-flop based sensors

Flip-flop based transducers [27–32] are a typical example of the advance in recent years in the field of sensors, including magnetometers. The priority given to this class of digital circuits over electronic oscillator-based devices (current-to-frequency converters, square-wave oscillators, ring oscillators, frequency-ratio output devices) is due to their universality, flexibility, the possibility of batch production and very high sensitivity.

Flip-flop transducers are, in fact, a version of stochastic analog-to-digital converters which use a comparator and a threshold-voltage generator to produce a random waveform with a constant probability–density distribution [33]. In the comparator the measured signal is compared with a random signal the result of which is that a random pulse train occurs at the output of the comparator. It presents an equivalent picture of the analog input signal. High sensitivity and noise immunity can be achieved if the comparison is frequently repeated. This is possible only in the case of small variations or invariability of the input signal. The electrical noise of flip-flop circuits can be used for random-signal generation, and owing to this fact these devices can perform satisfactory comparator functions. The operation of flip-flop sensors is based on the following principle: any physical system brought into an unstable state is highly sensitive to perturbations. Standard flip-flops consist of two transistors and two resistors each (Fig. 9.13), and possess two stable states. Under certain conditions, the flip-flop can be brought into an unstable state as, for instance, when the supply current is switched on or off. The flip-flop will be subsequently brought to one of the two stable states. The external and the internal

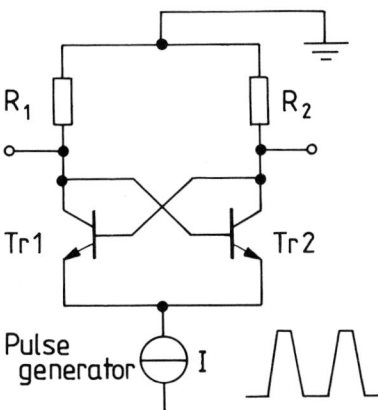

Fig. 9.13. The circuit diagram of a standard flip-flop, which includes two transistors and two resistors.

excitations, whose magnitudes may be very small, determine which stable state will be chosen by the device.

To create a sensing device from a standard flip-flop, the existing circuit components are replaced by elements that are sensitive to certain physical or chemical parameters and magnetic fields in particular. The external field B will then generate an excitation in the magnetosensitive element, thus bringing the flip-flop into a certain stable state. By observing which stable state is chosen, the presence or absence of the magnetic field can be determined. However, the magnitude of this parameter cannot be inferred. A reference is required in order to quantify the field B. The threshold-sensitivity of any given measurement is set by the total system noise, which is always present. The total system noise has therefore been used as a reference for the measurements aimed at low threshold sensitivity. In the unstable state, it is not only the external parameters that can generate a perturbation in the flip-flop, but the total system noise can also create an excitation with random characteristics. To infer the field B from the noise, the flip-flop measurements are repeated many times, and the number of "ones" and "zeros" in the output is counted. If the parameter to be measured has a zero magnitude, the number of ones will account for 50% of the number of unstable states, because the noise in this case is symmetrically distributed with a zero mean. When the field B is present, the resulting number of "ones" will not be 50% of the total unstable states, because of the asymmetry caused by the field B, but instead will depend on the magnitude of the field B. In other words, the number of the resulting "ones" is the measure of the magnetic field [29,32]. In short, the operation of a magnetic flip-flop sensor is based on: replacement of some components by magnetosensitive elements; bringing the flip-flop into an unstable state and detecting the final stable state; and repeating the sensing action many times [32].

Hall transducers with orthogonal or parallel-field excitation can be used in flip-

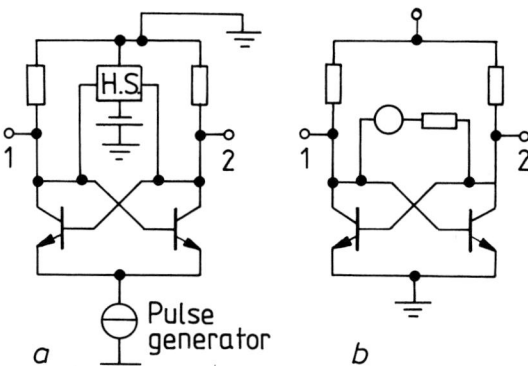

Fig. 9.14. (a) Circuit of Hall-plate flip-flop sensor; (b) the equivalent circuit [32].

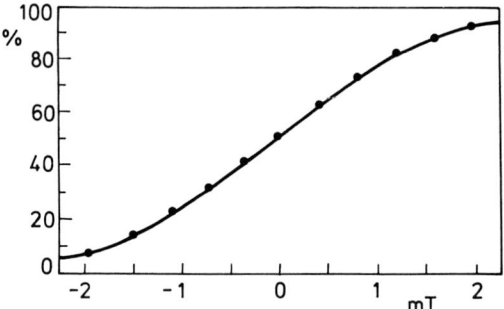

Fig. 9.15. Probability percentage of "ones" as a function of the magnetic field B measured by the flip-flop Hall sensor [31,32].

flop devices by connecting the two output terminals of the sensor to the two sides of the flip-flop [28,32]. The connections of the Hall plate's two bias nodes are then grounded and coupled to a bias voltage, separately (Fig. 9.14a). The equivalent circuit of a flip-flop Hall device is shown in Fig. 9.14b. The Hall voltage, which is applied to the two base contacts of the transistors, will switch the flip-flop from an unstable state to one of the two stable states. The devices can be integrated onto one chip by means of bipolar technology. The dependence of the output of the flip-flop sensor on the magnetic field B is illustrated in Fig. 9.15 [28,32]. The characteristic was recorded at a sampling frequency of 100 kHz, i.e. 10^5 unstable states per second.

Another magnetic flip-flop sensor with a NMOS split-drain MT (Fig. 9.16) is described in [32]. The switching on and the operation of this device are analogous to those of the flip-flop Hall version. Differential BMTs can also be used as sensitive elements of flip-flop circuits. In addition to magnetic-field registration, this promising class of digital sensors has been tested for the detection of other physical quantities, e.g. flip-flops with piezo-resistors [29], photodiodes and phototransistors

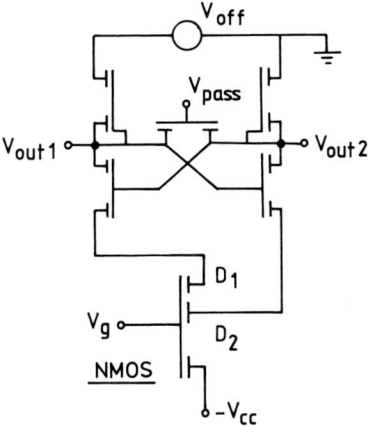

Fig. 9.16. N-MOS magnetic field flip-flop sensor using a split-drain device as a sensing element [32].

[31,32], thermal transistors and thermocouples and other types of physical and chemical transducers [32].

9.4. Interfacing solid-state magnetic sensors with digital systems

Digital systems show a dramatically increased capacity for gathering information from sensors by means of controlling and signal processing algorithms. The basic design criteria used to achieve this goal are the following: the binary signals are completely immune to errors associated with channel nonlinearity and induced noise; microprocessors tend to come armed with an independent timer (counter), whereby time and frequency encoding becomes a promising trend; the individual addressability of sensors is easily realizable; data multiplexing is relatively straightforward in a serial digital channel; the application of modern microelectronic technologies, etc. [34] (§1.6).

In modern measurement and control systems, a large number of sensors, magnetic-field transducers included, are connected to a central computer. The computer uses this input information to calculate relevant data about the process and, in accordance with the particular task, sends data back along the bus system to actuators such as valves, heaters, etc. An important feature is related to the most desirable format of the information sent along the bus data lines. A system could also be based on a setup in which the analog magnetosensitive signal is first converted into a digital form and this digital signal is subsequently transmitted to the central processor. Microprocessors with sophisticated functions, op-amps, A/D converters and other devices are currently available, and their prices are steadily dropping.

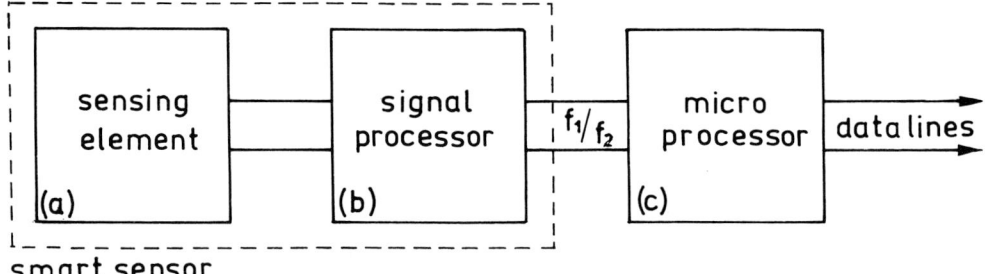

Fig. 9.17. Schematic diagram of a smart sensor with a microprocessor.

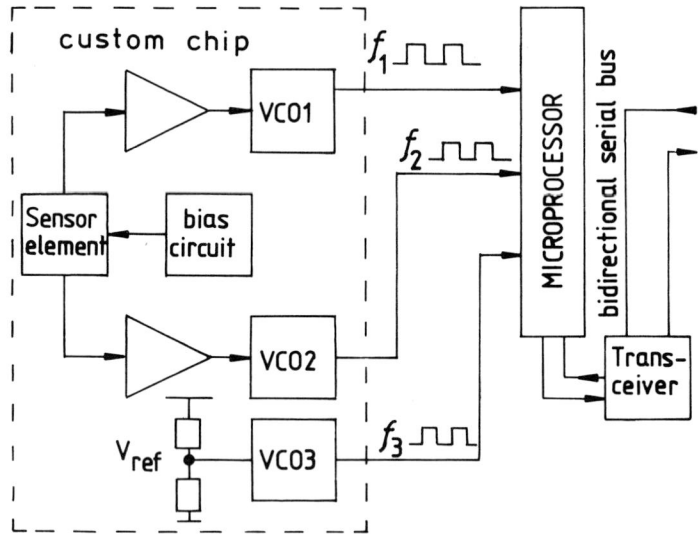

Fig. 9.18. Outline of the final design of the smart magnetic sensor [36,37].

A microprocessor-enhanced magnetosensor might be composed of the following: (a) a magnetosensitive element (a Hall device, magnetoresistor, MD, CDM, etc.); (b) a signal-processing part, generating a frequency ratio; (c) a microprocessor (Fig. 9.17) [35]. Parts (a) and (b) can be integrated onto a single chip. This is recommended when Si IC technology is used. The third part (c) of this smart solution should not necessarily be located close to the first two (§1.6). This idea has been successfully applied in a smart magnetic sensor, as well. Its circuit is presented in Fig. 9.18 [34,36,37]. It is advisable to use a symmetric structure such as a Hall device, a split-drain MOSFET or a differential BMT. This choice results from the need for initial minimization of some sensor defects like offset, nonlinearity, drift, etc. The influence of the common-mode errors is balanced out by means of design symmetry. This is termed structural compensation. Also, there is a tailored compensation, in which the individual device is "trimmed" to optimize its behavior (e.g. the addition of positive or negative temperature-coefficient resistors to the "bridge"). Internal electronics in the silicon chip allows both classes of compensation to be included. Tailored parameters can be stored in an internal ROM. The two sensor outputs, the on-chip amplifiers and the voltage-controlled oscillators VCO1 and VCO2 will tend to track each other with supply-voltage variations and externally induced noise. The effect of changes in the absolute value of frequency with temperature is eliminated, in particular, by taking the difference of frequency as the coded signal. There will, however, be residual cross-sensitivity problems associated with chip fabrication. There will also be second-order effects, such as the change in sensitivity of the voltage-controlled oscillators with temperature, which are not eliminated by the differencing process. It is here that tailored compensation becomes important. For this reason a third voltage-controlled oscillator VCO3 with fixed input is incorporated into the chip design. This oscillator acts as a temperature sensor, making possible the storage of individual calibration tables for different operating temperatures in reference tables in the microprocessor ROM, the programme also provides temperature compensation and linearization at the sensing head. The sensor chip is mounted on a thick-film substrate together with a microcontroller and a few minor components. As a result of these design procedures and precautions, the system designer is presented with a sensor that is apparently insensitive to temperature and produces a direct digital output that is an accurate linear measure of magnetic-field strength [36].

Another opportunity for the generation of an adequate temperature signal for VCO3 (Fig. 9.18) is the use of a functional multisensor (§8.1.1). This device, moreover, is structurally compensated. Another task the microprocessor could do is to perform periodic calibrations and tests of the magnetosensor. A detailed analysis of the software techniques for sensor compensations is given in [38]. The modern approach to smart transducer technology and hardware is discussed in [39].

The new generation of intelligent sensors can have the compensation parameters set up during the manufacturer's testing procedures by the process of programming the internal ROM. Thus, in future, the system designer will be freed from problems of sensor defects and information transmission which have been so burdensome in the past.

9.5. Improvement of solid-state magnetosensor characteristics

9.5.1. Offset reduction

All solid-state magnetosensors considered in the previous chapters belong to the modulating type. The problem arising in all differential transducers with an external power supply is that in general an electrical and/or structural imbalance leading to asymmetry of the energy flux occurs (§3.3.2). As a result, an offset is generated at the output, which varies from chip to chip. If the characteristics of both the sensor and the magnetic field are unknown, the offset and the useful signal cannot be distinguished. This is why the offset is described as an interfering input signal which produces additive errors in the sensor output:

$$V = SB + V_0, \tag{9.1}$$

where V is the differential output signal, $V_0 \equiv V_{\text{off}}$ is the offset in a field $B = 0$, and S is the magnetosensitivity. The output signal can be a voltage, current or frequency, and the offset signal is necessarily of the same type [40]. The offset V_0, which is measured in the units of the output signal, should be distinguished from the equivalent offset $B_{\text{eq,off}}$, measured in units of magnetic induction. The relationship between the two parameters is:

$$B_{\text{eq,off}} = \frac{V_0}{S}. \tag{9.2}$$

Therefore the equivalent offset $B_{\text{eq,eff}}$ is the value which will be displayed by the registration system as the magnetic induction when the real field is $B = 0$. According to (9.2), the value of $B_{\text{eq,off}}$ is not determined by the offset signal V_0 alone. It depends also on the sensitivity S. The behavior of $B_{\text{eq,off}}$ and V_0 as functions of the environment and variations in device parameters is not necessarily one and the same. If both the magnetic field and the offset signal are unknown quantities, it follows from (9.1) that it is impossible to distinguish what portion of the output signal V is generated by the real magnetic induction B and what part is associated with the offset. In this way a considerable error is introduced into the measured results owing to the fact that there is only one equation (9.1) describing the sensor output, which, however, includes two unknown variables: the offset signal V_0 and the induction B [41,42]. The offset problem can have serious adverse effects on sensor applications, e.g. brushless d.c. motors, contactless switches, permanent registration of the geomagnetic field, etc.

(a) Methods of offset reduction. The main causes for the occurrence of the offset in Hall sensors and differential BMTs have been analyzed in §4.2.4.2b and §6.4.3. Several methods for the reduction of this error have been generally accepted. The most important of them will be briefly discussed below.

– *Improvement in the fabrication technology and sensor design.* Imperfections in the fabrication technology, such as nonuniform diffusion, epitaxy, oxidation, etc., as well as inaccurate mask alignment, especially in sensors based on diffused Wheatstone-bridge-like structures, can contribute considerably to the offset. A small process-induced error in such devices may lead to great imbalance of the two or four resistors of the bridge and, therefore, to a substantial offset. Unfortunately the continuous improvement of technology leads to the devices becoming more expensive. Sensors with undiminished sensitivity and reduced offset can be realized by a careful selection of the geometrical ratio. The influence of mask-rotation errors can be reduced, for example, by the proper choice of the length to width ratio in Hall sensors. This method, however, does not lead to a satisfactory minimization of the offset in all kinds of magnetic-field transducers [42].

– *Offset adjustment and calibration.* Before the measurement is started, one can isolate the measuring system from the magnetic field (or set the magnetic field to a known value). In this way one of the two unknown variables in (9.1) is eliminated. By means of additional adjustable electronic circuitry, the output of the magnetosensor can be adjusted to zero. Because of the zero drift, this adjustment must be repeated many times. A further disadvantage is that in many cases it is rather difficult, or at least very inconvenient, to set the field B to zero, the cost of the system therefore inevitably increases [42].

– *Compensation of offset.* Compensation is the technique of offsetting the effects of changes in the environment on the input–output relationship of an instrument or a device. The operating principle in this case is either to prevent changes in the environment or to make the instrument insensitive to the changes that do occur. This method can be employed by using matched sensor elements integrated onto the same silicon chip. The method is based on the observation that neighboring elements on a silicon wafer often exhibit comparable imperfections, which makes it probable that neighboring sensors will also exhibit almost the same sensitivity and offset values. One of the magnetosensors can then be used as an offset reference, providing it is well shielded from any magnetic field [42].

– *Sensitivity-variation offset-reduction method.* The recently proposed sensitivity-variation offset-reduction method is a serious advance in solving the offset problem. It has been successfully applied to BMTs of the type shown in Fig. 6.9c and to Hall devices [40–42]. It is based on the introduction of a second equation which, together with (9.1), can give information about the offset V_{off} and the magnetic induction B. For this purpose, the sensitivity S of one and the same sensor is changed by means of some controllable parameter at a time $t = t_2$ in such a way that the variation of this parameter does not influence the offset signal. This approach does not encounter the problem of the offset varying with position on the chip, since one and the same device is used at different times. A system of two equations can thereby be obtained with the same unknown quantities as equation (9.1):

$$V_1 = S_1 B + V_0, \text{ at time } t = t_1, \tag{9.3}$$

$$V_2 = S_2 B + V_0, \text{ at time } t = t_2. \tag{9.4}$$

Equations (9.3) and (9.4) yield:

$$B = \frac{V_1 - V_2}{S_1 - S_2}. \tag{9.5}$$

Therefore the value of the magnetic induction is extracted from the source equations (9.3) and (9.4). If we keep the time interval t_1–t_2 short, and if the

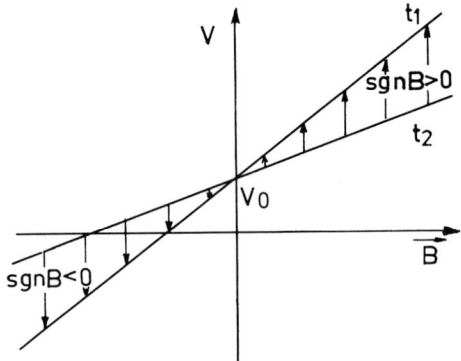

Fig. 9.19. Graphical representation of the sensitivity–variation method (ideal case) [41].

measurement repetition frequency is high enough, the slow drift of the offset signal V_0 is negligible. The magnitude of V_0 is not important in the sensitivity-variation method, so that whether the sensor exhibits a large spread in the offset signal from chip to chip is irrelevant. As a consequence, the process tolerances become less demanding in this respect. Only one sensor is used, but at two different times, which excludes errors that may occur if two devices are used whose offset signals are not exactly the same. More fundamentally, it can be said that the measurement has been expanded in time in the sensitivity-variation offset-reduction method and not in space as in the compensation techniques [41,42]. The sensitivity-variation method can be illustrated graphically as in Fig. 9.19. In such applications as the brushless electromotor, only the sign of the field is important (sign sensor). In that case, one need only know the sign of $S_1 - S_2$ and not its magnitude:

$$\operatorname{sgn} B = \operatorname{sgn}(S_1 - S_2)\operatorname{sgn}(V_1 - V_2). \tag{9.6}$$

The case represented by equation (9.6) is depicted in Fig. 9.19 for $S_1 - S_2 > 0$.

In the nonideal case, when the offset signal V_0 does change, Fig. 9.20 applies. ΔV_0 represents the amount of change in V_0. We can see from Fig. 9.20 that the sign sensor yields erroneous values of $\operatorname{sgn} B$ if $0 \leq B \leq B_r$. The value B_r is the new residual offset of the measurement, expressed in the units of the measured variable. In this nonideal case:

$$B = \frac{V_1 - V_2}{S_1 - S_2} + B_r, \quad B_r = \frac{\Delta V_0}{S_1 - S_2}. \tag{9.7}$$

B_r should be kept as small as possible [41]. If, however, ΔS is zero, while ΔV_0 is nonzero, the residual offset becomes infinite. This is certainly the most undesirable possibility.

The main advantages of the sensitivity-variation offset-reduction method are: small variations in the offset signal because of temperature variations; mechanical strain and other phenomena do not influence the measurements; the method is

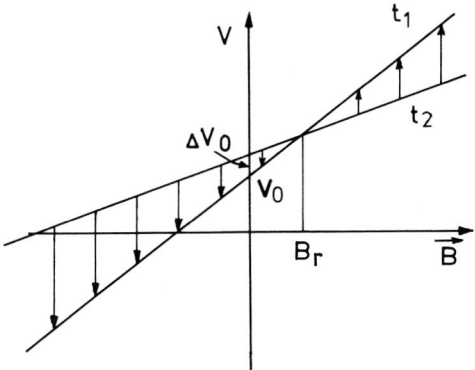

Fig. 9.20. Sensitivity–variation method in nonideal cases: both the offset V_0 and S change with the varying parameter; B_r is the residual offset expressed in units of the magnetic field [41].

a modern IC solution which does not demand that the value of the magnetic induction be known.

(b) Offset correction circuits. A simple and effective way to achieve zero offset in Hall microdevices is the formation of additional control electrodes in the active sensor area; they are located near the supply contacts [43]. The current I_k flowing in a control electrode produces an offset voltage at the Hall terminals given by $V_{\text{off}}(I_k) = kI_k$, where k is the control factor. The value of k is a function of the physical location of the control electrode in a somewhat complex manner, relating to the spreading resistance in a finite sheet with partially shortened edges. In general, the value of the k-factor increases with increasing distance between the nearest current contact and the control electrode, and it increases with the distance of the control electrode from the centerline. It has opposite signs on the opposite sides of the centerline. The offset at the Hall terminals when $B = 0$ can be completely eliminated by combining the initial offset and the offset due to the current I_k of the respective control electrode, provided they have opposite signs.

The circuits most widely used for offset compensation in Hall devices are presented in Fig. 9.21 [44,45]. The circuits in Fig. 9.21a, b and d are convenient for use with four-terminal Hall sensors, while the other circuits are designed for operation with five-terminal versions. The circuit in Fig. 9.21d seems to be the most convenient for practical applications since it does not cause loss of supply power and power loss at the output, neither is an additional biasing source necessary. The offset in parallel-field Hall devices can be successfully eliminated by the arrangement shown in Fig. 4.26e.

Since in the general case $V_0 \equiv V_{\text{off}}$ increases linearly with the supply current I_s, the circuit presented in Fig. 9.21h is convenient for certain applications. If the d.c. supply current is forced through a suitable coil placed close to the Hall sensor,

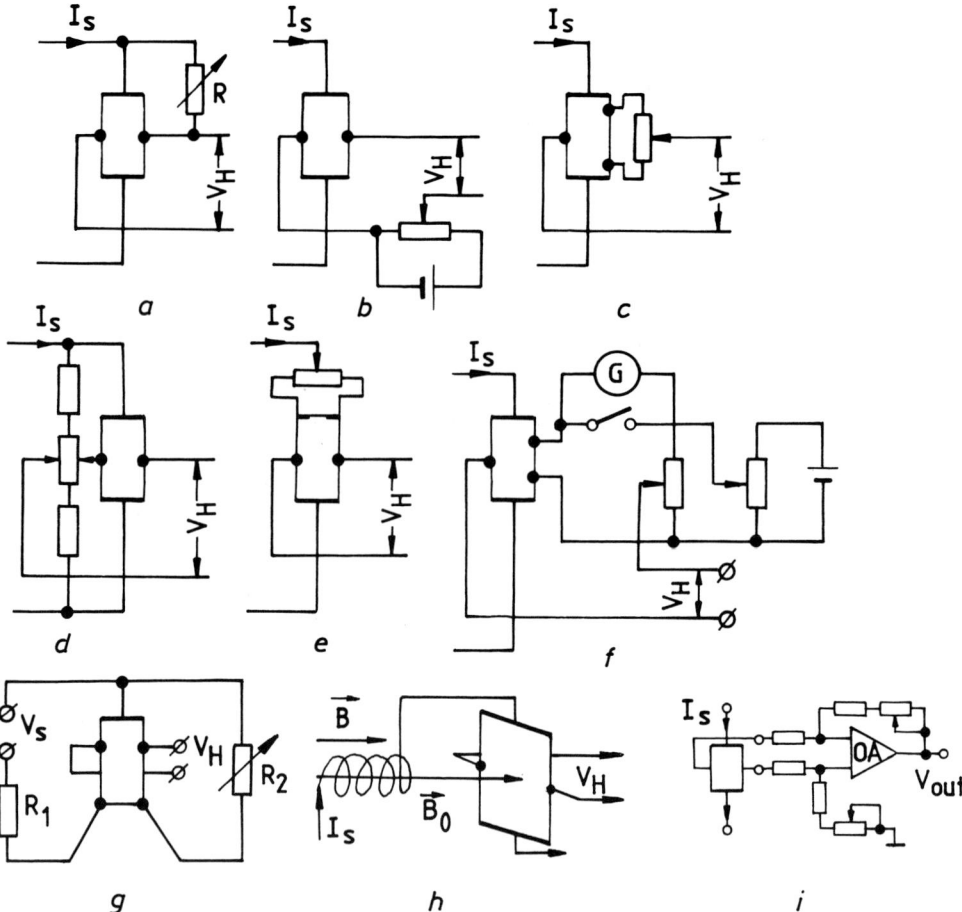

Fig. 9.21. Offset-compensation circuits.

a magnetic field of magnitude B_0 will be generated. The vector $\boldsymbol{B}_0$ has such a direction so as to neutralize the initial offset V_0, i.e. $V_H(B_0) + V_0 = 0$. With an optimized coil configuration the variation of $V_0(I)$ will always be compensated by the additional magnetic polarization $B_0(I)$ and the output will register a signal $V_H(\boldsymbol{B})$. An op-amp can be used for the offset-voltage correction as shown in Fig. 9.21i.

One of the main causes of offset in Hall microsensors are geometrical lithographic errors, due to misalignment between different masks in the IC processing, and etching randomness. However, stresses can easily cause an even larger offset than geometrical errors can. As the stress in the Hall plate depends very much on the way the device is mounted and, moreover, because stress appears to change with time and temperature, it is not possible to compensate for offset due to stress

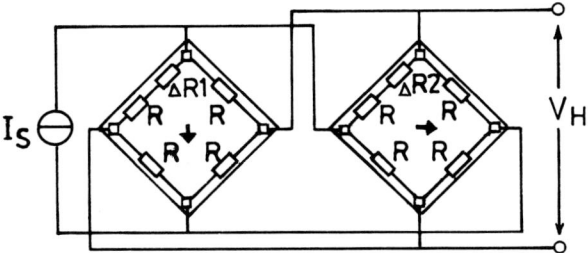

Fig. 9.22. Orthogonally switched on-chip Hall plates with current biasing to minimize the offset errors.

at the end of the production process. The lowest offset of about 2 mT is obtained by the incorporation of two or four identical square-shaped Hall plates (Fig. 4.1p), placed closely together in an orthogonal coupling arrangement (cf. Fig. 9.22). By interchanging the bias and the Hall contacts, the sign of the offset signal is reversed, while the sensitivity or the response to the magnetic field remains the same. If the output signals are added, the offsets will cancel, while the sensitivity will be doubled. This compensation technique which reduces both the offset and the thermal offset drift uses the reciprocity of symmetrical Hall plates, i.e. the exchangeability of the Hall and bias contacts without changing the magnetic-field sensitivity [46,47]. However, the matched Hall plates are never exactly identical. Differences in the local stress and geometrical errors are unavoidable and will result in an offset.

The insulating dams applied near the Hall contacts (Fig. 4.1g) are an improvement in the fabrication technology that reduces the equivalent offset caused by misalignment.

An ingenious and promising approach for the minimization of the influence of temperature and mechanical stress on the offset, the so-called low-offset spinning-current Hall plate, has been suggested in [48]. A four terminal Hall plate can be viewed as a Wheatstone bridge in which geometrical errors and the piezo-resistance effect result in a bridge imbalance [49]. It can easily be seen that geometrical errors cancel out when the Hall plate is switched orthogonally to itself. The variation of the piezo-resistance coefficients with the direction of the current flow prevents the stress from canceling out when switched orthogonally.

To extend the offset reduction in time, using the reciprocity principle, the direction of the current in a single, symmetrical Hall plate can be made to spin with steps of $\pi/2$ (cf. Fig. 9.23). However, it appears that orthogonal switching alone is not sufficient to cancel the offset completely. Therefore, a new symmetrical Hall plate was devised with 16 point contacts distributed uniformly along the border of the plate (cf. Fig. 9.24). The electronic circuitry commutates the contacts of the Hall plate in such a way that the direction of the current is made to spin with steps of $2\pi/16$. The Hall voltage is always measured at the pair of contacts positioned perpendicularly to the direction of current flow. An integrator averages

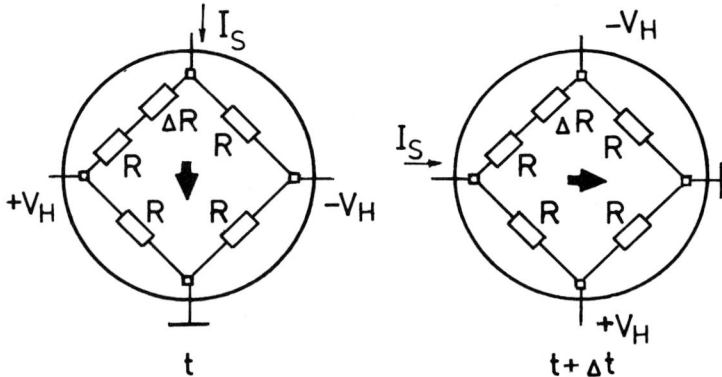

Fig. 9.23. Orthogonal switching of the Hall plate. At time t the current flows downwards, and at time $(t + \Delta t)$ the current flows perpendicularly to the initial direction of current flow. The Hall voltage is measured at the pair of contacts which is perpendicularly oriented to the current flow.

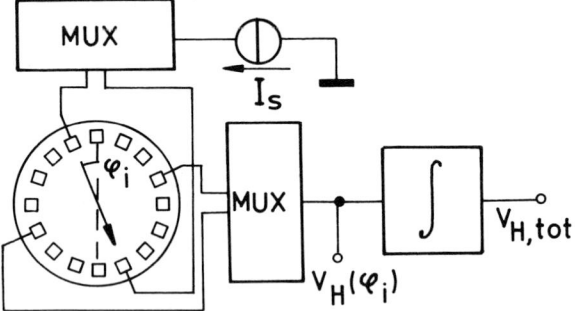

Fig. 9.24. Diagram of the implementation of the spinning- current offset-reduction method. One multiplexer changes the direction of the current flow while the other multiplexer selects simultaneously the proper Hall contact pair. The integrator averages the consecutively available Hall voltages [48].

the consecutive output signals which correspond to the different directions of the current flow.

The experimental results indicate that, by means of the low-offset spinning-current Hall plate, the offset is reduced from 50 mT to about 50 μT, i.e. by three orders of magnitude, at the cost of a reduced bandwidth [48].

There are two solutions to the problem of applying the sensitivity-variation offset-reduction method in Hall devices. The first solution is illustrated by Fig. 9.25 [42]. The sensitivity variation is accomplished by changing the impurity concentration of the active structure of the Hall plate. The well-known dependence of carrier mobilities $\mu_n(n)$ and $\mu_p(p)$ in silicon on the concentration of doping impurities (cf. also Fig. 2.4) [50], is used for this purpose. A concentration gradient in the active region is needed along with a means of altering the sensor volume.

The active structure of the plate is formed by two layers with different impu-

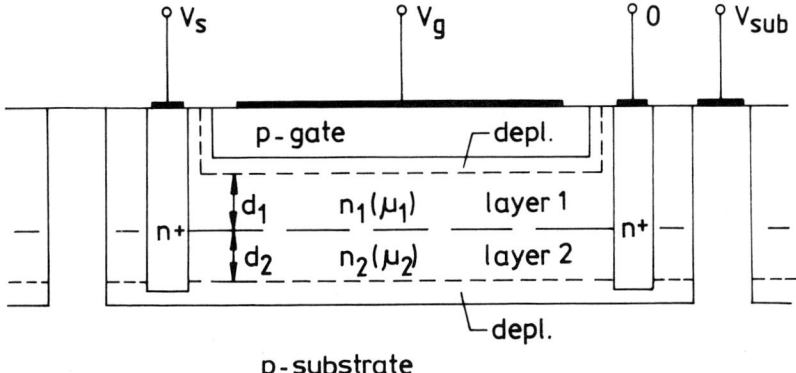

Fig. 9.25. Cross-section of a two-layer n-type Hall plate with a p-type gate used for the selection of the active region [42].

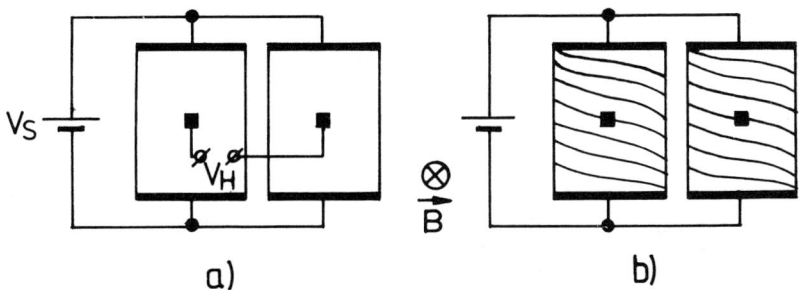

Fig. 9.26. (a) Switched Hall plate with separate active regions; (b) equipotential lines at $\boldsymbol{B} \neq 0$ [42].

rity concentrations (n_1 and n_2) and, therefore, with different carrier mobilities. The p-type gate is included as a means of selecting the active structure of the device, by means of a depletion layer formed by the reverse-biased $p-n$ junction. By fully depleting layer 1 of the platelet, layer 2 becomes the active region, which determines the sensitivity of the Hall plate. Conversely, by fully depleting the second layer (by applying a large enough reverse bias over the substrate–epilayer junction) layer 1 determines the sensitivity. Under the constant voltage bias of the Hall plate, the offset voltage V_0 remains unaltered when the sensitivity is changed in the above-mentioned way. The conditions necessary for offset reduction according to the sensitivity-variation method are now fulfilled [42]. Unfortunately, no experimental evidence can be presented of the double-layer Hall plate because of technological problems. Nevertheless, the innovative value of this solution is obvious.

The second possibility concerns switched Hall plates [42]. The basic principle is illustrated in Figs. 9.26 and 9.27. The structure shown in Fig. 9.26a consists of two isolated regions which are biased by a common source V_s, while the Hall contacts are placed in the middle of each region.

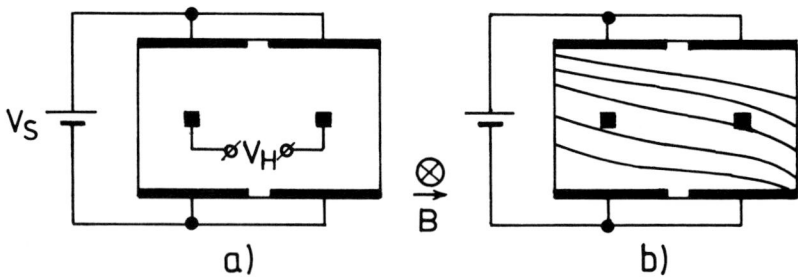

Fig. 9.27. (a) Switched Hall plate with lifted barrier; (b) equipotential lines at $B \neq 0$ [42].

The offset voltage will be nonzero. Figure 9.26b illustrates the equipotential lines in a nonzero, perpendicular magnetic field. If the Hall contacts are close to the center of their respective regions, they will also be on nearly the same equipotential lines. As a consequence, the response of this split structure to the magnetic field will be almost zero. If the barrier between the two regions is removed to a certain extent so that they are able to form a single device, the structure of Fig. 9.27 is obtained, which represents a slightly modified Hall plate. The magnetic sensitivity in this case is smaller compared to a platelet with Hall contacts on the sides. Nevertheless, it is significantly larger than the sensitivity of the split Hall plate of Fig. 9.26. The offset signals in the two cases should be almost equal since neither the mechanical stress, the geometrical misalignment and the power consumption, nor the current distribution has changed markedly. By switching the sensor between the two states, the conditions for offset reduction according to the sensitivity-variation method are fulfilled.

Switching Hall plates can be manufactured either by a CMOS process or by means of bipolar technology. A switched Hall plate fabricated by bipolar technology, for instance, when the sensitivity-variation offset-reduction method is applied, allows an 88% decrease in the original offset and the achievement of an extremely low offset value $\sim 10\ \mu T$ [42].

The offset compensation in differential BMTs is analogous to that in Hall sensors. The easiest and most widely used method is trimming by means of an additional resistor connected to the collector resistors (Fig. 9.7b). This arrangement, however, does not eliminate the temperature dependence of the offset signal. A certain reduction of the offset temperature drift together with its compensation in structures similar to the one in Fig. 6.9h can be achieved by means of a trimmer connected between the two base contacts B_1 and B_2 with $R_{C1} = R_{C2}$, (Fig. 8.5). The main advantage of this circuit is the possibility of improving, to a certain extent, the internal electrical symmetry of the BMT.

An elegant solution of the offset problem in the BMT sensor in Fig. 6.9j is suggested in [51]. In view of the confinement of the carrier injection to the center of the emitter–base junction, there is the possibility of shifting the center of the injection toward one of the collectors. This can be done by applying different

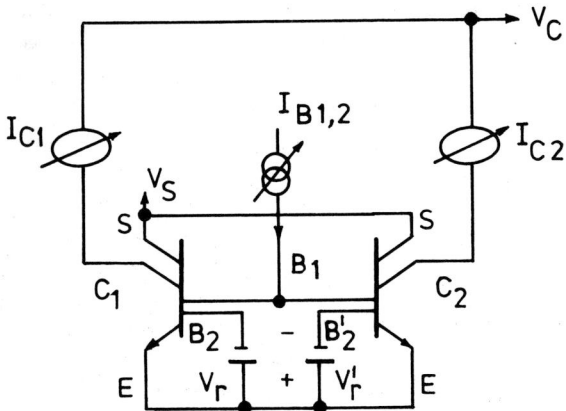

Fig. 9.28. Biasing circuitry for offset elimination in the BMT shown in Fig. 6.9j. The different negative potentials are applied to B_2 and B_2' in order to annul the offset.

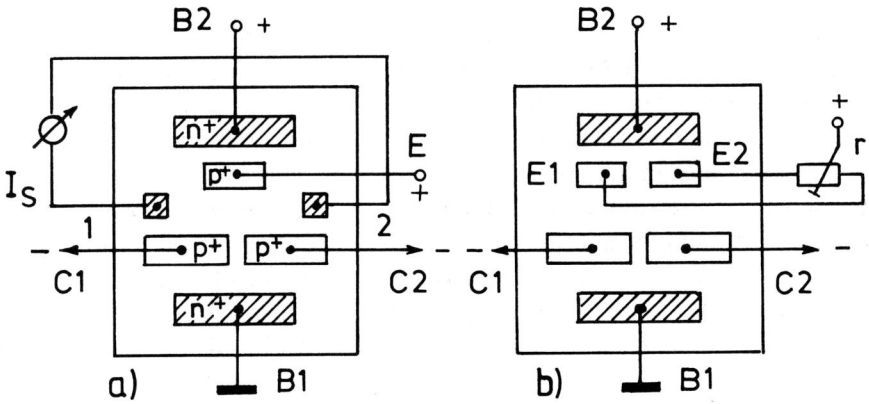

Fig. 9.29. Two versions of offset compensation in drift-aided BMTs.

negative potentials to each p^+-stripe, B_2 and B_2' (Fig. 9.28). The procedure allows adjustment of both collector currents until they are equal, thus annulling the offset.

Two other opportunities for BMT offset compensation are illustrated in Fig. 9.29. In orthogonal BMTs, similar to the one in Fig. 6.8c, two additional ohmic contacts 1 and 2 are formed with symmetrical positions with respect to the emitter E. They can be used together with the supply source I_s for the electrical control of the lateral flow of injected carriers, thus correcting a zero offset (Fig. 9.29a) [52]. The arrangement in Fig. 9.29b is suitable for the device structure shown in Fig. 6.8e. The two emitter currents are balanced by the trimmer r, and equal collector currents $I_{C1}(0) = I_{C2}(0)$ are established [53].

The application of the sensitivity-variation offset-reduction method to the parallel-field BMT with buried collector contacts shown in Fig. 6.9c, with the collector–

base bias V_{CB} as a parameter varying the sensitivity, yields very promising results. The use of V_{CB} for the control of the device's transducing efficiency is associated with the appearance of a high electric field in the depletion-layer regions, which changes the Hall angle (cf. §2.3.3). The magnetosensitivity decreases with increasing V_{CB}, whereas the offset remains nearly constant in a large V_{CB} interval. The difference in the response of the sensitivity and the offset signal to sinusoidal excitation is used to reduce the offset [54]. Experiments confirm a 95% reduction in the offset. The complete measurement circuit is described in detail in [54]. The collector–base bias can also be used as a varying parameter in the orthogonal BMT in Fig. 6.8c, when the substrate serves as an active sensor region [55]. The collector space-charge regions increase their widths as the voltage V_{CB} is increased, thus confining the majority-carrier current $I_{B1,2}$ to the bulk. A Hall field with a transverse orientation relative to the elongated collectors C_1 and C_2 is generated as a result. Magnetosensitivity is enhanced by the voltage V_{CB} while the offset, to a first approximation, remains the same. For instance, the sensitivity is changed by about 70% as the voltage V_{CB} changes by 20 V.

The direct application of the sensitivity-variation offset-reduction method to many types of differential BMT is blocked mainly by the absence (lack of knowledge so far) of a suitable variable parameter which does not disturb the offset. For example, the variation of the current I_{EB} in the device in Fig. 6.9h leads to an unpredictable change in the offset in each particular sample.

The threshold offset-frequency problems in some CDMs can be overcome by the use of a small biasing magnet mounted close to the sensor and producing a field of about 0.4 to 0.5 T, so that the device might operate in its linear region well above the threshold. Then the magnetic field to be measured adds to or subtracts from this biasing field, thus causing a proportional deviation from the bias output frequency.

Magnetodiodes and magnetoresistors have, by their very nature, a large quiescent signal. The response to the magnetic field is measured with respect to that quiescent signal [56]. If these elements are used in a bridge-like circuit, the offset (quiescent) signal is reduced, but asymmetries of the bridge are a new cause of the offset.

9.5.2. Cross-sensitivity reduction

In spite of attempts to optimize the sensor action and detect only the magnetic field, all magnetotransducers are subjected to the influence of a number of variables: temperature, pressure, radiation, etc. In addition to the adverse metrological effect, these cross-sensitivities often hamper the interpretation of the output signals.

The most important cross-sensitivity in solid-state devices, in magnetosensors in particular, is by far that of temperature [2]. Unfortunately the problem of temperature compensation is far from fully analyzed in the literature. The temperature drift of the offset and the temperature coefficient of magnetosensitivity are the main problems considered. The information concerning the particular techniques for overcoming these sensor shortcomings is quite often insufficient. An attempt is made below to consider in detail the minimization of the influence of temperature on the characteristics of magnetosensitive devices in compliance with the existing data and the author's experience.

(a) Compensation of the temperature drift of the offset. The structural compensation by means of geometrical and electrical device symmetry is a well-known technique for overcoming this type of cross-sensitivity. In this respect Hall sensors and differential BMTs can generally be regarded as bridge-format structures. Therefore the temperature drift of the offset, instead of being determined by the temperature variations of the "bridge" resistances R_1 to R_4, is determined by the temperature variation of the inevitable difference $\Delta R(T)$ between the respective resistors. According to the bridge model (Fig. 4. 10) the four resistors R_1 to R_4 of the Hall elements are incorporated in the plate itself, while in differential BMTs only two of the resistors are distributed in the active sensor region, while the other two have fixed values and are connected to the collector (drain) terminals. The temperature drift of the offset of these two types of magnetotransducers with symmetric structures can be minimized by passive and/or active circuits. Some of the methods and arrangements for offset-compensation discussed in §9.5.1 can successfully deal with and compensate the temperature dependence of the offset. In Hall devices the temperature-dependent offset is caused by the temperature sensitivity of the material's resistance.

Figure 9.30 shows a passive offset-temperature-compensation network, which consists of a Hall sensor, a thermistor $R_T(T)$ and two resistors — R_1 coupled in parallel with $R_T(T)$ and in series with R_2. The compensation Hall voltage V_H is measured across R_2 [57].

The necessary value of the network resistivity T.C. as well as the bridge balance

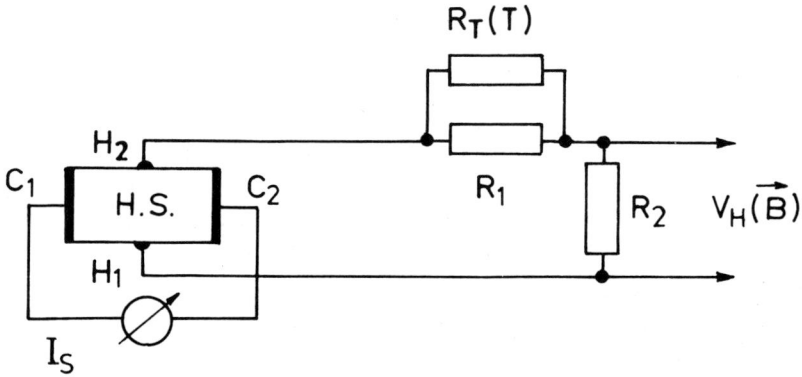

Fig. 9.30. Thermistor–resistor temperature compensation of the offset in a Hall sensor.

are ensured by the proper choice of R_1 and R_2. The thermistor $R_T(T)$ should be placed close to the Hall device in order to be under the same temperature conditions. The change in the Hall voltage with the field B tends to be more linear as the network impedance approaches the Hall output resistance. If a low-resistance thermistor is used, the components R_1 and R_2 must also have low resistances to improve linearity. Errors arise in this inexpensive circuit because of inaccuracies in the temperature coefficients and the different thermal masses of the Hall sensor and the compensating resistors. The power dissipated within the Hall element itself also causes a localized rise in temperature. It is also possible to operate two Hall devices in a differential configuration in an attempt to cancel the effect of their temperature sensitivity [58].

The active circuits for temperature compensation of the offset include op-amps and a temperature sensor T.S.. The compensation is accomplished at the next stage of signal processing. Such a circuit is shown in Fig. 9.31. The output signals of the Hall element and the T.S., after enhancement by op-amps $OA1$ and $OA2$, are subtracted in $OA3$. The gains of $OA1$ and $OA2$ are chosen in such a way that equal values of the temperature-dependent signals are achieved at the input of $OA3$ and the signal at its output is a function of the magnetic field B only. The comments concerning the registration errors in the circuit in Fig. 9.30 are valid for the arrangement in Fig. 9.31, also. In addition, it is necessary that the offset and the output of the temperature sensor should show identical temperature dependencies. In practice this cannot be achieved in a wide temperature range. The voltage developed between terminals C_1 and C_2 of the Hall device due to the current has been used as a temperature-dependent signal for the $OA1$ in Fig. 9.31 [59]. In this case the Hall sample should be long ($l \gg w$) to eliminate disturbance of the voltage $V_{C1,2}(T)$ by geometrical magnetoresistance. The practical implementation of this solution by means of a GaAs Hall structure manifests an accurate temperature offset compensation within ±1% of the full-scale sensitivity at temperatures over $\Delta T = 50°C$, while without compensation the offset is about 65% [59].

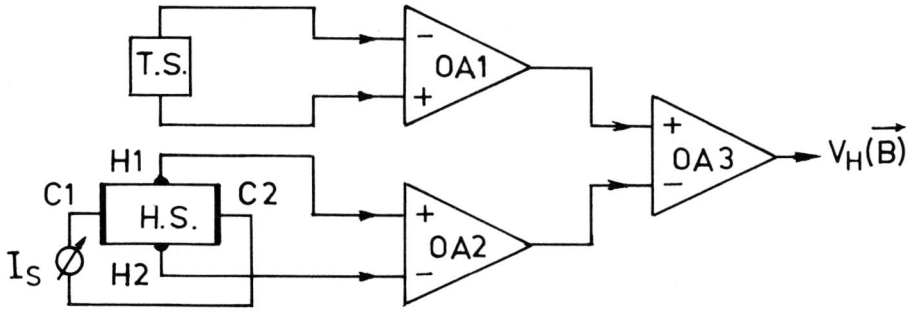

Fig. 9.31. Circuit for active thermocompensation of the offset in a Hall sensor, including op-amps and thermosensor T.S.

The use of heating elements together with environmental thermal feedback techniques is another approach to overcoming the temperature drift of the Hall-sensor offset by means of active circuitry [60]. Optimum results can be achieved by means of this arrangement when the Hall device, the temperature sensor and the heating element are integrated on one chip. The control of the "microthermostat" is accomplished by a circuit which receives the output signal of the temperature transducer and, thereby, determines the supply power for the heater. The temperature sensor and the heater can be combined into a single component whenever necessary. The operating temperature of the chip is usually kept 1 to 2°C higher than the maximum temperature determined by the particular application. By means of such circuits the temperature drift of the offset can be limited to within $\sim \pm 0.002\%$ °C^{-1} with a 1° accuracy of the substrate temperature stabilization in the ± 50°C interval, the time for a transition from $T = 25$°C to $T_{max} = 150$°C being about 100 ms [61]. A bipolar transistor can be used as a heating element. The problem of the temperature dependence of the magnetosensitivity is also solved by such a circuit. High power consumption and the large dimensions of the measurement module are the limitations of this method. Temperature stability can also be achieved by means of a cryogenic coolant, for instance by immersing the sensor in liquid nitrogen ($T = 77$ K).

The thermal drift of the offset of differential BMTs appears to be largely dependent on the thermal-expansion coefficients of the chip, the carrier and the covering epoxy resin. If this top material is omitted, the average flux-density equivalence of the drift is about 10^{-4} T °C^{-1}. The temperature drift of the BMT offset is also a function of the induction B. The symmetry of the temperature picture in the structure is disturbed by the redistribution of currents due to the magnetic field [53]. For instance, the exposure of the BMT in Fig. 6.9h for 10 minutes to a magnetic field $B = 1.5$ T changes the initial offset by 8 to 10%. After some time the initial offset is restored. The compensation of the temperature-dependent drift of the offset of BMTs is analogous to that of Hall sensors. The circuit in Fig. 9.31 is to be preferred, in which the output magnetotransistor signal $\Delta V_{C1,2}$ is transferred to the op-amp $OA2$. The microthermostatic arrangement can also be used for certain applications. In both cases the comments concerning the limitations associated with Hall sensors are valid. The use of the voltage $V_{EB}(T)$ as a thermometric signal at $I_{EB} = $ const is a satisfactory solution of the problem of BMT temperature-dependent offset drift in a wide temperature interval. A circuit suitable for that purpose is shown in Fig. 9.34.

The temperature drift of the offset in Hall devices and differential BMTs depends on the rate of variation of the temperature. The offset drift caused by abrupt changes of T can exceed by 40 to 50% the value observed when the samples are gradually heated. The mechanical strain in the chip and the carrier due to the thermal shock is the most probable reason for this adverse effect.

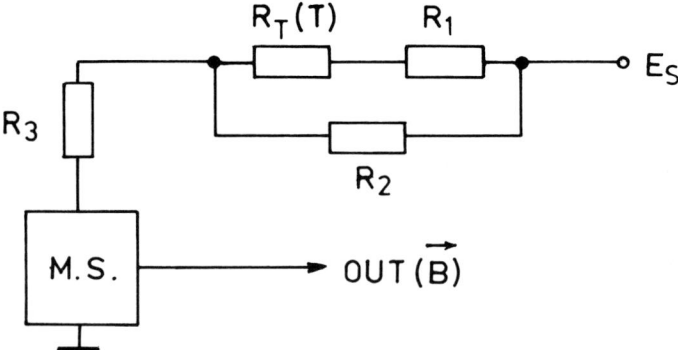

Fig. 9.32. Circuit including a thermistor for the compensation of the T.C. of sensitivity of magnetosensors.

(b) Compensation of the temperature dependence of magnetosensitivity. The variation with temperature of the magnetosensitivity can usually be minimized by the connection of compensation circuits to the input and output terminals of the sensor. This task is made easier by the fact that the T.C. of magnetosensitivity remains almost constant in a wide temperature range. There are active and passive compensation arrangements depending on the components included.

Figure 9.32 presents a passive circuit for compensation of the temperature dependence of magnetosensitivity in Hall devices, BMTs, magnetodiodes and other related magnetosensors (M.S.). The circuit includes a thermistor which is connected in series with the voltage source E_s. Like in the case in Fig. 9.30, the components R_1, R_2 and R_3 added to the thermistor provide for a linear $R_T(T)$ dependence [62]. The compensation of the temperature dependence of magnetosensitivity is achieved by varying the supply voltage with temperature with a coefficient K that has the same magnitude as the sensitivity T.C. but an opposite sign. This arrangement has the following disadvantages: time instability; insufficient linearity of the corrected $R'_T(T)$ dependence of the semiconductor thermistor; complicated adjustment of the compensation circuit; the difficulty of forming thermistors with negative temperature coefficients in ICs; obstacles to ensuring equal temperatures for the magnetosensor and the thermistor in the case of a circuit with discrete components, mainly owing to their large dimensions. The linearity of the circuit in Fig. 9.32 is improved if a forward bias p–n junction is used at a current $I = $ const instead of the thermistor. Silicon diodes are known to exhibit a $V_{pn}(T)$ dependence with a high degree of linearity in a wide temperature interval: $-250°C \leq T \leq +130°C$ [63]. The best results are achieved when the temperature transducer is integrated together with the magnetosensor on the same chip.

The active circuits used for the T.C. of magnetosensitivity include op-amps and suitable thermosensitive components which control the biasing of magnetodevices

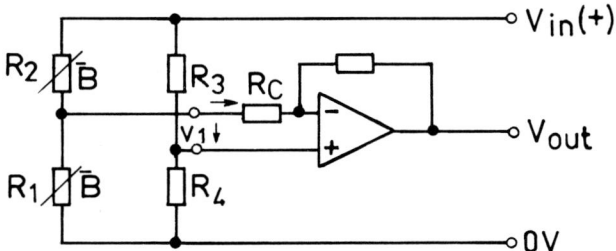

Fig. 9.33. Circuit for temperature compensation of a differential magnetoresistor [64].

or their output signals. The solutions are analogous to the network in Fig. 9.31. Such an arrangement used for the compensation of the T.C. sensitivity of magnetoresistors in a differential circuit is shown in Fig. 9.33 [64].

In the simplest case the temperature characteristic of the internal resistance itself is used as a compensation device. The magnetoresistors $R_1(B)$ and $R_2(B)$ are used in a magnetic potentiometric scheme such that the resistance of one of them increases and the resistance of the other one decreases. The op-amp equates the output current from the bridge $i = V_1/R_i$, where V_1 is the open-circuit output voltage and R_i the internal resistance of the bridge, i.e. $R_i = R_1 R_2/(R_1 + R_2) + R_3 R_4/(R_3 + R_4)$. Resistors R_3 and R_4 are independent of temperature, so that the variation with temperature of R_i is determined only by $R_1(B)$ and $R_2(B)$. As the open-circuit output voltage V_1 and the resistance R_i decrease with increasing temperature, the quotient i remains approximately constant. The output voltage V_{out} can be stabilized over a specified range by the resistor R_c. The optimal value of R_c is $(R_1 + R_2)/2$. An arrangement analogous to the one described above can be used with magnetodiodes connected in a bridge circuit.

The T.C. of sensitivity of the differential BMT in Fig. 9.8 is compensated by an on-chip epilayer resistor R_t for the I/V conversion [9]. The new functional possibilities of the differential BMTs described in §8.1.1 reveal prospects for reducing the T.C. of sensitivity with the aid of the thermometric parameter $V_{EB}(T)$ [65].

Figure 9.34 illustrates the electronic-instrumentation circuitry for reducing the T.C. of magnetosensitivity. The thermodependent output signal of the magnetotransistor, $\Delta I_{C1,2}(B, T) \sim V(B, T)$, is supplied to the input of an op-amp OA with a controllable gain coefficient. A linear thermodependent bias $V_{EB}(T)$ is fed to its controlling input. The influence of temperature on the sensor output voltage $V(B, T)$ is compensated by the respective variation of the gain coefficient control of the op-amp proportional to the temperature T. As a result, the output signal of the op-amp $V_{out}(B)$ is a function of the external magnetic field only. This method provides for the complete and precise reduction of the temperature dependence of the magnetosensitivity within a wide temperature range. A significant advantage of

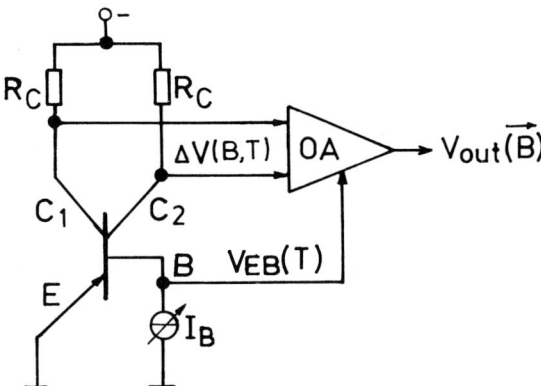

Fig. 9.34. Circuitry for reduction of the T.C. of sensitivity for differential BMTs using the signal $V_{EB}(T)$.

this method is the fact that the thermometric control signal $V_{EB}(T)$ to the op-amp OA is taken directly from the same crystal and reflects very well its actual temperature T. Thus, no additional forming of a thermodependent element in the structure is needed. This is, actually, a completely integrated solution. The maximum thermocompensation error of about 2% has been observed in experiments with silicon p^+-n-p^+ parallel-field BMT in Fig. 6.9h in a magnetic field $0 \le B \le 1$ T in a wide temperature interval $0 \le T \le 80°C$, which is a very promising result. The same network can be used with long Hall sensors, and the voltage between their supply contacts serves as a thermometric signal.

One possible digital method of temperature compensation of the T.C. of CDMs would be to sense the chip temperature by measuring the voltage across an on-chip diode junction with an analog to digital converter (A/D). A microprocessor could look up the required single calibration constant for a particular temperature in a large ROM-based table and use this to calculate B from the CDM frequency it also measures. A simpler method is proposed in ref. [66], which avoids the need for either an A/D or a ROM reference table, resulting in a completely self-contained temperature-compensated magnetic-field sensor. The change in frequency due to temperature variation can be compensated by an opposite change induced by varying biases. If one or more of the electrical biases could be varied in a suitable manner as the temperature of the CDM chip changes, the change in f_{CDM} (in a constant magnetic field) could be minimized. The chip temperature T can be measured using the voltage across a forward-biased diode fabricated on the CDM chip. Bipolar transistor bias-generating circuits, whose outputs are a function of this voltage, could also be laid out around the CDM, to make a very compact and potentially low-cost transducer. After the use of this phase-locking technique, the CDM output frequency can be compensated to remain within ±2% of a central value over the range −18°C to +102°C at a sensitivity of 57 kHz T^{-1}. A detailed description of the circuit of this useful thermocompensation arrangement can be found in [66].

9.5.3. Compensation of nonlinearity

The causes for the nonlinearity of the Hall device's static output characteristics have been described in §4.2.4.2f. In a number of practical cases, Hall sensors are connected to circuits whose input impedance does not considerably exceed the output impedance of the transducer. The value of the effective load resistance R_L must be matched to the sensor output in order to achieve a maximum efficiency. In this case the dependence of the Hall voltage V_H on the load R_L connected to the output terminals H_1 and H_2 is given by the expression [44]:

$$V_H = \frac{S_0}{1 + R_{\text{out}}/R_L}, \tag{9.8}$$

where S_0 is the static magnetosensitivity of the Hall device with $R_L \gg R_{\text{out}}$, and R_{out} is the output resistance. In (9.8), the parameters S_0 and R_{out} depend on the magnetic field B: S_0 grows with B due to the finite dimensions of the Hall contacts, while R_{out} grows because of the finite value of the ratio l/w. The increase in the internal sensor resistance R_{in} in a magnetic field can be compensated by a proper choice of the dimensions of terminals H_1 and H_2. This is particularly valid for semiconductors with high carrier mobility, such as InSb, InAs, InP, GaAs, etc. When $R_L/R_{\text{out}} \sim 1$, the $V_H(B)$ characteristic shows a high degree of linearity. If the sensitivity S_0 is assumed to be independent of the magnetic induction B, the $V_H(B)$ relationship would be sublinear [44,45]. Therefore the increase of $S_0(B)$ can be used for the linearization of $V_H(B)$, provided a load R_L has been connected between H_1 and H_2, although the Hall-device output resistance R_{out} increases with increasing B. In the presence of a load R_L, the magnetosensitivity S_0^* can be written in the form:

$$S_0^* = \frac{S_0}{1 + R_{\text{out}}/R_L}. \tag{9.9}$$

The parameter S_0^* does not depend on the supply current. It has been established from experiments with an n-type InAs Hall sensor that a minimum nonlinearity of $V_H(B)$ is observed when the load resistance R_L is seven times as high as the device's internal resistance R_{in} at $B = 0$, and three times the value of R_{in} at $B = 1$ T. The magnetosensitivity S_0^* is 75% of the value of S_0, and the output power of the sensor is 50% of the maximum [45]. The approach described above to the compensation of Hall-sensor nonlinearity by the choice of R_L is efficient only at a fixed temperature T. Unfortunately, the variation of the parameter T entails a nonlinear variation of the resistivity of the material. This particularly refers to InSb, InAs, GaAs, etc. The linearity of the output characteristic $V_H(B)$ is disturbed because the R_{out}/R_L ratio cannot be kept constant.

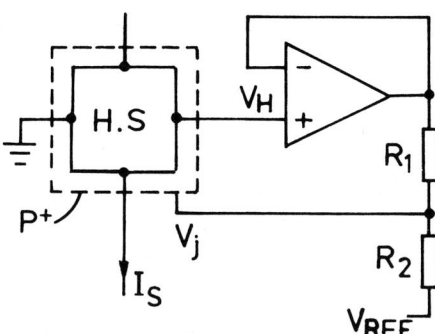

Fig. 9.35. A nonlinearity compensation circuit for a Hall microsensor [67].

An advance in the compensation of nonlinearity caused by the junction field effect in Si Hall microsensors isolated from the substrate by a heavily doped p^+-jacket (p^+–n junction) has been achieved by making use of the dependence of the magnetosensitivity of the voltage across the p^+–n junction (§4.2.4.2f) [67]. To this end, one should make the jacket voltage, V_j, dependent on V_H in such a way that the effective thickness of the plate could stay essentially constant. A simple circuit of feedback compensation that performs this operation is shown in Fig. 9.35. If $R_1 = R_2$ is chosen, $V_j = 1/2(V_{\text{Ref}} + V_H)$ (V_{Ref} being a negative reference voltage) and the mean plate thickness is maintained, to a first approximation, independently of V_H. Moreover, by making $V_j = V_{\text{Ref}} + AV_H$, where A denotes an appropriate amplification factor, one can, at least partially, compensate for some other nonlinearity effects.

In spite of the high efficiency of the circuit in Fig. 9.35, the best compensation of the different types of nonlinearity is achieved when the temperature is kept constant and the concentration of doping impurities in the active sensor region is $N_D \sim 7 \times 10^{14}$ cm^{-3} [67].

The problem of nonlinearity in BMTs can be most adequately resolved by optimizing the sensor mechanism and the device structure. A high degree of linearity is manifested by differential BMTs and split-drain versions whose operation is entirely based on the Lorentz deflection. Linearity of magnetodiode sensors is achieved by the differential structures in Fig. 5.1c and appropriate bridge circuits at low values of the magnetic induction B.

Digital signal processing employing a microprocessor is a sophisticated approach to the compensation of nonlinearity in solid-state magnetosensors (§9.4). At present this method is costly, but it outlines the future trend towards the overcoming of nonlinearity and other sensor defects.

9.6. Concluding remarks

The methods and circuits described in this chapter improve the main characteristics of solid-state magnetosensors and are of primary importance for their performance. Obviously, the satisfactory operation of the signal-processing circuits depends on the reproducibility and long-term stability of the device's parameters. They are directly connected with the maturity of chip-fabrication technology and the proper choice of the chip carriers. A review of modern magnetosensor technology is included in [35]. The on-chip integration of the sensor with the correction components considerably increases the latter's effectiveness. The shortcomings of integrated sensors, offset for instance, can be reduced if some critical processing steps are strictly controlled, e.g. mask alignment [68]. The on-chip integration of the transducing element with the appropriate circuitry for signal processing or performance correction may lead to some unexpected results. The circuitry itself may be adversely affected by the sensor. This is why in certain cases special measures should be taken to prevent the disturbance of the circuitry by the magnetotransducer. This may require some extravagant solutions concerning design of the layout [69].

Functional microtransducers offer a unique opportunity for a simultaneous generation of the useful and the correcting signals. This is a prerequisite for the creation of refined instruments of magnetometry.

The substitution of mechanical trimming by the process of downloading coefficients into the ROM of an intelligent magnetosensor may be considered a major advance, as it replaces one of the most expensive stages of sensor production. Attention is drawn here to the advantages of an optimization technique in sensor compensation, and the illustration of its application to frequency-response correction shows that it can be a powerful and versatile approach. The software element in the sensor subsystems is likely to become increasingly dominant as the smart concept penetrates into more areas of magnetosensor application.

References

[1] R.S. Popović, Sensors and Actuators, 5 (1984) 253–262.
[2] K. Maenaka and M. Maeda, Sensors and Materials, 5 (5) (1994) 265–284.
[3] Ch.S. Roumenin, Planar magneto-switch, Bulg. Patent 36 173 (May 3, 1983); Magneto-switch, Bulg. Patent 96536 (June 29, 1992).
[4] I.M. Vikulin, L.F. Vikulina and V.I. Stafeev, Galvanomagnitniye pribori, Radio i Sviaz, Moskwa, 1983 (in Russian).
[5] J.M. Huijsing, Sensors and Actuators, 10 (1986) 219–237; A, 30 (1992) 167–174.
[6] H.P. Baltes and R.S. Popović, Proc. IEEE, 74 (1986) 1107–1132.
[7] K. Matsui, S. Tanaka and T. Kobayashi, Proc. 1st Sensor Symp., Tokyo, Japan (1982) 37–40.
[8] A.G. Andreou, The Hall effect and related phenomena in microelectronic devices, Ph. D. dissertation, Johns Hopkins University, Baltimore, Maryland, USA, 1986; A.G. Andreou and C.R. Westgate, Proc. IEEE Lett., 73(3) (1985) 498–490.

[9] V. Zieren and B.P.M. Duyndam, IEEE Trans. Electron Devices, ED-29 (1982) 83–90.
[10] V. Zieren, Integrated silicon multicollector magnetotransistors, Ph. D. dissertation, Delft Univ. Techn., The Netherlands, 1983.
[11] A.W. Vinal, IBM J. Res. Develop., 25(3) (1981) 196–201.
[12] K. Maenaka, H. Okada and T. Nakamura, Sensors and Atuators, A, 21–23 (1990) 807–811.
[13] E.A. Vittoz, IEEE J. Solid-state Circuits, SC-18 (1983) 273–279.
[14] S. Takamiya and K. Fujikawa, IEEE Trans. Electron Devices, ED-19 (1972) 1085–1090.
[15] R.S. Popović and H.P. Baltes, IEEE J. Solid-state Circuits, SC-18 (1983) 426–428.
[16] R.S. Popović, Hall Effect Devices, Adam Hilger, Bristol, 1991.
[17] T. Nakamura and K. Maenaka, Sensors and Actuators, A, 21–23 (1990) 762–769.
[18] R. Popović, Electrical circuit which is linearly responsive to changes in magnetic field intensity, US Patent 4683 429 (1987).
[19] J. Burghartz and W. Von Münch, Sensors and Actuators, 11 (1987) 91–98.
[20] T.R. Lepkowski, G. Shade, S.P. Kwok, M. Feng, L.E. Dickens, D.L. Laude and B. Schoendube, IEEE Electron Device Lett., EDL-7 (1986) 222–224.
[21] A.R. Cooper and J.E. Brignell, J. Phys., E, Sci. Instrum., 17 (1984) 627–628.
[22] A. Nathan, I.A. McKay, I.M. Filanovsky and H.P. Baltes, IEEE J. Solid-state Circuits, SC-22 (1987) 230–232.
[23] K. Holzlein and J. Larik, Sensors and Actuators, A, 25–27 (1991) 349–355.
[24] S.V. Gumenjuk, B.I. Podlepetsky, V.V. Kremliov and N.A. Polyanskykh, Sensors and Actuators, A, 28 (1991) 231–234.
[25] Ch.S. Roumenin, S. Nihtianov and T. Minkova, Magnetosensitive device, Bulg. Patent 88 252 (April 26, 1989).
[26] S. Nihtianov and T. Minkova, Sensors and Actuators, A, 30 (1992) 101–104.
[27] S. Middelhoek and W.J. Lian, Flip-flop sensor, Dutch Patent 84 02 314.
[28] W. Lian and S. Middelhoek, 3rd Int. Conf. on Solid-state Sensors and Actuators, Philadelphia, USA (1985), 46–48.
[29] W.J. Lian and S. Middelhoek, IEEE Electron Device Lett., EDL-7 (1986) 238–240.
[30] W.J. Lian and S. Middelhoek, Sensors and Actuators, 9 (1986) 259–268.
[31] S. Middelhoek, P.J. French, J.H. Huijsing and W.J. Lian, Sensors and Actuators, 15 (1988) 119–133.
[32] W. Lian, Integrated silicon flip-flop sensors, Ph.D. dissertation, Delft Univ. Techn., The Netherlands, 1989.
[33] D. Seitzer, G. Pretzl and N.A. Hamdy, Electronic Analogue-to-Digital Converters, Wiley, Chichester, 1983.
[34] J.E. Brignell, J.Phys., E, Sci. Instrum., 18 (1985) 559–565.
[35] S. Middelhoek and S.A. Audet, Silicon Sensors, Academic Press, London, 1989, Chapter 8.
[36] A.R. Cooper and J.E. Brignell, Sensors and Actuators, 7 (1985) 189–198.
[37] J.E. Brignell, in: Sensors, Vol. 1, T. Granke and W. Ko (eds), VCH, Weinheim, 1989, 331–353.
[38] J. Brignell, Sensors and Actuators, A, 25–27 (1991) 29–35.
[39] W.H. Ko and C.D. Fung, Sensors and Actuators, 2 (1982) 239–250.
[40] Yi-Zi Xing, S. Kordić and S. Middelhoek, J. Phys., E, Sci. Instrum., 17 (1984) 657–663.
[41] S.Kordić, V. Zieren and S. Middelhoek, Sensors and Actuators, 4 (1983) 55–61.
[42] S. Kordić, Offset reduction and three dimensional field sensing with magnetotransistors, Ph.D. dissertation, Delft Univ. Techn., The Netherlands, 1987.
[43] R.J. Braun, IBM J. Res. Develop., 19(4) (1975) 344–352.
[44] A. Kobus and J. Tuszynski, Hallotrony i Gaussotrony, Wydawnictwa naukowo-techniczne, Warszawa, 1966 (in Polish).
[45] H. Weiss, Structure and Applications of Galvanomagnetic Devices, Pergamon Press, Oxford, 1969.
[46] No author given, Electron. Weekly, April 29 (1985) 59–61.
[47] P. Daniil and E. Cohen, J. Appl. Phys., 53 (1982) 8257–8259.
[48] P.J.A. Munter, Sensors and Actuators, A, 21–23 (1990) 743–746; A, 31 (1992) 206–209, with D.C. van Duyn.
[49] Y. Kanda, M. Migitaka, H. Yamamoto, H. Morozumi et al., IEEE Trans. Electron Devices, ED-29 (1982) 151–154.

[50] S.M. Sze, Physics of Semiconductor Devices, John Wiley and Sons, New York, 1981.
[51] Lj. Ristić, T. Smy and H.P. Baltes, Sensors and Materials, 2 (1988) 83–92.
[52] S.V. Jaskolski, H.P. Schutten, G.B. Spellman, J.K. Sedivy and M.W. Jensen, Columnated and trimmed magnetically sensitive semiconductor, US Patent 4 516 144 (May 7, 1985).
[53] Ch.S. Roumenin, Sensors and Actuators, A, 24 (1990) 83–105.
[54] S. Kordić and P.C. M van der Jagt, Sensors and Actuators, 8 (1985) 197–217.
[55] Ch.S. Roumenin, C.R. Acad. Bulg. Sci., 40 (5) (1987) 57–60.
[56] U. Dibbern, Sensors and Actuators, 10 (1986) 127–140.
[57] E.F. Vozenilek, Anal. Chem., 48 (1976) 1654–1656.
[58] D. Wedlake, Electron. Ind., 4(12) (1978) 33.
[59] B. Wilson and B.E. Jones, J. Phys., E, Sci. Instrum., 15 (1982) 364–366.
[60] B.S. Grainger and K.A. Rubinson, J. Phys., E, Sci. Instrum., 10 (1977) 773–775.
[61] J. Bretschi, Feinwerktechn. und Messtechn., 87 (1976) 335–338.
[62] E.S. Levshina and P.V. Novitsky, Izmeritelniye preobrazovately, Energoatomizdat, St. Petersburg, 1983 (in Russian).
[63] A.I. Krivonosov, Thermodiody i thermotriody, Energia, Moskwa, 1970 (in Russian).
[64] Sensoren-Magnetfeldhalbleiter, Teil 1, Datenbuch, Siemens, 1982/83.
[65] Ch.S. Roumenin and P.T. Kostov, Sensors and Actuators, 8 (1985) 307–318.
[66] S. Kirby, Sensors and Actuators, 3 (1982/83) 373–384; 4 (1983) 25–32.
[67] R.S. Popović and B. Halg, Solid-state Electron., 31 (1988) 681–688.
[68] N. Georgiev and P. Stamenova, C.R. Acad. Bulg. Sci., 47(1) (1994).
[69] N. Georgiev, C.R. Acad. Bulg. Sci., 46(9) (1993).

Chapter 10

Applications of solid-state magnetosensors

By virtue of their widely varied characteristics, solid-state magnetic sensors possess the remarkable property of being universal instruments for the measuring of almost all physical quantities. Suffice it to list the properties of the output signal, determined by five basic galvanomagnetic effects:
– The Hall effect is linear within the range of magnetic field $10^{-8} \leq B \leq 1$ T.
– Magnetoresistance is a square function of B in the range $10^{-8} \leq B \leq 0.3$ T, and linear or square in the interval $0.3 \leq B \leq 10$ T.
– The magnetoconcentration and magnetodiode effects are strongly nonlinear. The output signal of the respective device in the form of a positive-slope section of the characteristic can be moved along the axis of the magnetic field by deliberately changing the surface recombination rate.
– The output of the Schottky magnetodiode is nonlinear with a well-pronounced slope around point $B \approx 0$.

Two of these effects show monotonous variation of their electrical parameters with the field B (the Hall effect and Schottky magnetodiodes). The rest are symmetrical with respect to the field $B \approx 0$ (magnetoresistance), or with respect to a given fixed value B_0 (magnetoconcentration and MD effect). On the other hand, SQUIDs exhibit oscillatory behavior of the output.

The hierarchy of the possible applications of magnetotransducers is well known. The first level is the direct registration of the magnetic field itself (§1.4a–c). The second level includes the indirect methods (§1.4d, e, f, ...). Each one of these applications can be used as a component in a system or an instrument, such as an integrated silicon compass. Each, in its turn, may be a part of a yet larger system, such as a process controller in a submarine or satellite navigation system, etc. Given the continuous sophistication of information technologies, it becomes more and more difficult to predict the possible ultimate applications of magnetic sensors. It is for this reason that in this chapter the main attention is devoted to these two levels of use of solid-state magnetotransducers, which, in the author's view, hold especially high promise.

10.1. Measurement of the magnetic field

Hall devices, magnetoresistors, MDs, BMTs, CDMs and SQUIDs can be used as instruments for the accurate measurement of d.c. and a.c. magnetic fields. The selection of the particular sensor, however, depends on many factors and it is not possible to give a straightforward recommendation as to what type of transducer should be used for a specific purpose [1]. The most fundamental points to be considered are the range of the magnitudes of the field and the range of operating frequencies. The values of the magnetic induction, important for the magnetometry vary from about 1 fT (in biomagnetism and space research) to 100 T (in particle accelerators); this is a difference of 17 orders of magnitude. The range of frequencies for practical applications is also very wide: e.g. in geophysical and space research the lowest frequencies are about 10^{-8} Hz, whereas in domains like a.c. measurements they are several GHz. With the exception of SQUIDs, which are irreplaceable in low-field magnetometry at this stage, the boundaries of the operational regions for the rest of the solid-state sensors cannot be strictly defined, in view of the fact that in them the signal-to-noise ratio and the minimum detectable magnetic field are close in value. Various modifications of magnetometers have been successfully developed and are, in the majority, commercially available. They are based on Hall devices, magnetoresistors, differential magnetotransistors, magnetodiodes, etc.

The sensitivity of these magnetotransducers can be substantially enhanced by the use of flux concentrators which increase the magnetic field in the active sensor region [2–6]. The corresponding magnetic gain, A_M, is a simple straight concentrator in a field given by $A_M = B_M/B_0 = K(L_i/L_a)$, where K is an empirical constant $= 0.2$ for a concentrator with permeability $\gg 1$, L_i is the length of the concentrator and L_a the length of the air gap. Kovar (15%Co : 31%Ni : 54%Fe), already used in integrated-circuit packages as it has a coefficient of thermal expansion compatible with that of glass, thus allowing an effective seal, is a suitable magnetic material for flux concentration and makes it possible to incorporate flux concentrators in integrated-circuit packages without great difficulty. These are particularly useful when enhanced sensitivity is required and the distortion of the field is not important [5]. For instance, used with a Hall sensor, two flux-concentrators enhance its transducer efficiency by about 400 times [3]. In addition, the signal-to-noise ratio is improved by a factor of 100. Reference [7] presents an original flux-concentrator with a "butterfly" shape, based on $Co_{66}Mo_2Si_{16}B_{12}Fe_4$, attached to an orthogonal GaAs Hall device for the measurement of the Earth's field. In a field $B = 0.5G$, the output signal without a concentrator is 15 μV, and with a concentrator it reaches 1.5 mV. Whenever necessary, positive feedback can be introduced by the placing of suitable coils on the two parts of the flux concentrator, biased by the sensor's output signal itself (Fig. 10.1) [3,4,6]. Such an arrangement further enhances the sensitivity of the system. Besides Hall elements, this approach can successfully be applied to differential magnetotransistors. Another method for

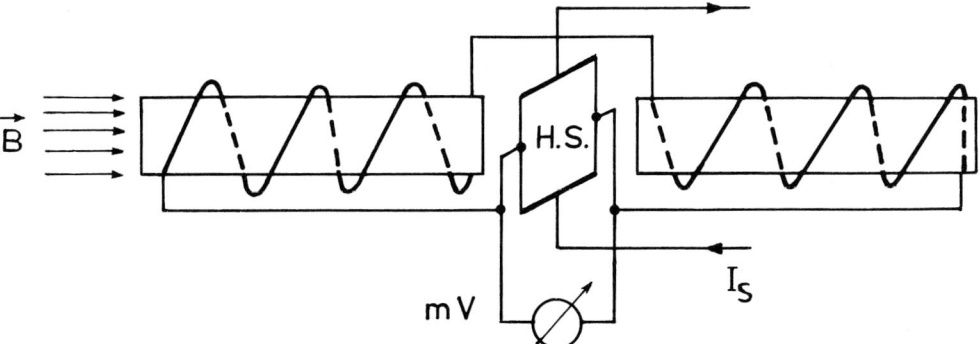

Fig. 10.1. Increasing the sensitivity of a Hall device by means of a fluxconcentrator and positive feedback.

enhancing the magnetosensitivity is the cooling of the magnetotransducer, by liquid nitrogen, for instance, ($T = 77$ K).

There are teslameters, varying in complexity and in calibration procedures, with analog and digital outputs, applicable to scientific and industrial use. Obviously, a detailed consideration of their performance is not possible here. A few typical examples are given, for instance, in [2,3,6,8,9]. Figure 10.2 shows the basic design of

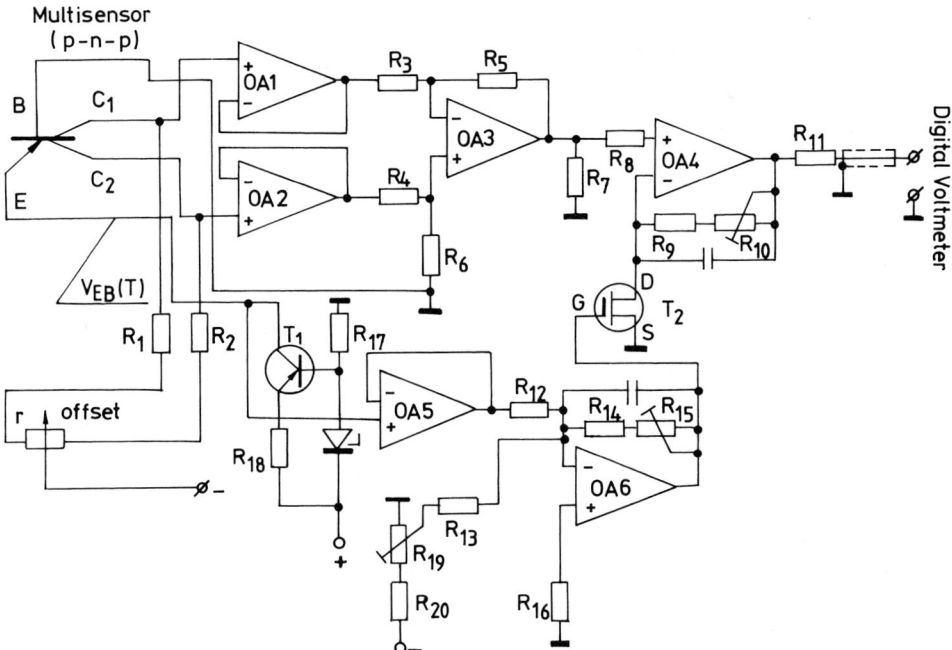

Fig. 10.2. Teslameter using a multisensor for magnetic field and temperature.

a teslameter, using a multisensor for magnetic field and temperature. It measures d.c. magnetic fields within the interval $0 \leq B \leq 1$ T, and a.c. fields in the frequency range $0 \leq f \leq 10$ kHz, at temperatures in the interval $0 \leq T \leq 70°$.

With the help of a transistor T_1, a voltage–current convertor is formed to bias the emitter–base circuit by a constant current. The output differential signal of the multisensor is amplified by op-amps OA_1, OA_2 and OA_3 and is calibrated by the op-amp OA_4. After suitable amplification by op-amps OA_5 and OA_6 and transistor T_2, the linear thermometric voltage $V_{EB}(T)$ is used to control the gain coefficient of op-amp OA_4. In fact, the op-amps OA_1 to OA_6 form a differential amplifier with a controllable gain coefficient. The output signal of the teslameter is registered by a conventional digital voltmeter. The basic error in the measurement of the d.c. and a.c. magnetic fields is less than 1% at $B \leq 0.6$ T. A calibrating solenoid has been incorporated in this instrument, generating a magnetic field of the order of 0.1 T, known with an accuracy of $\leq 1\%$ and highly homogeneous over a volume of several mm^3.

A modern trend in the application of integrated MDs whose sensitivity is in the 10^{-12} T or 10^{-3} γ range is the registration of very small anomalies in the geomagnetic field. They help in surveying for minerals and/or metals of low magnetic permeability, and this is a very powerful method in geology. Such magnetometers can be successfully used for the prediction of volcanic eruptions and earthquakes by detecting the variations of the geomagnetic field. Apart from the forecasting of such phenomena, these sensors are convenient tools of modern archaeology. SQUIDs may also be used for these applications.

Vector magnetometry has become a promising trend, especially after the invention of integrated vector magnetosensors for the simultaneous and/or successive determination of the components B_x, B_y and B_z, on the basis of the Hall effect and BMTs (§8.3). In addition to analyzing the topology of the magnetic field, these devices are also used in the design of electronic compasses. An example of this modern application is the monolithic silicon compass (Fig. 10.3), using two crossed

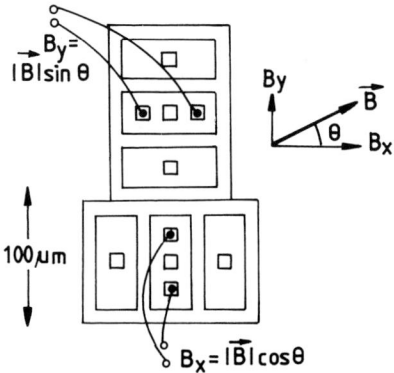

Fig.10.3. Two-dimensional magnetic detector using two parallel-field Hall cells from Fig. 4.26g [10].

parallel-field Si Hall microsensors (Fig. 4.26g) [10]. This integrated device detects the in-plane magnetic vectors, B_x and B_y, and contains also a signal-conversion circuit which calculates the direction $\Theta = \tan^{-1}(B_y/B_x)$, using translinear circuit technology. This magnetometer has not only the direction output but also the output for the intensity (the absolute value) of the magnetic field. The test device was fabricated using standard bipolar analog IC technology. As a result, the direction of the applied magnetic field was obtained with a maximum error of $\pm 2\%$ of full-scale deflection [10].

An implemented arrangement has been described [11] for measuring the absolute value of a magnetic field B with the help of three magnetoresistors, connected in series and placed at right angles. An advance in determining the gradient of the magnetic field has been made by the use of SQUIDs (§7.7), Hall devices (§8.2.1), and differential BMTs (§8.2.2).

Apart from the well-known thin-film magnetoresistors, the fundamental considerations of the sensor characteristics of solid-state magnetodevices like size, resolution, sensitivity, signal-to-noise ratio and frequency response have shown that such devices can successfully be used in the new read/write-head techniques for vertically stored magnetic data with a density of up to 10^7 bits per cm^2 [12]. One of the possible designs of such a read head, including a chip with a solid-state magnetosensor, is shown in Fig. 10.4. Passing the magnetic tape, containing the data, by a gap modifies the magnetic flux density, that generates a signal at the output of a sensor (a Hall device or a BMT) that does not depend on the speed of the tape. Ultramicro Hall sensors, made of a single crystal of n-type InSb with a sensitive area of 5×5 μm^2 and a high sensitivity of 140 μV mT^{-1} at a supply current of 0.1 mA are also suitable for this purpose. This transducing efficiency is sufficient for the detection of magnetic bubble domains [13]. Such novel read-head techniques can be applied in magnetic read memories of audio, video and data

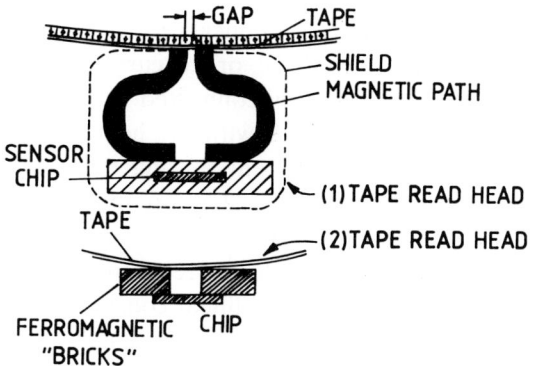

Fig. 10.4. Solid-state head, principle of operation: (*1*) adaptation of a conventional head, and (*2*) miniaturized chip form.

storage devices like magnetic cards, Winchester disks, coin validation systems, etc.

Aspects of the internal design of solid-state magnetosensors, the shape and location of contacts in particular, as well as lead paths are often matched to their use in magnetometry. For example, axial probes are constructed to detect axial fields in radial holes; tangential and extremely thin probes are used for measuring in very small air gaps, the small active area giving almost point registration of the field B, etc.

Integrated Si MD sensor arrays have recently found application in the noninvasive measurements necessary for the evaluation of the quality of soft and hard magnetic materials. The operation of these sensors is based on the change in the concentration of carriers in the base regions of the separate MDs due to the influence of the nonhomogeneous magnetic field $B(x, y)$, generated by the material being investigated. The electric voltage thus generated is transformed, by means of addressing electronic circuitry, into a video signal on the screen of a monitor, corresponding to the magnetic *map* of the ferromagnetic sample and characterizing its structure. A higher amplitude of the output signal of this sensor array can be obtained by the use of magnetotransistors.

The precise measurement of the parameters of the magnetic field can yield information about the properties of permanent magnets, the dynamic characteristics of electrical machines, the commutation processes and a great number of other important data concerning their performance, which are of interest to science and industry [2,3,6].

The registration of the Meissner effect in superconductive materials is a new application of Hall devices and BMTs. The multisensors are especially suitable for this purpose as they measure simultaneously the critical magnetic field B_c and the temperature of the phase-transition T_C.

10.2. Measurement of the current, power and related electrical quantities

One of the most promising indirect applications of magnetosensors is in the measuring of electrical quantities. Each d.c. or a.c. current I generates a respective d.c. or a.c. magnetic field, unambiguously connected with the current I, $(I \rightarrow B)$. This is why the measuring of this field B with the help of a solid-state magnetic sensor yields information about the current flowing through a given conductor. The main advantage of this approach is that no interruption of the current-carrier circuit is necessary and the measurement is noninvasive [1–3,6]. A simple tangential field measurement produces a linear relationship between the current I and the induction B, when there are no nonlinear constituents like iron cores. Hall sensors and magnetotransistors having linear output signals are usually used for this purpose. Figure 10.5 illustrates the principle of tangential measurement of the field around a conductor of circular cross-section. In the case of a loop enclosing the

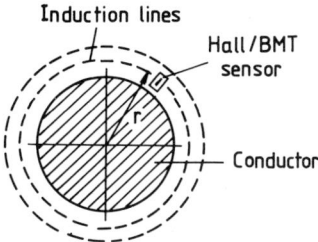

Fig. 10.5. Principle of contactless current measurement in a circular conductor by measuring the tangential field at a distance r.

conductor, what is obtained in accordance with Ampere's law is the following:

$$I = \oint H dl \; ; \quad I = 2H\pi r \; ; \quad H = \frac{I}{2\pi r} \quad \text{and} \quad B = \frac{\mu_0 I}{2\pi r}, \tag{10.1}$$

where μ_0 is the permeability of vacuum.

According to (10.1), the relationship between the current I and the induction B is linear, and in addition the arrangement is simple (Fig. 10.5). The method of noninvasive measuring of the current described above has certain limitations. Owing to the absence of a suitable flux concentrator, the measured magnetic induction B, and also the output signal have small magnitudes. That is why external parasitic fields and ferromagnetic masses exert an adverse influence. Besides, according to (10.1), the parameter H (or B) is inversely proportional to the distance r. Therefore, the transducer has to be placed as close to the surface of the electrical conductor as possible. As a result, the magnetosensitive device is in a field, B, with a strong gradient and the precision of the measured current I is a function of the location of the sensor. These drawbacks can be overcome by using a magnetic conductive core around the conductor, the transducer being located in an air gap in the core [14]. This approach can also provide for high sensitivity of the measuring system. Figure 10.6 shows an arrangement of a contactless current probe using an iron core [15]. When there is multiple winding, according to Fig. 10.6 the following relationships are valid:

$$In = \oint H dl \; ; \quad In = H_L \delta + H_{\text{Fe}} l_{\text{Fe}} \; ; \tag{10.2}$$

where n is the number of windings, l_{Fe} is the mean path length of the iron core, δ is the air gap in which a Hall or BMT sensor has been placed and H_L is the magnetic strength of the air gap.

If the negative influence of high magnetic fields is disregarded, then $B_{\text{Fe}} = B_L$ and the following relationships are obtained:

$$B_{\text{Fe}} = \mu_r \mu_0 H \; ; \quad In = \frac{B_L \delta}{\mu_0} + \frac{B_L l_{\text{Fe}}}{\mu_r \mu_0}. \tag{10.3}$$

Then the following expression is valid for the induction in the air gap:

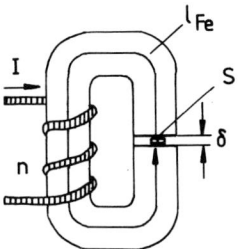

Fig.10.6. Arrangement for contactless current measurement using an iron core and a Hall or BMT sensor (S).

$$B_L = \frac{\mu_0 I n}{\delta + l_{Fe}/\mu_r}; \quad B_L = \frac{\mu_0 I n}{\delta}; \tag{10.4}$$

where the permeability μ_r of the core material has been assumed to be very large, i.e. $\mu_r > 10^3$.

The linear dependence from (10.4) is preserved at small currents I. At high values of the product In, nonlinearity occurs in the characteristics of the iron core along with remanence effects [1]. The typical values of the current which can be measured by the arrangement in Fig. 10.6 are from 10 A to several kA. Reference [16] presents a description of a contactless current probe using the approach from Fig. 10.6, in which for the first time a silicon functional multitransistor for magnetic field and temperature is used as a transducer (§8.1.1). This instrument can measure d.c. and a.c. currents of up to 300 A in a temperature interval $0 \leq T \leq 70°C$ with a sensitivity of 10 mV A^{-1} and error less than 0.1%.

The compensation method illustrated in Fig. 10.7 [1,15] can be used for extremely precise measurement of small currents without interrupting the current circuit. The iron core is kept free from induction B by an opposite magnetic field generated by the windings n_2 of the compensation coil, i.e. $I_1 n_1 + I_2 n_2 = 0$. The magnetosensor (as GaAs, InSb or Si Hall device, or a Si differential BMT) acts to indicate the null point $B = 0$. The output voltage is given by $V_{out} = -RI_1 n_1/n_2$, whereby the dependence between the current I_1 and V_{out} is highly linear.

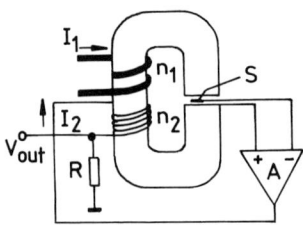

Fig. 10.7. Arrangement for indirect contactless current measurement using the compensation principle. The magnetosensor S in the air gap serves as a null indicator.

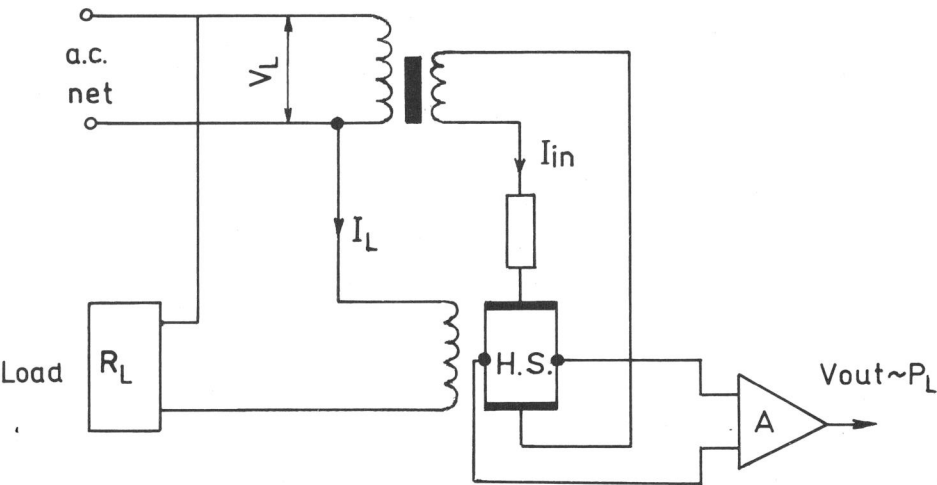

Fig.10.8. The basic arrangement for a.c.-power measurement using a Hall device.

The recent discovery of high-T_C superconductivity has brought forth the problem of contactless measuring of the current in superconducting cables, located in a cryogenic medium, e.g. at $T = 77$ K. An suitable solution in this case is the use of a multisensor for magnetic field and temperature, subjected directly to a temperature $T = 77$ K [17,18]. In the case of a failure that leads to the abrupt and uncontrollable rise of the temperature T, the information provided by the device about T leads to a command to turn off of the current in the main line, thereby preventing damage to expensive equipment.

The contactless measuring of the electrical power is based on the multiplicative nature of Hall sensors ($V_H \sim IB$) and of linear differential BMTs ($\Delta I_{C1,2} \sim I_E B$). Unlike all other electronic multipliers, which utilize the nonlinear transfer characteristics, the Hall-effect multiplier is the only true or ideal product-forming four-quadrant instrument, $\pm V_H = (\pm I)(\pm B)$. The basic circuit for a.c.-power measurement is shown in Fig. 10.8 [1–3,6]. By means of a transformer the load voltage is converted into a proportional control current from the a.c. line supply: $I_L \sim B$, and $V_L \sim I_{\text{in}}$. When the formula for the Hall voltage (4.29) is applied, the result is the following expression:

$$V_H = \left(\frac{R_H}{t}\right) I_{\text{in}} B = M V_L I_L = M P_L, \tag{10.5}$$

where M is the transduction constant.

The arithmetic mean value is derived in the signal-conditioning circuit, i.e. in a low-pass filter, so that the offset, V_{off}, of the Hall device is eliminated [1]. For example, the following relationships are valid for a resistive load and an a.c. line:

$$V_L(t) = \overline{V}_L \cos wt, \tag{10.6}$$

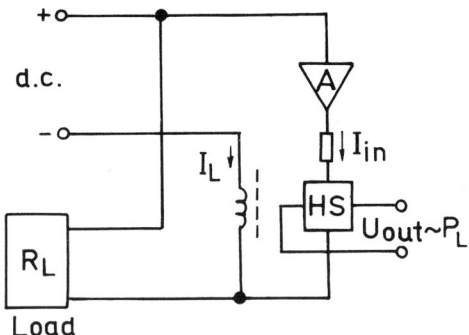

Fig. 10.9. Circuit diagram for d.c. power measurement using a Hall sensor.

$$I_L(t) = \overline{I}_L \cos wt, \tag{10.7}$$

$$V_{\text{off}}(t) = \overline{V}_{\text{off}} \cos wt. \tag{10.8}$$

Then the Hall voltage can be presented as:

$$V_H = M V_L(t) I_L(t) + V_{\text{off}}(t). \tag{10.9}$$

Taking into account (10.6) and (10.8), the voltage V_H is given by

$$V_H = 0.5 M \overline{V}_L \overline{I}_L (1 + \cos 2wt) + V_{\text{off}} \cos wt. \tag{10.10}$$

The term $V_{\text{off}} \cos wt$ does not contribute to the arithmetic mean, and finally the following relationship between V_H and the resultant active power P is obtained:

$$V_H = 0.5 M \overline{P} (1 + \cos 2wt). \tag{10.11}$$

The principle of determining the d.c. power using a Hall sensor is analogous to the a.c. measurement presented in Fig. 10.9 [1–3,6]. A linear differential BMTs, including multisensors, can also be successfully applied in a.c. and d.c. power registration.

However, the practical Hall multipliers suffer from a number of limitations such as: a relatively small output signal, which requires amplification; offset and faults of the conventional Hall-voltage amplification methods; and fluctuation in the offset and reduced dynamic range for the amplifier caused by a common-mode voltage [19]. The current and watt meter presented in ref. [19] uses a common-mode voltage rejection circuit (Fig. 9.2), and GaAs Hall sensor. This solution overcomes some of the limitations listed above: a calibration error smaller than ±0.2% has been achieved at a current ranging from 1.0 to 50A. The arrangement is like that in Fig. 10.9.

A magnetoresistive element (MR) can also perform a multiplication of the current and the magnetic flux density [11,20] (Fig. 10.10). In a relatively high magnetic field, the increase in magnetoresistance in a sample with $l = w$ is linear with respect to B, and the sensitivity is just equal to that of a Hall element of

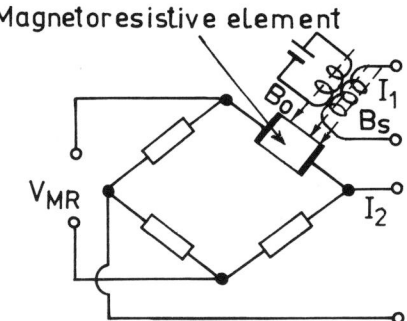

Fig.10.10 General principle of MR multiplier; $V_{MR} = SB_s I_2 \sim I_1 I_2$ [20].

the same material and with the same thickness t (§4.4.2). In order to make use of the linear characteristic, a bias magnetic flux density B_0 together with a signal from B_s should be applied to the MR device. The output voltage is then given by $V_{MR} = SB_s I_2 \sim I_1 I_2$.

Hall sensors and MR devices can be successfully used as microwave wattmeters [2,3,11,20]. In this case the Hall instruments do not need an external supply power.

Amidst the huge variety of applications of solid-state magnetosensors for measuring electrical quantities, the most important cases should be mentioned: linear and square detectors, gyrators, isolators, circulators, modulators, demodulators, amplifiers, frequency-spectrum analyzers, magnetic switches, digital-to-analog converters, and multipliers of vector quantities. For detailed information on these and other instruments cf., for instance, [2,3,6,11,21], as well as the references listed therein.

10.3. Measurement of nonelectrical and nonmagnetic quantities

10.3.1. Principles of operation

Another promising field of the indirect application of solid-state magnetotransducers is the registering of nonelectrical and nonmagnetic quantities, whereby the information is magnetically encoded and the output sensor signal becomes the carrier of the original quantities, such as mechanical displacements, accelerations and vibrations, linear and angular positions, stress, force, weight, pressure, etc. The techniques used for the contactless measuring of mechanical quantities are based on moving the galvanomagnetic device (Hall element, BMT, magnetoresistor, MD, etc.) along a straight line in a nonuniform magnetic field, turning the device around in a uniform field, or, in general, moving it along a given trajectory in space where the distribution of the field $\boldsymbol{B}$ has been properly designed, $B = f(x)$. The sensitivity K of such an arrangement, e.g. measured in volts per mm of displacement, is defined by the expression $K = S_V(I_s) \Delta B / \Delta x$, where S_V is the voltage magnetosensitivity of the transducer, $\Delta B/\Delta x$ is the gradient of the field B along the x-axis, and I_s is the supply current. The theoretical estimation of the value of the detectable displacement by the use of an InAsP Hall sensor, for instance, with $S_V = 10^{-4}$ V G^{-1} and a field gradient $\Delta B/\Delta x = 10^4$ G mm^{-1} has shown that the sensitivity K has an unusually high value of 1 V mm^{-1} = 1 mV μm^{-1} or 0.1 μV Å^{-1} [2]. When the Hall device is replaced by a BMT, whose sensitivity S may be 10 times as high, the value 1 μV Å^{-1} is obtained for the parameter K. The advantages of the contactless measuring of mechanical quantities with the help of galvanomagnetic transducers are the following: the sensitivity K can be easily changed, by varying the current I_s and the gradient of the field B, to adequately tune the device for the specific application; an output signal of any shape can be easily produced (e.g. sine wave, d.c. or pulse voltage); a harmonic analysis of variations can be performed and several components of a motion can be measured simultaneously; the force required to move the sensor is small; there is practically no reaction force, such as attraction of the core in inductive transducers; the magnitude of the output signal does not depend on the speed of the motion, and extremely low values of the operating frequency, including $f = 0$, are possible; the moving system is much simpler than in the case of the inductive or capacitive transducers; the system has low power consumption; the working temperature may be raised up to 200°C depending on the packaging and the semiconductor material used (GaAs); there is stability with regard to external impacts like humidity, radiation, chemical corrosion, etc.

10.3.2. Magnetic sensor systems

The task of designing contactless galvanomagnetic instruments for mechanical quantities implies the development of magnetic circuits with a gradient $\Delta B/\Delta x$ suitable for the particular applications. Various configurations including a galvanomagnetic transducer and magnetic systems are classified in Fig. 10.11 [1–3,6,22–25]. The arrangements in Fig. 10.11a1, b1, e1, e2, f1, g1, g2, g5–g9, and h3 are actuated by a permanent magnet, while those from Fig. 10.11a2, b2, c1–c3, d1, g3, g4, h1, and h2 are magnetically biased and actuated by magnetically conductive cores. For instance, the configurations a1, b1, b2, c1–c3, e1, and e2 are suitable for use in conjunction with electronic switching circuits to produce digital position sensors, limit and proximity switches, keyboard switches, etc. The arrangements in Fig. 10.11d1, f1, h1, and h2 are the foundation of brushless DC motors, angle detectors and decoders, synchroresolvers, tachometers, crankshaft position sensors in car ignition control, etc. The linear gradient systems in Fig. 10.11g1–g9 are of particular interest as they are suitable for the designing of linear instruments for displacement registration within the 2- to 3-mm range. The magnetic circuits shown in Fig. 10.11g6–g9 [22,24,25] exhibit the highest transducing efficiency, i.e. the highest linear gradient. The output signal is similar to a sinusoidal wave (d1, h1–h3) digitized as a square wave by a comparator [1].

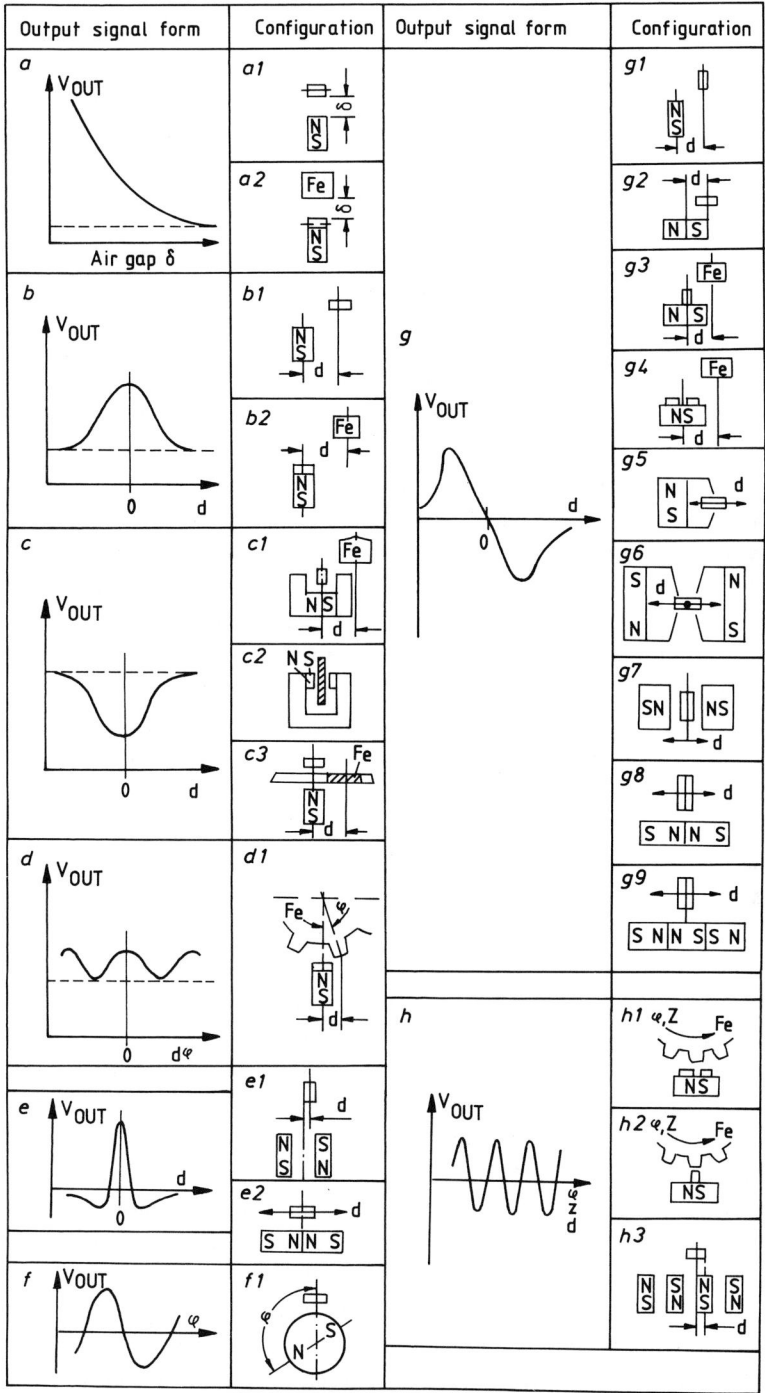

Fig. 10.11. Magnetic circuits with galvanomagnetic sensors: configurations and forms of output signals.

In the arrangements shown in Fig. 10.11, the excitation permanent magnets are made of AlNiCo-type alloys, platinum–cobalt alloys, iron–cobalt–vanadium (chromium) alloys, chromium–iron–cobalt alloys, rare-earth–cobalt alloys, rare-earth–iron alloys, hard ferrites like Ba- and Sr-ferrites, etc. When the permanent magnet is properly stabilized with respect to temperature and demagnetizing forces, the only magnet error of significance results from reversible temperature changes. Temperature cycling of the magnet to 200°C provides the necessary stabilization against irreversible temperature effects with a permanent loss in the flux of less than 2%. The reversible temperature changes follow a nearly straight line of $\sim -0.008\%$ °C^{-1} [2]. This shift is compensated by the same network that corrects the temperature coefficient of the magnetosensor.

The magnets usually have a cubic or cylindrical shape and the magnetic energy density depends on the volume of the material. The value of the induction measured directly at the poles is 0.1 to 1.2 T. The distance d between the sensor and the magnet is of considerable importance for the performance of this class of instruments, because the field intensity drops very rapidly with increasing distance from the magnetic circuit, the dependence being approximately $B \sim 1/d^2$. Therefore, in order to generate a maximum output signal the sensor and the magnet must be located as close to each other as possible. In linear displacement transducers, the precision of the measurements depends not only on S, but on the constancy of the air gap as well, i.e. the relative movement should be strictly parallel to the magnetic system. This condition sets the demand for stability of the moving part of the instrument. With modern electronics and with the availability of microprocessors for signal conditioning, nonlinearity characteristics do not limit analog signals, since they can be made linear. Moreover, the usable range of displacements can be significantly increased. Stability and reproducibility of the characteristics of the system are the only conditions in this case.

Depending on their application, the magnetic systems use galvanomagnetic sensors with differing parameters. Hall devices and differential BMTs are best suited to linear position transducers, as their output signals are linear. Differentially connected magnetoresistors with a built-in permanent magnet to provide for a linear output produce a large output signal. The arrangement of differentially coupled MDs is analogous. In digital position detection, the linearity of the sensor output is not necessarily required and MDs, some types of the single-collector BMTs and other types of transducers offer an adequate choice.

If the measuring system operates at high temperatures, GaAs sensors are indispensable. Under these adverse conditions, the reduced magnetosensitivity can be enhanced by the building of a magnetic flux-concentrator into the package [26]. The influence of the temperature can best be minimized by the functional multisensors for magnetic field and temperature and the scheme from Fig. 10.2. The combined influence of the temperature in the linear arrangements using Hall devices can be eliminated by means of two identical and stationary fixed sensors displaced at some

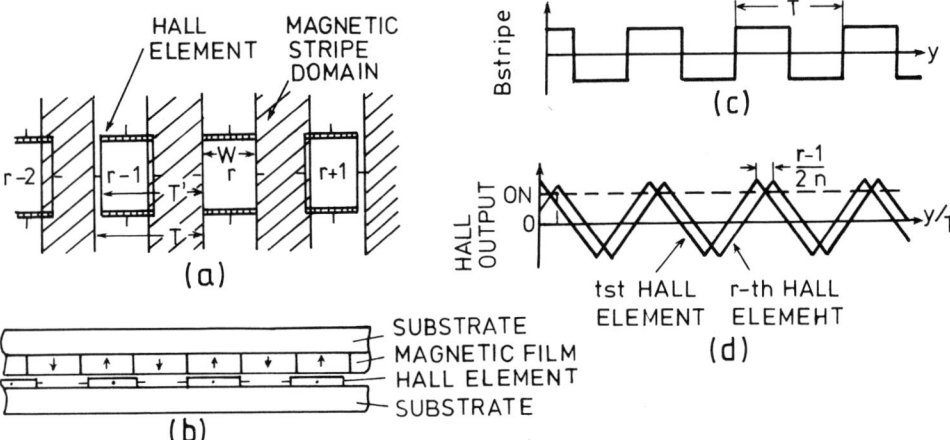

Fig.10.12. Principle of the digital displacement transducer: (a) plan view; (b) cross-sectional view; (c) stray magnetic field distribution generated by domains; and (d) signals given by the first and the rth Hall sensor; dashed line shows threshold level of the comparator.

distance along the direction of travel [23]. For this purpose, the sum of the two outputs, having been sensitivity matched at one temperature, will have to be divided by their difference, which would be only temperature dependent in the linear range, to yield a stable output.

The well-developed IC technology provides the perfect opportunity to incorporate a magnetosensor along with analog or digital signal electronics onto one chip. Practically, there are promising trends in the development of instruments making use of periodically arrayed parallel stripe domains in a bubble film, generating stray magnetic fields with a regularly periodic distribution of period T. Figure 10.12 illustrates the principle underlying such a digital displacement instrument (cf. Fig. 4.21) [13]. The $(2n + 1)$ Hall elements are integrated onto the substrate with a periodic separation of T'. The relation between the arrangement of the Hall devices and the stripe domains is that $(2n - 1)T = 2nT'$. If the Hall output of the rth Hall sensor is the maximum, the distance between the first stripe domain and the first Hall element is $(r - 1)T/2n$. The resolution of the displacement is given by $T/2n$. Since it is possible to realize magnetic domains and Hall sensors with dimensions of less than a micron, a fine digital displacement transducer with a submicron resolution is feasible [13].

10.3.3. Selected contactless mechanical sensing devices

This paragraph deals with what, in the author's view, are some of the most interesting and promising fields of application of the contactless techniques for the measurement of mechanical quantities.

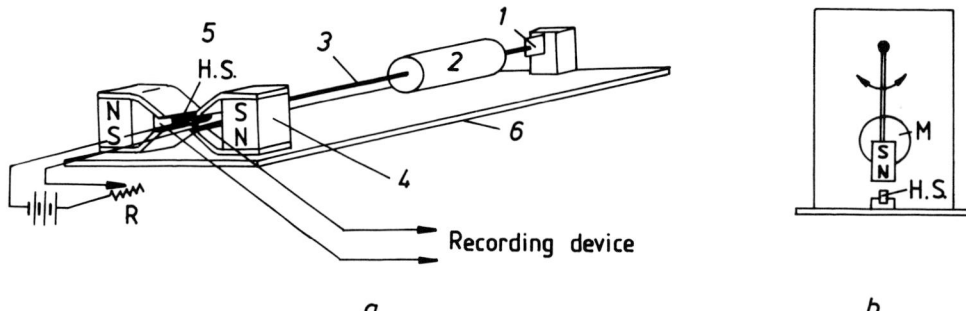

Fig. 10.13. Principle of operation of: (a) galvanomagnetic seismograph; (b) tiltmeter.

Galvanomagnetic seismograph and related instruments. The operating principle of a seismograph utilizing a Hall sensor or a linear differential BMT is shown in Fig. 10.13a [27]. The mass 2 is suspended on spring 1. By means of arm 3, the linear magnetosensor 5 connected with mass 2 is placed in the inhomogeneous magnetic field, with a linear gradient, generated by circuit 4 (Fig. 10.11g6). The magnets are rigidly fixed to the seismograph base 6. Any movement of the seismograph base 6 due to oscillations of the earth will cause a displacement of the transducer 5. The output signal is registered by a suitable recording device. The measurements described in [27] with an instrument using a Hall sensor show that the sensitivity is 100 times higher than in an electrodynamic seismograph with the same dimensions for arm 3. If the Hall sensor H.S. is replaced by a linear differential BMT (Fig. 6.9h), operating in a multisensor regime for B and T, and the scheme from Fig. 10.2 is used, the sensitivity of the instrument will be at least 5×10^2 times higher than that of a conventional seismograph. A similar device (Fig. 10.13b) can also be used as a tiltmeter. If, owing to the horizontal tilting of the ground, the seismograph base 6 tilts at an angle α, the pendulum will be moved from its initial position at an angle $\beta \sim \alpha$. The use of this approach makes it possible to register tilts of 0.001 seconds of arc [2,27]. A contactless accelerometer, similar to the design in Fig. 10.13a, can register accelerations within the range 10^{-4} to 10^{-3} g (g is the free-fall acceleration). The Hall sensor is displaced in a magnetic field with gradient 10^4 G mm^{-1}. The movable part of this instrument is sunk in a special liquid to achieve the necessary damping.

Pressure cans and detection of linear displacement. A magnetosensor with a linear output is actuated in a pressure can by a permanent magnet mounted on a pressure-sensitive diaphragm (Fig. 10.11g1). The diaphragm flexes under pressure and thereby moves the magnet relative to the transducer (Fig. 10.14a). The differential pressure, ΔP, can be measured by the arrangement in Fig. 10.14b. Hall devices, differential BMTs, MDs operating in a linear mode, etc. can be used as

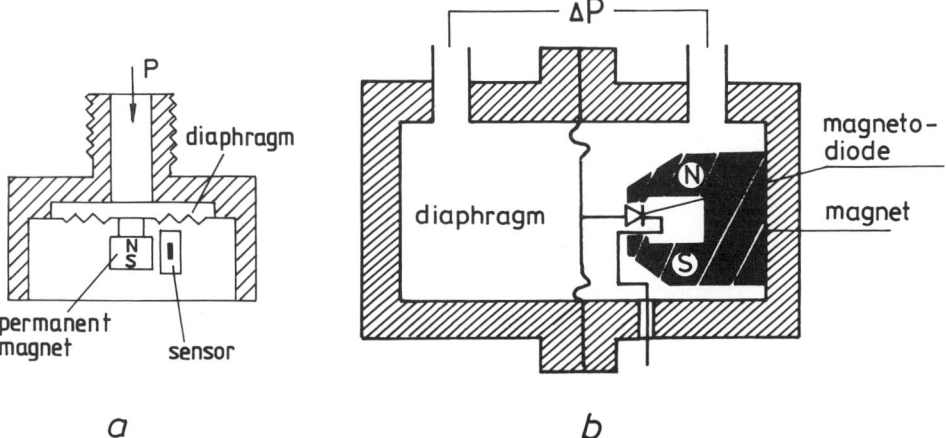

Fig.10.14. Construction of a pressure can using a Hall sensor (a), and arrangement for measuring differential pressure ΔP (b).

galvanomagnetic elements in this application. If the pressure measuring can is to be used for automotive applications, GaAs is preferable as the sensor material. This type of electromechanical instruments is a cost-effective alternative to the piezoresistive silicon pressure transducers [1,28]. Besides the magnetic circuit shown in Fig. 10.14, the arrangements in Fig. 10.11g8, in which the gradient of the field B is substantially higher, can be successfully used in a pressure can. The sound-measuring microphone shown in Fig. 10.15 has the same principle of operation as the device in Fig. 10.14.

The magnetic configurations in Fig. 10.11g8 and g9 are the basis of an instrument recently developed for the measurement of linear displacements in geotechnical experiments [29]. A multisensor for magnetic field and temperature and the circuit in Fig. 10.2 have been used. The overall error is less than ± 1 μm within the interval ± 2 mm and at temperatures in the range $0 \leq T \leq 60°C$. This is a highly promising result in the study of vertical and horizontal deformations of test bodies. A contactless instrument using a Hall device has also been developed for the measuring of normal stress, shear stress and force for applications in geotechnical engineering [30].

A linear position transducer using a magnet and a Hall sensor and designed for clinical dental research laboratories, its purpose being to measure the movements of the jaw, is described in [31]. A galvanomagnetic micrometer based on an InAs Hall device and the magnetic circuit shown in Fig. 10.11g6 is described in [25]. Within the interval of ± 500 μm, the instrument exhibits a resolution of 5×10^{-3} μm which is one order of magnitude higher than that of the inductive transducers.

The large mismatch between the thermal expansion coefficients of silicon and the available plastic encapsulants introduces severe stresses on the chip surface,

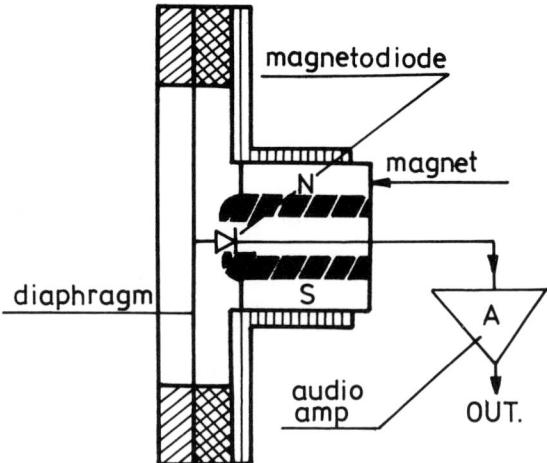

Fig. 10.15. Galvanomagnetic microphone using a MD.

which might lead to stress-induced displacements of metallization paths. A sophisticated solution for on-chip metal-deformation measurements using Hall elements is presented in [32]. A local magnetic field is introduced onto the chip surface by leading a current through the central conductor. Two Hall sensors symmetrically integrated along this conductor measure the magnetic field. The difference in the output voltages of the two Hall devices is a measure of the displacement of the center of the current and hence of the central conductor. This approach shows good linearity for moderate displacements (≤ 5 μm).

The strong dependence of the offset voltage of Hall devices on mechanical strain without the application of a magnetic field is well known [33]. An IC strain sensor using this effect has been proposed for square, rectangular and circular diaphragms. This unique pressure transducer is attractive for miniature biomedical applications. The idea has been implemented by the one-chip combination of a module for the measurement of the magnetic field B and the pressure.

Brushless DC motors. A brushless DC motor uses a diametrically magnetized permanent-magnet rotor (Fig. 10.11f1) with coils in the stator [1,15]. The position of the rotor relative to the stator coils is indicated by two magnetosensors (Hall devices, BMTs, MDs, etc.); the four stator coils W_1, W_2, W_3 and W_4 are powered via the resultant output voltages from the transducers H_1 and H_2, so that a torque is produced by the magnetic field of the rotor (Fig. 10.16). As a result of the polarity dependence of the Hall voltage on the field B, the two sensors H_1 and H_2, displaced by 90°, suffice to control the four coils W_1, W_2, W_3 and W_4. The two transducers are located directly in the air gap between the rotor and the stator and are activated by the permanent-magnet rotor, or they are placed outside the motor

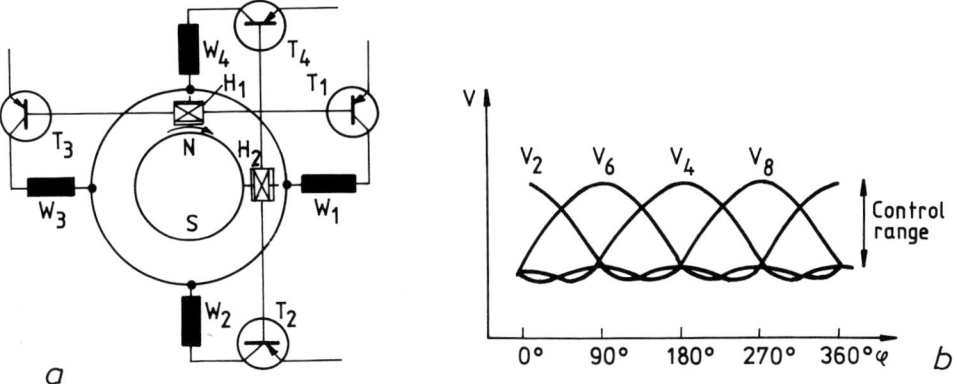

Fig.10.16. Principle of operation of a brushless electromotor (a); (b) $V_2 \ldots V_8$ are typical characteristics of the control voltages for the coils $W_1 \ldots W_4$ belonging to the Hall voltages [1,15].

windings and are acted upon by a control magnet mounted onto the motor shaft [1,34].

The advantages of brushless DC motors are their long life, low level of noise, the absence of disruptive brush sparking and their very good controllability.

Angular transducers. If a linear output sensor is rotating round its own axis in a homogeneous magnetic field, the output voltage is a regular sinusoidal curve. Owing to difficulties associated with the connection of the wires to the terminals of a moving transducer, it is preferable that the magnet generating the field should rotate, while the sensor should be stationary (Fig. 10.11f1). Iron sectors are used to enhance the sensitivity. Figure 10.17 illustrates the operating principle of an angular transducer using two equal linear magnetosensitive devices [2,35]. When the permanent-magnet rotor is turned through an angle α, two voltages $V_1 = V_{\max} \sin \alpha$ and $V_2 = V_{\max} \cos \alpha$ are induced in the two sensors displaced by 90°. The remaining

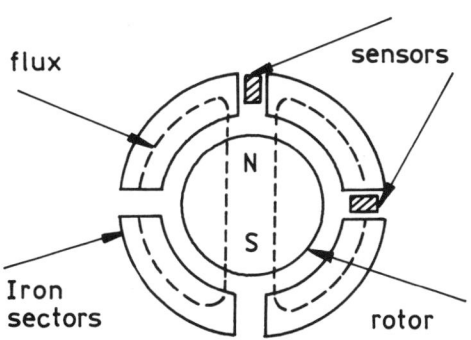

Fig. 10.17. Angular transducer using magnetosensors.

two air gaps are appropriately shaped to eliminate the asymmetry in the magnetic system. An instrument has been developed on this principle for the measurement of the angle of rotation with an accuracy of 0.1° [2].It is preferable that Hall devices and differential BMTs are used in this arrangement. The transducer in Fig. 10.17 has universal applicability as a function generator of sin or cos waveforms, in the conversion of various mechanical angular displacements into electrical signals, and in synchroresolvers, tachometers, contactless potentiometers (cf. Figs. 4.35 and 4.37), etc.

Contactless counters, interrupters and keyboard switches. A revolution counter, based on the configuration shown in Fig. 10.11h1 and differentially connected magnetoresistors, has been developed [1,15]. The optimal signal amplitude is achieved when the distance between the centers of the individual magnetoresistors corresponds to exactly half the tooth period. The sinusoidal output waveform from the Wheatstone-bridge configuration, which includes two MRs, is usually digitized by a comparator and the pulses are counted over a set period of time [1]. Rotation sensors based on MRs for automotive applications are available. A gear-tooth counter, using two Hall probes integrated together with digital circuitry for rough environmental conditions, is described in [36]. A special layout of the Hall elements minimizes their offset and offset drift. An active filter increases the maximum possible distance between the sensor IC and the toothed wheel over a wide temperature range. Full protection against peak pulses and electromagnetic influences is provided.

Digital sensors which latch on reaching a predetermined flux level and which have a built-in hysteresis so that they may release at a lower flux level are especially suitable for position detection in automotive applications [1,37]. The monolithic, integrated electronic module of these promising transducers consists of a constant voltage supply, a magnetosensor (usually a Hall device), an amplifier, a Schmitt trigger, an output stage with an open collector and protection circuitry. The operating principle of such a digital proximity detector for measuring a rotational speed is shown in Fig. 10.18. The devices shown in Fig. 9.1 can be successfully used for such applications.

Another version of an IC interrupter with a Hall device is the magnetic switch with a memory. Once turned on by a magnetic field, it remains in this state after the field B has been removed. It is turned off by a magnetic field with the opposite polarity, i.e. by $-B$. The indicator of the direction of rotation, shown schematically in Fig. 10.19 [38], is a promising application of this operating mode of the switch. When the disc rotates in a clockwise direction, sectors SN, "–", SN, "–", etc. turn up in succession to occupy a position below the magnetic switch. As the first N-pole passes, the switch is turned on and a signal corresponding to the logical "zero" appears at the output. This state is preserved as long as the part of the disc that contains no magnets is passing by the switch. A signal corresponding to the logical

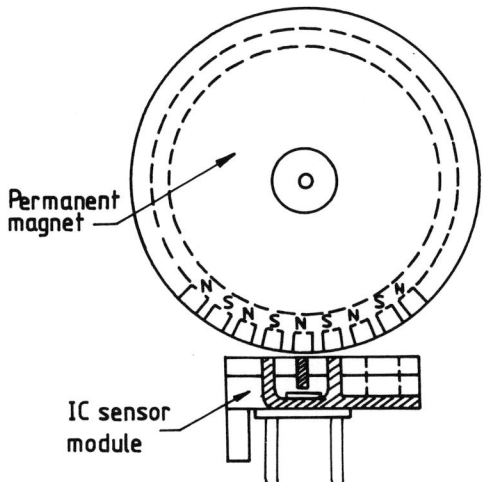

Fig.10.18. Magnetic circuit and IC magnetosensor module used to detect rotation or to measure the rotational speed of a wheel.

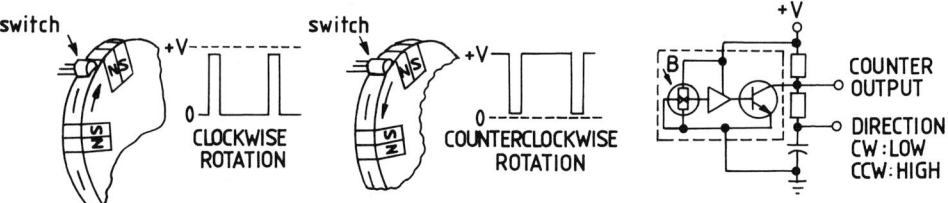

Fig. 10.19. Arrangement of an indicator for the direction of rotation. The indicator uses a magnetoswitch [38].

"one" appears at the moment the S-pole arrives. In this way when the disc turns in a clockwise direction, a succession of short positive pulses is generated at the output. Most of the time, the output is in the "zero" state. When the disc moves in the counterclockwise direction, the sequence under the switch is as follows: NS, "–", NS, "–", etc. In this case most of the time the output is in the state of "1" and short negative pulses are generated. The magnetosensitive IC determines the direction of rotation of the monitored object by the shape of the output signal. The electronic circuit also provides information about the frequency of rotation. Together with the Si Hall devices, silicon-based BMT sensors, magnetodiodes, etc. can successfully be used in the magnetic switch shown in Fig. 10.19. The contactless instruments described above hold special promise for applications in automobile electronics.

Figure 10.20 illustrates one of the versions of a keyboard switch [6,15,39]. It helps to eliminate the typical shortcomings of mechanical switches, like contact burning

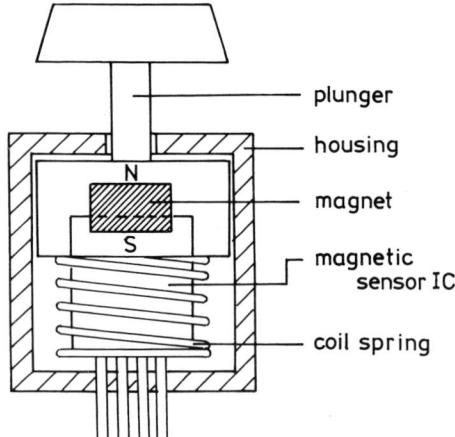

Fig.10.20. Version of a contactless switch used in solid-state keyboards. A permanent magnet on the plunger operates the sensor.

and erosion, slow response, sensitivity to vibrations and shocks, etc. Hall devices, MDs, BMTs and other transducers can be used as sensor elements in solid-state keyboard switches.

High sensitivity to a low magnetic field (of the order of 1 mT) may be achieved by the use of the so-called flux-gate magnetometers [40–43]. It is difficult at present to fabricate fully integrated versions of these sensors [41], but they have a number of applications, for instance, in small distance measurements, in earth magnetic-field topography, in medicine [42], etc.

Detection of gravitational waves. According to the basic conclusions of Einstein's general theory of relativity, a mass engaged in accelerated motion is expected to give out gravitational waves, which propagate with the speed of light. The collapse and birth of new stars, collisions of black holes, and rapidly moving twin-stars can be sources of gravitational radiation. For the detection of this fundamental physical phenomenon SQUID-based instruments, capable of measuring extremely small displacements, can be used [44,45]. Longitudinal oscillations of extremely small amplitudes are expected to arise in large, freely attached cylinders under the impact of gravitational radiation. The registration of the oscillations necessarily requires a DC SQUID measuring technique. A version of an antenna for the receiving of gravitational waves is shown schematically in Fig. 10.21 [46]. The instrument is an aluminium cylinder of length 3 m and weight 4800 kg, hung in a vacuum at a temperature $T = 4.2$ K. The frequency of the basic longitudinal mode of oscillation is $w_\alpha/2\pi \simeq 842$ Hz with a Q factor $\simeq 5 \times 10^6$. A circular diaphragm of niobium is pressed all along its perimeter to one of the ends of the Al cylinder, and Nb coils are mounted on both ends of the cylinder. These coils are

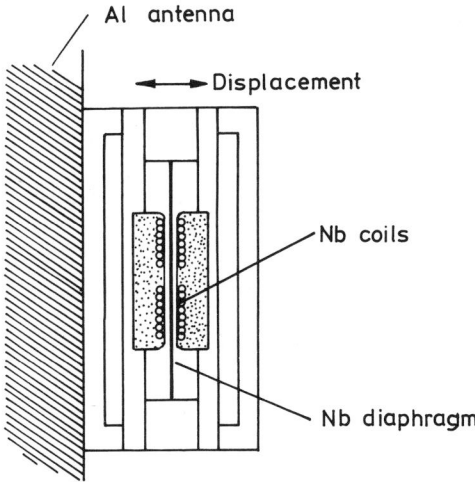

Fig. 10.21. Superconducting antenna for the detection of gravitational waves.

coupled in parallel and connected with the input coil of a DC SQUID. The whole arrangement is superconductive. A superconducting current, whose magnetic field generates a force counteracting the deformation of the diaphragm circulates in the closed loop. This current can be controlled to provide coincidence of the resonant frequency of the diaphragm's oscillations with the frequency of oscillation of the cylinder. The longitudinal variations of the Al cylinder, induced by the gravitational waves, cause variations of the diaphragm with respect to the two coils, whose inductances are modulated as a result. Owing to magnetic-flux quantization, a part of the superconducting current is directed to the input coil of the DC SQUID and thereby the oscillations are registered. This antenna has a mean-square value of the sensitivity to deformations of $\langle \delta l^2 \rangle^{1/2}/l \simeq 10^{-18}$, where l is the length of the cylinder, and δl is the longitudinal displacement [46]. But even this unsurpassed sensitivity, limited by the thermal noise of the cylinder, makes it possible to register only gravitational pulses originating from our own Galaxy. As the collapse of a star is a rare phenomena, the transducing efficiency of the antenna needs to be considerably increased. Notwithstanding the existing difficulties, apart from the very fact of the fundamental character of the possibility of detecting gravitational waves, these experiments encourage the research and development of SQUID technology. The measuring of the gradient of the gravitational force by means of SQUIDs is based on an analogous principle. These advanced experiments employ magnetosensors as tools to gain additional knowledge about the surrounding world.

At the end of this short survey of the fields of application of solid-state magnetic sensors, the author would like to underline two trends. The first is associated with the increasingly wider use of magnetotransistors and magnetodiodes together with Hall devices and magnetoresistors in data acquisition systems. A dominant position is likewise occupied by low-field and vector magnetometry, using SQUIDs and BMTs. The second trend concerns the sophistication of contactless instruments themselves, including the digital processing of the magnetosensitive signal, combined with a dramatic improvement of sensor performance and the expansion of the measuring functions.

References

[1] R. Popović and W. Heidenreich, in Sensors, Vol. 5, R. Bolk and J. Overshott (eds.), VCH, Weinheim, 1989, 43–96.
[2] A. Kobus and J. Tuszynski, Hallotrony i Gaussotrony, Wydawnictwa naukowo-techniczne, Warszawa, 1966 (in Polish).
[3] H. Weiss, Physik und Anwendung galvanomagnetischer Bauelemente, Vieweg, Braunschweig, 1969.
[4] N.P. Milligan and J.P. Burgess, Solid-State Electron., 7 (1964) 323–333.
[5] J.E.L. Hollis, Measurement and Control, 6 (1973) 38–40.
[6] T. Takov and V. Minchev, Poluprovodnikovi datchitchi, Technika, Sofia, 1986, Chapter 3 (in Bulgarian).
[7] P. Extance and G.D. Pitt, Proc. of the Int. Conf. on Solid-State Sensors and Actuators — Transducers' 85, Boston, Massachusetts, USA (1985) 304–307.
[8] P. Daniil and E. Cohen, J. Appl. Phys., 53 (1982) 8257–8259.
[9] G. Torzo, A. Sconza and R. Storti, J. Phys., E, Sci. Instrum., 20 (1987) 260–262.
[10] K. Maenaka, M. Tzukahara and T. Nakamura, Sensors and Actuators, A, 21–23 (1990) 747–750.
[11] S. Kataoka, Recent Development of Magnetoresistive Devices and Applications, Circular of the Electrotechnical Laboratory, No. 182 (Tokyo, Japan) 1974.
[12] A.W. Vinal, IEEE Trans. Magn., MAG-20 (1984) 681–686.
[13] Y. Sugiyama, Fundamental Research on Hall Effect in Inhomogeneous Magnetic Fields, Res. of the Electrotechnical Laboratory, No. 838 (Tokyo, Japan) 1983.
[14] F. Kuhrt and K. Maaz, Elektrotechn. Z., 77 (1956) 487–490.
[15] Sensoren-Datenbuch, Magnetfeldhalbleiter, Teil 1, Siemens, 1982/83.
[16] Ch.S. Roumenin and P.T. Kostov, Electroprom. i priborostr., 12 (1986) 30–32 (in Bulgarian).
[17] Ch.S. Roumenin, C.R. Acad. Bulg. Sci., 41(5) (1988) 71–74.
[18] Ch.S. Roumenin, Sensors and Actuators, A, 24 (1990) 83–105.
[19] K. Matsui, S. Tanaka and T. Kobayashi, Proc. 1st Sensor Symp., 1981, Tokyo, Japan (1982) 37–40.
[20] S. Kataoka, Proc. IEE, 111 (1964) 1937–1947.
[21] W.E. Bulman, Solid-State Electron., 9 (1966) 361–372.
[22] Ch.S. Roumenin, C.R. Acad. Bulg. Sci., 41(4) (1988) 25–28.
[23] Y. Netzer, Proc. IEEE, 69 (1981) 491–492.
[24] M. Nalecz and Z.L. Warsza, Solid-State Electron., 9 (1966) 485–495.
[25] G.I. Kotenko, Galvanomagnitniye Preobrazovately i ih Primeneniye, Energoatomizdat, St. Petersburg, 1982 (in Russian).
[26] W. Pattenpaul, J. Huber, H. Weidlich, W. Flossmann and U. Borcke, Solid-State Electron., 24 (1981) 781–786.
[27] M. Nalecz and I. Zawicki, Bull. Seismol. Soc. of America, 52(2) (1962) 439–454.
[28] Ch.J. Everett, Pressure transducer device using Hall element, US Patent 4 484 173 (Nov. 20, 1984).
[29] Ch.S. Roumenin, Engineer. Geol. and Hydrogeol., 17 (1987) 41–50; 20 (1991) 23–32 (in Bulgarian).
[30] A.V.D. Bica and C.R.I. Clayton, J. Phys., E, Sci. Instrum., 22 (1989) 548–551.
[31] W.D. McCall and E.J. Rohan, IEEE Trans. Instrum. and Measurem., IM-26 (1977) 133–136.
[32] A. Bossche and J.R. Mollinger, Sensors and Actuators, A, 21–23 (1990) 754–757.
[33] Y. Kanda and A. Yasukawa, Sensors and Actuators, 2 (1982) 283–296.
[34] I.M. Vikulin, L.F. Vikulina and V.I. Stafeev, Galvanomagnitniye pribori, Radio i Sviaz, Moskwa, 1983 (in Russian).

[35] R.S. Davidson and R.D. Gourlay, Solid-State Electron., 9 (1966) 471–484.
[36] H. Jasberg, Sensors and Actuators, A, 21–23 (1990) 737–742.
[37] P. Kleinschmidt and F. Schmidt, Sensors and Actuators, A 31 (1992) 35–45.
[38] J.D. Spencer, EDN, 25(3) (1980) 151–156.
[39] S. Kordić, Sensors and Actuators, 10 (1986) 347–378.
[40] D.B. Dimitrov, Sensor and Actuators, A, 39 (1993) 45–47.
[41] D.B. Dimitrov, I. Halianov, J. Kassabov et. al., J. Phys, Condens. Matter, 5 (1993) 1257–1260.
[42] K. Havada, Y. Sunouchi and H. Sakamoto, IEEE, Trans. on Magn., MAG-25 (1989) 3399–3401.
[43] D.B. Dimitrov and J. Kassabov, Magnetosensitive Device, Bulg. Patent, 95 576 (Jan. 13, 1992).
[44] S. L. Shapiro, R.F. Stark and S. J. Teukolsky, Am. Sci., 73 (1985) 248–257.
[45] P.F. Michelson, J.C. Price and R.C. Taber, Science, 237 (1987) 150–157.
[46] J. Clarke, Proc. IEEE, 77 (1989) 1208–1223.

Conclusion

Unlike other transducers of physical and chemical quantities, magnetosensors have the unique ability to reveal realities that cannot be perceived by the human senses. Hence our knowledge of magnetic fields comes solely from these devices, combining in themselves different laws of Nature. The solid-state sensors discussed in this book currently pretend to be among the most precise instruments for the acquisition of magnetic data, while SQUIDs are the absolute record-holders in terms of sensitivity. The main trend in the development of these "magic" crystals is in their miniaturization and the ceaseless improvement of their performance and resolution, markedly influenced by sophisticated IC technologies. This is why the data gathered with the help of microsensors do not change the picture of the magnetic field being investigated. The on-chip integration of more than one sensor function is the most suitable approach to the development of smart magnetometers and their wide range of applications. Despite the occurrence of different transducing mechanisms associated with the type of conductivity, device structure, operating conditions, etc., the Hall effect underlies the action of most magnetosensors. It is amazing, indeed,, how such a clear and seemingly simple phenomenon could govern in a universal manner the performance of a whole range of complex electronic devices, being at the same time a tool for studying the charge transport in solid-state materials.

The reasonable question arises: how far have solid-state magnetotransducers gone in their evolution? Might this be a scientific field whose capacity for development is already exhausted? The author seriously doubts that this class of sensors could have reached the limits of their potentialities. The Hall arrangement has been known for more than a century now, but it has been only during the past few years that the sophisticated 2-DEG superlattice and parallel-field Hall devices have been investigated and developed; moreover this is, just the beginning of the new generation of microsensors. Integrated magnetodiodes, magnetotransistors and carrier-domain magnetometers are more the generators of new physical effects and ideas than well-proven engineering solutions. With the help of the new materials of extremely high carrier mobility and by the use of low temperatures as a operational environment, the magnetoresistors have been a subject of continuous improvement.

High-T_C superconductivity promises unexpected and "unpredicted" meetings with SQUIDs. In view of the availability of supercomputers and much varied software, the fact that the modeling of solid-state magnetosensors is in its initial phase, coupled with a lack of satisfactory knowledge concerning their figures of merit, is yet another indicator that these electronic devices are far from "maturity".

It will not perhaps be long before the first synthesized magnetobiosensors appear, which, by the way, Nature "invented" long ago and has been using ever since. The relation between biological and solid-state versions, together with the experience accumulated so far in this "magnetician" field of research, indeed, encourage us on our way towards the unknown. The development of solid-state magnetic sensors continues...

Appendices

Appendix I: Magnetic terms and units

Term, quantity, symbol	MKSA unit	Subunits	CGS unit	Conversion factors
Magnetic field strength, H	A m^{-1}	$1\,\text{A cm}^{-1} = 10^2\,\text{A m}^{-1}$ $1\,\text{mA cm}^{-1} = 0.1\,\text{A m}^{-1}$ $1\,\text{kA m}^{-1} = 10^3\,\text{A m}^{-1}$	Oe (Oersted)	$1\,\text{Oe} = 79.58\,\text{A m}^{-1}$ $= 0.796\,\text{A cm}^{-1}$
Magnetization, M	A m^{-1}			
Magnetic induction, B (flux density)	$\text{T (Tesla)} = \text{V·s m}^{-2}$	$1\,\text{mT} = 10^{-3}\,\text{T}$ $1\,\mu\text{T} = 10^{-6}\,\text{T}$ $1\,\text{nT} = 10^{-9}\,\text{T}$ $1\,\text{pT} = 10^{-12}\,\text{T}$ $1\,\text{fT} = 10^{-15}\,\text{T}$ $1\,\gamma = 1\,\text{nT}$	G (Gauss)	$1\,\text{G} = 10^{-4}\,\text{T}$ $1\,\text{kG} = 0.1\,\text{T}$ $1\,\text{mG} = 10^{-7}\,\text{T}$ $1\,\gamma = 10^{-5}\,\text{G}$
Magnetic flux, Φ	$\text{Wb (Weber)} = \text{V·s}$		Mx (Maxwell)	$1\,\text{Mx} = 10^{-8}\,\text{Wb}$
Magnetic polarization, J				
Permeability absolute, μ	T·m·A^{-1}		G Oe^{-1}	
Permeability of vacuum, μ_0	$4\pi \times 10^{-7}\,\text{T·m·A}^{-1}$	$0.4\pi \times 10^{-4}\,\text{T·cm·A}^{-1}$	1	

Appendix II: Important quantities and constants used in galvanomagnetism

No.	Physical quantity and constants	Symbol	SI unit	CGS unit and conversion
1	Conductivity	σ	S m^{-1}	$10^{-2} \times \Omega^{-1}\cdot\text{cm}^{-1}$
2	Resistivity	ρ	$\Omega\cdot\text{m}$	$10^{2} \times \Omega\cdot\text{cm}$
3	Hall coefficient or constant	R_H	$\text{m}^3\,\text{C}^{-1}$	$10^{6} \times \text{cm}^3\cdot\text{C}^{-1}$
4	Carrier concentration	n, p	m^{-3}	$10^{-6} \times \text{cm}^{-3}$
5	Carrier mobility and Hall mobility	$\mu_n, \mu_p; \mu_{Hn}, \mu_{Hp}$	$\text{m}^2\cdot\text{V}^{-1}\cdot\text{s}^{-1}$	$10^{4} \times \text{cm}^2\cdot\text{V}^{-1}\cdot\text{s}^{-1}$
6	Elementary charge	q	1.602×10^{-19} C	4.8×10^{-10} statcoulombs
7	Electron rest mass	m_0	9.109×10^{-31} kg	9.109×10^{-28} g
8	Boltzmann constant	k_B	1.380×10^{-23} J K^{-1}	1.380×10^{-16} erg °C^{-1}
9	Thermal energy ($T = 300$ K)	$k_B T$	4.14×10^{-21} J	4.14×10^{-14} erg
10	Thermal voltage at $T = 300$ K	$K_B T/q$	0.0259 V	
11	Planck's constant; reduced constant	h $\hbar = h/2\pi$	6.626×10^{-34} J·s 1.054×10^{-34} J·s	6.626×10^{-27} erg·s 1.054×10^{-27} erg·s
12	Dielectric constant (permittivity of vacuum)	ε_0	8.854×10^{-12} F·m^{-1}	1

The data for Appendices I and II have been taken mainly from "Dictionary of Physics", H.J. Gray and A. Isaacs (eds.), Longman UK Ltd., Third Edition, 1990.

Subject index

Actuator 1,2
Ambipolar diffusion length 56, 59
Ambipolar diffusion constant 66
Analog-digital converter 8
Angular transducers 401
Applications of solid-state magnetosensors 9, 383
 – direct 9
 – indirect 9
Avalanche bipolar magnetotransistor (AMBT) 246
Average time 20
Average velocity 20

Basic equations of
 – conductivity in crossed electric and magnetic fields 18–22
 – Hall effect 23–27, 31–34, 42–45
 – magnetoconcentration effect 55–57
 – magnetodiode effect 58–63, 65–69
 – magnetoresistance effect 46–48, 50–52
 – superconductivity 76–82
Biasing circuits of
 – BMT sensors 236, 343, 345, 349
 – CDMs 268, 271, 343, 344
 – flip-flop sensors 355–357
 – Hall sensors 168, 343, 345–347
 – MDs 347, 348
 – magnetoresistots 183, 184, 186, 343
 – magnetosensitive multivibrator sensor 354
 – multisensors 314, 322, 326, 329, 331, 339
 – SQUIDs 88, 286, 307
 – UJMTs 344, 345
Bipolar magnetotransistor sensors (BMT) 229
 – base-region magnetosensitivity of 234
 – device structures of
 – – orthogonal 244–247, 256
 – – parallel-field 248–256
 – electrical control of the polarity of magnetosensitivity in 236
 – emitter-injection modulation in 239, 241

 – emitter region magnetosensitivity of 232
 – figures of merit of 255
 – filament magnetosensitivity in 239
 – frequency response of 260
 – linearity of 259
 – Lorentz deflection in 234
 – magnetodiode effect in 235
 – magnetosensitivity of collector-base depletion region 241
 – magnetosensitivity of 255
 – noise of 257
 – offset of 258
 – principle of operation of 245–247, 249–255
 – – at cryogenic temperatures 247, 248, 253
 – temperature coefficient of magnetosensitivity of 259
Boundary conditions 29, 32
Bridge circuit model of a Hall sensor 133
Brushless DC motor 400
Bulk bipolar Hall sensor 153–160
Buried Hall sensor 156

Cardiomagnetography 308
Carrier-domain magnetometer 195, 266
 – four-layer 270
 – three-layer 195, 266
Circulating current 39, 73
Coherent phenomena in superconductors
 – flux quantization 75, 76
 – infinite conductivity 75, 76
 – Josephson effect 75, 76
 – Meissner effect 75, 76
Coherence length 77
Collisions 20
Conductivity 18
 – bipolar 44, 52, 53
 – in crossed electric and magnetic fields 18
 – infinite 75, 76
 – intrinsic 44, 45, 54
 – monopolar 18
 – tensor of 19, 21

Conformal mapping   117–119, 166
Contactless counters, interrupters and switches   402, 403
Contactless potentiometer   183, 186
Conversion of magnetic energy by sensors   2–4
Cooper pairs   75, 78
Corbino disk   30, 32, 176
Critical current in superconductor   80
Critical temperature in superconductor   75
Cube of sensor effects   5, 6
Cyclotron frequency   19
Cyclotron radius   19

de Broglie wave   89
Dember field   71
Depletion-layer BMT   241
Detection of gravitational waves   404
Differential amplification magnetic sensor (DAMS)   243, 350, 351
Diffusion coefficient   55
Diffusion length   59, 62
Digital displacement transducer   397
Directional figure of merit (DFM)   110, 158
Doping impurities   38
Drift-aided BMT   236, 241
Drift mobility   18
Drift velocity   18, 36

Effective electron temperature   35
Effective lifetime   67, 68
Effective mass   19
Einstein special theory of relativity   7
Electric force   7
Electromotive force   23
Electronic compass   386

Faraday coils   12
Figures of merit of solid-state magnetic sensors   97–99
  – accuracy   106
  – creep   110
  – cross-sensitivity   109
  – detection limit   107, 108
  – directivity   103
  – drift   110
  – electrical excitation   111
  – error   105
  – excitation by magnetic fields   103, 114
  – frequency response   104
  – hysteresis   106
  – input and output impedance   111
  – magnetosensitivity   99
  – – absolute   99
  – – relative   99, 100
  – noise   106–108
  – nonlinearities   100–102
  – offset   108, 109
  – operating life   110
  – output form   106
  – range of magnetic fields   103
  – reliability   110
  – repeatability   106
  – resolution   104, 105
  – response time   111
  – room conditions   111
  – stability   110
  – temperature coefficient of sensitivity   109
Filament (or domain)   192–195, 266, 270
Flux concentrator   384
Flux-gate magnetometer   404
Flux quantization   75, 76, 78
Flux transformer   283, 284
Functional gradiometer sensors   325
  – BMT   328
  – Hall effect   326
Functional magnetic-field sensors   313

Galvanomagnetic coefficients   26
Galvanomagnetic quantities and constants   412
Galvanomagnetic microphone   399, 400
Galvanomagnetic phenomena   17
Galvanomagnetic seismograph   398
Galvanomagnetic tiltmeter   398
Geomagnetic field   10, 11, 15
Geometrical correction factor   37, 38
Geometrical magnetoresistance
  – mixed conductivity   51
  – monopolar conductivity   47

Hall   23
  – angle   20, 25, 35, 36
  – coefficient   26, 27, 42, 44, 45
  – – of n-type conductivity   25
  – – of p-type conductivity   25
  – factor   26
  – – of n-type silicon   27
  – – of p-type silicon   27
Hall current   30, 32–34
  – in short structure   30, 31
  – mode of operation   31, 139
Hall effect   23, 28
  – at high electric fields   34
  – in diodes and transistors   73, 74
  – in ferromagnetic materials   30
  – in inhomogeneous conditions   37–41, 72
  – in long samples   23, 45
  – in medium-length samples   37
  – in metals   27

– in mixed conductivity   42
– in nonequilibrium conditions   71
– in short samples   31
Hall field   23
Hall mobility   25, 36, 42, 44
Hall sensor   115
– basic characteristics of   128
– efficiency of   129
– frequency response of   138
– geometrical correction factor of   120–124
– materials, fabrication technology and packaging of   141–145
– noise of   133
– nonlinearities of   137
– offset voltage of   130
– orthogonal   115
– output voltage of   128
– parallel-field   165
– – device structures and characterization of   168, 169–174
– sensitivity of   136, 140, 147
– shapes   115–119
– signal-to-noise ratio of   135
– temperature coefficient of magnetosensitivity of   138
Hall voltage   23, 125
– in long samples   23, 24
– mode of operation   125
Hooge parameter   213

Ideality factor in MD effect   62
Improvement of solid-state magnetosensors   359
– compensation of nonlinearity   377
– compensation of the temperature dependence of magnetosensitivity   374
– compensation of the temperature drift of the offset   371
– cross-sensitivity reduction   370
– offset reduction   359
Inertial system   7
Infinite sample   30
Information-processing system   1
Injection phenomena in semiconductor devices   53
Interface of solid-state magnetosensors   343, 357
– signal-processing electronics   346

JFET Hall sensor   174, 175
Josephson effect   75, 76
– nonstationary   80, 81
– stationary   80

Josephson tunnel junctions   79, 81, 279
– current-voltage curves of   81, 83, 85

Keyboards   403, 404

Law of conservation of energy   4
Linear displacement detection   398
Locally inverted magnetic fields   161
London penetration depth   75
Lorentz force   7, 19, 23

MAGFET   145, 149, 152
Magnetic field   6, 7
– high   21
– low   21
– mapping   388
– vector of   6
Magnetic field strength   9
Magnetic-field vector sensors   329
– 2-D and 3-D BMT   333, 338
– 2-D and 3-D Hall effect   331
Magnetic flip-flop sensors   354
Magnetic force   19
Magnetic flux   9
Magnetic flux quantization   75, 78
Magnetic induction   9
Magnetic permeability   9
Magnetic recording head   387
Magnetic sensor   9
Magnetic sensor systems   394
Magnetic terms and units   411
Magnetically controlled oscillator (MCO)   353
Magnetization   9
Magnetoconcentration effect   54–58
Magnetodiode effect   52, 58
– in long structures with one injecting contact   58
– in structures with two injecting contacts   64
Magnetodiode regimes   67
– diffusion   68
– insulator   68
– ohmic   67
– semiconductor   67
Magnetodiode sensors   191
– current–voltage characteristics of   202–204
detection limit of   211
– device parameters of   191–200
– device structures of   191–200
– frequency response of   210
– integrated   197–200, 223–225
– magnetosensitivity of   204–207
– materials of   192, 196–200
– noise of   211
– operation of   191–200

– orthogonal   192–200, 213–218
– parallel-field   192–200, 218–225
– S-type   192–194
– Schottky   198–200, 223, 224
– silicon on sapphire (SOS)   197–200, 222
Magnetogradient effect (MGE)   328
Magneto-operational amplifier (MOP)   351
Magnetoresistance effect   46
  – geometrical   47–49
  – in mixed conductivity   50
  – in monopolar conductivity   46
  – physical   46
Magnetoresistance mobility   26, 47, 48
Magnetoresistance sensors   164, 176
  – characterization of   180
  – design and materials of   176
  – multiterminal   183
Magnetosensitive device technology   11
Magnetosensitive multivibrator sensor   353
Magnetothyristors   274, 275
Measurement of
  – current   388
  – magnetic field   384
  – nonelectrical and nonmagnetic quantities   393
  – power   391
Measurement process   8
Methods and applications of magnetic-field measurements   8
Minority carrier lifetime   55
Modulating magnetosensor   97
MOSFET Hall sensors   145
  – split-drain   140, 148, 150, 151
Multiplier   391–393
Multisensing   14, 313
Multisensors
  – for magnetic field and temperature   314
  – for magnetic field, temperature and light   319

Nature of the magnetic field and the Lorentz force   6
Narrow bandgap semiconductors   49
Negative magnetoresistance   49
Negative differential resistance   192–195
Noise   106
Nonelectrical quantities   1, 2
Nuclear magnetic resonance (NMR)   12
Nyquist relation   134

Offset   108
Ohm's law   18
Optoelectronic magnetosensitive device   11

Physical magnetoresistance
  – in mixed conductivity   50
  – in monopolar conductivity   46
Piezo-Hall effect   144
Plasma   53
Pressure cans   398
Principles and technologies for the acquisition of magnetic data   1

Recombination rate   54, 66
Reference system   6, 7
Relaxation time   30, 35
Resistively shunted junction (RSJ) model   84
RMS noise voltage   106, 107

Scattering mechanisms   26, 27
Schwarz-Christoffel's transformation   118
Sensitivity-variation offset reduction method   361
Sensor   1
Short structure   30
Signal conversions   3
Signal processor   1
Signals   2
  – chemical   2
  – electrical   2
  – magnetic   2
  – mechanical   2
  – radiant   2
  – thermal   2
Smart sensors   13, 358
Solid-state magnetosensor   97, 98
SQUID   82, 85
  – DC   89
  – – noise of   294
  – – structures of   291
  – RF   86
  – – noise of   296
  structures of   295
SQUID gradiometers   305
  – basic configuration of   305
  – characteristics of   307
SQUID sensors   291, 295
  – DC   291, 297
  – high-T_c 302–304
  – noise characteristics and sensitivity of   288, 294, 296
  – periphery of   285–288
  – read-out schemes of   88, 284, 286
  – RF   295, 301
  – signal-input coupling of   283
  – system   279
Stochastic A/D converters   354

Submicron Hall sensors   160
Suhl effect   73
Superconducting antenna   404, 405
Superconducting bridge contacts   281, 292
Superconducting condensate   80
Superconducting galvanometer   300
Superconducting materials   75
Superconductivity   75
  – high-T_c 77
  – low temperature   75
Superlattice   163, 164
  – energy-band diagram of   164
Super magnetoresistor   304
Surface recombination   54, 58

Tandem conversions   9, 10
Teslameter   385
Thin film magnetosensitive device   12

Transducer
  – input   1, 2
  – output   1
Transit time   35
Two-dimensional electron-gas Hall device   163
Types of sensors (or converters)
  – modulating   4, 5
  – self-generating   3, 5

Unijunction magnetotransistor sensor   260, 264
  – device structures of   261
  – operation of   261

Van der Pauw dual theorem   34, 139, 140
Velocity saturation   35

Weak link   79, 80
Wheatstone bridge   184, 348